www.wadsworth.com

www.wadsworth.com is the World Wide Web site for ThomsonWadsworth and is your direct source to dozens of online resources.

At *www.wadsworth.com* you can find out about supplements, demonstration software, and student resources. You can also send email to many of our authors and preview new publications and exciting new technologies.

www.wadsworth.com
Changing the way the world learns®

www.wadsworth.com

www.wadsworth.com is the World Wide Web [illegible]

[illegible]

[illegible]

Sex, Self, and Society

The Social Context of Sexuality

TRACEY L. STEELE
Wright State University

Australia • Canada • Mexico • Singapore • Spain
United Kingdom • United States

Acquisitions Editor: *Bob Jucha*
Assistant Editor: *Stephanie Monzon*
Editorial Assistant: *Melissa Walter*
Technology Project Manager: *Dee Dee Zobian*
Marketing Manager: *Matthew Wright*
Marketing Assistant: *Tara Pierson*
Advertising Project Manager: *Linda Yip*
Project Manager, Editorial Production: *Megan E. Hansen*
Art Director: *Maria Epes*
Print/Media Buyer: *Emma Claydon*
Permissions Editor: *Sarah Harkrader*
Production Service: *Peggy Francomb, Shepherd, Inc.*
Copy Editor: *Cindy Blum*
Cover Designer: *Yvo Riezebos*
Cover Image: *Illustration by csaimages.com*
Compositor: *Shepherd, Inc.*
Printer: *Malloy Incorporated*

Printed in the United States of America
1 2 3 4 5 6 7 08 07 06 05 04

For more information about our products, contact us at:
Thomson Learning Academic Resource Center
1-800-423-0563
For permission to use material from this text or product, submit a request online at http://www.thomsonrights.com.
Any additional questions about permissions can be submitted by email to thomsonrights@thomson.com

Library of Congress Control Number: 2004109854

ISBN 0-534-52943-7

Thomson Wadsworth
10 Davis Drive
Belmont, CA 94002-3098
USA

Asia
Thomson Learning
5 Shenton Way #01-01
UIC Building
Singapore 068808

Australia/New Zealand
Thomson Learning
102 Dodds Street
Southbank, Victoria 3006
Australia

Canada
Nelson
1120 Birchmount Road
Toronto, Ontario M1K 5G4
Canada

Europe/Middle East/Africa
Thomson Learning
High Holborn House
50/51 Bedford Row
London WC1R 4LR
United Kingdom

Latin America
Thomson Learning
Seneca, 53
Colonia Polanco
11560 Mexico D.F.
Mexico

Spain/Portugal
Paraninfo
Calle Magallanes, 25
28015 Madrid, Spain

For Kasey

Contents

CHAPTER 2

Constructing and Critiquing Sexual Categories 50

CHAPTER 3

Conflated Constructs: Sex, Gender, and Sexuality 74

CHAPTER 4

What's Love Got to Do with It? Constructions of Desire, Love, and Intimacy 97

Preface

This book is about sultry, seductive, hot, naughty, sweaty, bawdy, steamy, wild, animalistic, S-E-X!

Just picture wet luscious quivering lips; sleek, well-oiled, taut bodies; tight, firm, round buttocks; bodacious ta-tas, and pulsating thrusting, pounding . . .

I'm assuming I now have your undivided attention. Let's face it, the old adage is true: sex sells, and a preface is designed to help sell books. So, I could probably stop here and be assured you will give this book at least more than a cursory glance because, short of traffic accidents and mass murder, no other topic seems to grab and hold our attention quite so well as sex. But, all pandering aside, that is precisely why sex deserves serious academic scrutiny. It is a vital force that runs throughout our public and our private lives, but typically it is treated solely as a private concern. In point of fact, there is much we can learn about and from sex. It is the lifeblood of much of our mental, physical, and social energies. Sex lies at the nexus of a complex array of emotions, desires, biological processes, physiological responses, political aims, cultural conditioning, and social control. Indeed, no understanding of individuals or society is complete without a thorough analysis of sex and how it is constituted.

Sex, Self, and Society will expose you to a wide variety of important and provocative readings that speak to the meaning of sex and sexuality in contemporary America. The selections should provide you with a better understanding of the dynamic interplay between what society teaches us about sex and what we, as individuals, interpret, define, and experience it to be.

This book is divided into three key sections which each speak to different aspects of contemporary American understandings of sexuality. The first section begins with readings that are designed to make visible the subterranean conceptual framework and social forces that undergird modern conceptions of sexuality. Each of its five chapters reveals the power that culture has to shape and structure what we define and experience as sexual or erotic. The second section of the text examines the *mechanisms* through which culture creates and constrains sexuality—social institutions. The five chapters in this section target a combined seven social institutions to demonstrate how they order and manage the social-sexual landscape. This section also explores the tension that exists between institutional sexual imperatives (what we are *supposed* to think, feel, and do) and individual conformity and resistance (what we *actually* think, feel, and do). The final section of the text, "hot topics," looks at five unsettled and unsettling sex-related subjects that continue to provoke wide public debate. An examination of these topics reveals the active contestation of public claims makers and their struggles to determine who will have dominion over normative definitions of the realm of the sexual. Furthermore, these debates bring to the surface many of the normative assumptions that might otherwise remain unseen. Together, these readings should provide you with an acute appreciation of the social construction of sexuality, the role of institutions in creating and maintaining sexual constructions, and the costs and social consequences of retaining the existing sexual order.

KEY FEATURES

Anthologies can be useful educational tools, but because they generally employ a topical/shotgun approach to the material, readers often come away from the experience without a holistic appreciation of the subject matter. For that reason, I have included many features (described below) which should help to provide this book with the continuity and coherence that edited volumes often lack.

- The use of the theoretical paradigm ***social constructionism*** provides an overarching, thematic structure to the book and offers a valuable mechanism for active, critical analysis of American sexuality.
- An embedded framework of ***introductions*** creates a cognitive roadmap of the book and establishes and reinforces the overall organization of the text.
- ***Article prologues*** alert readers to important substantive concepts, connect selections to the constructionist perspective, and situate readings within the text's larger themes.
- ***Integrative*** critical ***discussion questions*** build upon and reinforce ideas and issues discussed in previous readings. These questions also challenge readers to look at the material in new ways and to explore how important concepts, issues, and insights may connect to their own lives.

Below are several other significant features that guided the design of this anthology.

- The readings in this anthology were selected to be ***inclusive*** and to represent a mixture of ***diverse voices*** representing a variety of ***social locations.*** Attention to issues of diversity are crucial because our ideas about sex are affected by factors such as race, class, physical and mental ability, gender, and sexual identity. These elements of our biographies interact, compete, and combine in complex ways to profoundly affect our sexual expectations, desires, and behaviors.

- The inclusion of ***classical and contemporary readings*** provides readers with exposure to critical foundational works as well as innovative and incisive present-day contributions that advance our understanding of sexual matters.
- ***Multiple levels of social analysis*** are utilized in the text including entries that address issues relevant at the individual, group, and social-structural levels.
- As an added benefit, this reader includes several easily accessible ***InfoTrac College Edtion® bonus readings*** that modify and extend the material covered in selected chapters.

I hope you find the material in this reader to be both engaging and thought-provoking. Enjoy!

ACKNOWLEDGMENTS

Writing is a humbling experience and never more so than when it comes time to try to give due credit (and sincere thanks) to those whose ideas and input have helped make a work possible. Though it is only my name that appears on the cover of this volume it is, in fact, the product of the hard work and graciousness of so many others. In recognition of this, I would like to acknowledge my debt and extend my gratitude to the many friends, colleagues and reviewers who have provided encouragement, wise counsel, and inspiration as this anthology has progressed. I would also like to thank the many helpful Wadsworth employees who have been such a pleasure to work with. So, thanks to Jeanne Ballantine, John Batacan, Jim Beers, Anna Bellasari, Jackie Bergdahl, Dana Britton, Maureen Clinton, Kasey Coleman, Nicole Foran, Sarah Harkrader, C. Lee Harrington, Eve Howard, Patricia Gagne, Michael Goslin, Janice Irvine, Bob Jucha, Michael Kimmel, Lin Marshall, Dawne Moore, David Orenstein, Barbara Risman, Julie Wilgen, and Norma Wilcox.

Introduction

"Shhhhh," we are told in embarrassed whispers, "We don't talk about things like that." Sex belongs in the bedroom, not the boardroom, and certainly not in the hallowed halls of academia. "Sex is personal." "Sex is private." It is a tacky topic that appeals only to deviants, radicals, and perverts. We don't discuss sexual matters openly within the family or, for that matter, in any "proper" social context. Yet, as philosopher Michel Foucault has noted, sex and discussions about sex are virtually ubiquitous (1980). In fact, sex permeates most aspects of our daily lives.

From advertising, family-leave policies, and the wearing of wedding rings to separate restroom facilities for men and women, and even inter-office harassment, flirtation and innuendo, sex is fundamentally entrenched in human relations and social organization. And, like so many other dimensions of identity and categories of human experience, we find that a critical examination of sex and the sexual reveals much about the distribution of privilege and power within society. It also can reveal much about the ways in which we come to understand ourselves as human beings and the particular ways in which we express sexual desires.

SEX AS A SOCIAL ENTERPRISE

This reader is designed to explore the many ways that social forces and interaction construct and situate our understanding and experiences of sex. Though many may consider sex and sexual expression to be principally physiological experiences inherently tied to raging hormones, insistent impulses, and unruly genitalia, this reader will show that, at its very foundation, sex is a social experience grounded in interpersonal relations, social scripts, and cultural norms and values. It will also show that far from being part of our "natural" programming as human beings, sex is a social act that is shaped and affected by social forces and is learned through interaction with others. It will demonstrate that, sexually speaking, what we view as "natural," "normal," and invariant is socially produced, reproduced, and contested. Sex varies across both temporal and geographic boundaries. In other words, sex—including our sexual activities and desires, our understanding of what sex is, and what sex means to us as individuals, depends—it depends on such things as when and where you were born, your sex, age, ethnicity, social class, and marital status, who your friends and family are, what kind of job you have (or plan to have), the status of the economy, and even the god, gods or goddesses that you worship.

The pages in this volume will reveal a wide and incredibly diverse set of factors that define and shape "sex." They will demonstrate that sex is a vital part of the self and the social order and that a true understanding of sexuality is not possible without an active and critical examination of the relationship of sex to each.

THE EMERGING FIELD OF SEXUALITY

Despite its pervasiveness and integral importance to human social relations, it is only within recent years that academia has directed a focused gaze upon the realm of the sexual. Sex-research chroniclers trace the origins of sex research to the early 1890s through the 1930s when research on sexual matters tended to be rather myopic in scope and generally took a moralistic "social problems" approach to the study of sexual behavior (Bullough 1994; Epstein 1994). Early sex research typically took the form of surveys explicitly designed to help government officials "educate" the public about "vices" such as masturbation and venereal disease. Other early contributions included comparative and cross-cultural ethnographies of the sexual lives of people living in foreign and exotic locales (Best 1981). By and large however, both types of these early efforts tended to be small in scale and overly concerned with unusual or marginalized sexual behaviors (Epstein 1994; Stein 1992). They did not reflect any sustained interest or concern about sexuality by academics, or society as a whole, nor did this body of research treat sex as an integral part of social systems or as a central feature of individual identity.

The beginning of modern sex research is generally associated with the work of Alfred Kinsey and other behavioral scientists who began studying human sexual behavior in the 1940s (Epstein 1994). The bulk of these works were primarily concerned with documenting, describing, and categorizing sexual behaviors. These efforts included significant and methodologically rigorous national social surveys of sexual behavior conducted by researchers such as Kinsey (1948, 1953) and Masters and Johnson (1966, 1970).

A second strain of modern sex research derived from qualitative analyses conducted by deviance researchers. These researchers attempted to provide a glimpse into the lives and activities of those engaging in sexually "deviant" acts. Such works included Humphries' (1961) research on "*tearooms*" (public restrooms used by men for same-sex sexual encounters), an analysis of the ways that nudists negotiate respectability (Weinberg 1958), the social organization and identity formation of male youth street hustlers (Reiss 1961), and the techniques used by madams to train house prostitutes (Heyl 1979).

Both strains of these significant works were quite useful in exemplifying the diversity of sexuality that existed in American society in the mid- and mid-to-late twentieth century. Data from the quantitative surveys provided a much needed baseline for documenting and describing Americans' sexual proclivities. Works from the deviance literature helped identify the rich diversity of human sexual behaviors and to demonstrate how individuals attempted to manage marginalized "deviant" sexual behaviors. However, as a whole, both strands of research tended to treat sexuality as something completely apart from the social processes shaping it. Statistics on topics such as the frequency of adult coitus, the duration of sexual activities, and the average age of first sexual encounters were taken as ends in themselves. Questions about social context were left unexamined. For example, *why* does the frequency and duration of sexual encounters matter in this society? Why might they not matter in other societies? What effect did World War II have on the sexual and procreative patterns of adults during the 1940s (the era of Kinsey's research)? Are patriarchal societies more likely than matriarchal societies to define sex from the perspective of the male (e.g., male orgasm)? Does the economic base of a country shape the average age of first intercourse? And so on. In short, the impact of social forces on the rates and diversity of sexual behaviors remained hidden from public view.

In a similar way, works deriving from the deviance tradition (informally dubbed the "nuts and sluts" approach), tacitly accepted existing standards of normality and abnormality as unquestioned truths. They failed, for example, to examine how and why ideas about "normal" or "typical" sexuality were determined in the first place or whose interests such definitions might serve. They also failed to recognize or address how sexuality was embedded structurally in the fabric of society. Society's role in the production of sex and the sexual remained critically unexamined. To a large extent, the field of sexuality remained a marginalized and underutilized field of study.

In recent years, however, academic inquiry into the field of sexuality and sexual behavior has seen exponential expansion. Sexuality research has gained increased recognition from the academic community as a central component of self and an indelible feature of social structure and organization.

Much of the increased interest in sexuality has emerged as an outgrowth of research in the areas of gender, feminist, and women's studies that surfaced in the 1960s and became a staple of academic research by the late 1970s and early 1980s. Researchers working in this vein revealed that biological ***sex*** (whether one is a male or female) was a crucial source of differentiation omnipresent in social relations and the very fabric of society. These analyses demonstrated that differentiation by sex was of essential importance in shaping the distribution of social goods and services. In other words, these researchers showed that being male or female greatly influenced the opportunities, responsibilities, and behaviors considered socially "appropriate" for each sex. In addition, research in gender, women, and feminist studies revealed the salience of ***gender*** (differences between men and women that are culturally, rather than biologically based, i.e., masculinity and femininity) to identity-formation, self-concept, locus-of-control, career aspirations, and countless other concerns produced and mediated at the individual level.

In modern American society, gender and sexuality are issues that are closely intertwined. In fact, they are often ***conflated*** or treated as cognitive synonyms. So, the successful accomplishment of a heterosexual identity (the dominant ***normative*** [typical/standard], expectation) is largely equated with, and contingent upon, the successful accomplishment of gender, despite the fact that the gender and sexuality are *not* mutually deterministic identities. For example, muscular, aggressive, hockey-playing men are generally assumed to be heterosexual, while flannel-wearing, tobacco-chewing females are labeled lesbians. The reality is that "macho" men can be (and often are) gay and "butch" women can be (and often are) straight. Nonetheless, assumptions about sexual identity tend to be aligned with how well individuals measure up to contemporary gender norms rather than their actual sexual behaviors or preferences. And, these distinction matter. As with sex, gender, race, class, and other axes of human differentiation, being assigned to a non-dominant sexual category (e.g., non-heterosexual) means a loss or denial of social privilege and often a marginalized sense of self. It means you will likely have farther to go with fewer resources and greater burdens to carry with you in life's journeys.

Because gender and sexuality are vital aspects of social and individual identity that can affect individual life chances, and because they are so culturally conflated, academic explorations of maleness and femaleness as well as femininity and masculinity lend themselves to similar considerations of sexual behavior and identity. For example, contemporary researchers have begun to explore the ways in which sexuality, like gender, operates as an organizing principle throughout, across, and within social systems (Martin and Collinson 1999). Similarly, contemporary sex research is going beyond issues of structure to explore issues of process (how cultural understandings of sex and sexuality come to be) and the role of individual sexual ***agency*** (the extent to which people make conscious and active choices about sex and sexual behaviors). In addition, many of the issues of key concern and contention for gender and feminist theorists, such as rape, prostitution, sexual harassment, and pornography, are matters that are ultimately linked to the social organization of sexuality. For these reasons, and more, feminist and gender studies have been instrumental in the emergence of the field of sexuality.

A second, and equally crucial, area of study to contribute to the emergence of sexuality as a legitimate site of scholarly discourse is the field of gay and lesbian studies. Researchers in this arena have made invaluable empirical and theoretical contributions to the academic exploration of sexuality, rethinking, and challenging (a.k.a. **"*queering*"**) the boundaries of dominant sexual categories (Seidman 1992).

Prior to the establishment of gay and lesbian studies, researchers typically found it sufficient to treat homosexuality as an anomalous and deviant human condition. Much analytic capital was spent in attempts to determine its prevalence, explain its etiology, and describe the norms and environs of the cultural milieus in which same-sex sexual activities took place. Some researchers sought to portray its purported evils, others to defend and humanize the category's denizens. But it was only with the emergence of gay and lesbian studies that heterosexuality, and the categorization of sexual identity itself, began to be fully critiqued and challenged. Rather than treating heterosexuality as the unspoken and "proper" or "natural" norm, academic inquiry began to address multiple forms of sexual behavior and identity (including heterosexuality, bisexuality, asexuality, and transgender identity) and treat them as equally significant subjects of analysis.

Spawned during the politically turbulent and socially volatile decade of the 1960s, gay, lesbian, and women's studies created an intellectual space in which sexuality could be viewed as a valid and crucial field of study devoid of the stigma it had often been ladened with in eras past. Moreover, important social changes including the increased sexual agency of women, the spread of AIDS, innovations in fertility, increased media attention, and the ready availability of contraceptives, have resulted in a social order that today views sex not just as an enjoyable extra-curricular activity but also as an important topic that is an integral part of social life worthy of serious academic attention.

This expansion is reflected in increased research, scholarly publications, and the development or expansion of social-science-based sexuality courses. This reader is designed to speak to this proliferation by offering a selection of readings that focus on the social bases of human sexuality using the theoretical perspective known as social constructionism.

THE SOCIAL CONSTRUCTIONIST PERSPECTIVE

Social constructionism is a theoretical perspective that holds that reality is created and sustained through human social interaction. The roots of this perspective are grounded in the sociological perspective known as symbolic interactionism, but the social-constructionist perspective owes its greatest intellectual debts to Peter Berger and Thomas Luckmann's landmark work *The Social Construction of Reality* (1966) as well as Malcolm Spector and John Kitsue's influential *Constructing Social Problems* (1977).

Works in the constructionist tradition contend that our understanding of the world around us, including our values, norms, attitudes, and knowledge systems, derive from our social relations. It is society, social constructionists argue, that will fashion what is right and wrong, just and unjust, moral and immoral, normal and deviant, fact, fantasy, and falsehood. For constructionists, meaning is amorphous, inchoate; it takes particular shapes within particular social environments. Adherents explain that we view and come to know the world through a cultural lens; therefore, our experiences are contextualized and given meaning within particular social systems. We learn from society how we are to cut up and organize the world, what features of social existence we will categorize, order, and give meaning to. Take the case of a person who is killed after being injected with a lethal dose of deadly chemicals. What is the meaning of this occurrence? If it is the demise of a serial killer who has been sentenced to death by the state, in some societies this act would be considered justice; however, other societies would deem this to be pure barbarism. Similarly, a constructionist perspective would reveal that if the deceased were an elderly patient with an excruciatingly painful terminal disease, geographical boundaries might well determine whether the death is viewed as mercy, murder, or of any social consequence at all. A constructionist perspective reveals that in each of these cases, a life has been taken through the deliberate actions of another, yet the act is evaluated quite differently based on the cultural criteria that "matters" and defines acceptable behavior within a given social order.

The social-constructionist perspective is typically contrasted with essentialism. ***Essentialism*** holds that there is a core, fundamental essence that inheres in the objects and features of our world. This essence is fixed, unchangeable, definitive, and has an objective reality. This essence, essentialists contend, exists independently of culture and human discernment. Often, essentialist claims appeal to "biology" or "nature." For example, the assertion that there are *natural* differences between men and women in the willingness to stop and ask directions when lost is essentialist, as are the contentions that women have a *genetic* predisposition to nurture and that white men can't jump.

For social constructionists, however, the social order is not part of the nature of things. It exists only as a product of human activity (Berger and Luckmann 1966). A constructionist would argue, for example, that categorizations such as male and female, white and nonwhite, are social rather than natural in origin and can vary across cultural and temporal boundaries. Further, there are many individuals who would not fit neatly into these dicholonized categories. Where does one place the ***intersexed*** (individuals with both male and female sex organs or ambiguous genitalia)? Children of mixed racial ancestry? Who counts as white? Nonwhite? And who makes these decisions?

Issues of definition are of particular concern for social constructionists because they are said to serve the interests of the powerful. The privileged members of a social order have the most control over how reality is constructed, and therefore, they have a greater ability to delineate and circumscribe the world in ways that are beneficial to them. Accordingly, a constructionist might ask what groups benefit from classifying and defining reality in specific ways or why particular classifications and definitions are enforced during some historical epochs but not others.

A SOCIAL CONSTRUCTIONIST APPROACH TO SEX

For well over the past 20 years, the social constructionist paradigm has emerged as one of the foremost theoretical perspectives applied to the study of sex in the social sciences. Though it is most often associated with the work of Michel Foucault, the constructionist approach to sexuality found some of its earliest expression in the work of sociologists (Epstein 1992; Jackson 1999; Stein and Plummer 1994).

One of the first and most significant sociological works that heralded the emergence and prominence of the constructionist approach to sexuality was

John Gagnon and William Simon's 1973 treatise titled *Sexual Conduct: The Social Sources of Human Sexuality* (see also Simon and Gagnon 1967, 1969). A classic refutation of essentialism and psychological drive theories, Simon and Gagnon argued that sex is not natural but has been *naturalized* historically. They explained that people *learn* to be sexual and that they do so in the same ways they learn everything else, "without much reflection, they pick up directions from their social environment" (1973:14).

Another landmark work presaging the emergence of the constructionist perspective and exposing the complicity of social processes in sexual matters was Mary McIntosh's article "The Homosexual Role" (1968). This work openly critiqued essentialist dualistic models of sexual identity (i.e., the homosexual versus heterosexual classification system) and pointed to the social origins of "the homosexual role."

Both of these innovative works challenged prevailing assumptions about the biological "nature" of sex by exposing the fact that sexual behaviors and identity do not emerge fully formed at birth but, rather, take shape across the life-course assiduously guided by social scripts and normative expectations circumscribing the boundaries and even the potentialities of sex within a given socio-cultural domain. Formal applications of social constructionism began soon thereafter, and by the 1980s constructionist approaches densely populated the academic landscape. Today, its popularity endures, and it remains one of the most dominant and valuable approaches to sexuality available.

Constructionist works on sexuality cover a wide spectrum of topics and vary greatly in subject matter. The bulk of constructionist approaches focus on the social construction of sexual identity (e.g., homosexual, heterosexual, bisexual, virgin, queer) though a good number are also concerned with the construction of specific social problems or particular aspects of sexuality (e.g., prostitution, rape, sexual consent, infertility, and love).

Another facet of variation among social constructionists is the role of biology in sexual matters. Most constructionist works grant biological processes at least some role in the enactment or actualization of sexuality (albeit often a minor or quite minimal role). However, the "purest" and most radical constructionist approaches contend that sexuality itself is socially constructed. This strand of social constructionism rejects the very existence of biological drives and argues instead that what exists is the *potential* for human consciousness, behavior, and physical experience. These theorists argue that social forces, namely definition, regulation, organization, and categorization are required to incite this potential (Foucault 1980).

Despite these variations, constructionist analyses of sexuality proceed along common trajectories. The communal methodology of the social constructionist is to deconstruct ***hegemonic*** (dominant) sexual definitions and categorizations such as rape, prostitution, and homosexuality by documenting their historical or cultural variations (or to indicate the potential for such variation) thereby exposing the instrumentality and situated interests of humans and social systems that are involved in their creation. Sexual constructionists also bring attention to the immense forces of social control that are often brought to bear to enforce these social constructs. They also point to variations in when, how, and upon whom prevailing definitions are enforced. In short, to deconstruct sexual definitions and categories is to reveal that the existing definitions of the social order are variant products of cultural systems that serve particularized interests rather than natural, essential, ontological categories of human existence.

And what do constructionist analyses of sexuality tell us? We learn from social constructionists that sex is a social product rather than a presocial, natural act. Our subjective sense of sex is given shape and substance within the confines of a carefully constructed cultural canopy of sexual mores and social strictures disseminated through social institutions and negotiated, challenged, and perpetuated by individuals. As many of the readings in this book will show, sex may involve bodies, touch, and sensations, but it is our culture that ultimately determines which gestures, movements, and tactile sensations will be experienced as sexual or count as sex because virtually anything can be sexualized (or desexualized) through social definition and regulation (Weeks 1985). In fact, evidence from monkeys raised in isolation suggests that culture, social interaction, and socialization are so important to our perceptions of sex that without them we wouldn't even know how to reproduce (Harlow 1965).

Social constructionist perspectives reveal that who and what we find erotic and sexy (e.g., males or females, the young or the old, blondes, brunettes, or redheads, breasts, backsides, nostrils, or toes), the specific behaviors deemed appropriate for the expression of sexual feelings (e.g., touching, holding hands, gripping, groping, kissing, masturbation, making love, oral sex, cyber sex, anal sex), and the pool from which we are "allowed" to select our sexual partners (e.g., from any human to those of a specific biological sex, a specific race, a specific social class or rank, a certain religion, a specific age range to anything with a pulse) are just a few of the ways in which culture constructs and regulates sexuality. In short, we learn that sex is contingent, culturally manufactured, and socially engineered to take particular forms during particular historical epochs. Constructionist accounts reveal that sex is "fluid and changeable, the product of human action and history rather than the invariant result of the body, biology or an innate sex drive" (Vance 1989:160).

The constructionist perspective is not without its critics or shortcomings. For example, by critiquing categories such as homosexuality and heterosexuality constructionist analyses must use and may, therefore, indirectly ***reify*** (treat a socially constructed concept as real or concrete) the very concepts they are deconstructing. Further, social constructionism cannot fully determine the role of biology in sexual behavior (for discussion of these and other critiques see Stein 1992; Epstein 1987; and Vance 1989). Nonetheless, the constructionist perspective is unique in its ability to reveal the complex, nuanced, and often hidden social forces that shape, enable, and constrain sexual potentialities within given cultural contexts, thereby illuminating prospective paths of social change. It is also one of the few perspectives that recognizes the involvement and complicity of social actors in accepting, resisting, and reproducing hegemonic sexual forms. In sum, social constructionism provides a unique and critical approach to the study of sexuality that puts in highest relief the intricate, dialectical relationship between sex, self, and society.

THE GOALS AND ORGANIZATION OF THIS VOLUME

In this volume I provide a broad and critical examination of issues of sex and sexuality in American society. I have included both classic and contemporary readings that address multiple levels of social analysis ranging from individual interaction and identity to the impact of social-structural forces and social institutions. While not every reading in this volume will be an explicitly constructionist work, each piece speaks to the constructionist mission in a meaningful way. Many of the included readings examine the ways in which broad social forces fashion and control matters concerning sex and the erotic. Other selections explore the multiple ways that individuals internalize, negotiate, replicate, resist, and challenge dominant sexual constructions and normative standards regarding sexuality.

Another defining feature of this text is the use of an integrative framework focusing on issues of race, class, gender, and physical ability. Each of these domains contextualizes and conditions the social constructions of sexuality. For example, social strictures regarding African-American sexuality may differ significantly from those governing Caucasian sexuality in much the same way that normative standards of sexuality and sexual behavior differ for males and females. Further, inclusion of these additional sources of social differentiation help reveal how social identities are further stratified into dominant and subordinate forms. Similarly, rather than treating them as "special" or "deviant" topics, non-hegemonic forms of sexual identity (e.g., transgenderism, homosexuality, and bisexuality) are incorporated throughout the text. Consistent with the constructionist framework, the readings assess and challenge sexual categorizations, recapture their history, and address questions of power and subordination, resources and privilege. I have selected readings that I hope will be engaging and challenging, yet will remain highly readable.

The book is divided into 15 chapters divided equally between three substantive sections. The sections have been designed to move from a broad understanding of the social context and construction of sexual behavior (Section 1), through an assessment of the ways in which social institutions organize and disseminate cultural constructions of sexuality (Section 2), to the exploration of specific divisive and controversial social issues (Section 3) while taking into account multiple axes of social identity. As such, I hope that this reader will serve as a stimulus for a critical and focused examination of contemporary American sexuality.

SOURCES

Berger, Peter L. and Thomas Luckman. 1966. *The Social Construction of Reality*. New York: Doubleday.

Best, Joel. 1981. "The Social Organization of Deviance," *Deviant Behavior* 2:231–259.

Bullough, Vern. 1994. *Science in the Bedroom: A History of Sex Research*. New York: Basic Books.

Epstein, Steven. 1987. "Gay Politics, Ethnic Identity: The Limits of Social Constructionism," *Socialist Review* 17:9–51.

Epstein, Steven. 1994. "A Queer Encounter: Sociology and the Study of Sexuality," *Sociological Theory* 12:2.

Foucault, Michel. 1980. *The History of Sexuality Volume I: An Introduction*. New York: Vintage Books.

Gagnon, John and William Simon. 1973. *Sexual Conduct: The Sources of Human Sexuality*. Chicago: Aldine.

Harlow, Harry. 1965. "The Affectional Systems," in Allan Schrier et al. Eds., *Behavior of Nonhuman Primates: Modern Research Trends*. New York: Academic Press.

Heyl, Barbara Sherman. 1979. *Madam as Entrepreneur: Career Management in House Prostitution*. New Brunswick, N.J.: Transaction Books.

Hooker, Evelyn. 1958. "Male Homosexuality in the Rorschach," *Journal of Projective Techniques* 22:33–54.

Humphries, Laud. 1970. *Tearoom Trade: Impersonal Sex in Public Places*. Chicago: Aldine.

Jackson, Stevi. 1999. *Heterosexuality in Question*. London: Sage.

Kinsey, A., W. Pomeroy and C. Martin. 1948. *Sexual Behavior in the Human Male*. Philadelphia: W. B. Saunders.

Kinsey, A., W. Pomeroy, C. Martin and P. Gebhard. 1953. *Sexual Behavior in the Human Female*. Philadelphia: W. B. Saunders.

Martin, Patricia Yancey and David Collinson. 1999. "Gender and Sexuality in Organizations," in *Revisioning Gender,* Myra Marx Feree, Judith Lorber, and Beth Hess eds. Thousand Oaks, CA: Sage.

Masters, W. H. and V. E. Johnson. 1966. *Human Sexual Response*. Boston: Little, Brown.

Masters, W. H. and V. E. Johnson. 1970. *Human Sexual Inadequacy*. Boston: Little, Brown.

McIntosh, Mary. 1968. "The Homosexual Role," *Social Problems* 16:182–192.

Reiss, Albert J. 1961. "The Social Integration of Peers and Queers," *Social Problems* 9:102–120.

Seidman, Steven. 1996. *Queer Theory/Sociology*. Cambridge, Mass.: Blackwell.

Simon, William and John H. Gagnon. 1967. "Homosexuality: The Formulation of a Sociological Perspective," *Journal of Health and Social Behavior* 8:177–185.

Simon, William and John H. Gagnon. 1969. "On Psychosexual Development," in *Handbook of Socialization Theory and Research,* D. A. Goslin ed. Chicago, IL: Rand McNally.

Spector, Malcolm and John I. Kitsue. 1977. *Constructing Social Problems*. Menlo Park, CA: Cummings.

Stein, Arlene. 1989. "Three Models of Sexuality: Drives, Identities and Practices," *Sociological Theory* 7(1):1–13.

Stein, Arlene and Ken Plummer. 1994. " 'I Can't Even Think Straight': 'Queer' Theory and the Missing Sexual Revolution in Sociology" *Sociological Theory* 12:178–187.

Stein, Edward. 1992. *Forms of Desire: Sexual Orientation and the Social Constructionist Controversy*. New York: Routledge.

Vance, Carole S. 1998. "Social Construction Theory: Problems in the History of Sexuality," in Peter M. Nardi and Beth E. Schneider eds. *Social Perspectives in Lesbian and Gay Studies: A Reader*. New York: Routledge.

Weeks, Jeffrey. 1985. *Sexuality and Its Discontents: Meanings, Myths, & Modern Sexualities*. London: Routledge.

Weinberg, Martin S. 1976. "The Nudist Management of Respectability," in *Sex Research: Studies from the Kinsey Institute,* Martin S. Weinberg ed. New York: Oxford University Press.

PART I

Sexual Constructions

INTRODUCTION

The works in this section of the text explore how our understandings about sex and the realm of the sexual come to be—how culture molds and shapes both the meaning and experience of sexuality. These selections also examine how individuals negotiate sexual interaction and identity given these cultural expectations. The first chapter "*Constructing Sex, the Sexual, and the Erotic*" introduces the constructionist perspective and exposes the multitude of social forces that produce, shape, and constrain conceptualizations of sex and sexual behaviors. In short, it reveals that the act of sex is not "natural," but is, in fact, a *social* enterprise. Further, it uncloaks many of the hidden assumptions we make about sexuality and desire (e.g., what is "hot," what is "inappropriate") and reveals these to be social products as well. The chapter should challenge you to re-examine your ideas about what is "natural" and should also help you begin to appreciate how social forces are crucial in producing and regulating the world of sex.

This section's first article will provide much of the theoretical glue for this chapter, as well as the entire volume. It includes a discussion of the functions of sex, introduces the constructionist perspective, and explores and challenges the cultural meanings of sex. The chapter's other selections extend the constructionist critique of sexuality to show that even ideas about what is and what is not sexual or erotic (e.g., breasts, a kiss, a touch or caress), are shaped by social and cultural influences such as race, class, and reproductive status.

The section's second chapter, "*Constructing and Critiquing Sexual Categories,*" moves from exposing the social roots of sexuality to an examination of the formation and maintenance of the categories through which our society ascribes sexual meaning and identity (e.g., heterosexuality, homosexuality). These readings reveal not only that these categories are socially constructed, but also that the way they are constructed profoundly shapes sexual experiences, erotic possibilities, and even experiences of social subordination and social privilege. Works included in this chapter are designed to challenge you to critically explore and evaluate existing hegemonic categorizations of sexuality and the consequences of organizing sexuality in these ways. Selections include Steven Epstein's discussion of queer theory in "A Queer Encounter" as well as Jonathan Katz' compelling analysis of the history and invention of heterosexuality.

The third chapter, "*Conflated Construct: Sex, Gender, and Sexuality,*" integrates two additional social constructs, sex (i.e., male and female) and gender (i.e., femininity and masculinity), into the analysis of sexuality and reviews their conflation with and complex interrelationship to cultural formations of sexuality and sexual identity. Each piece should help you to better understand the distinctions between these concepts and to better appreciate some of the problems that are created by treating these constructs as mutually reinforcing or interchangeable indicators of identity. Selections include John Stoltenberg's "How Men Have (a) Sex," and Judith Lorber's "Beyond the Binaries: Depolarizing the Categories of Sex, Sexuality, and Gender."

Chapter 4 is titled "*'What's Love Got to Do With It?' Constructions of Desire, Love, and Intimacy.*" This chapter takes up the constructionist mantle by demonstrating cultural and temporal variations in notions of love, intimacy, and desire, and specifically examines the relative importance of romantic love in sexual relationships over time. Selections include Francesca Cancian's important contributions on the feminization of love as well as Steven Seidman's influential analysis of the sexualization of love.

The culminating chapter of the section, "*Constructing the Sexual Self: The Negotiation and Actualization of Sexual Identity and Behavior,*" explores the issue of sexual identity—how individuals negotiate and mediate the impact of culture, socialization, biology, and biography to attain a sense of themselves as sexual beings. This chapter's works speak to the active tension between the social and the self—between cultural hegemony and individual conformity, resistance, and adaptation. The selections also highlight how social factors such as gender, class, and race influence decisions about sexual identity and behavior. These selections include Michael Messner's autobiographical account of "Becoming 100 percent Straight" as well as Holly Devor's cogent analysis of the negotiation of sexual identity for female transsexuals.

1

Constructing Sex, the Sexual, and the Erotic

'Doing it': The Social Construction of S-E-X

TRACEY STEELE

What is sex? And by this question I do not mean the dichotomous morphological division of the species into male and female. Rather, I mean what is "sex," **S-E-X**?[1] Is it the sacred joining of two souls on a spiritual and physical plane? A commodity bought, sold, and traded in the cultural marketplace? A way to sustain the species? A delightful muscle spasm? A conjugal obligation? A weapon used to objectify, humiliate, and subjugate?

Sex is all of these things and many more. Its meanings and functions vary from epoch to epoch and culture to culture. It can simultaneously serve a variety of private motives and social aims, which themselves are shaped by the vicissitudes of personal desire and the sexual possibilities articulated through existing social conventions. In short, the meaning of sex is a product of both individual and social factors. What often goes unnoticed, however, is that while the purpose and ultimate significance of individual sexual interactions may differ, the underlying sense that we know what sex *is,* generally, does not. I may be in it for love, and you may be in it for money, pleasure, exercise, companionship, procreation, revenge, pity, ego, or countless other reasons, but we both "*know*" what "*it*" is. After all, it's sex, it's natural, and *everybody* knows what sex is.

But do we? Curiously, we seem to forget, or wholly disregard, the fact that the road to adulthood is routinely paved with intense insecurity, doubt, and angst concerning that great mystery of mysteries—sex. Playing "doctor" with the next-door neighbor, furtive scrutiny of purloined pornography, tentative adolescent fumblings, as well as others' world-wise double entendres that induce our hesitant laughter, silent confusion, and nervously feigned sophistication all testify to a decided uncertainty and ***lack*** of knowledge regarding sex. The truth is sex is *not* something innate, something we instinctively know; it is something we learn about, sometimes slowly, sometimes painfully, as we grow to adulthood.

In the following pages I will utilize what is known as the social-constructionist perspective to explore the meaning of sex in contemporary American society. This perspective posits that sex, rather than being natural and instinctual, is,

Steele, T. (ed). 2005. Sex, Self & Society. Wadsworth

in fact, principally learned and produced within specific cultural contexts. I begin by ***deconstructing*** the meaning of sex in contemporary American society. Deconstructing sex involves examining the hidden assumptions built into how sex is defined and understood. In addition, the deconstruction of sex will reveal both the possibilities and the limits of existing definitions (Rubin 1984). It will also bring to light how existing definitions work to privilege some social groups at the expense of others. Together these discussions will show that sex, rather than being a wholly instinctual, innate phenomenon, is largely shaped and defined by factors external to the individual. More to the point, these discussions will show that sex is, fundamentally, a *social* enterprise.

THE SOCIAL CONSTRUCTIONIST PARADIGM

Modern Western societies are not the first to explore sexuality as a topic of intellectual inquiry. Philosophers, scholars, and clergy of countless generations have pondered its significance and place and have come to radically different conclusions about how best to define, express, circumscribe, and manage libidinal energies and behaviors.

Lynne Segal (1994) has identified three historical intellectual traditions that have dominated Western thinking about sex. These frameworks are the spiritual, the biological, and the social. In pre-industrial Europe, societal views about sexuality were largely shaped by spiritual and religious beliefs. With the emergence of the industrial revolution, however, the dominance of these views were supplanted by models grounded in science and scientific thinking. In these early stages, modern science was heavily influenced by Darwinian logic; thus, scientific explorations of human behavior tended to be flavored with a decidedly biological bent. It is within this historical moment Segal argues that the notion that sex is a biologically-based phenomenon became an accepted truism in Western thought (Segal 1994).[2]

Generally, such *essentialist* approaches to sexuality hold that sex is a matter of biological essence, that it is a product of biological force "seeking expression in ways that are preordained" (Epstein 1987:15).[3] For essentialists, sexuality is fundamentally a product of organic and biochemical processes, a function of our innate nature. Essentialists consider sexual identities to be "cognitive realizations of genuine, underlying differences" (Epstein 1992:241). For example, essentialists maintain that individuals are *by nature* heterosexual or homosexual. We are *born* straight or gay. Men are *naturally* more sexually aggressive and promiscuous than women (ostensibly because they have been *biologically* programmed to spread their seed and maximize their procreative potential) (Buss 1998). All of these claims are grounded in the fundamental notion that sex is a category of human existence dictated by nature that is eternally "unchanging, asocial and transhistorical" (Rubin 1984:275).

The association of sex with nature in these models is not unintentional. By emphasizing something occurs in nature, a sense of validity is conferred upon a subject (Tiefer 1995). In other words, by invoking nature, the *claims* of scientists regarding the etiology of sex and sexual behavior are elevated to apparent *facts* and larger *truths,* conveniently eliding the role of social forces in their creation.

In recent years, a formidable challenge to essentialist models of sex and sexual behavior has emerged. Though biological models still hold considerable sway in both popular and scientific thinking, new thinking about sex, sexuality, and the erotic assert that each is, in fact, socially constructed and not simply a matter of biological mandates.

Social constructionism is the theoretical perspective which argues that our perception of what is real is defined only by the meaning that we attribute to a given situation (Berger and Luckman 1967; Blumer 1969). These meanings are derived in a social context and are created through social interaction that occurs in particular social environments. Simply put, it is through our interaction with others that we learn how to interpret and evaluate the world around us. Constructionists also assert that social hierarchies play an important role in ascription of meaning; individuals and groups at the top of social hierarchies are those most likely to have their definitions and views imposed and enforced *as* reality.

Applied to sexuality, this perspective holds that sexuality has no "inherent essence" (Harding, 1998:9). Ideas about sexuality are not hard-wired or "natural"; they arise in particular social-historical contexts (Foucault 1980). For constructionists, sexuality is an arrangement of cultural norms, values, and expectations which themselves are fundamentally shaped by hierarchies and matrices of social power relations (Foucault 1980; Gagnon and Parker 1995; Harding 1998; Vance 1984).

Constructionists argue that if sex and the realm of the sexual *were* natural, we would not have to be taught what sex was and it would take consistent forms across societies (Tiefer 1995). But, as we know, sex is learned, and it is expressed in an astonishing variety of ways. Robert Padgug observes: "The forms, content, and context of sexuality always differ. There is no abstract and universal category of 'the erotic' or 'the sexual' applicable without change to all societies" (1992:54). Sex and the erotic are highly malleable constructs expressed in innumerable modes across the globe—in fact, we can (and do) attach erotic desires to almost anything. For example, for nearly one thousand years, small feet were considered sexually appealing in Chinese women; for the Mangbettu of Central Africa, it is elongated cone-shaped heads that are particularly attractive; and in modern American culture it is women's breasts that are supposed to make (heterosexual) men 'hot'. In one culture, the term 'sex' may be reserved exclusively for heterosexual intercourse while in another, 'sex' may refer to any sexual encounter that produces an orgasm in at least one of the participants. The bottom line is—sex varies.[4]

The definition of sex, who can engage in it, at what age, for how long, and from what position are just a few examples of sites of sexual variation. Our culture provides the template upon which erotic desires are channeled and shaped. Because we experience the same cultural sexual indoctrination as other members of society, we generally share the same sexual beliefs, values, and desires as those around us. So, for all intents and purposes, sex *appears* to be invariant and natural when it is not.

The implications of this perspective challenge many contemporary notions about sexuality. For example, a constructionist perspective asserts that male sexual aggression is learned rather than inborn, that women may not naturally wish to bear children, and that public nudity, promiscuity, and premarital sex are not inherently wrong or sinful. Rather, constructionists contend, ideas and values surrounding sexual reality are learned, and they can, and do, change. In the following section, I adopt a constructionist framework to critically examine, identify, explore, and deconstruct cultural assumptions about S-E-X that are embedded within contemporary American society. Though the social construction of sexuality encompasses far more than the just the delineation of activities that constitute the act of sex in our society, this construct is essential to examine because it serves as a primary focal point for most scientific inquiry and public debate—for most of us, sexuality is primarily about S-E-X. A critical examination of the social construction of S-E-X can therefore provide considerable conceptual entré into a broader and less culturally-bound analysis of the sexual realm.

ISSUES OF DEFINITION

I begin my course on sexuality by having students write their definition of sex on a small index card. "Let's say someone from Mars came to Earth and asked you to explain what this thing called sex is that is so often the focus of movies, songs, and barroom conversations," I begin. "What would you say? What is it? What is it for? How do you know when you've had it? How would you explain to this ultimate outsider what this activity 'sex' is all about?"

Quickly they set to the task, most of them confident and eager. A few finish quickly. "The man puts his penis in the woman's vagina." "Intercourse" "It is something adults do when they are in love." "It's the way we reproduce the species." A few take longer to complete the task. Slowly, steadily, frustration grows. Brows wrinkle, heads tilt, lips purse. Many struggle to put into words something they believe they know, something natural, something that, as they put pen to paper, suddenly does not seem so simple, or so automatic. Typically, clarification is requested. "Do we need to discuss gay sex?" "Does the Martian know what genitalia are?" "Should I limit it to 'vanilla' sex?" "Sex like in the movies or 'real' sex?" "Should I explain about how men want it all the time?" "The Martian knows NOTHING about sex," I reply. "Start from ground zero." Invariably, someone remarks, "This is hard!"

And indeed it is. When we speak of sex and matters sexual we may be referring to a single specific act, a group of acts, or something that can encompass an enormously wide variety of thoughts, emotions, and behaviors. What is sex? Does it inhere in certain affective states (e.g., love), physiological responses (e.g., excitement, orgasm) specific actors (e.g., males with females), or specific acts (e.g., vaginal penetration by an erect penis)?

The ambiguity inherent in these questions was acutely illustrated in 1997 with the eruption of a sexual political scandal involving former President Bill Clinton and White House Intern Monica Lewinsky. Amid intense political pressure and media scrutiny, the president made a public address in which he vehemently denied "having sex" with Ms. Lewinsky. However, later, the public learned that the President and the intern had been involved in *oral* sex. While many members of the general public may have felt understandably deceived, the President was by no means alone in making this kind of categorical distinction.

Several recent studies indicate that the American public *does* differentiate between oral sex and what it considers to be "real" sex (Bogart, Cecil, Wagstaff, Pinkerton, and Abramson 2000; Bogart, Pinkerton, Myaskovsky, Wagstaff, and Abramson 1999; Sanders and Reinish 1999). The work of Bogart et al. (2000), for example, indicates that while a vast majority of the college students they sampled (97 percent) considered heterosexual vaginal intercourse to be sex, slightly fewer (93 percent) were willing to label anal heterosexual sex as sex, and fewer than half (44 percent) considered that oral sex constituted real sex.[5]

An interesing extension of such views is reflected in national opinion surveys which indicate that youth may be engaging in significantly higher rates of oral sex and other sexual behaviors believing that by doing so they have not had sex. Further, many indicate that by engaging in this particular form of erotic expression they have not technically lost their virginity (Sanders and Reinisch 1999; Indigo 2000). In short, despite the existence of a wide variety of socially identified sexual behaviors, in modern American society, the designation of having had sex is typically limited to a quite narrow range of erotic expression.

In the following pages, I present a limited deconstruction of the term "sex" as it is understood in modern American culture. A critical examination of S-E-X reveals that the term represents a particular constellation of socially determined sexual norms and expectations. In other words, sex, *real* sex, means something very specific in this culture, but the content of that meaning lies cloaked beneath layers of taken-for-granted assumptions that are so embedded in our cultural framework that they are rendered conceptually invisible. So, what *is* the content of S-E-X? What exactly *does* S-E-X mean? I raise this and subsequent questions not in search of definitive answers or solutions, but rather to stimulate awareness and provoke discovery—to peel back the layers of the social fabric to expose both the limits and possibilities of human sexual interaction.

SO, S-E-X IS . . .

To discern the basic assumptions that go into defining sex in this culture I find it quite useful to strip away rhetoric and discursive distractions. Though there are several ways to do this, let's try pantomimes. Take a moment and, using only your hands, make a gesture for sex. Go ahead, it won't work nearly as well if you don't play along. C'mon

O.K. Finished? Great! Now, repeat this gesture and pay close attention as you do it. What did you notice? Note the shapes your fingers form, the types of movements they make, the direction and intensity of your hands' movements. Each of these will likely speak to core cultural assumptions about sex—that is, who does what to whom, where, and how. The most common gesture for the sex act involves one hand or finger actively breaching the passive boundaries of the other in a mock penetrative motion. This deceptively simple gesticulation goes quite a long way towards capturing many of the essential elements that constitute the modern conceptualization of S-E-X. This is the "sex" that will be explored in the paragraphs below.

Penetration and Male Agency

As symbolized by the preceding exercise, active penile penetration is one of the most essential components of the social construction of S-E-X in our culture. Its centrality is well illustrated in the research conducted by Bogart and her colleagues described above (Bogart et al. 2000). In both of the cases involving penile penetration, (anal and vaginal), a majority of the respondents concluded that the activities constituted sex. This was not the case for the nonpenetrative activity described in the scenarios (oral sex) (Bogart et al. 2000).[6] Similar findings were echoed in Sanders and Reinisch's 1999 study, which indicated that 99.5 percent of their Midwestern college student sample defined penile-vaginal intercourse as sex, while somewhat fewer (81%) reported that penile-anal intercourse would qualify. However, only about 40 percent of the respondents, indicated that oral sex constituted having sex (1999).

These findings point to the cultural privileging of sex as penile penetration, revealing a conceptualization of sexuality that is tacitly connected to male agency.[7] Sex in America is a male purview; women *can* have sex, but to be a "real man," men *must* have sex, "real," aggressive, penetrative sex (Stoltenberg 1990). In our culture's ***hegemonic*** (culturally dominant) articulation of sex it is men who are defined as the sexual actors, as those who seek sex, and as those around whom notions of sexuality are invariably constructed.[8]

Heterosexuality

Another important assumption built into our cultural system is that S-E-X is a *heterosexual* activity—it is penetrative, male agentic, and it takes place between a man and a woman. But, *must* S-E-X involve a man and a woman? If so, are we asserting that gay men and lesbians do not have sex? Why then all the concern about homo**sex**uality if we don't really count same-sex sexual activities as "real" sex? What *would* we call these activities? Near sex? Pseudo sex? Sex-like? And what of bisexuals? Does it only count as S-E-X when a bisexual participates in penetrative, male agentic sexual activity with someone of the "opposite" biological sex?

Do we consider mutual digital genital stimulation (and subsequent orgasm) between two lesbians sex? If not, this begs the question, can lesbians ever "really" have sex? If so, then why would we not consider this same activity performed between a man and a woman "sex"? In grade-school vernacular this is merely "third base," clearly short of making it "home" or "all the way." How can we say Veronica had sex with

Betty but Archie only got to third if they engaged in identical sexual practices?

Continuing in this same vein, we might ask if there is a need to call gay male sexual activities something entirely different than lesbian sexual activities. After all, gay men have penises, they can engage in penile penetrative activity. But, herein lies the problem of constructing a definition of sex as contingent on the actors involved; identical acts may be evaluated differently depending upon the biological sex of the participants. For example, research indicates that it is typically considered to be sex when a man penetrates a woman anally with his penis (93 percent believed as much in Bogart et al.'s [2000] research and 81 percent in Sanders and Reinisch's [1999]). Would as many people consider that this same act counted as sex if it were conducted by a man, upon a man? Perhaps not. However, even if Americans' tacitly heterosexist definitional criterion were expanded to include same-sex sexual expression, valid and important alternative conceptualizations would still be excluded. For example, must the erotic activities involve *two* people? What about masturbation? Is this sex? Multiple partners? Non-human partners?

It is also important to note that our culture gives us important cues as to the culturally legitimated forms of sex. We do not typically use the phrase "heterosecual sex" when we discuss S-E-X. Heterosexuality is presumed. We have not need to "Sally, Sally, guess what! Bob and I had *heterosexual* sex last night!!" Specification of heterosexuality is conceptually redundant because S-E-X is culturally defined as a heterosexual activity.

The presence of a qualifier or hyphen differentiates subordinated sexual forms from the hegemonic: hence the denotations of *lesbian sex* and *gay-male sex* rather than simply "sex." Linguistic devices such as this signal that the modified forms are considered to be qualitatively different from, and inferior to, the presumed norm. They operate in a similar manner as when they are used to qualify occupations in gendered ways such as 'male nurse,' 'female firefighter,' and the **'W'**NBA.

Vaginal Specification

Another essential component of the hegemonic construction of sex in contemporary American society follows logically from the three preceding presumptive elements—if sex is constituted as a penetrative heterosexual male activity, what, we may ask, is the site of this activity? Where is all this effort supposed to be directed? To answer this, we can again gain considerable insight by looking at the linguistic cues; take for example, anal sex and oral sex. Both are sex acts that specify the orifice or bodily site of sexual expression. Both modify the presumptive case. What orifice is involved in traditional vanilla S-E-X? What is the presumed site of erotic expression when we say he, she, or they had sex? The answer is the vagina. The vagina disappears from view linguistically because its presence is presupposed in much the same way that the heterosexuality of the participants in S-E-X remains unacknowledged.[9]

Why such a focused concern surrounding the penetration of a vagina by a penis? Modern conceptualizations of S-E-X are derivate of historical constructions which held that sex was an activity directed toward procreation (Seidman 1996). In other words, our understanding of what sex is has evolved over time from definitions of sex tied to human reproduction. *Real* sex was what could get you pregnant. Historically, this has necessitated vaginal penile penetration. However, contemporary constructions are not linked to reproductive ends—the majority of Americans today do not have sex for procreative purposes. Most typically, sex is a recreational activity pursued for pleasure rather than procreation (Seidman 1996; Vance 1984).[10] In fact, for the sexually active (particularly non-Catholic women) considerable energy (and expense) is typically spent trying to *avoid* pregnancy. This shift away from reproductively oriented constructions of sex renders the heterosexual imperative embedded in contemporary constructions particularly anachronistic.

Further, medical technology has now made it possible for the human species to reproduce without any direct contact between male and female generative parts. In fact, technology has so far removed procreation from sex that a woman can give birth to a child harvested from another woman's eggs that have been fertilized with a complete stranger's sperm. There is no sex (as we define it) in this reproductive equation. The issue of human cloning moves the reproduction of the species even farther into the laboratory and away from the embodied and imperfect process of human sexual intercourse.[11]

Orgasm, Pleasure, and Love

Another issue that is closely tied to the issue of reproduction is orgasm. What is the role of orgasm in S-E-X? Typically, orgasm is viewed as the culmination of a sexual encounter. However, not all orgasms are created equal. Male orgasm in heterosexual coital relations is the event that is most closely associated with human reproduction (Laumann, Gagnon, Michael and Michaels 1994; Masters and Johnson 1966), and both males and females indicate that it is chiefly *male* orgasms that signal whether a particular sexual interaction qualifies as sex (Segal 1994; Bogart et al. 2000). And, although many females are capable of multiple orgasms,[12] it is typically the number of male climaxes that heterosexual partners count when describing how many times they have had sex. So accustomed are we to conceiving of sex in this manner that it is difficult to imagine alternate conceptualizations. Consider Marilyn Frye's discussion of the number-of-times question for lesbians:

> Some might have counted a two- or three-cycle evening as one "time" they "had sex"; some might have counted it as two or three "times." Some may have counted as "times" only the times both partners had orgasms; some may have counted as "times" occasions on which at least one had an orgasm; those who do not have orgasms or have them far more rarely than they "have sex" may not have figured orgasms into the calculations; perhaps some counted as a "time" every episode in which both touched the other's vulva more than fleetingly and not for something like a health examination. For some, to count every reciprocal touch of the vulva would have made them count as "having sex" more than most people with a job or work would dream of having time for; how do we suppose those individuals counted "times?" Is there any good reason why they should *not* count all those as "times?" Does it depend on how fulfilling it was? Was anybody else counting by occasions of fulfillment? (1990:308)

What of fulfillment? What about love? Unlike the factors mentioned previously, neither of these concepts is intrinsically bound up in prevailing constructions of sex. Though often desired, neither is generally considered to be a conditional prerequisite of the claim of having had sex. It may be the cultural ideal and the most culturally *legitimated* form, but sex that occurs between two people who are in love is by no means the *hegemonic* norm. S-E-X is about physical pleasure and is typically measured (male) orgasm by (male) orgasm.

A few considerations remain. For example, what do we call cases where there is orgasm between two people who have no actual physical contact? Is that sex? Is, for example, phone sex "sex"? Is cyber sex "sex"? If the informal student polls I have taken in my sexuality courses are any indication, it appears the answer is "not really"—S-E-X is, apparently, a contact sport. Ironically, however, when these same students are asked whether these activities should be considered "cheating" if engaged in by someone in a committed relationship with someone outside of that relationship, most say "yes." While these acts will not typically qualify as S-E-X, for many of us they feel like betrayal. Structurally, S-E-X and romantic love may be conceptually defined as separate and distinct categories of human activity but at the interpersonal level such normative distinctions appear to dissolve quite readily.

Consent

Finally, we must address the issue of consent. Even in cases that fit all of the criteria tacitly embedded in our cultural construction of S-E-X, one would hope that most Americans would be loathe to consider that a forced sexual assault could be construed as sex. I'm certain most survivors of sexual assault don't. However, even though S-E-X is generally constructed as a pleasurable pursuit, sexual pleasure is more strongly associated with its male participants. Because sexual agency is ascribed to males in contemporary conceptualizations, S-E-X **does not** necessarily require enjoyment or consent on the part of the sexual "recipient" (be they male or female).[13] Further, not only are issues of power, domination, and force not precluded, some authorities contend that contemporary definitions of sexuality proceed from a foundation of male domination and aggression. From this perspective, S-E-X *is* violence within patriarchal regimes (Dworkin 1981; MacKinnon 1987; Stoltenberg 1990).

The dangers inherent in contemporary constructions of S-E-X begin to emerge quite clearly here. S-E-X is something one "gets," something (or some*one*) one "has," the subjective needs and desires of the "object of affection," are largely irrelevant. S-E-X is, simply put, about the pleasure of *the* actor. Nor is this the only example of how the social construction of S-E-X manifests and propagates social inequality. The examples discussed above demonstrate quite clearly that our culture privileges some forms of sexuality (e.g., penetrative, male agentic, heterosexual) while simultaneously marginalizing others (e.g., non-penetrative, female-agentic, gay, lesbian, transgendered). From this we can begin to discern not only the benefits that accrue to those members of society whose beliefs and activities fall within prescribed expectations, but also the costs of our sexual constructions for subordinated groups—those whose sexual values and practices fall outside normative boundaries. Vance notes:

> Our ability to think about sexual difference is limited . . . by a cultural system that organizes sexual differences in a hierarchy in which some acts and partners are privileged and others are punished. Privileged forms of sexuality . . . are protected and rewarded by the state and subsidized through social and economic incentives. Those engaging in privileged acts, or pretending to do so, enjoy good name and good fortune. Less privileged forms of sexuality are regulated and interdicted by the state, religion, medicine, and public opinion. Those practicing less privileged forms of sexuality . . . suffer from stigma and invisibility although they also resist (Vance 1984:19).

The benefits of conformity are as enriching as the price of nonconformity is costly. Yet, as Vance notes, many do resist the confines of social strictures and normative boundaries. And, it is in resistance, struggle, and defiance that we find both societies' limits and its potential. Social constructions are far from static: as ideological silhouettes they shift and change, ebb and flow, remaining ever-powerful, yet, as historically ephemeral as the social structures they serve.

The examination I have presented here is by no means meant to be perceived as exhaustive; there are many more aspects of S-E-X that remain for analysis. We could, for example, discuss the construction of S-E-X as shameful, S-E-X as sin, S-E-X as signifier of adulthood, and so on, as well as the social inequalities that each of these categorizations engenders. Nonetheless, it is my hope that this brief analysis demonstrates the importance of critically examining the hidden assumptions about sex that are deeply rooted within our cultural framework, and fosters an appreciation of how the existence of alternate understandings of sex and sexual behaviors across a diversity of cultural contexts may help us better understand our own.

NOTES

1. I use the terms "sex," "real" sex, and S-E-X interchangeably in this essay to refer to hegemonic definitions of sex (the act) in contemporary American society.

2. That is not to imply that one tradition replaces another; multiple perspectives (e.g. spiritual, scientific, and constructionist accounts) can (and do) co-exist—the issue is which attains cultural dominance as the chief explanatory mechanism.
3. Though not always biological in form, essentialism contends that there is a fundamental underlying essence, reality, or truth that inheres in objects of study.
4. For examples of sexual variation across the globe see Ford and Beach (1951), Gregersen (1983), LeVine (1959), Marshall (1971), Ortner and Whitehead (1981), Sanday (1996), and Tiefer (1995) (Reading 2 this volume).
5. Nearly all (99.5 percent) of Sanders and Reinisch's (1999) 599 college-student sample determined that penile-vaginal intercourse counted as "sex," though 81 percent felt that penile-anal intercourse also qualified. Approximately 40% considered oral sex sufficient to claim they had engaged in sex, and two percent even contended that deep kissing could be considered sex.
6. It is unclear how scenarios involving homosexual participants would have been labeled.
7. It is also significant that all qualitative characteristics of penetration are generally considered irrelevant. For example, normatively speaking, the relative temporal duration of penile penetration (how long the penetration or stroke lasts) is not generally considered a matter of concern. The question is not the depth, frequency, or duration of penetration, but whether penile penetration has occurred at all.
8. It is unclear how the use of an artifical penis (e.g. dildo) by a woman would be evaluated.
9. It is important to point out that heterosexuality is not a sufficient precondition to require the presumption of penile penetration and vaginal reception. Presuming the participation of a male and a female does not necessarily dictate that sex will equate with traditional intercourse. For example, alternate conceptualizations of S-E-X that also included a heterosexual presumption could include penile-mammary interactions, female engulfment of a male member, or any other interactional activity defined as sexual within a given social order.
10. This move away from a procreative-based standard may also help to explain why anal sex often counts as sex in contemporary social surveys. Except for vaginal specification (and particularly because it involves active penile penetration), anal sex is consistent with much of the normative criteria for real sex and the enactment of hegemonic masculinity.
11. The 1993 film *Demolition Man* provides a useful example of the separation of sexuality and reproduction through science. In a particularly compelling scene, the protagonist (played by Sylvester Stallone) is asked by the female lead (Sandra Bullock) to have sex. The protagonist is quite surprised and dismayed to find that no bodily contact is involved. *Sex,* he is to learn, occurs between two parties through the use of mechanical devices placed on the participating parties' heads that stimulate the erotic centers of the brain. In fact, Bullock's character is shocked and repulsed when the protagonist dares to suggest that they lock lips and "exchange saliva."
12. For an historical analysis of scientific views on the significance and role of female orgasms (particularly clitorally-based orgasms) vis à vis sex see Lynne Segal's "The Liberated Orgasm" Chapter 2 in *Straight Sex: The Politics of Pleasure.*
13. The issue of consent is particularly noteworthy in the case of **statutory rape.** Statutory rape is committed when an individual engages in sexual intercourse with a younger person who is under the legal age of consent, even if that person agrees to the sexual encounter. Statutory rape laws hold that minors are less able than adults to appreciate the meaning and consequences of their actions and are, therefore, unable to give true, legally binding, consent. However, the actual age of consent is determined by state legislatures and therefore varies from state to state.

SOURCES

Berger, Peter and Thomas Luckman. 1967. *The Social Construction of Reality: A Treatise in the Sociology of Knowledge.* Garden City, NY: Anchor Books.

Blumer, Herbert. 1969. *Studies in Symbolic Interaction.* Englewood Cliffs, NJ: Prentice Hall.

Bogart, Laura M., Heather Cecil, David A. Wagstaff, Steven D. Pinkerton, and Paul R. Abramson. 2000. "Is It 'Sex'?: College Students' Interpretations of Sexual Behavior Terminology," *The Journal of Sex Research* 37(2).

Bogart, L. M., S. D. Pinkerton, H. Cecil, L. Myaskovsky, D. A. Wagstaff, and P. R. Abramson. 1999. "Attitudes Toward and Definitions of Having Sex" [Letter], *Journal of the American Medical Association* 282:1917–1918.

Buss, D. M. 1998. "Sexual Strategies Theory: Historical Origins and Current Status." *Journal of Sex Research* 35(1).

Dworkin, Andrea. 1981. *Pornography: Men Possessing Women*. London: Women's Press.

Epstein, Steven. 1987. "Gay Politics, Ethnic Identity: The Limits of Social Constructionism." *Socialist Review* 93/94.

Epstein, Steven. 1992. "Gay Politics, Ethnic Identity: The Limits of Social Constructionism," in Edward Stein (ed.) *Forms of Desire: Sexual Orientation and the Social Constructionist Controversy*. New York: Routledge.

Ford, Clellan S. and Frank A. Beach. 1951. *Patterns of Sexual Behavior*. New York: Harper and Row.

Foucault, Michel. 1980. *The History of Sexuality Volume I: An Introduction*. New York: Vintage Books.

Frye, Marilyn. 1990. "Lesbian 'Sex'," In Jeffner Allen (ed.) *Lesbian Philosophies*. Albany, NY: Statue University of New York Press.

Gagnon, John H. and Richard G. Parker. 1995. "Conceiving Sexuality," in Richard G. Parker and John H. Gagnon (eds.) *Conceiving Sexuality: Approaches to Sex Research in a Postmodern World*. New York: Routledge.

Gregersen, Edgar. 1983. *Sexual Practices. The Story of Human Sexuality*. London: Mitchell Beazley.

Harding, Jennifer. 1998. *Sex Acts: Practices of Femininity and Masculinity*. London: Sage.

Henig, Robin Marantz. 1996. "The Price of Perfection." *Civilization* (April 1996).

Indigo, Susannah. 2000. "Blow Jobs and Other Boring Stuff." *Salon.com*. www.salon.com/sex/feature/2000/12/14/teens/index.html. Published December 14, 2000. Accessed January 6, 2001.

Laumann, E.J. Gagnon, R. Michael, and S. Michaels. 1994. *The Social Organization of Sexuality*. Chicago: University of Chicago Press.

LeVine, R. A. 1959. "Gusii Sex Offenses: A Study in Social Control." American Anthropologist 61.

Marshall, D. 1971. "Sexual Behavior on Mangaia," in D. Marshall and R. Suggs (eds.) *Human Sexual Behavior*. New York: Basic Books.

Masters, W.H., V.E. Johnson. 1966. *Human Sexual Response*. Boston: Little, Brown.

MacKinnon, Catherine. 1987. *Feminism Unmodified: Discourses on Life and Law*. Cambridge, MA: Harvard University Press.

Ortner, S.B. and H. Whitehead. 1981. *Sexual Meanings: The Cultural Construction of Gender and Sexuality*. Cambridge, MA: Cambridge University Press.

Padgug, Robert. 1992. "Sexual Matters: On Conceptualizing Sexuality in History," in Edward Stein (ed.) *Forms of Desire: Sexual Orientation and the Social Constructionist Controversy*. New York: Routledge.

Rubin, Gayle S. 1984. "Thinking Sex: Notes for a Radical Theory of the Politics of Sexuality," in *Pleasure and Danger: Exploring Female Sexuality*, Carole Vance (ed). Boston: Routledge & Kegan Paul.

Sanday, Peggy Reeves. 1996. "The Socio-cultural Context of Rape: A Cross-Cultural Study," in *Culture and Sexuality*, Lois J. McDermott (ed). Needham Heights, MA: Simon and Schuster.

Sanders, S. A. and J. M. Reinisch. 1999. "Would you say you 'had sex' if . . . ?" *Journal of the American Medical Association 281*(2).

Seidman, Steven. 1996. "The Sexualization of Love" in Steven Seidman (ed.) *Queer theory/sociology*. Cambridge, MA: Blackwell.

Segal, Lynne. 1990. *Slow Motion: Changing Masculinities, Changing Men*. New Brunswick, NJ: Rutgers University Press.

Segal, Lynne. 1994. *Straight Sex: The Politics of Pleasure*. London: Virago.

Stoltenberg, John. 1990. "How Men Have (a) Sex." From *Refusing to Be a Man*. Meridian Books.

Tiefer, Leonore. 1995. *Sex is Not a Natural Act and Other Essays*. Boulder, CO: Westview Press.

Vance, Carole. 1984. *Pleasure and Danger: Exploring Female Sexuality*. Boston, MA: Routledge and Kegan Paul.

QUESTIONS FOR DISCUSSION

1. What is sex? What are the constituting elements of your own definition of S-E-X?
2. How are definitions of sex socially stratified? That is, what kinds of groups might have definitions of sex that differ from the hegemonic norm (e.g., do males differ from females, Americans from Italians, parents from children, etc.)?
3. Which, if any, aspect of the contemporary American construction of sexuality do you expect to change first? How do you think it will change? When? Why? Which element do you think is most in need of change? Why?
4. Do you agree or disagree with those who argue that sex is violence within patriarchal societies? Why?

Unnatural Acts

LEONORE TIEFER

Anthropologists catalogue an extraordinary diversity of sexual behaviors and practices across the globe. Examples include variations in the duration and frequency of sexual encounters, norms of sexual permissiveness, and even the "normal" positions assumed by participants engaged in sexual acts. In the following excerpt, psychologist Leonore Tiefer explores the cultural variation of one specific practice closely associated with sex and romance in American culture—kissing—to argue that kissing, like sex, is far from a natural act.

I have a T-shirt that reads, "Sex is a natural act." I used to think it was at least amusing, at best profound. If people would only relax and let their natural reactions flow, I thought, sex would be more of a pleasure and less of a Pandora's box.

I'm wiser now. I think the sentiment of the T-shirt distorts the truth. The urge to merge may be natural for birds and bees, but the biological takes a back seat in our own species. We humans are the only ones with a sex drive that isn't solely related to procreation.

Originally, the message that sex is natural was meant to relieve guilt feelings—you can't be blamed for doing what is healthy and normal. Such permission was extremely useful for a time. It enabled many people to break free from choking inhibitions.

But the message was taken too literally. I now meet people who believe hormones control their sex life. They feel no pride when sex is good and have no idea what to do when it is not. Letting Mother Nature do the driving sounds like the lazy person's dream; actually it makes you feel powerless and ignorant.

Belief that sexuality comes naturally relieves our responsibility to acquire knowledge and make choices. You don't have to teach your kids anything special—when the time comes, they'll know what to do. You don't have to talk with your partner about your love life—it'll all just happen automatically.

What happens automatically is often brief, routine, and more in the category of scratching an itch than indulging a beautiful expression. Such a sexual style may satisfy a person for whom sex has a low priority. It is unreasonable to expect mutual pleasure, variety, or emotional intimacy without some information and a lot of practice. If all you need for fulfilling sex comes already built in, then any difficulties must be due to physical breakdown. Many couples seek medical help when what they need is a course on communication.

Most important, the attitude that sex is a natural act implies that great sex occurs early in a relationship and stays constant throughout. A dynamic vision of continuing change and adjustment is more realistic—it's not failing memory that leads some older couples to report that sex keeps getting better.

High school and college classes dwell at length on statistics, plumbing, and contraception. Students rarely read about connections between sexuality and feelings. Nor do they discuss what influences sexual attraction or how psychological needs are met through sex. Often students enter a course and leave it still thinking that love will guide the way to sexual happiness.

From Sex Is Not a Natural Act and Other Essays, 2nd ed. by Leonore Tiefer, Boulder, Colorado: Westview Press, 2004. Reprinted by permission of the author.

Limiting instruction to issues like birth control and venereal disease prevention may promote public health goals, but it does little to enrich the quality of sexual experience. Techniques of pregnancy prevention don't work to prevent sexual disappointment.

Natural sex, like a natural brassiere, is a contradiction in terms. The human sex act is a product of individual personalities, skills, and the scripts of our times. Like a brassiere, it shapes nature to something designed by human purposes and reflecting current fashion.

THE KISS

Nothing seems more natural than a kiss. Consider the French kiss, also known as the soul kiss, deep kiss, or tongue kiss (to the French it was the Italian kiss). Western societies regard this passionate exploration of mouths and tongues as an instinctive way to express love and to arouse desire. To a European who associates deep kisses with erotic response, the idea of one without the other feels like summer without sun.

Yet soul kissing is completely absent in many cultures of the world, where sexual arousal may be evoked by affectionate bites or stinging slaps. Anthropology and history amply demonstrate that, depending on time and place, the kiss may or may not be regarded as a sexual act, a sign of friendship, a gesture of respect, a health threat, a ceremonial celebration, or a disgusting behavior that deserves condemnation.

Considering the diversity of kissing customs, it astonishes me that so few social scientists have given the kiss any attention. Kissing is usually relegated to an occasional footnote, if authors bother to mention it at all. My computer search of kissing references in *Psychological Abstracts* and the *Index Medicus* turned up some papers on mononucleosis ("the kissing disease"), one article on a fish known as the "kissing gourami," and unrelated work by people with names like Kissing and Kissler. Even sex researchers are uninterested. *Sex* now refers to intercourse, not kissing or petting. In textbooks on human sexuality, kissing rarely appears in the index.

Anthropology and Kissing

I became fascinated by the remarkable cultural and historical variations in styles and purposes of kissing, given how "natural" it seems to pursue whatever customs each of us has grown accustomed to. Clellan Ford and Frank Beach (1951) compared the sexual customs of the many tribal societies that were recorded in the Human Relations Area Files at Yale. Few field studies mentioned kissing customs at all. Of the 21 that did, some sort of kissing accompanied intercourse in thirteen tribes—the Chiricahua, Cree, Crow, Gros Ventre, Hopi, Huichol, Kwakiutl, and Tarahumara of North America; the Alorese, Keraki, Trobrianders, and Trukese of Oceania; and the Lapps in Eurasia. There were some interesting variations: the Kwakiutl, Trobrianders, Alorese, and Trukese kiss by sucking the lips and tongue of their partners; the Lapps like to kiss the mouth and nose at the same time.

But sexual kissing is unknown in many societies, including the Balinese, Chamorro, Manus, and Tinguian of Oceania; the Chewa and Thonga of Africa; the Siriono of South America; and the Lepcha of Eurasia. In such cultures, the mouth-to-mouth kiss is considered dangerous, unhealthy, or disgusting, the way Westerners might regard a custom of sticking one's tongue into a lover's nose. Ford and Beach reported that when the Thonga first saw Europeans kissing, they laughed, remarking, "Look at them—they eat each other's saliva and dirt."

Deep kissing apparently has nothing to do with degree of sexual inhibition or repression in a culture. On certain Polynesian islands, women are orgasmic and sexually active, yet kissing was unknown until Westerners and their popular films arrived. Research in parts of Ireland, by contrast, where sex was considered dirty and sinful, and, for women, a duty, shows that the Irish also were oblivious to tongue kissing until recent decades.

Many tribes across Africa and elsewhere believe that the soul enters and leaves through the mouth and that a person's bodily products can be collected and saved by an enemy for harmful purposes. In these societies, the possible loss of saliva would cause a kiss to be regarded as a dangerous gesture. There, the "soul kiss" is taken literally. (It was taken figuratively in Western societies; recall Christopher Marlowe's "Sweet Helen, make me immortal with a kiss! Her lips suck forth my soul.")

Although the deep kiss is relatively rare around the world as a part of sexual intimacy, other forms of mouth or nose contact are common—particularly the "oceanic kiss," named for its prevalence among cultures in Oceania but not limited to them. The Tinguians place their lips near the partner's face and suddenly inhale. Balinese lovers bring their faces close enough to catch each other's perfume and to feel the warmth of the skin, making contact as they move their heads slightly. Another kiss, as practiced by Chinese Yakuts and Mongolians at the turn of the century, has one person's nose pressed to the other's cheek, followed by a nasal inhalation and finally a smacking of lips.

The oceanic kiss may be varied by the placement of the nose and cheek, the vigor of the inhalation, the nature of the accompanying sounds, the action of the arms, and so on; it is used for affectionate greeting as well as for sexual play. Some observers think that the so-called Eskimo or Malay kiss of rubbing noses is actually a mislabeled oceanic kiss; the kisser is simply moving his or her nose rapidly from one cheek to the other of the partner, bumping noses en route.

Small tribes and obscure Irish islanders are not the only groups to eschew tongue kissing. The advanced civilizations of China and Japan, which regarded sexual proficiency as high art, apparently cared little about it. In their voluminous display of erotica—graphic depictions of every possible sexual position, angle of intercourse, and variation of partner and setting—mouth-to-mouth kissing is conspicuous by its absence. Japanese poets have rhapsodized for centuries about the allure of the nape of the neck, but they have been silent on the mouth; indeed, kissing is acceptable only between mother and child. (The Japanese have no word for kissing—though they recently borrowed from English to create *kissu.*) In Japan, intercourse is "natural"; a kiss, pornographic. When Rodin's famous sculpture *The kiss* came to Tokyo in the 1920s as part of a show of European art, it was concealed from public view behind a bamboo curtain.

Among cultures of the West, the number of nonsexual uses of the kiss is staggering: greeting and farewell, affection, religious or ceremonial symbolism, deference to a person of high status. (People also kiss icons, dice, and other objects, of course, in prayer, for luck, or as part of a ritual.) Kisses make the hurt go away, bless the sacred vestments, seal a bargain. In story and legend a kiss has started wars and ended them, awakened Sleeping Beauty and put Brunnhilde to sleep.

Classifying Kisses

Efforts to sort all of these kisses into neat categories apparently began centuries ago. According to Christopher Nyrop (1901), a Danish linguist who wrote a history of the kiss, the ancient rabbis recognized three kinds: greeting, farewell, and respect. The Romans also distinguished three kinds of kisses: *oscula* (friendly kisses), *basia* (love kisses), and *suavia* (passionate kisses). The most imaginative system, was proposed in 1791 by an Austrian writer, W. von Kempelen, who divided kisses according to their sound: the *freundschaftlicher hellklatschender Kerzenskuss* (the affectionate clear-ringing kiss coming from the heart), the acoustically weaker discreet kiss, and the *ekeljafter Schmatz* (a loathsome smack). Von Kempelen's categories did not gain widespread use.

When Nyrop wrote his book, he reported no fewer than 30 different German words to indicate types of kisses, from *Abschiedskuss* (farewell kiss) to *Zuckerkuss* (sweet, or "candy" kiss). The structure of the language permits composite nouns, but even so, German shows a remarkable linguistic richness in its variety of kisses. Today, *abkussen* means to give many little kisses all over; *erkussen* is (slang) for getting a gift or favor by kissing ("sucking up"?); *fortkussen* is to kiss away tears; and *wiederkussen* is to return a kiss you have been given.

The Germans are not the only ones to classify their kisses. Allan Edwardes, in *The Jewel in the Lotus* (1959), described the Hindu science of kissing: There is *sootaree-sumpoodeh* (the probing, tongue-sucking kiss), to be distinguished from *jeebh-juddh* (tongue-tilting), *jeebhee* (tongue scraping), and *hondh-chubbow* (lip biting). Of course, the question sexologists cannot yet answer—because research to date has been more descriptive than subjective—is to what extent Germans and Hindus actually have more diverse experiences than the French, who struggle along monolinguistically with *embrasser,* the Spanish, who have only *besar,* or the Russians, with *tselovat.*

Ceremonial Kisses

Classifications are entertaining but not especially illuminating. Types of kisses overlap and change over time. For example, St. Paul may not have known what would come of his simple advice to Christians to "salute one another with an holy kiss," a brief admonition in Romans 16:16 that is repeated in Corinthians I and II. Over the centuries, the "holy kiss" was interpreted and reinterpreted; it found expression in baptism, marriage, confession, and ordination. The *osculum pacis,* or kiss of peace, supposed to represent God's kiss of life and Christ's kiss of eternal blessing, was exchanged in some locations between priest and congregant, in others between clergy only. It passed out of common practice after the Reformation but is enjoying a renaissance in modern Catholic, Anglican, and Episcopalian congregations.

The famous kiss between bride and groom that concludes a wedding ceremony was actually part of ancient pagan rites and signified that legal bonds were being assumed. I've always thought the clergy's injunction to the groom "I now pronounce you husband [or 'man'] and wife, you may kiss the bride" represented the clergyman's quasi-parental permission to the new couple to be sexual (though it can't be an accident that the permission has typically been extended to the new husband!).

There are many more stories, about holy and profane kisses, social and ceremonial kisses, changing customs (e.g., how European customs of kissing changed to bowing and hat lifting during the time of the plague), but by now I have made my point that an act like kissing cannot only be choreographed in very different ways but can serve many functions and carry many different meanings, all depending on the customs of a particular era. And, amazingly, each social group, each generation, feels its kisses are the normal, the natural ones.

Importance of Kissing

Why is kissing so popular, and why does it adapt to so many meanings? There have been some theories. Desmond Morris, following Freud, noted that a baby experiences its earliest joys, gratifications, and frustrations through its mouth, which becomes a site of emotional associations. In many cultures infants are lavishly touched, cuddled, and kissed all over their bodies, not only by their mothers but also by other relatives and friends. The infant learns that touching something soft with the mouth is a calming and pleasurable sensation. Adult kisses recall some of this infant gratification. Kissing symbols for luck, Morris argued, is emotionally reassuring—it's not just a random gesture to appease the gods or fate.

It certainly is true that the lips, mouth, and tongue are among the most exquisitely sensitive parts of the body. The tongue itself is sensitive to pressure, temperature, taste, smell, and movement. Lips, tongue, and mouth detect and transmit to the brain a range of incoming sensations; and the brain, in turn, devotes a disproportionate amount of its resources to processing their messages and linking them up to behavioral reactions and psychological functions. The space devoted to messages, from and to the lips alone is far greater than that devoted to sensory or motor function for the entire torso.

Opportunities for kissing to develop multiple social meanings also arise because of variations in elements such as posture and facial expression—an especially important factor in communicating emotion. Through processes of social learning, trial and error, imitation, and reward and punishment, kisses acquire their multiple meanings and intense associations. Rules for social and sexual kisses vary not only across cultures but even among social classes and subcultures in the United States, making research into the social scripting of kissing a fertile area for those interested in how sexual choreography develops and what the various components mean to participants. The "naturalness" of kissing, as with so many other aspects of social life, turns out to be a biological potential shaped and cultivated by the real human nature—culture.

REFERENCES

Edwardes, A. (1959). *The Jewel in the Lotus: A historical survey of the sexual culture of the East.* New York: Julian Press.

Ford, C. S., and Beach, F. A. (1951). *Patterns of Sexual Behavior.* New York: Harper and Row.

Nyrop, C. (1901). *The Kiss and Its History,* trans. by W. F. Harvey. London: Sands and Co.

DISCUSSION QUESTIONS

1. Is love sufficient to guide the way to sexual happiness? If pleasure were not the dominant goal of sexual relations (e.g., if procreation were the primary reason individuals engaged in sex) would sexual happiness matter? Should it?
2. Does the information Tiefer presents about worldwide variation in the practice of kissing definitively support the essentialist or the social constructionist perspective? Why or why not?
3. Construct your own classification system for kissing.
4. What other aspects of sex and romance in the United States may be culture bound?

Breastfeeding and the Good Maternal Body

CINDY A. STEARNS

Ideas about what is erotic and/or sexual are strongly shaped by cultural norms and values—in other words, they are socially constructed. And we can attach erotic desire to virtually anything—while one society may eroticize lips, another may sexualize feet, buttocks, or ear lobes. In modern American society, one of the primary sites of erotic desire is the female breast. However, the female breast is also the location of breastfeeding, a practice linked to the revered realm of motherhood. Stearns' article demonstrates the malleability of the sexual by exploring ways in which American women actively negotiate competing constructions of the breast—first as a site of erotic desire and second as a symbol of maternal love and devotion.

From "Breastfeeding and the Good Maternal Body" by Cindy A. Stearns, *Gender and Society* 13(3): 308–325, 1999. Used by permission of Sage Publications.

This article focuses on one aspect of the sociology of breastfeeding: how women experience and negotiate the act of breastfeeding in front of others. The performance of breastfeeding is complicated by conflicting cultural beliefs about women's breasts. In American culture, breasts are viewed primarily as sexual objects, although the sexual breast is not common historically or cross-culturally (Dettwyler 1995). For example, in *A History of the Breast,* Yalom describes breasts in American society as "the crown jewels of femininity" (1997, 3). For American women, breasts are part of the "doing" of femininity and the presentation of heterosexuality (West and Zimmerman 1987). Like hair or makeup, breasts are displayed as part of women's appearance work and are sometimes "improved" either through implants or special garments that are typically designed to make breasts larger and more noticeable.

The prominence of the sexualized breast poses a problem for breastfeeding women and their maternal bodies. The good maternal body is not commonly believed to be simultaneously sexual, despite the obvious facts of human reproduction (Davis-Floyd 1992; Newton 1977). The sexual aspects of women and the maternal aspects of women are expected to be independent of each other. Thus, breastfeeding raises questions about the appropriate uses of women's bodies, for sexual or nurturing purposes. Young theorizes that breastfeeding and "(b)reasts are a scandal because they shatter the border between motherhood and sexuality" (1998,132–133). Given the strong cultural preference for sexualized breasts, women who breastfeed are transgressing the boundaries of both the good maternal body and woman-as-(hetero)sexual object. Taken to an extreme, transgression might also mean losing one's child, as in the representative but not unique case of "Karen Carter." Carter, the mother of a nursing 2-year-old, called a crisis line with concerns about feelings of sexual arousal while breastfeeding (Umansky, 1998). As a result of her phone call, Carter's child was taken from her by the Department of Social Services for more than one year. Thus, the construction of the good maternal body as being at all costs *not* sexual is taken very seriously by both the culture and the law.

Increasingly, women are confronted with the dilemma of the sexual or the nurturing, maternal breast. Breastmilk is currently recommended as the optimal nutrition for babies by the medical community and breastfeeding advocates. Breastmilk and breastfeeding provide a number of health benefits, including reduction in gastrointestinal and ear infections, among others (for a recent summary of the medical advantages of breastfeeding for infants, see Stuart-Macadam 1995). A variety of recent health promotion efforts, including those for women receiving welfare benefits, are geared toward getting more women to breastfeed. However, while breastfeeding has become the medical gold standard for infant feeding, it is still not the typical form of feeding. Research indicates that while most physicians suggest breastfeeding as an alternative to their patients, physicians lack the training, skills, and knowledge necessary to successfully support a breastfeeding relationship (Freed et al. 1995; Howard, Schaffer, and Lawrence 1997). The United States remains primarily a bottle-feeding culture. Only 53 percent of women initiate breastfeeding and only "1 in 5 infants receives any breastmilk *at all* at 20 weeks of age" (Baumslag and Michels 1995, xxi). Those babies who are breastfed typically receive a few weeks of breastmilk, followed by many months of bottled formula. Therefore, while the slogan "breast is best" is advocated to pregnant women and new mothers, the act of breastfeeding is still unusual.

Given the cultural emphasis on the sexualization of women's bodies, and their breasts in particular, accompanied by the low incidence of breastfeeding, the ways in which women go about breastfeeding should tell us a great deal about how the maternal body is actively constructed through both discourse and behaviors. In doing breastfeeding in front of others, women negotiate the definitions of their nursing behaviors as sexual or nurturing. Breastfeeding women must manage their own ideas about the appropriateness of breastfeeding as well as their perceptions of how others actually respond or might respond to their breastfeeding. In discussing how, where, and when they will breastfeed, breastfeeding women reveal how they both accommodate and resist the cultural expectations and barriers they encounter in a culture in which the meaning and place of women's breasts is contested. The analysis reveals that fundamental aspects of the work and experience of breastfeeding and mothering are shaped by cultural beliefs about how women should use their bodies.

METHODOLOGY

This research involves in-depth interviews with 51 women in Sonoma County, California, from January 1996 through April 1998. The tape-recorded interviews lasted an average of 90 minutes and were transcribed verbatim.

The mostly snowball sample was composed of women who referred other women for possible inclusion in the study. However, to create a sample more diverse in ethnicity and class, strategic contacts were made with the government-subsidized Women, Infants and Children program (WIC), mothers' clubs, a teen parenting program, and lactation consultants, all of whom were very helpful in publicizing the study and referring potential participants to the researcher. The sample is very diverse in terms of class, ranging from a physician and a lawyer to several women living on welfare.

The requirements for participation in the study were that the woman was currently breastfeeding or had stopped breastfeeding within the past six months. As it turned out, 84 percent of the women were currently breastfeeding when interviewed and 16 percent had weaned within the past six months (three of these within one month of the interview).

FINDINGS

Breastfeeding raises many possibilities for public performance. The average newborn nurses about every two hours. Unless a woman stays at home for several months and is able to only breastfeed at home, and in private, most women must think about how they will go about breastfeeding in front of others.

All of the women in this study had given significant thought to how to breastfeed in public. Women had a great deal to say about how they breastfeed in front of others, where and when breastfeeding would be inappropriate, and how they perceive others do or might react to their breastfeeding. Women were keenly aware that the activity of breastfeeding in public might result in negative feedback, or worse yet, legal action. During the time of the study, there were a few publicized cases in which breastfeeding women were asked to leave public establishments because their behavior was viewed as inappropriate public activity.

California passed legislation in 1997 clarifying that breastfeeding in public is legal behavior, not public nudity. At least 13 other states have passed breastfeeding legislation in the past five years (Baldwin and Friedman 1998). The various legislative efforts do not legalize an illegal activity—they simply clarify that the act of breastfeeding is not public nudity (a criminal activity). Many of the women, however, understood the legislation to mean that breastfeeding in public was currently illegal or had been illegal and adapted their behavior accordingly.

The media stories of breastfeeding as well as the legislation contributed to the women's anticipation of an environment hostile to breastfeeding in public. None of the women in this study had been asked to leave a public establishment because she was breastfeeding. Nonetheless, the perception of a hostile public environment was clearly in women's thinking as they planned their breastfeeding behavior. Indeed, women report many instances of receiving direct and indirect feedback from friends, family, and others concerning the appropriateness of their breastfeeding in front of others. Consequently, women proceeded with their breastfeeding as though it were deviant behavior, occurring within a potentially hostile environment. Women both resisted and accommodated this hostile environment as they went about breastfeeding in front of others.

Discretion

In describing how they breastfeed, women uniformly emphasized the importance and/or necessity of learning to breastfeed discreetly. Discretion for these women, typically refers to not showing the breast—and especially, the nipple—in public. For many, achieving this was difficult, especially with the first child or with a child who had nursing problems. Breastfeeding is a learned activity for both mother and baby. Trying to get a baby positioned correctly on the breast while simultaneously hiding all parts of the breast is not always easy for the new (or sometimes even the established) mother.

Being able to be discreet in public was a skill that women valued highly. Jody describes her frustrations with her firstborn:

> Well, I always felt like the biggest accomplishment that you could make (was) if no one even knew you were breastfeeding. I always felt like I could never do that. And so I always, I remember feeling very flushed, my face would get very red and I'd be embarrassed trying to get it out, get her up right, um so I always tried to manage, you know, to say "Let me nurse her first and then we'll go."

Being an invisible breastfeeding mother was the goal for many women. Women would describe wearing special clothing, draping a receiving blanket over the baby, sitting in obscure places, and other means of hiding the breastfeeding. Women would speak with pride about no one even knowing what they were doing, when, in fact, they were really breastfeeding.

Discretion was important even to those women most adamant about women's right to breastfeed anywhere and everywhere. For example, Marian reported that she would always breastfeed in public and that "if anyone wanted to arrest me, they could go ahead and arrest me. I didn't care. I was doing what was best for my child." Yet, she also describes how she breastfeeds:

> If it offended somebody because I was nursing my child in the mall, and I had a blanket over my shoulder or a big shirt that was completely covered where all you maybe saw was this little toe, if somebody got offended by that, as far as I'm concerned that's their problem. And they're the ones that have the problem, not me.

Discretion, though much desired, was not always achieved. Many women had stories to tell about breastfeeding "gone wrong." Laura's tale of her first outing is a humorous example:

> I remember going out to pizza, our first like major breastfeeding in public and nursing her on this side and C sitting across from me and I said, joking, "See anything?" And he thinks a little bit of breast tissue, not the nipple area, and he's like "Oh, no. Its fine." I said "OK." And then we were having a conversation and I lifted her up and was burping her and C and I are talking and all of a sudden he says, "Um, now you can." And my whole breast was like on the table like staring at the table right over there and here he was like "Um, um." And I'm, "What?" looking around and my breast was "BOING!" and the waiter came at that point in time and said, "Do you want your CHECK" . . . It was so embarrassing.

Many of the respondents viewed breastfeeding in public as a potential problem if others might feel uncomfortable. Embarrassment occurred if the woman perceived that the verbal or body language of others indicated they felt the breastfeeding was inappropriate to the situation. Women tried to avoid situations that might lead to negative definitions, as many women felt the potential for public censure was always present.

Avoiding Some Places and Claiming Other Spaces

One way that women deal with the potential stigma of breastfeeding is to avoid being seen breastfeeding. In fact, many of the women in this study went to great lengths to make sure that they were not seen breastfeeding outside their homes. Women would nurse in department store dressing rooms and cars and even stay home exclusively to avoid public breastfeeding. Women also described breastfeeding their children before leaving home and timing their trips out accordingly. This worked better with older babies, as infants eat much more frequently.

In addition, there were some institutional settings in which some women felt breastfeeding simply wasn't appropriate. Several women mentioned work settings as creating breastfeeding problems. A working mother describes an informal work event in which she brought her newborn baby and nursed: "I remember being with people from work at this shower that I mentioned and unbuttoning and feeling really exposed."

Other women would not breastfeed at all in particular work situations. For example, Patty, a part-time college instructor, said that she would "breastfeed just about anywhere" but not in front of the chair of her department, who had never commented on her pregnant state or birth.

One woman describes that she will breastfeed in most places with her third child, whereas she felt too intimidated with her first and somewhat more liberated with the second. Still, Theresa, a Catholic, won't nurse in church and comments, "I don't know why that would bother me, it shouldn't bother me. (Laughs) God created nursing, I don't think he has a problem with me nursing!"

Women receive a lot of unsolicited advice about where and when they should nurse. Their own ideas about appropriate locations were frequently influenced by others' preferences.

Family members were most likely to provide unsolicited commentary on the appropriateness of breastfeeding in various situations. In one case even a video of a new baby, mailed to family members in the Midwest, was the subject of censure. A niece reported back to the baby's mother about the family's viewing of the "the boob segments." She described how grandma "fast forwarded through the nursing scenes" and commented, "Well, I think we've seen enough of that."

Women expressed strong beliefs about claiming certain spaces as appropriate for breastfeeding. Sometimes women would note spaces where they absolutely felt they had a right to breastfeed (usually their home) or not be breastfeed (usually bathrooms).

The bathroom was frequently noted as the bottom line for women on their willingness to accommodate. Katie [remarks]:

> That's how I feed him, you know, and I don't like going and sitting in bathrooms to feed him because, you know, I don't think I'd like to go in there and eat lunch so I don't see why he should have to, plus I don't like sitting in there and smelling it.

La Leche League was sometimes mentioned as a safe place for breastfeeding mothers, with many women exclaiming that you could even breastfeed a "4-year-old" at one of their meetings. Mothers' clubs were also highlighted as gatherings where breastfeeding could more openly and easily take place.

Monitoring the Male Gaze

Women frequently noted that breastfeeding in front of some categories of men created significant issues. Male strangers in a public place posed a problem for several women. For example, one mother describes her dilemma when her crying baby demanded a feeding and the only available spot was in front of male strangers: "I was like I don't feel like turning anybody on while I'm doing this (laughing) you know. It was kind of like, let's just do it quick here." Theresa similarly explains why she tries not to breastfeed in front of men she does not know: "Because I mean our society is so puritanical and has so many screwed-up ideas about sexuality, plus there's a lot of weird people out there too, you know."

Breastfeeding women fear that the exposure of their breasts will he misread as a sexual invitation to male strangers and they fear potential consequences of that misreading. Some women would define the men who were looking as having problems, specifically that the observer was being sexually inappropriate. For example, Lynette observed:

> Yeah, well that's their problem, I feel if my baby is hungry and I'm eating at a restaurant, he should have the right too. You know, everybody else is and so, I mean, cause I feel, this is how I say it, if you don't like it, don't look, you know. You're not supposed to look as a woman breastfeeds anyways, you know, what's wrong with you.

Weirdo and *pervert* were words frequently used to describe men perceived to be leering rather than looking.

Breastfeeding in front of male family and friends was also an issue for many women. Most frequently, the concern was with fathers and fathers-in-law.

Many women could not articulate why they felt it was inappropriate to nurse in front of particular men but would say it just didn't feel right. The mother of a 13-week-old girl describes why she would not breastfeed in front of her father-in-law: "I don't know why I felt that way; I just felt like I shouldn't be flashing my breasts in front of him."

The use of the word *flashing* is notable here and was used by more than one woman in describing inappropriate breastfeeding. Flashing describes inappropriate public nudity and a sexual displays—breastfeeding women wish to avoid that label for their activity.

Many times women are strongly signaled by others that they need to be discreet in front of the male family member. Rita, a 41-year-old mother of two boys (each nursed for more than three years), describes her experience: "My husband's parents, when they were visiting, I sat down to nurse one time and they were here and she, his mother, got up and got a blanket and put it over me." Other women experienced similar "coaching" from mothers and mothers-in-law (many of whom had never breastfed) about the conduct of their breastfeeding.

All of the women in this study reported feeling comfortable breastfeeding in front of their male partners. However, sometimes fathers were concerned with the mothers' possible public exposure.

A partner who worried about the public breastfeeding could be a real deterrent to women's desire to breastfeed at all. Recent research indicates that a supportive father is an important factor in the choice to breastfeed, as well as in the duration of breastfeeding (Bar-Yam and Darby 1997).

Redefining Breasts

Some women had thought a great deal about "what breasts were really for" and how to redefine their breasts and breastfeeding for themselves and for others. Often this discussion brought up how to cope with sexual imagery. Jeanette, the 24-year-old mother of a 10-month-old girl, described her frustrations about others viewing her breastfeeding as sexual:

> I've read a lot of articles on nursing in public and some people say that, well, a woman baring her breast in public is, is taboo and it's a sexual thing and, well, I don't get turned on when I'm nursing my daughter. You know, I'm giving her something that she's telling me that she needs. I'm feeding her. It's like taking a bottle out of my diaper bag and giving it to her. In my eyes there is no difference. Milk is milk. I'm not doing it out of a sexual need.

Jeanette's notion that "milk is milk" and that her breastmilk is the same as giving a bottle is a way of saying this is not really a sexual act. In this use of language, shared by a few other women, breastfeeding is not an embodied experience but instead simply a food service.

In contrast, other women redefined breastfeeding as being a natural use of their bodies:

> You know, it's like you want us to look at, to gawk at, and that, but when we're doing

> what comes natural and what we are supposed to do for our child and the next generation, you're telling us it's wrong. And that really made me angry.

The natural perspective is an embodied description of breastfeeding. Breasts are made to produce milk and providing it to a child is "natural," not sexual.

Code Words

Women who breastfed older children faced additional issues. Most American children are not breastfed beyond the first few weeks. A common public belief reported (but not always shared) by those in the study was that breastfeeding beyond infancy or babyhood was not nutritionally necessary. Thus, the motivations for breastfeeding an older child were often questioned. Breastfeeding women were aware that breastfeeding an older baby or child in public was often met with increased suspicion by strangers, family, and friends. Furthermore, children who are older can often talk and this raises problems in controlling the disclosure of breastfeeding. In contrast to a baby who signals hunger with a cry that might mean bottle or breast, a child who can ask for its mother's breast alerts those present that there is a breastfeeding relationship.

A few of the women did not mind their child publicly asking for breastfeeding. However, most viewed this as a dilemma to be managed. Some prevented public problems by creating a rule for their toddlers that breastfeeding would only happen at home or even in a certain chair at home. Another response to the anticipation of negative reactions to an older breastfeeding child is the development of intentionally defined "code words" to be used by the older child and others in the family. The secret words, such as *milky* and *nummies,* would allow the mother to talk about nursing in public without others knowing she is talking about it. Mothers might say, "OK, let's go in another room and have "milky" or "You can have juice now and nummies when we get home."

The use of code words to conceal extended breastfeeding is a form of women's resistance because mothers must actively resist prevailing cultural norms about the appropriate duration of breastfeeding. In this study, women reported receiving increased pressure to end breastfeeding as the child got older, most notably when children passed certain milestones, including first birthdays, walking, and talking. Women used code words to maintain the breastfeeding relationship. Although the code words do not attempt to change other people's opinions about breastfeeding, they do give the mother and child license to resist and change their definitions of what is an appropriate breastfeeding relationship.

Using Medical Guidelines to Inform Discourse

In resisting cultural pressures and defending their right to breastfeed, women frequently relied on the discourse of medical authority. Sometimes they reported telling inquiring relatives, friends, and others that the "breast was best," according to the latest medical research. Ironically, many of their mothers, older friends, and relatives had received the opposite advice when making feeding decisions for their own children. Manufactured infant formula was promoted as scientifically superior to breastfeeding by many physicians in previous decades (Apple, 1987).

Medical authority was used by women to support their own breastfeeding decisions. For example, a mother of a 3-month-old girl describes why six months of breastfeeding is her goal: "Well, it's six months just more or less because that's what I keep reading in books. It's best to keep breastfeeding until they're six months old." In contrast, a 21-year-old mother of an 18-month-old describes the medical and scientific reasons she uses to repeatedly justify to her family why she is still nursing her son:

> All this energy I put into him, all this, everything I give him, you know, I want to give him everything that I know that I, you know, nutritionally, you know, and now I've seen test results the longer you nurse, the higher their IQ and less chances for cancer and asthma and all of these benefits. I mean, how can you not want to give your child those benefits?

To some extent women use the medical guidelines and medical research selectively as a justification for their own individual choices about breastfeeding. They invoke medical authority to explain to family members why they are "still" breastfeeding at eight weeks or 18 months.

Changes over Time

Women report that their comfort level with breastfeeding grows over time and with successive children. For example, Davida began nursing by hiding in her old bedroom when visiting her parents' home. By the fourth child, Davida describes still being uncomfortable in front of her father but choosing to breastfeed in front of him anyway.

With their second child, many women described having better feelings about their breastfeeding bodies and regrets about their body image and their nursing with the first child. A 30-year-old white/Native American mother is a good example. During the time of the interview she was nursing two of her children, a 1-year-old boy and a 3-year-old girl. She weaned her oldest child at two months because of family pressure and looked back with regret: "I just wish that I had been OK about my own breasts and my own body and what I was doing, as I am now. And I wasn't then." A common theme among the women was to breastfeed the second child longer than the first. Many women indicated this was because they felt more comfortable about their bodies and breastfeeding in general, and especially breastfeeding in front of family, friends, and in public.

CONCLUSIONS

Women engage in a variety of behaviors to try to make their breastfeeding fit into a hostile environment. In doing so, women actively create the good maternal body before an audience that is more familiar and comfortable with the sexualized breast than the nurturing breast. The construction of the good maternal body involves constant vigilance to how the breastfeeding is viewed by others. The major concern of women is that their breastfeeding is perceived as maternal and not sexual behavior. To transgress the precarious boundaries of the good maternal body is to risk being labeled a bad mother and/or sexually inappropriate or deviant.

While women overwhelmingly accommodated to perceived cultural demands, especially with the perceived need to practice discretion in nursing, breastfeeding women also expressed irritation and anger with the necessity to be careful, secretive, and/or watchful. Women wished that the culture were more accommodating to the fullness of their maternal bodies. Sometimes women reported that the experience of breastfeeding was often greatly enjoyed when done alone or with supportive family or friends, while the demands of the tricky public performance of breastfeeding muted that pleasure. Often women would lament that in other cultures, their nursing would be viewed as natural and a part of everyday life, rather than experienced as a problem.

It is important to consider women's agency, how they resisted the labeling of their breastfeeding as deviant. Dellinger and Williams accurately note that "[w]hat is too often missing in the theoretical debates about femininity and the body is the complicated relationship between powerful hegemonic ideologies and women's agency reflected in their actual lived experiences" (1997, 152). Breastfeeding women do resist the prevailing definitions. They redefine their breasts, drawing on medical guidelines to make their case. They use code words to conceal and continue long-term breastfeeding and they also claim certain spaces as their own and as appropriate for breastfeeding. Women who breastfeed for longer periods of time and who nurse multiple children often feel more entitled to breastfeed when and where they want. The very act of breastfeeding, particularly prolonged breastfeeding, is itself a form, of resistance to the sexualized image of the breast and the good maternal body.

Certainly the breastfeeding experience is a lens through which to better understand the place of women in society more generally. An abundance of feminist and sociological research has highlighted the centrality of bodily experience to understanding the constraints and possibilities for the transformation of women's lives (Bordo, 1993; Weitz, 1998). The carefully managed and often secretive nature of much breastfeeding reveals volumes about women's status. As Bartky writes, "woman's body language speaks eloquently, though silently, of her subordinate status" (1998, 36). Women accomplish the breastfeeding of their children with constant vigilance to location, situation, and observer. Women breastfeed in anticipation of, and reaction to, the male gaze and the possibility of inappropriate responses or censure. Women spend a lot of time hiding the breastfeeding, attempting to be discreet, and being careful around situations where there are men and therefore the possibility of their breastfeeding being misread as sexual. Women expend a good deal of effort trying not to offend others. The actual labor of breastfeeding is increased because women must constantly negotiate and manage the act of breastfeeding in every sector of society—in public and in the home.

While this article describes how women balance competing ideas about how their breasts should be used, it is also important to pose the possibility that women might experience their breasts as sexual and nurturing simultaneously. In other words, the maternal body might also be sexual. Bordo, in describing the frequent use of the term *bra burners* to describe feminists, raises an important point:

> But whether or not bras were burned, the uneasy public with whom the image stuck surely got it right in recognizing the deep political meaning of women's refusal to "discipline" our breasts, culturally required to be so exclusively "for" the other—whether as instrument and symbol of nurturing love, or as erotic fetish. (1993, 20)

As long as women's breasts are defined exclusively as "for the other" women will likely feel the need to negotiate their breastfeeding carefully, And few women may even choose to breastfeed at all because of consequent fears of embarrassment and censure that come from defining breasts as only sexual and the act of breastfeeding as private behavior (Dettwyler 1995).

The social consequences for women of the limitations and constraints on their breastfeeding are numerous. Breastfeeding is work; work that is not shared and work that is rendered invisible by the way it is required to be hidden. Breastfeeding is the superior form of infant nutrition, but it is also often an all-encompassing and time-consuming form of mothering that can only be done by the mother. To be expected to hide breastfeeding is to hide much of the early work of mothering.

The perceived need to hide breastfeeding and to proceed with discretion also effectively keeps some women at home and out of public life more than they would be otherwise. It is this very fear that breastfeeding will exclude women from public life, will essentialize women's otherness from, difference from men, that may explain why breastfeeding has not been constructed as a feminist issue with much frequency (Kahn, 1989 makes this argument). But certainly a feminist social agenda would include allowing women to have the choice to use their bodies in ways they wish. Despite the perception of a hostile environment for breastfeeding, the women in this study overwhelmingly reported enjoying the experience of breastfeeding, describing it as calming, providing a special closeness to the child and emotionally gratifying. And, medical research is beginning to investigate the physical health benefits to the mother of breastfeeding as well.

REFERENCES

Apple, Rima. 1987. *Mothers & medicine: A social history of infant feeding, 1890–1950.* Madison: University of Wisconsin Press.

Baldwin, E., and K. Friedman. 1998. A current summary of breastfeeding legislation in the U.S. years. Online.

Available: http://www.lalecheleague.org/lawbills.html. Accessed 10 April 1998.

Bar-Yam, N. B., and L. Darby. 1997. Fathers and breastfeeding: A review of the literature. *Journal of Human Lactation* 13:45–50.

Bartky, S. L. 1998. Foucault, femininity, and the modernization of patriarchal power. In *The politics of women's bodies: Sexuality, appearance, and behavior,* edited by R. Weitz. New York: Oxford University Press.

Baumslag, Naomi, and Dia Michels. 1995. *Milk, money and madness: The culture and politics of breast-feeding.* Westport, CT: Bergin & Garvey.

Bordo, Susan. 1993. *Unbearable weight: Feminism, Western culture and the body.* Berkeley: University of California Press.

Davis-Floyd, Robbie. 1992. *Birth as an American rite of passage.* Berkeley: University of California Press.

Dellinger, K., and C. L. Williams. 1997. Makeup at work: Negotiating appearance rules in the work-place. *Gender & Society* 11:151–77.

Dettwyler, Katherine. 1995. Beauty and the beast: The cultural context of breastfeeding in the United States. In *Breastfeeding: Biocultural perspectives,* edited by P. Stuart-Macadam and K. Dettwyler. New York: Aldine.

Freed, G. L., S. J. Clark, J. Sorenson, J. A. Lohr, R. Cefalo, and P. Curtis. 1995. National assessment of physicians' breastfeeding knowledge, attitudes, training, and experience. *Journal of the American Medical Association* 273:472–76.

Goffman, Erving. 1963. *Stigma: Notes on the management of spoiled identity.* New York: Simon & Schuster.

Howard, C., S. Schaffer, and R. Lawrence. 1997. Attitudes, practices, and recommendations by obstetricians about infant feeding. *Birth* 24:240–246.

Kahn, Robbie Pfeufer. 1989. Mother's milk: The "moment of nurture" revisited. *Resources for Feminist Research* 18:29–33.

———. 1995. *Bearing meaning: The language of the birth.* Chicago: University of Illinois Press.

Newton, Niles. 1977. *Maternal emotions: A study of women's feelings toward menstruation, pregnancy, childbirth, breastfeeding, infant care and other aspects of their femininity.* New York: Paul B. Hoeber.

Stuart-Macadam, Patricia. 1995. Biocultural perspectives on breastfeeding. In *Breastfeeding: Biocultural perspectives,* edited by P. Stuart-Macadam and K. Dettwyler. New York: Aldine.

Umansky, Lauri. 1998. Breastfeeding in the 1990's: The Karen Carter case and the politics of maternal sexuality. In *"Bad mothers": The politics of blame in twentieth century America,* edited by M. Ladd-Taylor and L. Umansky. New York: New York University Press.

Weitz, Rose. 1998. A history of women's bodies. In *The politics of women's bodies: Sexuality, appearance, and behavior,* edited by R. Weitz. New York: Oxford University Press.

West C., and D. Zimmerman. 1937. Doing gender. *Gender & Society* 1:125–151.

Yalom, Marilyn. 1997. *A history of the breast.* New York: Knopf.

Young, Iris Marion. 1998. Breasted experience: The look and the feeling. In *The politics of women's bodies: Sexuality, appearance, and behavior,* edited by R. Weitz. New York: Oxford University Press.

Boundary Lines: Labeling Sexual Harassment in Restaurants

PATTI A. GIUFFRE AND CHRISTINE L. WILLIAMS

As the previous articles reveal, notions about what is considered sexual are ultimately social. The same object, person, or behavior that is considered attractive, erotic, or sexy in one context may be deemed mundane or even sacrosanct in another. Culture ultimately determines what is sexual and, by the same token, what is not. For example, a gentle pat on our posteriors may be considered a sign of affection, innocent fun, or an unwelcome trespass upon our bodily integrity depending on who is touched, who does the touching, and under what conditions the touching occurs. In the following reading, Giuffre and Williams take up this issue and demonstrate that like the construction of the erotic, the construction of the sexually inappropriate (in this case sexual harassment), is socially determined.

Sexual harassment occurs when submission to or rejection of sexual advances is a term of employment, is used as a basis for making employment decisions, or if the advances create a hostile or offensive work environment (Konrad and Gutek, 1986). Sexual harassment can cover a range of behaviors, from leering to rape (Ellis, Barak, and Pinto, 1991; Pryor, 1987; Reilly et al., 1992; Schneider, 1982). Researchers estimate that as many as 70 percent of employed women have experienced behaviors that may legally constitute sexual harassment (MacKinnon, 1979; Powell, 1986); however, a far lower percentage of women claim to have experienced sexual harassment. Paludi and Barickman write that "the great majority of women who are abused by behavior that fits legal definitions of sexual harassment—and who are traumatized by the experience—do not label what has happened to them 'sexual harassment' " (1991, 68).

It is difficult to label behavior as sexual harassment because it forces people to draw a line between illicit and "legitimate" forms of sexuality at work—a process fraught with ambiguity. Whether a particular interaction is identified as harassment will depend on the intention of the harasser and the interpretation of the interchange by the victim, and both of these perspectives will be highly influenced by workplace culture and the social context of the specific event.

This article examines how one group of employees—restaurant workers—distinguishes between sexual harassment and other forms of sexual interaction in the workplace. We conducted an in-depth interview study of waitpeople and found that complex double standards are often used, in labeling behavior as sexual harassment: identical behaviors are labeled sexual harassment in some contexts and not others. Many respondents claimed that they *enjoyed* sexual interactions involving co-workers of the same race/ethnicity, sexual orientation, and class/status backgrounds. Those who were offended by such interactions nevertheless dismissed them as natural or inevitable parts of restaurant culture.[1] When the same behavior

From "Boundary Lines: Labeling Sexual Harassment in Restaurants," by Patti A. Guiffre and Christine L. Williams, *Gender & Society* 8(3), 1994. Used by permission of Sage Publications.

occurred in contexts that upset these hegemonic heterosexual norms—in particular, when the episode involved interactions between gay and heterosexual men, or men and women of different racial/ethnic backgrounds—people seemed willing to apply the label sexual harassment.

We argue that identifying behaviors that occur only in counterhegemonic contexts as sexual harassment can potentially obscure and legitimate more insidious forms of domination and exploitation. As Pringle points out, "Men control women through direct use of power, but also through definitions of pleasure—which is less likely to provoke resistance" (1988, 95). Most women, she writes, actively seek out what Rich (1980) termed "compulsory heterosexuality" and find pleasure in it. The fact that men and women may enjoy certain sexual interactions in the workplace does not mean they take place outside of oppressive social relationships, nor does it imply that these routine interactions have no negative consequences for women. We argue that the practice of labeling as "sexual harassment" only those behaviors that challenge the dominant definition of acceptable sexual activity maintains and supports men's institutionalized right of sexual access and power over women.

METHODS

The occupation of waiting tables was selected to study the social definition of sexual harassment because many restaurants have a blatantly sexualized workplace culture (Cobble, 1991; Paules, 1991). Unremitting sexual banter and innuendo, as well as physical jostling, create an environment of "compulsory jocularity" in many restaurants (Pringle, 1988, 93). Sexual attractiveness and flirtation are often institutionalized parts of a waitperson's job description; consequently, individual employees are often forced to draw the line for themselves to distinguish legitimate and illegitimate expressions of sexuality, making this occupation an excellent context for examining how people determine what constitutes sexual harassment. In contrast, many more sexual behaviors may be labeled sexual harassment in less highly sexualized work environments.

Eighteen in-depth interviews were conducted with male and female wait staff who work in restaurants in Austin, Texas. Respondents were selected from restaurants that employ equal proportions of men and women on their wait staffs. Co-worker sexual harassment is perhaps the most common form of sexual harassment (Pryor 1987; Schneider 1982); yet most case studies of sexual harassment have examined either unequal hierarchical relationships (e.g., boss-secretary harassment) or harassment in highly skewed gender groupings (e.g., women who work in nontraditional occupations) (Benson and Thomson 1982; Carothers and Crull 1984; Gruber and Bjorn 1982). This study is designed to investigate sexual harassment in unequal hierarchical relationships, as well as harassment between organizationally equal co-workers.

The sample was generated using "snowball" techniques and by going to area restaurants and asking waitpeople to volunteer for the study. The sample includes eight men and ten women. Four respondents are Latina/o, two African-American, and 12 white. Four respondents are gay or lesbian; one is bisexual; 13 are heterosexual. (The gay men and lesbians in the sample are all "out" at their respective restaurants.) Fourteen respondents are single; three are married; one is divorced. Respondents' ages range from 22 to 37.

FINDINGS

Respondents agreed that sexual banter is very common in the restaurant: staff members talk and joke about sex constantly. With only one exception, respondents described their restaurants as highly sexualized. This means that 17 of the 18 respondents said that sexual joking, touching, and fondling were common, everyday occurrences in their restaurants. For example, when asked if he and other waitpeople ever joke about sex, one waiter replied, "about 90 percent of [the jokes] are about sex." According to a waitress, "at work . . .

[we're] used to patting and touching and hugging." Another waiter said, "I do not go through a shift without someone . . . pinching my nipples or poking me in the butt or grabbing my crotch. . . . It's just what we do at work."

These informal behaviors are tantamount to "doing heterosexuality," a process analogous to "doing gender" (West and Zimmerman 1987). By engaging in these public flirtations and open discussions of sex, men and women reproduce the dominant cultural norms of heterosexuality and lend an air of legitimacy—if not inevitability—to heterosexual relationships. In other words, heterosexuality is normalized and naturalized through its ritualistic public display. Indeed, although most respondents described their workplaces as highly sexualized, several dismissed the constant sexual innuendo and behaviors as "just joking," and nothing to get upset about. Several respondents claimed that this is simply "the way it is in the restaurant business," or "just the way men are."

With the one exception, the men and women interviewed maintained that they enjoyed this aspect of their work. However, in a few instances, sexual conduct was labeled as sexual harassment. Seven women and three men said they had experienced sexual harassment in restaurant work. Of these, two women and one man described two different experiences of sexual harassment, and two women described three experiences.

In general, respondents labeled their experiences sexual harassment only if the offending behavior occurred in one of three social contexts: (1) if perpetrated by someone in a more powerful position, such as a manager; (2) if perpetrated by someone of a different race/ethnicity; or (3) if perpetrated by someone of a different sexual orientation.

Powerful Position

In the restaurant, managers and owners are the highest in the hierarchy of workers. Generally, they are the only ones who can hire or fire waitpeople. Three of the women and one of the men interviewed said they had been sexually harassed by their restaurants' managers or owners. In addition, several others who did not personally experience harassment said they had witnessed managers or owners sexually harassing other waitpeople. This finding is consistent with other research indicating people are more likely to think that sexual harassment has occurred when the perpetrator is in a more powerful position (e.g., Ellis et al., 1991).

Carla describes being sexually harassed by her manager:

> One evening, [my manager] grabbed my body, not in a private place, just grabbed my body, period. He gave me like a bear hug from behind a total of four times in one night. By the end of the night I was livid. I was trying to avoid him. Then when he'd do it, I'd just ignore the conversation or the joke or whatever and walk away.

She claimed that her co-workers often give each other massages and joke about sex, but she did not label any of their behaviors sexual harassment. In fact, all four individuals who experienced sexual harassment from their managers described very similar types of behavior from their co-workers, which they did not define as sexual harassment. For example, Cathy said that she and the other waitpeople talk and joke about sex constantly: "Everybody stands around and talks about sex a lot . . . Isn't that weird? You know, it's something about working in restaurants and, yeah, so we'll all sit around and talk about sex." She said that talking with her co-workers about sex does not constitute sexual harassment because it is "only joking." She does, however, view her male manager as a sexual harasser:

> My employer is very sexist. I would call that sexual harassment. Very much of a male chauvinist pig. He kind of started [saying] stuff like, "You can't really wear those shorts because they're not flattering to your figure. . . . But I like the way you wear those jeans. They look real good. They're tight." It's like, you know [I want to say to him], "You're the owner, you're in power. That's evident. You know, you

> need to find a better way to tell me these things." We've gotten to a point now where we'll joke around now, but it's never ever sexual, ever. I won't allow that with him.

Cathy acknowledges that her manager may legitimately dictate her appearance at work, but only if he does so in professional—and not personal—terms. She wants him "to find a better way to tell me these things," implying that he is not completely out-of-line in suggesting that she wear tight pants. He "crosses the line" when he personalizes his directive, by saying to Cathy "*I like* the way you wear those jeans." This is offensive to Cathy because it is framed as the manager's personal prerogative, not the institutional requirements of the job.

One of the men in the sample, Frank, also experienced sexual harassment from a manager:

> I was in the bathroom and [the manager] came up next to me and my tennis shoes were spray-painted silver so he knew it was me in there and he said something about, "Oh, what do you have in your hand there?" I was on the other side of a wall and he said, "Mind if I hold it for a while?" or something like that, you know. I just pretended like I didn't hear it.

Frank also described various sexual behaviors among the waitstaff, including fondling, "joking about bodily functions," and "making bikinis out of tortillas." He said, "I mean, it's like, what we do at work. . . . There's no holds barred. I don't find it offensive. I'm used to by now. I'm guilty of it myself." Evidently, he defines sexual behaviors as "sexual harassment" only when perpetrated by someone in power over him.

Two of the women in the sample also described sexual harassment from customers. We place these experiences in the category of "powerful position" because customers do have limited economic power over the waitperson insofar as they control the tip (Crull 1987). Cathy said that male customers often ask her to "sit on my lap" and provide them with other sexual favors. Brenda, a lesbian, described a similar experience of sexual harassment from women customers:

> One time I had this table of lesbians and they were being real vulgar towards me. Real sexual. This woman kind of tripped me as I was walking by and said, "Hurry back." I mean, gay people can tell when other people are gay. I felt harassed.

In these examples of harassment by customers, the line is drawn using a similar logic as in the examples of harassment by managers. These customers acted as though the waitresses were providing table service to satisfy the customers' private desires, instead of working to fulfill their job descriptions. In other words, the customers' demands were couched in personal—and not professional—terms, making the waitresses feel sexually harassed.

It is not difficult to understand why waitpeople singled out sexual behaviors from managers, owners, and customers as sexual harassment. Subjection to sexual advances by someone with economic power comes closest to the quid pro quo form of sexual harassment, wherein employees are given the option to either "put out or get out." Studies have found that this type of sexual harassment is viewed as the most threatening and unambiguous sort (Ellis et al., 1991; Fitzgerald, 1990; Gruber and Bjorn, 1982).

Some women are reluctant to label blatantly offensive behaviors as sexual harassment. For example, Maxine, who claims that she has never experienced sexual harassment, said that customers often "talk dirty" to her:

> I remember one day, about four or five years ago when I was working as a cocktail waitress, this guy asked me for a "Slow Comfortable Screw" [the name of a drink]. I didn't know what it was. I didn't know if he was making a move or something. I just looked at him. He said, "You know what it is, right?" I said, "I bet the bartender knows!" (laughs). . . . There's another one, "Sex on the Beach." And there's another one

called a "Screaming Orgasm." Do you believe that?

Maxine is subject to a sexualized work environment that she finds offensive; hence her experience could fit the legal definition of sexual harassment. But because sexy drink names are an institutionalized part of restaurant culture, Maxine neither complains about it nor labels it sexual harassment: Once it becomes clear that a "Slow Comfortable Screw" is a legitimate and recognized restaurant demand, she accepts it (albeit reluctantly) as part of her job description. In other words, the fact that the offensive behavior is institutionalized seems to make it beyond reproach in her eyes. This finding is consistent with others' findings that those who work in highly sexualized environments may be less likely to label offensive behavior "sexual harassment" (Gutek, 1985; Korrrad and Gutek, 1986).

Only in specific contexts do workers appear to define offensive words and acts of a sexual nature as sexual harassment—even when initiated by someone in a more powerful position. The interviews suggest that workers use this label to describe their experiences only when their bosses or their customers couch their requests for sexual attentions in explicitly personal terms. This way of defining sexual harassment may obscure and legitimize more institutionalized—and hence more insidious—forms of sexual exploitation at work.

Race/Ethnicity

The restaurants in our sample, like most restaurants in the United States, have racially segregated staffs (Howe, 1977). In the restaurants where our respondents are employed, men of color are concentrated in two positions: the kitchen cooks and bus personnel (formerly called busboys). Five of the white women in the sample reported experiencing sexual harassment from Latino men who worked in these positions. For example, when asked if she had ever experienced sexual harassment, Beth said:

Yes, but it was not with the people . . . it was not, you know, the people that I work with in the front of the house. It was with the kitchen. There are boundaries or lines that I draw with the people I work with. In the kitchen, the lines are quite different. Plus, it's a Mexican staff. It's a very different attitude. They tend to want to touch you more and, at times, I can put up with a little bit of it but . . . because I will give them a hard time too but I won't touch them. I won't touch their butt or anything like that.

[Interviewer: So sometimes they cross the line?]

It's only happened to me a couple of times. One guy, like, patted me on the butt and I went off. I lost my shit. I went off on him. I said, "No. Bad. Wrong. I can't speak Spanish to you but, you know, this is it."

Beth reported that the waitpeople joke about sex and touch each other constantly, but she does not consider their behavior sexual harassment. Like many of the other men and women in the sample, Beth said she feels comfortable engaging in this sexual banter and play with the other waitpeople (who were predominantly white), but not with the Mexican men in the kitchen.

Part of the reason for singling out the behaviors of the cooks as sexual harassment may involve status differences between waitpeople and cooks. Studies have suggested that people may label behaviors as sexual harassment when they are perpetrated, by people in lower status organizational positions (Grauerholz, 1989; McKinney, 1990); however, it is difficult to generalize about the relative status of cooks and waitpeople because of the varied and often complex organizational hierarchies of restaurants (Paules 1991, 107–110).

Because each recounted case of sexual harassment occurring between individuals of different occupational statuses involved a minority man sexually harassing a white woman, the racial context seems equally important. For example, Ann also said that she and the other waiters and waitresses joke about sex and touch each other "on the butt" all the time, and when asked if she had ever experienced sexual harassment, she said,

> I had some problems at [a previous restaurant] but it was a communication problem. A lot of the guys in the kitchen did not speak English. They would see the waiters hugging on us, kissing us and pinching our rears and stuff. They would try to do it and I couldn't tell them, "No. You don't understand this. It's like we do it because we have a mutual understanding but I'm not comfortable with you doing it." So that was really hard and a lot of times what I'd have to do is just sucker punch them in the chest and just use a lot of cuss words and they knew that I was serious. And there again, I felt real weird about that because they're just doing what they see go on everyday.

According to Brenda:

> They're mostly Mexican, not even Mexican-American. Most of them, they're just starting to learn English.
>
> [Interviewer: What do they do to you?]
>
> Well, I speak Spanish, so I know. They're not as sexual to me because I think they know I don't like it. Some of the other girls will come through and they will touch them like here [points to the lower part of her waist]. . . . I've had some pretty bad arguments with the kitchen.
>
> [Interviewer: Would you call that sexual harassment?]
>
> Yes. I think some of the girls just don't know better to say something. I think it happens a lot with the kitchen guys. Like sometimes, they will take a relleno in their hands like it's a penis. Sick!

Each of these women identified the sexual advances of the minority men in their restaurants as sexual harassment, but not the identical behaviors of their white male co-workers; moreover, they all recognize that they draw boundary lines differently for Anglo men and Mexican men: each of them willingly participates in "doing heterosexuality" only in racially homogamous contexts. These women called the behavior of the Mexican cooks "sexual harassment" in part because they did not "have a relationship" with these men, nor was it conceivable to them that they *could* have a relationship with them, given cultural and language barriers—and, probably, racist attitudes as well. The white men, on the other hand, can "hug, kiss, and pinch rears" of the white women because they have a "mutual understanding"—implying reciprocity and the possibility of intimacy.

The importance of this perception of relationship potential in the assessment of sexual harassment is especially clear in the cases of the two married women in the sample, Diana and Maxine. Both of these women said that they had never experienced sexual harassment. Diana, who works in a family-owned and -operated restaurant, claimed that her restaurant is not a sexualized work environment. Although people occasionally make double entendre jokes relating to sex, according to Diana, "there's no contact whatsoever like someone pinching your butt or something." She said that she has never experienced sexual harassment.

Similarly, Maxine, who is Colombian, said she avoids the problem of sexual harassment in her workplace because she is married:

> The cooks don't offend me because they know I speak Spanish and they know how to talk with me because I set my boundaries and they know that . . . I just don't joke with them more than I should. They all know that I'm married, first of all, so that's a no-no for all of them. My brother used to be a manager in that restaurant so he probably took care of everything. I never had any problems anyway in any other jobs because,

like I said, I set my boundaries. I don't let them get too close to me.

> [Interviewer: You mean physically?]

> Not physically only. Just talking. If they want to talk about, "Do you go dancing? Where do you go dancing?" Like I just change the subject because it's none of their business and I don't really care to talk about that with them . . . not because I consider them to be on the lower levels than me or something but just because if you start talking with them that way then you are just giving them hope or something. I think that's true for most of the guys here, not just talking about the cooks. . . . I do get offended and they know that so sometimes they apologize.

Both Maxine and Diana said that they are protected from sexual harassment because they are married. In effect, they use their marital status to negotiate their interactions with their co-workers and to ward off unwanted sexual advances. Furthermore, because they do not view their co-workers as potential relationship "interests," they conscientiously refuse to participate in any sexual banter in the restaurant.

The fact that both women speak Spanish fluently may mean that they can communicate their boundaries unambiguously to those who only speak Spanish (unlike the female respondents in the sample who only speak English). For these two women, sexual harassment from co-workers is not an issue. On the other hand, at least from Maxine's vantage point, racial harassment is a bigger problem in her workplace than is sexual harassment. When asked if she ever felt excluded from any groups at work, she said:

> Yeah, sometimes. How can I explain this? Sometimes, I mean, I don't know if they do it on purpose or they don't but they joke around you about being Spanish. . . . Sometimes it hurts. Like they say, "What are you doing here? Why don't you go back home?"

Racial harassment—like sexual harassment—is a means used by a dominant group to maintain its dominance over a subordinated group. Maxine feels that, because she is married, she is protected from sexual harassment (although, as we have seen, she is subject to a sexualized workplace that is offensive to her); however, she does experience racial harassment where she works, and she feels vulnerable to this because she is one of very few nonwhites working at her restaurant.

One of the waiters in the sample claimed that he had experienced sexual harassment from female co-workers, and race may have also been a factor in this situation. When Rick (who is African-American) was asked if he had ever been sexually harassed, he recounted his experiences with some white waitresses:

> Yes. There are a couple of girls there, waitpeople, who will pinch my rear.

> [Interviewer: Do you find it offensive?]

> No (laughs) because I'm male. . . . But it is a form of sexual harassment.

> [Interviewer: Do you ever tell them to stop?]

> If I'm really busy, if I'm in the weeds, and they want to touch me, I'll get mad. I'll tell them to stop. There's a certain time and place for everything.

Because of the race difference, Rick may experience their behaviors as an expression of racial dominance, which probably influences his willingness to label the behavior as sexual harassment.

In sum, the interviews suggest that the perception and labeling of interactions as "sexual harassment" may be influenced by the racial context of the interaction. If the victim perceives the harasser as expressing a potentially reciprocal relationship interest, they may be less likely to label their experience sexual harassment. In cases where the harasser and victim have a different race/ethnicity and class background, the possibility of a relationship may be precluded because of racism, making these cases more likely to be labeled "sexual harassment."

This finding suggests that the practices associated with "doing heterosexuality" are profoundly racist. The white women in the sample showed a great reluctance to label unwanted sexual behavior sexual harassment when it was perpetrated by a potential (or real) relationship interest—that is, a white male co-worker. In contrast, minority men are socially constructed as potential harassers of white women: any expression of sexual interest may be more readily perceived as nonreciprocal and unwanted. The assumption of racial homogamy in heterosexual relationships thus may protect white men from charges of sexual harassment of white women. This would help to explain why so many white women in the sample labeled behaviors perpetrated by Mexican men as sexual harassment, but not the identical behaviors perpetrated by white men.

Sexual Orientation

There has been very little research on sexual harassment that addresses the sexual orientation of the harasser and victim (exceptions include Reilly et al. 1992; Schneider 1982, 1984). Surveys of sexual harassment typically include questions about marital status but not about sexual orientation (e.g., Fain and Anderton 1987; Gruber and Bjorn 1982; Powell 1986). In this study, sexual orientation was an important part of heterosexual men's perceptions of sexual harassment. Of the four episodes of sexual harassment reported by the men in the study, three involved openly gay men sexually harassing straight men. One case involved a male manager harassing a male waiter (Frank's experience, described earlier). The other two cases involved co-workers. Jake said that he had been sexually harassed by a waiter:

> Someone has come on to me that I didn't want to come on to me. . . . He was another waiter [male]. It was laughs and jokes the whole way until things got a little too much and it was like, "Hey, this is how it is. Back off. Keep your hands off my ass." . . . Once it reached the point where I felt kind of threatened and bothered by it.

Rick described being sexually harassed by a gay baker in his restaurant:

> There was a baker that we had who was really, really gay . . . He was very straightforward and blunt. He would tell you, in detail, his sexual experiences and tell you that he wanted to do them with you. . . . I knew he was kidding but he was serious. I mean, if he had a chance he would do these things.

In each of these cases, the men expressed some confusion about the intentions of their harassers—"I knew he was kidding but he was serious." Their inability to read the intentions of the gay men provoked them to label these episodes sexual harassment. Each man did not perceive the sexual interchange as reciprocal, nor did he view the harasser as a potential relationship interest. Interestingly, however, all three of the men who described harassment from gay men claimed that sexual banter and play with other *straight* men did not trouble them. Jake, for example, said that "when men get together, they talk sex," regardless of whether there are women around. He acceded, "people find me offensive, as a matter of fact," because he gets "pretty raunchy" talking and joking about sex. Only when this talk was initiated by a gay man did Jake label it as sexual harassment.

Johnson (1988) argues that talking and joking about sex is a common means of establishing intimacy among heterosexual men and maintaining a masculine identity. Homosexuality is perceived as a direct challenge and threat to the achievement of masculinity and consequently, "the male homosexual is derided by other males because he is not a real man, and in male logic if one is not a real man, one is a woman" (p. 124). In Johnson's view, this dynamic not only sustains masculine identity, it also shores up male dominance over women; thus, for some straight men, talking about sex with other straight men is a form of reasserting masculinity and male dominance, whereas talking about sex with gay men threatens the very basis for their masculine privilege. For this reason they may interpret the sex talk and conduct of gay men as a form of sexual harassment.

In certain restaurants, gay men may in fact intentionally hassle straight men as an explicit strategy to undermine their privileged position in society. For example, Trent (who is openly gay) realizes that heterosexual men are uncomfortable with his sexuality, and he intentionally draws attention to his sexuality in order to bother them:

> [Interviewer: Homosexuality gets on whose nerves?]
>
> The straight people's nerves . . . I know also that we consciously push it just because, we know, "Okay." We know this is hard for you to get used to but tough luck. I've had my whole life trying to live in this straight world and if you don't like this, tough shit." I don't mean like we're shitty to them on purpose but it's like, "I've had to worry about being accepted by straight people all my life. The shoe's on the other foot now. If you don't like it, sorry."
>
> [Interviewer: Do you get along well with most of the waitpeople?]
>
> I think I get along with straight women. I get along with gay men. I get along with gay women usually. If there's ever going to be a problem between me and somebody it will be between me and a straight man.

Trent's efforts to "push" his sexuality could easily be experienced as sexual harassment by straight men who have limited experience negotiating unwanted sexual advances. The three men who reported being sexually harassed by gay men seemed genuinely confused about the intentions of their harassers, and threatened by the possibility that they would actually be subjected to and harmed by unwanted sexual advances. But it is important to point out that Trent works in a restaurant owned by lesbians, which empowers him to confront his straight male co-workers. Not all restaurants provide the sort of atmosphere that makes this type of engagement possible. Clearly, not all gay men would be able to push their sexuality without suffering severe retaliation (e.g., loss of job, physical attacks).

In contrast to the reports of the straight men in this study, none of the women interviewed reported sexual harassment from their gay or lesbian co-workers.

Some heterosexual women claimed they feel *more* comfortable working with gay men and lesbians. For example, Kate prefers working with gay men rather than heterosexual men or women. She claims that she often jokes about sex with her gay co-workers, yet she does not view them as potential harassers. Instead, she feels that her working conditions are more comfortable and more fun because she works with gay men. Similarly, Cathy prefers working with gay men over straight men because "gay men are a lot like women in that they're very sensitive to other people's space." Cathy also works with lesbians, and she claims that she has never felt sexually harassed by them.

The gays and lesbians in the study did not report any sexual harassment from their gay and lesbian co-workers.

In sum, our interviews suggest that sexual orientation is an important factor in understanding each individual's experience of sexual harassment and his or her willingness to label interactions as sexual harassment. In particular, straight men may perceive gay men as potential harassers. Three of our straight male respondents claimed to enjoy the sexual banter that commonly occurs among straight men, and between heterosexual men and women, but singled out the sexual advances of gay men as sexual harassment. Their contacts with gay men may be the only context where they feel vulnerable to unwanted sexual encounters. Their sense of not being in control of the situation may make them more willing to label these episodes sexual harassment.

Our findings about sexual orientation are less suggestive regarding women. None of the women (straight, lesbian, or bisexual) reported sexual harassment from other female co-workers or from gay men. In fact, all but one of the women's reported cases of sexual harassment involved a heterosexual man. One of the two lesbians in the sample (Brenda) did experience sexual harassment from a group of lesbian customers (described earlier), but she claimed that sexual orientation is *not* key to her defining the situation as harassment. Other studies have shown that lesbian and bisexual women are routinely subjected to sexual harassment in the workplace (Schneider 1982, 1984); however, more research is needed to elaborate the social contexts and the specific definitions of harassment among lesbians.

The Exceptions

Two cases of sexual harassment were related by respondents that do not fit in the categories we have thus far described. These were the only incidents of sexual harassment reported between co-workers of the same race: in both cases, the sexual harasser is a white man, and the victim, a white woman. Laura—who is bisexual—was sexually harassed at a previous restaurant by a cook:

> This guy was just constantly badgering me about going out with him. He like grabbed me and took me in the walk-in one time. It was a real big deal. He got fired over it too. . . . I was in the back doing something and he said, "I need to talk to you," and I said, "We have nothing to talk about." He like took me and threw me against the wall in the back. . . . I ran out and told the manager, "Oh my God. He just hit me," and he saw the expression on my face. The manager went back there . . . then he got fired.

This episode of sexual harassment involved violence, unlike the other reported cases. The threat of violence was also present in the other exception, a case described by Carla. When asked if she had ever been sexually harassed, she said,

> I experienced two men, in wait jobs, that were vulgar or offensive and one was a cook and I think he was a rapist. He had the kind of attitude where he would rape a woman. I mean, that's the kind of attitude he had. He would say totally, totally inappropriate [sexual] things.

These were the only two recounted episodes of sexual harassment between "equal" co-workers that involved white men and women, and both involved violence or the threat of violence.[2]

Schneider (1982, 1991) found the greatest degree of consensus about labeling behavior sexual harassment when that behavior involves violence. A "victim of sexual harassment may be more likely to be believed when there is evidence of assault (a situation that is analogous to acquaintance rape). The assumption of reciprocity among homogamous couples may protect assailants with similar characteristics to their victims (e.g., class background, sexual orientation, race/ethnicity, age)—*unless* there is clear evidence of physical abuse. Defining only those incidents that involve violence as sexual harassment obscures—and perhaps even legitimatizes—the more common occurrences that do not involve violence, making it all the more difficult to eradicate sexual harassment from the workplace.

DISCUSSION AND CONCLUSION

We have argued that sexual harassment is hard to identify, and thus difficult to eradicate from the workplace, in part because our hegemonic definition of sexuality defines certain contexts of sexual interaction as legitimate. The interviews with waitpeople in Austin, Texas, indicate that how people currently identify sexual harassment singles out only a narrow range of interactions, thus disguising and ignoring a good deal of sexual domination and exploitation that take place at work.

Most of the respondents in this study work in highly sexualized atmospheres where sexual banter and touching frequently occur. There are institutionalized policies and practices in the workplace that encourage—or at the very least tolerate—a continual display and performance of heterosexuality. Many people apparently accept this ritual display as being a normal or natural feature of their work; some even enjoy this behavior. In the in-depth interviews, respondents labeled such experiences as sexual harassment in only three contexts: when perpetrated by someone who took advantage of their powerful position for personal sexual gain; when the perpetrator was of a different race/ethnicity than the victim—typically a minority man harassing a white woman; and when the perpetrator was of a different sexual orientation than the victim—typically a gay man harassing a straight man. In only two cases did respondents label experiences involving co-workers of the same race and sexual orientation as sexual harassment—and both episodes involved violence or the threat of violence.

These findings are based on a very small sample in a unique working environment, and hence it is not clear whether they are generalizable to other work settings. In less sexualized working environments, individuals may be more likely to label all offensive sexual advances as sexual harassment, whereas in more highly sexualized environments (such as topless clubs or striptease bars), fewer sexual advances may be labeled sexual harassment. Our findings do suggest that researchers should pay closer attention to the interaction context of sexual harassment, taking into account not only gender but also the race, occupational status, and sexual orientation of the assailant and the victim.

Of course, it should not matter who is perpetrating the sexually harassing behavior: sexual harassment should not be tolerated under any circumstances. But if members of oppressed groups (racial/ethnic minority men and gay men) are selectively charged with sexual harassment, whereas members of the most privileged groups are exonerated and excused (except in cases where institutionalized power or violence are used), then the patriarchal order is left intact. This is very similar to the problem of rape prosecution: minority men are the most likely assailants to be arrested and prosecuted, particularly when they attack white women (LaFree, 1989). Straight white men who sexually assault women (in the context of marriage, dating, or even work) may escape prosecution because of hegemonic definitions of "acceptable" or "legitimate" sexual expression. Likewise, as we have witnessed in the current debate on gays in the military, straight men's fears of sexual harassment justify the exclusion of gay men and lesbians, whereas sexual harassment perpetrated by straight men against both straight and lesbian women is tolerated and even endorsed by the military establishment, as in the Tailhook investigation (Britton and Williams, forthcoming). By singling out these contexts for the label "sexual harassment," only marginalized men will be prosecuted, and the existing power structure that guarantees privileged men's sexual access to women will remain intact.

Sexual interactions involving men and women of the same race and sexual orientation have a hegemonic status in our society, making sexual harassment difficult to identify and eradicate. Our interviews suggest that many men and women are active participants in the sexualized culture of the workplace, even though ample evidence indicates that women who work in these environments suffer negative repercussions to their careers because of it (Jaschik and Fretz 1991; Paludi and Barickman 1991; Reilly et al. 1992; Schneider 1982). This is how cultural hegemony works—by getting under our skins and defining what is and is not pleasurable to us, despite our material or emotional interests.

Our findings raise difficult issues about women's complicity with oppressive sexual relationships. Some women obviously experience pleasure and enjoyment from public forms of sexual engagement with men; clearly, many would resist any attempt to eradicate all sexuality from work—an impossible goal at any rate. Yet, it is also clear that the sexual "pleasure" many women seek out and enjoy at work is structured by patriarchal, racist, and heterosexist norms. Heterosexual, racially homogamous relationships are privileged in our society: they are institutionalized in organizational policies and job descriptions, embedded in ritualistic workplace practices, and accepted as legitimate, normal, or inevitable elements of workplace culture. This study suggests that only those sexual interactions that violate these policies, practices, and beliefs are resisted and condemned with the label "sexual harassment."

We have argued that this dominant social construction of pleasure protects the most privileged groups in society from charges of sexual harassment and may be used to oppress and exclude the least powerful groups. Currently, people seem to consider the gender, race, status, and sexual orientation of the assailant when deciding to label behaviors as sexual harassment. Unless we acknowledge the complex double standards people use in "drawing the line," then sexual domination and exploitation will undoubtedly remain the normative experience of women in the workforce.

NOTES

1. It could be the case that those who find this behavior extremely offensive are likely to leave restaurant work. In other words, the sample is clearly biased in that it includes only those who are currently employed in a restaurant and presumably feel more comfortable with the level of sexualized behavior than those who have left restaurant work.
2. It is true that both cases involved cooks sexually harassing waitresses. We could have placed these cases in the "powerful position" category, but did not because in these particular instances, the cooks did not possess institutionalized power over the waitpeople. In other words, in these particular cases, the cook and waitress had equal organizational status in the restaurant.

REFERENCES

Benson, Donna J., and Gregg E. Thomson. 1982. Sexual harassment on a university campus: The confluence of authority relations, sexual interest and gender stratification. *Social Problems* 29:236–251.

Britton, Dana M., and Christine L. Williams. Forthcoming. Don't ask, don't tell, don't pursue: Military policy and the construction of heterosexual masculinity. *Journal of Homosexuality*.

Carothers, Suzanne C., and Peggy Crull. 1984. Contrasting sexual harassment in female- and male-dominated occupations. In *My troubles are going to have trouble with me: Everyday trials and triumphs of women workers,* edited by K. B. Sacks and D. Remy. New Brunswick, NJ: Rutgers University Press.

Crull, Peggy. 1987. Searching for the causes of sexual harassment: An examination of two prototypes. In *Hidden aspects of women's work,* edited by Christine Bose, Roslyn Feldberg, and Natalie Sokoloff. New York: Praeger.

Ellis, Shmuel, Azy Barak, and Adaya Pinto. 1991. Moderating effects of personal cognitions on experienced and perceived sexual harassment of women at the workplace. *Journal of Applied Social Psychology* 21:1320–1327.

Fain, Terri C., and Douglas L. Anderton. 1987. Sexual harassment: Organizational context and diffuse status. *Sex Roles* 17:291–311.

Fitzgerald, Louise F. 1990. Sexual harassment: The definition and measurement of a construct. In *Ivory power: Sexual harassment on campus,* edited by Michele M. Paludi. Albany: State University of New York Press.

Grauerholz, Elizabeth. 1989. Sexual harassment of women professors by students: Exploring the dynamics of power, authority, and gender in a university setting. *Sex Roles* 21:789–801.

Gruber, James E., and Lars Bjorn. 1982. Blue-collar blues: The sexual harassment of women auto workers. *Work and Occupations* 9:271–298.

Gutek, Barbara A. 1985. *Sex and the workplace.* San Francisco: Jossey-Bass.

Howe, Louise Kapp. 1977. *Pink collar workers: Inside the world of women's work.* New York: Avon.

Jaschik, Mollie L., and Bruce R. Fretz. 1991. Women's perceptions and labeling of sexual harassment. *Sex Roles* 25:19–23.

Johnson, Miriam. 1988. *Strong mothers, weak wives.* Berkeley: University of California Press.

Konrad, Alison M., and Barbara A. Gutek. 1986. Impact of work experiences on attitudes toward sexual harassment. *Administrative Science Quarterly* 31:422–438.

LaFree, Gary D. 1989. *Rape and criminal justice: The social construction of sexual assault.* Belmont, CA: Wadsworth.

MacKinnon, Catherine A. 1979. *Sexual harassment of working women: A case of sex discrimination.* New Haven, CT: Yale University Press.

McKinney, Kathleen. 1990. Sexual harassment of university faculty by colleagues and students. *Sex Roles* 23:421–438.

Paludi, Michele, and Richard B. Barickman. 1991. *Academic and workplace sexual harassment.* Albany: State University of New York Press.

Paules, Greta Foff. 1991. *Dishing it out: Power and resistance among waitresses in a New Jersey restaurant.* Philadelphia: Temple University Press.

Powell, Gary N. 1986. Effects of sex role identity and sex on definitions of sexual harassment. *Sex Roles* 14:9–19.

Pringle, Rosemary. 1988. *Secretaries talk: Sexuality, power and work.* London: Verso.

Pryor, John B. 1987. Sexual harassment proclivities in men. *Sex Roles* 17:269–290.

Reilly, Mary Ellen, Bernice Lott, Donna Caldwell, and Luisa Deluca. 1992. Tolerance for sexual harassment related to self-reported sexual victimization. *Gender & Society* 6:122–138.

Rich, Adrienne. 1980. Compulsory heterosexuality and lesbian existence. *Signs* 5:631–660.

Schneider, Beth E. 1982. Consciousness about sexual harassment amont heterosexual and lesbian women workers. *Journal of Social Issues* 38:75–98.

———. 1984. The office affair: Myth and reality for heterosexual and lesbian women workers. *Sociological Perspectives* 27:443–464.

———. 1991. Put up and shut up: Workplace sexual assaults. *Gender & Society* 5:533–548.

West, Candace, and Don H. Zimmerman. 1987. Doing gender. *Gender & Society* 1:125–151.

DISCUSSION QUESTIONS

1. In addition to race/ethnicity, class/status, and sexuality, what other social characteristics affect our evaluation of the "appropriateness" of others' sexual behaviors?
2. The authors note that many restaurants have a "blatantly sexualized" workplace culture. How well might this research translate to less sexualized work settings? More sexualized settings? Outside of workplace settings?
3. Can workplace policies minimize and/or eliminate the sexual harassment of employees? If so, how? If not, why not?
4. What are some of the strategies that individuals can use to avoid or thwart harassing behaviors and comments? How reasonable, necessary, useful and/or just is it to expect potential targets to bear the responsibility of prevention?
5. How would you refute the argument that by dressing "sexy" some servers invite harassment?
6. The readings in this chapter provide examples of the social construction of sexuality in contemporary American society. How are elements of the social structure (e.g., patriarchy, heterosexuality, capitalism, etc.) reflected in these constructions? For example, how do heterosexual interests shape sex and sexual harassment? How do current constructions of breastfeeding and sexual harassment reflect the privileging of capitalism and patriarchy?

2

Constructing and Critiquing Sexual Categories

The Invention of Heterosexuality

JONATHAN NED KATZ

There are many possible ways to sort, label, and categorize sex and sexual behavior. Perhaps the most common modern categorization is to place people in groups based on the physiology (i.e., biological sex) of the actors involved. ***Heterosexual*** *refers to individuals who are attracted to or engage in sexual activities with those of the "opposite" biological sex, while* ***homosexual*** *refers to those who are attracted to or engage in sexual activities with those of the "same" sex. In this work, Katz traces the historical development of heterosexuality as a social category and reveals the important social conditions that gave rise to the ascendancy and primacy of this particular social-sexual distinction in the modern era.*

Heterosexuality is old as procreation, ancient as the lust of Eve and Adam. That first lady and gentleman, we assume, perceived themselves, behaved, and felt just like today's heterosexuals. We suppose that heterosexuality is unchanging, universal, essential: ahistorical.

Contrary to that common sense conjecture, the concept of heterosexuality is only one particular historical way of perceiving, categorizing, and imagining the social relations of the sexes. Not ancient at all, the idea of heterosexuality is a modern invention, dating to the late nineteenth century. The heterosexual belief, with its metaphysical claim to eternity, has a particular, pivotal place in the social universe of the late nineteenth and twentieth centuries that it did not inhabit earlier. This essay traces the historical process by which the heterosexual idea was created as ahistorical and taken for granted.

By not studying the heterosexual idea in history, analysts of sex, gay and straight, have continued to privilege the "normal" and "natural" at

From "The Invention of Heterosexuality" by Jonathan Ned Katz. Reprinted by permission.

the expense of the "abnormal" and "unnatural." Such privileging of the norm accedes to its domination, protecting it from questions. By making the normal the object of a thoroughgoing historical study we simultaneously pursue a pure truth and a sex-radical and subversive goal: we upset basic preconceptions. We discover that the heterosexual, the normal, and the natural have a history of changing definitions. Studying the history of the term challenges its power.

Contrary to our usual assumption, past Americans and other peoples named, perceived, and socially organized the bodies, lusts, and inter course of the sexes in ways radically different from the way we do. If we care to understand this vast past sexual diversity, we need to stop promiscuously projecting our own hetero and homo arrangement. Though lip-service is often paid to the distorting, ethnocentric effect of such conceptual imperialism, the category heterosexuality continues to be applied uncritically as a universal analytical tool. Recognizing the time-bound and culturally specific character of the heterosexual category can help us begin to work toward a thoroughly historical view of sex.

BEFORE HETEROSEXUALITY: EARLY VICTORIAN TRUE LOVE, 1820–1860

In the early nineteenth century United States, from about 1820 to 1860, the heterosexual did not exist. Middle-class white Americans idealized a True Womanhood, True Manhood, and True Love, all characterized by "purity"—the freedom from sensuality.[1] Presented mainly in literary and religious texts, this True Love was a fine romance with no lascivious kisses. This ideal contrasts strikingly with late-nineteenth and twentieth century American incitements to a hetero sex.*

Early Victorian True Love was only realized within the mode of proper procreation, marriage, the legal organization for producing a new set of correctly gendered women and men. Proper womanhood, manhood, and progeny—not a normal male-female eros—was the main product of this mode of engendering and of human reproduction.

The actors in this sexual economy were identified as manly men and womanly women and as procreators, not specifically as erotic beings or heterosexuals. Eros did not constitute the core of a heterosexual identity that inhered, democratically, in both men and women. True Women were defined by their distance from lust. True Men, though thought to live closer to carnality, and in less control of it, aspired to the same freedom from concupiscence.

Legitimate natural desire was for procreation and a proper manhood or womanhood; no heteroerotic desire was thought to be directed exclusively and naturally toward the other sex; lust in men was roving. The human body was thought of as a means towards procreation and production; penis and vagina were instruments of reproduction, not of pleasure. Human energy, thought of as a closed and severely limited system, was to be used in producing children and in work, not wasted in libidinous pleasures.

The location of all this engendering and procreative labor was the sacred sanctum of early Victorian True Love, the home of the True Woman and True Man—a temple of purity threatened from within by the monster masturbator, an archetypal early Victorian cult figure of illicit lust. The home of True Love was a castle far removed from the erotic exotic ghetto inhabited most notoriously then by the prostitute, another archetypal Victorian erotic monster. Between

* Some historians have recently told us to revise our idea of sexless Victorians: their experience and even their ideology, it is said, were more erotic than we previously thought. Despite the revisionists, I argue that "purity" was indeed the dominant, early Victorian, white, middle-class standard. For the debate on Victorian sexuality see John D'Emilio and Estelle Freedman, *Intimate Matters: A History of Sexuality in America* (New York: Harper & Row, 1988). p. xii.

1820 and 1860, only rarely was reference made to those other illicit erotic figures, the "sodomite" and "sapphist"—terms with no antonyms. State sodomy laws defined a particular, obscure act, referred to in a limited legalese, not a criminal, medical, or psychological type of person. Because the social organization of gender was dominant, not the mode of pleasure, the term "invert" was one of the early medical names for deviants from True Womanhood and True Manhood.[2] "Invert," stressing gender rather than erotic deviation, was one of those terms that did not take in the hetero- and homo-eroticized twentieth century.

LATE VICTORIAN SEX-LOVE: 1860–1892

Heterosexuality and "Homosexuality" did not appear out of the blue in the 1890s. These two eroticisms were in the making from the 1860s on. In late Victorian America and in Germany, from about 1860 to 1892, our modern idea of an eroticized universe began to develop, and the experience of a heterolust began to be widely documented and named. Playing a pioneering role in the titling and theorizing of sexual normalcy we may be surprised to find the early theorists and defenders of man-man love.

At the start of this era, the eloquent, embattled exponent of the new male-female lustiness was Walt Whitman whose third edition of *Leaves of Grass,* published in 1860, first included a section, "Children of Adam," frankly evoking the sexual-procreational intercourse of men and women. Whitman's women participated lustily and equally with men in the act of conceiving robust babies. Another of Whitman's new sections, "Calamus," vividly detailed acts of erotic communion between men. Whitman named and evoked hot "amative" relations between men and women, and sizzling "adhesive" intimacies between men. Whitman's titling of these amative and adhesive intimacies was an attempt to position male-female and male-male eroticisms together as a natural "healthy" division of human erotic responses. Though now perhaps better known as man-lover, Whitman was also the trailblazer of a previously silent and vilified male-female lust.

At about the same time, in 1863, in Germany, another pioneer defender of man-man love, Karl Heinrich Ulrichs, was also producing new sexual names to defend the love of the man who loved men, the "Uming," for the True Man. Uming's sex-love for a True Man, Ulrichs argued, was as natural as the "Dioning-love" of True Man and True Woman.

A few years later in Germany, in a letter of May 6, 1968 to Ulrichs, Karl Maria Kertbeny, is first known to have *privately* used two new terms coined by him: "heterosexuality" and "homosexuality"—the debut of the modern lingo! Kertbeny's "Heterosexualität" referred to a strong lust drive toward "opposite sex" intercourse (associated with numbers of morally reprehensible acts). It had the same sense as another of his new terms, "Normalsexualität." Kertbeny's term homosexual was first used *publicly* in Germany in 1880 in a defense of homosexuality. Kertbeny's coinage of the term heterosexual in the service of homosexual emancipation is considering the term's later use—one of sex history's grand ironies.

Heterosexual next made its public appearance in 1889, in the fourth German edition of Dr. Richard Krafft-Ebing's *Psychopathia Sexualis* where it was distinguished from homosexual. The homosexual emancipationist's word homosexual was appropriated by Krafft-Ebing and other late victorian German medical men (and later, by American doctors) as these Dr. Frankensteins' way of naming, condemning, and asserting their own right to regulate a group of homoerotic creatures just then emerging into sight in the bars, dance halls, and streets of their countries' larger cities.

But naming the specifying the sex deviant simultaneously delimited a sex norm—the new heterosexuality. The medical moralists' interest in a few powerless perverts would help to ensure the conformity of the majority to a new sex ethic, one that was congruent with the pursuit of consumer happiness and capitalist profit.

In the late nineteenth century United States, several social factors converged to cause the eroticizing of consciousness, behavior, emotion, and identity that became typical of the twentieth-century Western middle class. The transformation of the family from producer to consumer unit resulted in a change in family members' relation to their own bodies; from being an instrument primarily of work, the human body was integrated into a new economy, and began more commonly to be perceived as a means of consumption and pleasure. Historical work has recently begun on how the biological human body is differently integrated into changing modes of production, procreation, engendering, and pleasure so as to alter radically the identity, activity, and experience of that body.[3]

The growth of a consumer economy also fostered a new pleasure ethic. This imperative challenged the early Victorian work ethic, finally helping to usher in a major transformation of values. While the early Victorian work ethic had touted the value of economic production, that era's procreation ethic had extolled the virtues of human reproduction. In contrast, the late Victorian economic ethic hawked the pleasures of consuming, while its sex ethic praised an erotic pleasure principle for men and even for women.

In the late nineteenth century, the erotic became the raw material for a new consumer culture. Newspapers, books, plays, and films touching on sex, "normal" and "abnormal," became available for a price. Restaurants, bars, and baths opened, catering to sexual consumers with cash. Late Victorian entrepreneurs of desire incited the proliferation of a new eroticism, a commoditized culture of pleasure.

In these same years, the rise in power and prestige of medical doctors allowed these upwardly mobile professionals to prescribe a healthy new sexuality. Medical men, in the name of science, defined a new ideal of male-female relationships that included, in women as well as men, an essential, necessary, normal eroticism. Doctors, who had earlier named and judged the sex-enjoying woman a "nymphomaniac," now began to label women's *lack* of sexual pleasure a mental disturbance, speaking critically, for example, of female "frigidity" and "anesthesia."*

By the 1880s, the rise of doctors as a professional group fostered the rise of a new medical model of Normal Love, replete with sexuality. The new Normal Woman and Man were endowed with a healthy libido. The new theory of Normal Love was the modern medical alternative to the old Cult of True Love. The doctors prescribed a new sexual ethic as if it were a morally neutral, medical description of health. The creation of the new Normal Sexual had its counterpart in the invention of the late Victorian Sexual Pervert. The attention paid the sexual abnormal created a need to name the sexual normal, the better to distinguish the average him and her from the deviant it.

HETEROSEXUALITY: THE FIRST YEARS, 1892–1900

In the periodization of heterosexual American history suggested here, the years 1892 to 1900 represent "The First Years" of the heterosexual epoch, eight key years in which the idea of the heterosexual and homosexual were initially and tentatively formulated by US doctors. The earliest-known American use of the word "heterosexual" occurs in a medical journal article by Dr. James G. Kieman of Chicago, read before the city's medical society on March 7, 1892 and published that May portentous dates in sexual history.[4] But Dr. Kieman's heterosexuals were definitely not exemplars of normality. Heterosexuals, said Kieman, were defined by a mental condition, "psychical hermaphroditism." Its symptoms were "inclinations to both sexes." These heterodox sexuals also betrayed inclinations "to abnormal methods of

* This reference to females reminds us that the invention of heterosexuality had vastly different impacts on the histories of women and men. It also differed in its impact on lesbians and heterosexual women, homosexual and heterosexual men, the middle class and working class, and on different religious, racial, national, and geographic groups.

gratification," that is, techniques to insure pleasure without procreation. Dr. Kieman's heterogeneous sexuals did demonstrate "traces of the normal sexual appetite" (a touch of procreative desire). Kieman's normal sexuals were implicitly defined by a monolithic other-sex inclination and procreative aim. Significantly, they still lacked a name.

Dr. Kieman's article of 1892 also included one of the earliest known uses of the word homosexual in American English. Kieman defined "Pure homosexuals" as persons whose "general mental state is that of the opposite sex." Kieman thus defined homosexuals by their deviance from a gender norm. His heterosexuals displayed a double deviance from both gender and procreative norms.

Though Kieman used the new words heterosexual and homosexual, an old procreative standard and a new gender norm coexisted uneasily in his thought. His word heterosexual defined a mixed person and compound urge, abnormal because they wantonly included procreative and non-procreative objectives, as well as same-sex and different-sex attractions.

That same year, 1892, Dr. Krafft-Ebing's influential *Psychopathia Sexualis* was first translated and published in the United States.[5] But Kieman and Krafft-Ebing by no means agreed on the definition of the heterosexual. In Krafft-Ebing's book, "hetero-sexual" was used unambiguously in the modern sense to refer to an erotic feeling for a different sex. "Homo-sexual" referred unambiguously to an erotic feeling for a "same sex." In Krafft-Ebing's volume, unlike Kieman's article, heterosexual and homosexual were clearly distinguished from a third category, a "psycho-sexual hermaphroditism," defined by impulses toward both sexes.

Krafft-Ebing hypothesized an inborn "sexual instinct" for relations with the "opposite sex," the inherent "purpose" of which was to foster procreation. Krafft-Ebing's erotic drive was still a reproductive instinct. But the doctor's clear focus on a different-sex versus same-sex sexuality constituted a historic, epochal move from an absolute procreative standard of normality toward a new norm. His definition of heterosexuality as other-sex attraction provided the basis for a revolutionary, modern break with a centuries-old procreative standard.

It is difficult to overstress the importance of that new way of categorizing. The German's mode of labeling was radical in referring to the biological sex, masculinity or femininity, and the pleasure of actors (along with the procreant purpose of acts). Krafft-Ebing's heterosexual offered the modern world a new norm that came to dominate our idea of the sexual universe, helping to change it from a mode of human reproduction and engendering to a mode of pleasure. The heterosexual category provided the basis for a move from a production-oriented, procreative imperative to a consumerist pleasure principle—an institutionalized pursuit of happiness.

Despite the clarity of Krafft-Ebing's heterosexual/homosexual distinction, Dr. Kieman was not the only American medical writer to have difficulty understanding the hetero/homo pair as a normal/pervert duo. Perceived as ambivalent procreator, the heterosexual did not at first exemplify the quintessence of the normal. In 1893, for example, Dr. Charles Hughes assured his fellow medical men that, by medical treatment (hypnosis and sometimes surgery!), the mind and feelings could be "turned back into normal channels, the homo and hetero sexual changed into beings of natural erotic inclination, with normal impulsions."[6] The hetero, as person of mixed, procreative and nonprocreative, disposition, still stood with the nonprocreative homo as abnormal characters in the late nineteenth-century pantheon of sexual perverts.

Only gradually did doctors agree that heterosexual referred to a normal, "other-sex" eros. This new standard-model heterosex provided the pivotal term for the modern regularization of eros that paralleled similar attempts to standardize masculinity and femininity, intelligence, and manufacturing.[7] The idea of heterosexuality as the master sex from which all others deviated was (like the idea of the master race) deeply authoritarian. The doctors' normalization of a sex that was hetero proclaimed a new heterosexual separatism—an erotic apartheid that forcefully segregated the sex normals from the sex perverts. The new, strict boundaries made the emerging erotic world less polymorphous—safer for sex normals. However, the idea of such creatures as heterosexuals and homosexuals emerged from the narrow world of medicine to become a commonly accepted notion only in the early twentieth century. In 1901, in the comprehensive *Oxford English Dictionary,* "heterosexual" and "homosexual" had not yet made it.

THE DISTRIBUTION OF THE HETEROSEXUAL MYSTIQUE: 1900–1930

In the early years of this heterosexual century the tentative hetero hypothesis was stabilized, fixed, and widely distributed as the ruling sexual orthodoxy: The Heterosexual Mystique. Starting among pleasure-affirming urban working-class youths, southern blacks, and Greenwich Village bohemians as defensive subculture, heterosex soon triumphed as dominant culture.[8]

In its earliest version, the twentieth century heterosexual imperative usually continued to associate heterosexuality with a supposed human "need," "drive," or "instinct" for propagation, a procreant urge linked inexorably with carnal lust as it had not been earlier. In the early twentieth century, the falling birth rate, rising divorce rate, and "war of the sexes" of the middle class were matters of increasing public concern. Giving vent to heteroerotic emotions was thus praised as enhancing baby-making capacity, marital intimacy, and family stability. (Only many years later, in the mid-1960s, would heteroeroticism be distinguished completely, in practice and theory, from procreativity and male-female pleasure sex justified in its own name.)

The first part of the new sex norm—hetero—referred to a basic gender divergence. The "oppositeness" of the sexes was alleged to be the basis for a universal, normal, erotic attraction between males and females. The stress on the sexes' "oppositeness," which harked back to the early nineteenth century, by no means simply registered biological differences of females and males. The early twentieth century focus on physiological and gender dimorphism reflected the deep anxieties of men about the shifting work, social roles, and power of men over women, and about the ideals of womanhood and manhood. That gender anxiety is documented, for example, in 1897, in *The New York Times'* publication of the Reverend Charles Parkhurst's diatribe against female "andromaniacs," the preacher's derogatory, scientific-sounding name for women who tried to "minimize distinctions by which manhood and womanhood are differentiated."[9] The stress gender difference was a conservative response to the changing social-sexual division of activity and feeling which gave rise to the independent "New Woman" of the 1880s and eroticized "Flapper" of the 1920s.

The second part of the new hetero norm referred positively to sexuality. That novel upbeat focus on the hedonistic possibilities of male-female conjunctions also reflected a social transformation—a revaluing of pleasure and procreation, consumption and work in commercial, capitalist society. The democratic attribution of a normal lust to human females (as well as males) served to authorize women's enjoyment of their own bodies and began to undermine the early Victorian idea of the pure True Woman—a sex-affirmative action still part of women's struggle. The twentieth-century Erotic Woman also undercut nineteenth century feminist assertion of women's moral superiority, cast suspicions of lust on women's passionate romantic friendships with women, and asserted the presence of a menacing female monster, "the lesbian."[10]

From about 1900 on through the 1920s, a mixed bag of novelists, playwrights, sex educators, and profit-seeking publishers and play producers struggled to establish the legal right to discuss and distribute a new commodity, the explicit (for its time) heterosexual drama, novel, and advice book. The writers included James Branch Cabell, Theodore Dreiser, F. Scott Fitzgerald, Elinor Glyn, James Joyce, D. H. Lawrence, and sex educators like Mary Ware Dennett.[11]

In the perspective of heterosexual history, this early twentieth century struggle for the more explicit depiction of an "opposite-sex" eros appears in a curious new light. Ironically, we find sex-conservatives, the social-purity advocates of censorship and repression, fighting against the depiction not just of sexual perversity but also of the new normal heterosexuality. That a more open depiction of normal sex had to be defended against forces of propriety confirms the claim that heterosexuality's predecessor, Victorian True Love, had included no legitimate eros.

Before 1930 in the United States, heterosexuality was still fighting an uphill battle. As late as 1929, a federal court in Brooklyn found Mary Ware Dennett, author of a 21-page sex education pamphlet for young people, guilty of mailing this obscene essay. Dennett's pamphlet criticized other sex-education materials for not including a "frank, unashamed declaration that the climax of sex emotion is an unsurpassed joy, something which rightly belongs to every normal human being" *after* they fell in love and married. If it seemed "distasteful" that the sex organs were "so near . . . our 'sewerage system,'" Dennett assured America's youth that this offensive positioning of parts was probably protective ("At any rate, there they are, and our duty is . . . to take mighty good care of them."). The word heterosexual did not appear.[12]

THE HETEROSEXUAL STEPS OUT: 1930–1945

In 1930, in *The New York Times,* heterosexuality first became a love that dared to speak its name. On April 30th of that year, the word "heterosexual" is first known to have appeared in *The New York Times Book Review.* There, a critic described the subject of André Gide's *The Immoralist* proceeding "from a heterosexual liaison to a homosexual one." The ability to slip between sexual categories was referred to casually as a rather unremarkable aspect of human possibility. This is also the first known reference by *The Times* to the new hetero/homo duo.[13]

The following month the second reference to the hetero/homo dyad appeared in *The New York Times Book Review,* in a comment on Floyd Dell's *Love in the Machine Age.* This work revealed a prominent antipuritan of the 1930s using the dire threat of homosexuality as his rationale for greater heterosexual freedom. *The Times* quoted Dell's warning that current abnormal social conditions kept the young dependent on their parents, causing "infantilism, prostitution and homosexuality." Also quoted was Dell's attack on the "inculcation of purity" that "breeds distrust of the opposite sex." Young people, Dell said, should be "permitted to develop normally to heterosexual adulthood." "But," *The Times* reviewer emphasized, "such a state already exists, here and now." And so it did. Heterosexuality, a new gender-sex category, had been distributed from the narrow, rarified realm of a few doctors to become a nationally, even internationally cited aspect of middle-class life.[14]

HETEROSEXUAL HEGEMONY: 1945–1965

The "cult of domesticity" following World War II—the reassociation of women with the home, motherhood, and child-care; men with fatherhood and wage work outside the home—was a period in which the predominance of the hetero norm went almost unchallenged, an era of heterosexual hegemony. This was an age in which conservative mental-health professionals reasserted the old link between heterosexuality and procreation. In contrast, sex-liberals of the day strove, ultimately with success, to expand the heterosexual ideal to include within the boundaries of normality a wider-than-ever range of nonprocreative, premarital, and extra-marital behaviors. But sex-liberal reform actually helped to extend and secure the dominance of the heterosexual idea, as we shall see when we get to Kinsey.

The post-war sex-conservative tendency was illustrated in 1947, in Ferdinand Lundberg and Dr. Marnia Farnham's book, *Modern Woman: The Lost Sex.* Improper masculinity and femininity was exemplified, the authors decreed, by "engagement in heterosexual relations . . . with the complete intent to see to it that they do not eventuate in reproduction."[15] Their procreatively defined heterosex was one expression of a post-war ideology of fecundity that, internalized and enacted dutifully by a large part of the population, gave rise to the postwar baby boom.

The idea of the feminine female and masculine male as prolific breeders was also reflected in the stress, specific to the late 1940s, on the homosexual as sad symbol of "sterility"—that particular loaded term appears incessantly in comments on homosex dating to the fecund forties.

In 1948, in *The New York Times Book Review,* sex liberalism was in ascendancy. Dr. Howard A. Rusk declared that Alfred Kinsey's just published report on *Sexual Behavior in the Human Male* had found "wide variations in sex concepts and behavior." This raised the question: "What is 'normal' and 'abnormal'?" In particular, the report had found that "homosexual experience is much more common than previously thought," and "there is often a mixture of both homo and hetero experience."[16]

Kinsey's counting of orgasms indeed stressed the wide range of behaviors and feelings that fell within the boundaries of a quantitative, statistically accounted heterosexuality. Kinsey's liberal reform of the hetero/homo dualism widened the narrow, old hetero category to accord better with the varieties of social experience. He thereby contradicted the older idea of a monolithic, qualitatively defined, natural procreative act, experience, and person.[17]

Though Kinsey explicitly questioned "whether the terms 'normal' and 'abnormal' belong in a scientific vocabulary," his counting of climaxes was generally understood to define normal sex as majority sex. This quantified norm constituted a final, society-wide break with the old qualitatively defined reproductive standard. Though conceived of as purely scientific, the statistical definition of the normal as the sex-most-people-are-having substituted a new, quantitative moral standard for the old, qualitative sex ethic—another triumph for the spirit of capitalism.

Kinsey also explicitly contested the idea of an absolute, either/or antithesis between hetero and homo persons. He denied that human beings "represent two discrete populations, heterosexual and homosexual." The world, he ordered, "is not to be divided into sheep and goats." The hetero/homo division was not nature's doing: "Only the human mind invents categories and tries to force facts into separated pigeonholes. The living world is a continuum.[18]

With a wave of the taxonomist's hand, Kinsey dismissed the social and historical division of people into heteros arid homos. His denial of heterosexual and homosexual personhood rejected the social reality and profound subjective force of a historically constructed tradition which, since 1892 in the United States, had cut the sexual population in two and helped to establish the social reality of a heterosexual and homosexual identity.

On the one hand, the social construction of homosexual persons has lead to the development of a powerful gay liberation identity politics based on an ethnic group model. This has freed generations of women and men from a deep, painful, socially induced sense of shame, and helped to bring about a society-wide liberalization of attitudes and responses to homosexuals.[19] On the other hand, contesting the notion of homosexual and heterosexual persons was one early, partial resistance to the limits of the hetero/homo construction. Gore Vidal, rebel son of Kinsey, has for years been joyfully proclaiming:

> . . . there is no such thing as a homosexual or a heterosexual person. There are only homo- or heterosexual acts. Most people are a mixture of impulses if not practices, and what anyone docs with a willing partner is of no social or cosmic significance.
>
> So why all the fuss? In order for a ruling class to rule, there must be arbitrary prohibitions. Of all prohibitions, sexual taboo is the most useful because sex involves everyone . . . we have allowed our governors to divide the population into two teams. One team is good, godly, straight; the other is evil, sick, vicious.[20]

Though Vidal's analysis of our "wacky division" is persuasive, we may now go one step further and question not only the division into hetero and homo persons but the hetero/homo division itself.

Kinsey popularized the idea of a "continuum" of activity and feeling between hetero and homo poles. His liberal reform of the hetero/homo dualism did widen the narrow, old hetero category to accord with social experience, suggesting there were degrees of heterosexual and homosexual behavior and experience. But that famous continuum of erotic acts and feelings reaffirmed the idea of a sexuality divided between the hetero and homo. Kinsey's hetero/ homo rating scale, from zero to six, sounded neat, quantitative, and ever-so scientific; his influential sex-liberalism thus upheld and consolidated the hetero/homo polarity, giving it new life and legitimacy.

HETEROSEXUALITY QUESTIONED: 1965–1982

By the late 1960s, anti-establishment counterculturalists, fledgling feminists, and homosexual-rights activists had begun to produce an unprecedented critique of sexual repression in general, of women's sexual repression in particular, of marriage and the family—and of some forms of heterosexuality. This critique even found its way into *The New York Times.*

In March 1968, in the theater section of that paper, freelancer Rosalyn Regelson cited a scene from a satirical review brought to New York by a San Francisco troup:

> . . . a heterosexual man wonders inadvertently into a homosexual bar. Before he realizes his mistake, he becomes involved with an aggressive queen who orders a drink for him. Being a broadminded liberal and trying to play it cool until he can back out of the situation gracefully, he asks, "How do you like being a ah homosexual?" To which the queen drawls drily, "How do you like being ah whatever it is you are?"

Regelson continued:

> The Two Cultures in confrontation. The middle-class liberal, challenged today on many fronts, finds his last remaining fixed value, his heterosexuality, called into question. The theater . . . recalls the strategies he uses in dealing with this ultimate threat to his worldview.[21]

By the early 1970s, militant homosexuals and feminists were explicitly criticizing some institutional forms of heterosexuality. A "heterosexual dictatorship," enforcing that single erotic standard, had been named by Christopher Isherwood.[22] "Compulsory heterosexuality" was named in 1976 as one of the crimes against women by the Brussels Tribunal on Crimes Against Women.[23]

In 1979, Lillian Faderman coined the term "heterocentric" to condemn a world-view that made homosexuals (especially lesbians) invisible.[24] And in the summer of 1980, *Signs,* the feminist scholarly journal, published Adrienne Rich's "Compulsory Heterosexuality and Lesbian Existence." Taking off from earlier discussions among feminists in the movement press, Rich's was the first "respectable" work to so explicitly question the implicit assumption of heterosexuality. Rich's essay was important in beginning to legitimize the serious study of heterosexuality as social and political institution.[25]

Several years before herpes and AIDS became focal points for the anxiety of heterosexuals, their sex-love had begun to loose its old certainty, its unquestioned status. Masters and Johnson's *Crisis: Heterosexual Behavior in the Age of AIDS* in fact named an emergency predating AIDS.[26] Yet the media response to AIDS has created a major, unremarked, quantitative increase in references to heterosexuality and a historic, qualitative change in the public perception of heterosexuals as one of the endangered erotic species.

HETEROSEXUAL HISTORY: OUT OF THE SHADOWS

Our brief survey of the heterosexual idea suggests a new hypothesis. Rather than naming a conjunction old as Eve and Adam, heterosexual designates a word and concept, a norm and role, an individual and group identity, a behavior and feeling, and a peculiar sexual-political institution particular to the late nineteenth and twentieth centuries.

Because much stress has been placed here on heterosexuality as word and concept, it seems important to affirm that heterosexuality (and homosexuality) came into existence before it was named and thought about. The formulation of the heterosexual idea did not create a heterosexual experience or behavior; to suggest otherwise would be to ascribe determining power to labels and concepts. But the titling and envisioning of heterosexuality did play an important role in consolidating the construction of the heterosexual's social existence. Before the wide use of the word heterosexual, I suggest, women and men did not mutually lust with the same profound, sure sense of normalcy that followed the distribution of "heterosexual" as universal sanctifier.

According to this proposal, women and men make their own sexual histories. But they do not produce their sex lives just as they please. They make their sexualities within a particular mode of organization given by the past and altered by their changing desire, their present power and activity, and their vision of a better world. That hypothesis suggests a number of good reasons for the immediate inauguration of research on a historically specific heterosexuality.[27]

The study of the history of the heterosexual experience will forward a great intellectual struggle still in its early stages. This is the fight to pull heterosexuality, homosexuality, and all the sexualities out of the realm of nature and biology into the realm of the social and historical. Feminists have explained to us that anatomy does not determine our gender destinies (our masculinities and femininities). But we've only recently begun to consider that *biology does not settle our erotic fates.* The common notion that biology determines the object of sexual desire, or that physiology and society together cause sexual orientation, are determinisms that deny the break existing between our bodies and situation and our desiring. Just as the biology of our hearing organs will never tell us why we take pleasure in Bach or delight in Dixieland, our female or male anatomies, hormones, and genes will never tell us why we yearn for women, men, both, other, or none. That is because desiring is a self-generated project of individuals within particular historical cultures. Heterosexual history can help us see the place of values and judgments in the construction of our own and others' pleasures, and to see how our erotic tastes—our aesthetics of the flesh—are socially institutionalized through the struggle of individuals and classes.

The study of heterosexuality in time will also help us to recognize the *vast historical diversity of sexual emotions and behaviors*—a variety that challenges the monolithic heterosexual hypothesis. John D'Emilio and Estelle Freedman's *Intimate Matters: A History of Sexuality in America* refers in passing to numerous substantial changes in sexual activity and feeling: for example, the widespread use of contraceptives in the nineteenth century, the twentieth century incitement of the female orgasm, and the recent sexual conduct changes by gay men in response to the AIDS epidemic. It's now a commonplace of family history that people in particular classes feel and behave in substantially different ways under different historical conditions.[28] Only when we stop assuming an invariable essence of heterosexuality will we begin the research to reveal the full variety of sexual emotions and behaviors.

The historical study of the heterosexual experience can help us *understand the erotic relationships of women and men in terms of their changing modes of social organization.* Such modal analysis actually characterizes a sex history well underway.[29] This suggests that the eros-gender-procreation system (the social ordering of lust, femininity and masculinity, and babymaking) has been linked closely to a society's particular organization of power and production. To understand the subtle history of heterosexuality we need to look carefully at correlations between (1) society's organization of eros and pleasure; (2) its mode of engendering persons as feminine or masculine (its making of women and men); (3) its ordering of human reproduction; and (4) its dominant political economy. This General Theory of Sexual Relativity proposes that substantial historical changes in the social organization of eros, gender, and procreation have basically altered the activity and experience of human beings within those modes.[30]

A historical view locates heterosexuality and homosexuality in time, helping us distance ourselves from them. This distancing can help us formulate new questions that clarify our long-range sexual-political goals: what has been and is the social function of sexual categorizing? Whose interests have been served by the division of the world into heterosexual and homosexual? Do we dare not draw a line between those two erotic species? Is some sexual naming socially necessary? Would human freedom be enhanced if the sex-biology of our partners in lust was of no particular concern, and had no name? In what kind of society could we all more freely explore our desire and our flesh?

A new sense of the historical making of the heterosexual and homosexual suggests that these are ways of feeling, acting, and being with each other that we can together unmake and radically remake according to our present desire, power, and our vision of a future political-economy of pleasure.

NOTES

1. Barbara Welter, "The Cult of True Womanhood: 1820–1860." *American Quarterly,* vol. 18, (Summer 1966); Welter's analysis is extended here to include True Men and True Love.
2. Michael Lynch, "New York Sodomy, 1796–1873," "paper presented at the New York Institute for the Humanities, February 1, 1985.
3. See, for example, Catherine Gallagher and Thomas Laqueur, eds., "The Making of the Modern Body: Sexuality and Society in the Nineteenth Century," *Representations,* no. 14 (Spring 1986), (republished, Berkeley: University of California Press, 1987).
4. Dr. James G. Kieman, "Responsibility in Sexual Perversion," *Chicago Medical Recorder,* vol. 3 (May 1892), pp. 185–210.
5. R. von Krafft-Ebing, *Psychopathia Sexualis, with Especial Reference to Contrary Sexual Instinct: A Medico-Legal Study,* trans, Charles Gilbert Chaddock (Philadelphia: F. A. Davis, 1892), from the 7th and revised German ed. Preface, November 1892.
6. Dr. Charles H. Hughes, "Erotopathia—Morbid Eroticism," read at the Pan American Medical Congress, September, 1893; *Allenist and Neurologist,* vol. 14, no. 4 (October 1893), pp. 531–578.
7. For the standardization of gender see Lewis Terman and C. C. Miles, *Sex and Personality, Studies in Femininity and Masculinity* (New York: McGraw-Hill, 1936). For the standardization of intelligence see Lewis Terman, *Stanford-Binet Intelligence Scale* (Boston: Houghton Mifflin, 1916). For the Standardization of work, see "scientific management" and "Taylorism" in Harry Braveman, *Labor and Monopoly Capital: The Degradation of Work in the Twentieth Century* (New York: Monthly Review Press, 1974).
8. See D'Emilio and Freedman, *Intimate Matters,* pp. 194–201, 231, 241, 295–296; Ellen Kay Trimberger, "Feminism, Men, and Modern Love: Greenwich Village, 1900–1925," in *Powers of Desire: The Politics of Sexuality,* eds. Ann Snitow, Christine Stansell, Sharon Thompson (New York: Monthly Review Press, 1983), pp. 131–152; Kathy Peiss, "'Charity Girls' and City Pleasures: Historical Notes on Working Class Sexuality, 1880–1920," in *Powers of Desire,* pp. 74–87; and Mary P. Ryan, "The Sexy Saleslady: Psychology, Heterosexuality, and Consumption in the Twentieth Century," in her *Womanhood in America,* 2nd ed. (New York: Franklin Watts; 1979), pp. 151–182.
9. [Rev. Charles Parkhurst], "Woman. Calls Them Andromaniacs. Dr. Parkhurst So Characterizes Certain Women Who Passionately Ape Everything That is Mannish. Woman Divinely Preferred. Her Supremacy Lies in Her Womanliness, and She Should Make the Most of It—Her Sphere of Best Usefulness the Home," *The New York Times,* May 23, 1897, p. 16:1.
10. See Lisa Duggan, "The Social Enforcement of Heterosexuality and Lesbian Resistance in the 1920s," in *Class, Race, and Sex: The Dynamics of Control,* ed. Amy Swerdlow and Hanah Lessinger (Boston: G. K. Hall, 1983), pp. 75–92; Rayna Rapp and Ellen Ross, "The Twenties Backlash: Compulsory Heterosexuality, the Consumer Family, and the Waning of Feminism," in Swerdlow and Lessinger; Christina Simmons, "Companionate Marriage and the Lesbian Threat," *Frontiers,* vol. 4, no. 3 (Fall 1979), pp. 54–59; and Lillian Faderman, *Surpassing the Love of Men* (New York: William Morrow, 1981).
11. David Loth, *The Erotic in Literature: A Historical Survey of Pornography as Delightful as it is Indiscreet* (London: Secker & Warburg, 1962). Ch. IX, "The Bars Begin to Drop," pp. 145–170.
12. Mary Ware Dennett, *The Sex Side of Life, An Explanation for Young People* (Astoria, New York: Published by the Author, 1928).
13. Louis Kronenberger, review of Andrè Gide, *The Immoralist, New York Times Book Review,* April 20, 1930, p. 9.
14. Henry James Forman, review of Floyd Dell, *Love in the Machine Age* (New York: Farrar & Rinehart), *New York Times Book Review,* September 14, 1930, p. 9.
15. Ferdinand Lundberg and Dr. Marnia F. Farnham, *Modern Woman the Lost Sex* (NY: Harper, 1947).
16. Dr. Howard A. Rusk, *New York Times Book Review,* January 4, 1948, p. 3.
17. Alfred Kinsey, Wardell B. Pomeroy, Clyde E. Martin, *Sexual Behavior in the Human Male* (Philadelphia: W.B. Saunders, 1948), pp. 199–200.
18. Kinsey, *Sexual Behavior,* pp. 637, 639.

19. See Steven Epstein, "Gay Politics, Ethnic Identity: The Limits of Social Constructionism," *Socialist Review* 93/93 (1987), pp. 9–54.

20. Gore Vidal, "Someone to Laugh at the Squares With" [Tennessee Williams]. *New York Review of Books,* June 13, 1985: reprinted in his *At Home: Essays, 1982–1988* (New York: Random House, 1988), p. 48.

21. Rosalyn Regelson, "Up the Camp Staircase," *The New York Times,* March 3, 1968, Section II, p. 1:5.

22. I have not traced when Isherwood first referred to the "heterosexual dictatorship." Isherwood dates to 1929, at age 24, his becoming "suddenly, blindly furious that 'Girls are what the state and the church and the law and the press and the medical profession endorse, and command me to desire.'" *Christopher and His Kind* (New York: Avon Books, 1977), p. 12.

23. Diana Russell and Nicole van de Vens, eds., *Proceedings of the International Tribunal on Crimes Against Women* (Millbrae, California: Les Femmes, 1976), pp. 42–43, 56–57; cited by Adrienne Rich, "Compulsory Heterosexuality and Lesbian Existence," *Signs,* vol. 5, no. 4 (Summer 1980), pp. 653.

24. *The Body Politic* (December-January 1977), pp. 19, 21.

25. Lillian Faderman, "Who Hid Lesbian History?," *Frontiers: A Journal of Women Studies,* vol. 4, no. 3 (Fall 1979), p. 74.

26. William H. Masters, Virginia Johnson, Robert Kolodny, *Crisis: Heterosexual Behavior in the Age of AIDS* (New York: Grove Press, 1988).

27. In addition to those already cited in these notes, a few writers have begun the historicizing of heterosexuality. In 1987, GMP in London published *Heterosexuality,* a collection of essays by gay men and lesbians edited by Gillian E. Hanscombe and Martin Humphires; the essay by Jon Ward on "The Nature of Heterosexuality," is the most relevant to historical analysis. In April 1988, at a lesbian and gay studies conference at Brooklyn College, I spoke on that subject, and Henry Abelove and Randolph Trumbach spoke on a separate panel of the same title. In February 1989, Abelove, G. S. Rousseau, and Trumbach participated in a panel of the same title at the University of California, Berkeley. See Marv Shaw, "Inventing Straights: Homosexuality," *Bay Area Reporter,* March 9, 1989. Also see Abelov's "Some Speculations on the History of 'Sexual Intercourse' During the 'Long Eighteenth Century' in England," *Genders,* vol. 6 (November 1989).

28. D'Emilio and Freedman, *Intimate Matters,* pp. 57–63, 268, 356.

29. Ryan, *Womanhood:* John D'Emilio, "Capitalism and Gay Identity" in *Powers of Desire,* pp. 100–113; Jeffrey Weeks, *Coming out: Homosexual Politics in Britain from the Nineteenth Century to the Present* (London: Quartet Books, 1977); D'Emilio and Freedman, *Intimate Matters,* Katz, "Early Colonial Exploration."

30. This tripartite system is *intended* as a revision of Gayle Rubin's pioneering work on the social-historical organization of eros and gender. See "The Traffic in Women: Notes on the Political Economy of Sex," in *Toward an Anthropology of Women,* ed. Rayna R. Reiter (New York: Monthly Review Press, 1975), pp. 157–210, and "Thinking Sex: Notes for a Radical Theory of the Politics of Sexuality," in *Pleasure and Danger: Exploring Female Sexuality,* ed. Carole S. Vance (Boston: Routledge & Kegan Paul, 1984) pp. 267–329.

DISCUSSION QUESTIONS

1. What kind of social changes might result from a movement toward the creation of a society in which we could "all more freely explore our desire and our flesh?" How might life be different in such a society?

2. In this article, Katz demonstrates how classifications of sexuality have, in recent history, centered around issues such as procreation and the erotic pleasure principle. What other kinds of sexual classification systems can you envision? In other words, in what other ways might we group sexual acts and/or sexuality to create distinct social categories? What kinds of social conditions would increase the likelihood that these specific kinds of categorical distinctions would emerge?

3. Think about the different historical periods outlined by Katz. Do you think that the definition of sex (the act) changed across these time periods? How? Why? How would bisexuality and asexuality be viewed during each of the historical epochs Katz describes?

4. Think back to the Stearns article on breastfeeding and the Giuffre and Williams' article on sexual harassment. How do you think the division of the world into heterosexual and homosexual categories has affected definitions of what is erotic and the what/who of sexually "appropriate" behaviors?

5. What interests have been served by the division of the world into heterosexual and homosexual? Who benefits? Who suffers from dividing the world in this way?

'Bisexuality' in an Essentialist World: Toward an Understanding of Biphobia

AMANDA UDIS-KESSLER

In this selection, Udis-Kessler explores the negative attitudes some gays and lesbians have toward bisexuals. These negative views include the perception that bisexuals are "fence-sitters" who eschew commitment in order to retain heterosexual privilege. Udis-Kessler focuses on the essentialist/constructionist debate to argue that the discomfort gays and lesbians have with bisexuality largely stems from the specter of sexual choice that bisexuality (and social constructionism) supposedly represents.

Imagine this. You are at work on a lesbian/gay political or educational event. Or, possibly, you are involved in AIDS education or activism. You work as hard as anyone else, and you are as committed as anyone else. However, your sexuality is denied, if not actually disparaged. You are considered an oppressor and sometimes a spy; your concerns are thought to be different and threatening. Some of the people with whom you interact think you shouldn't be part of lesbian/gay/AIDS work at all. In their minds, you are one of the "fence-sitters, traitors, cop-outs, closet cases, people whose primary goal in life is to retain 'heterosexual privilege,' power-hungry cold-hearted seducers who use and discard their same-sex lovers like so many Kleenex."[1] If nothing else, you are responsible for the spread of AIDS. Imagine. All of this, simply because you identify as bisexual.[2]

There are two reasons why lesbian/gay communities would do well to examine, and ultimately unlearn, their biphobia. First, gaybashings and gay-related murders are increasing throughout the nation. The incidence of homophobia on college campuses has risen alarmingly. Sex and AIDS educators who insist on being honest about homosexuality have lost federal funding, and judges and officials around the country have made it clear that their bigotry will affect their judicial and legal decision-making. Gay religious groups are kicked out of churches; their clergy are sometimes forced to leave their orders, or are silenced. [Many] states have refused to pass antidiscrimination laws for their lesbian, gay, and bisexual citizens. Twenty years after Stonewall, 10 years after the assassination of Harvey Milk, there is a daunting amount of work to do, and lesbians and gays cannot afford to exclude anyone who wants to participate. This especially includes bisexuals, who share the same issues, and not just "half the time." We don't get half-gaybashed when we walk down the street with our same-sex lovers; we don't even get bashed half as often. We don't get half-fired from our jobs, or lose only half of our children in court battles. When HIV-positive, we don't progress to ARC and then stop, rather than

From "Bisexuality in an Essentialist World: Toward and Understanding of Biophobia" by Amanda Udis Kessler, in *Bisexuality: A Reader and Sourcebook*, ed. by Thomas Geller, p. 51-63.

getting full-blown AIDS. Lesbian/gay issues are our issues, and we want to work on them with lesbians and gay men. We can offer our strength, energy, and creativity. It's too dangerous a time to let biphobia get in the way of resources.

I said there were two reasons for unlearning biphobia. In the past few years, lesbian/gay communities have begun serious work on the racism and sexism which exists within every part of our culture. They have begun to acknowledge their power to oppress, as well as their experience of being oppressed. With the growth of coalitioning, they have taken up others' causes as others have taken up lesbian and gay causes. In this process, lesbians and gays are learning new ways to communicate and relate, ways that are not hierarchical and dualistic but that point forward to an egalitarian society. Looked at in this light, biphobia is just one more wall that needs to come down. Various arguments have been advanced as to why lesbians and gay men, no strangers to bigotry and persecution, should continue to malign anyone's sexuality. The claims are similar to some of those offered for homophobia: jealousy, fear of differences that one doesn't understand, fear that one might actually be bisexual. I want to consider a possible explanation which goes beyond these and draws on sexual identity theory, specifically the essentialist/constructionist debate.[3]

This debate appears in various disciplines, and concerns the nature of someone's identity; for our purposes, the debate concerns the nature of sexuality. Essentialists will describe sexuality as an essence, or ontological category, while constructionists (properly called social constructionists) will claim that sexuality does not exist as a category in and of its own right, only as a facet of specific human lives and experiences. To the extent that sexuality is a category for the constructionist, it is a socially constructed one.

Essentialists describe sexual orientation as innate: a basic, individually immanent part of one's sex drive. That is, there is a predetermined orientational "core of truth" in any person, regardless of his or her sexual behavior. Note the normative dimensions of this outlook: it is as unnatural and perverted for a homosexual to engage in heterosexual sex as it is for a heterosexual to engage in homosexual sex. To an essentialist, sexual orientation is deterministic. The very phrase "sexual orientation" is essentialist in nature.

Constructionism, in contrast, begins with empiricism: existence precedes essence. Sexual orientation, far from being a predetermined ontological category, is seen as a constructed descriptor of discrete acts. While the essentialist moved from the category to the specific example, the constructionist moves from the specific example to the category. In studying individuals and the societies in which they live, the constructionist encounters sexual scripts that vary with era and location. Moreover, even within a specific era and location, there can be variations on the sexual script. One's sexuality emerges throughout one's life in response to changing scripts and situations; it is relational. As Cliff Arnesen recently noted, "The process of socialization and sexual orientation are interwoven. They produce the many ways that men and women learn how and whom to love. In short, there is no universal 'map' on loving.[4] To a constructionist, sexual orientation is not deterministic; it is not simply the unfolding of a natural process. Rather, it is learned, contingent, unpredictable; the potential for change remains throughout every life. To a constructionist, sexuality is as sexuality does.

Consider the history of the sexual language we now use. The term "sexuality" did not appear in the Oxford English Dictionary until 1800. Although there has been homosexual behavior since animals became sexual, and certainly since people became sexual, the word "homosexuality" did not exist until 1869, and did not reach common usage until the 1890s. We talk now of "the love that dared not speak its name," but we would do well to remember that it had no name until about a century ago. The essentialist would argue that we discovered the category of homosexuality, much the same way scientists discovered oxygen and the heliocentric design of our solar system.

The constructionist would argue that we constructed the category because it became useful to have such a category. Further, the constructionist would argue that there is no hard and fast distinction between heterosexuality and homosexuality.

Before I move into the history of this debate within lesbian/gay liberation, perhaps I should make my own bias on the debate clear. I believe that sexuality is overdetermined and has both essentialist and constructionist roots. Nonetheless, I am primarily a constructionist. If we look at individuals across the span of their lives, we see the limitations of essentialism. At any given point in time, a person may experience what feels like an essential pull in a given direction, yet several years or decades later experience the exact reverse. The essentialist has no way to describe a woman who is happily heterosexual until age forty and happily homosexual thereafter, except to say that the first 40 years didn't count in some way, did not represent the woman's true self. This seems unreasonable as well as demeaning; it delegitimates forty years of sexual experience which may have been very meaningful. Essentialism has been of even less help in understanding sexuality in non-Western cultures.

Constructionism seems to me the better way of describing human sexual capacity and variety, as long as its limitations are also clarified. Most specifically, while constructionism does imply fluidity across time and place, it does not *necessarily* imply willful choice or intentionality, either in an individual's life or in a given society. Cultures are not as malleable as we might like to think, and individuals who undergo sexual changes in their lives may have no sense of having chosen them. The issue of choice is extremely important, given that the constructionist suggestion of its possibility is the grounds upon which many essentialists criticize constructionism. Both essentialism and constructionism also have political limitations, which we will consider later; for now let us consider the part which the sexual-identity debate played in the history of our communities and political struggles.

Before Stonewall, lesbian and gay groups such as the Mattachine Society and Daughters of Bilitis attempted to show heterosexual society how similar heterosexuals and homosexuals were, as a way of gaining acceptance in society at large. The idea of proudly and openly creating a counterculture, or indeed, of taking radical political stances, was not common in the decades before Stonewall. The changing politics following Stonewall involved a drastic identity shift: lesbians and gay men went from assimilationism to an "ethnic" model of oppression and counterculture.

In this ethnic model, lesbians and gay men came to see themselves as a specific oppressed minority, much like any racial or ethnic minority in a white, racist society. They drew on the civil rights movement and its language to describe their situation and their newfound resistance; George Weinberg's coinage of the term "homophobia" strengthened the parallel with racism. In coming to understand themselves as an ethnic minority, lesbians and gay men began to stress the difference between themselves and heterosexual people, and to ground that difference in biology.[5] In fact, the constructionist view of sexuality, with its fluidity and its connotation of choice, threatened lesbians and gay men as soon as it was proposed. Constructionism challenged the "oppressed ethnic minority" approach by arguing that sexuality could not be compared to skin color as a natural phenomenon. To many lesbians and gay men, constructionism took away their greatest asset, the ethnic self-conception, and did not offer a sound replacement. It could too easily be utilized by homophobic leaders as an argument that lesbians and gay men could change and should therefore be forced to do so.

Constructionists replied that this view distorted the role that choice and intentionality may play in constructionism, in which the personal sexuality changes addressed need be neither chosen nor intentional—and certainly not forced by outside pressure or self-hatred. They further explained that they were not proposing the idea of "sexual preference," as though object choice were equivalent to choice of breakfast cereal; in a society of compulsory heterosexuality, a phrase that connoted choosing between equally accepted sexual options was inadequate and inaccurate. Communities replied that since their homophobic opponents would happily twist constructionism to discredit them, they needed an essentialist identity to strengthen themselves. Accuracy in sexual-identity description simply was not as important as the politics of the community. While sexual theorists continued the essentialist/constructionist debate in academic journals and other settings, community members lived essentialist lives, creating separatist culture and politics, putting their primary energy into the community and their primary trust in other lesbians and gay men.[6] Steven Epstein, as late as 1987, noted that "while constructionist theories have been preaching the gospel that the hetero/homosexual distinction is a social fiction, gays and lesbians, in everyday life and in political action, have been busy hardening the categories."[7] In 1989, AIDS activists, gay male culturalists, and lesbian separatists demonstrate the continuation of essentialism. While constructionism may represent the better description of human sexuality, the very elements which make lesbian/gay communities strong today—perhaps which make them possible as communities at all—are essentialist.

This history has been told many times; these connections have been made before by theorists on both sides of the debate. My claim is that this history can be linked with biphobia in lesbian/gay communities. We will need to take a somewhat roundabout path to see how this can be, beginning with the relationship between biphobia and the fear of being bisexual.

Consider a lesbian who has gone through a hard coming-out process, who has taken a long time to arrive at a sense of her identity and to settle into her local community. Or consider a gay man who can look back on a history of pain and homophobia, who was beaten up throughout high school, frozen out by parents, thrown out by landlords. Each of these people may have a hard time with bisexuality, especially if it is the bisexuality within themselves. The lesbian, finally secure in a lesbian identity, may not want to remain in touch with her bisexuality; it may threaten her sense of self and her community location. Likewise, the gay man with the scars of homophobia may be horrified to discover his bisexuality. What, then, was this pain about? Do the experiences that shaped him mean nothing? Was there an easier way? And should he have taken it? Both of these people have gone through pain and soul-searching to reach their identities, which provide them with a sense of unity, a social location, and a political commitment; to see those identities fluctuate would be unnerving, and would threaten the meaning of their personal histories.

Likewise, lesbian/gay communities have gone through tremendous growing pains, indeed, are built on a great deal of shared pain. The communities and their struggles only make sense if the pain was in some way the inevitable product of being oneself in a homophobic and heterosexist society. Note that the essentialist view of sexuality seems to be required in this equation. Just as bisexuality would threaten the two people described above, the fluidity and connotation of choice within constructionism would seem to challenge both the history and the future of a community built on pain.

Now we can begin to see the threat posed by bisexuality somewhat more clearly. The charge that bisexuals retain "heterosexual privilege" does not need to come from a closet bisexual. The idea that bisexuality destroys the point of gay rights is not a product of fear of the unknown. Rather, the biphobia reflected in these statements arises from the fear of constructionism. Community-oriented individuals, protective of the essentialist view of sexuality that seems to give rhyme and reason to their communities, equate the fluidity and apparent choice-making of bisexuality with that of constructionism, and see only a threat to that which they hold dear. Bisexuality, then, signifies constructionism, and the assumed link between the two is not, in reality, entirely absent. Constructionism does posit that everyone has, if not the experience of living a bisexual life, than at least the potential to do so. Moreover, constructionism makes the claim that one's sexuality is not necessarily firmly set at age five, or even at age fifty. While it is basic to constructionism that no mode of sexuality or object choice is better than another, bisexuality occupies a place of importance in constructionism that it does not in essentialism.

If bisexuality signifies constructionism and constructionism is seen as a threat to some lesbians and gay men, bisexuals will be scapegoated sooner or later. In the presence of a walking example of constructionism, angers and fears quite unrelated to the individual have a chance to surface. Lesbians and gay men are as capable of prejudice as any heterosexual bigot. Moreover, because the bisexual threat involves fluidity, the response to it can be to stereotype in a particularly problematic way. To understand some of the misconceptions about bisexuals and bisexuality, we need to return to the social sciences and consider the idea of reification.

Reification, a constructionist concept, describes the hardening of a social construct into an ontological category in the worldview of a set of people.[8] As a socially constructed behavior or idea is institutionalized, society comes to forget its original social construction and perceives it as natural and without a beginning in time. That which is historically contingent comes to be seen as inevitable. A government gains power and rewrites history around itself. A religious practice which arose because of a practical and culture-specific need is touted as a divine revelation for all people.

Reification is a way of explaining how people give social constructs ontological validity. I noted before that the terms "sexuality" and "homosexuality" are recent developments in human sexual history, and that constructionists would argue that we created them when we needed them. Moreover, they would argue that the creation of those terms played a part in the reification of sexual acts, and in the artificial division between homosexuality and heterosexuality to which I alluded before. Herein lies an explanation for some of the stereotypes about bisexuality. Consider that heterosexuality and homosexuality would be easy to reify compared to bisexuality. They seem to make sense as categories. But how does one reify fluidity? How does one make a category of the potential to have either kind of relationship? The essentialist answer is to change bisexuality from a potential-for-either to a requirement-for-both identity, and this, in fact, is what happened. The lore which developed described bisexuals as people who could not be satisfied with either sex, but who had to be involved with both, usually at the same time. Bisexuals became stereotyped as swingers who eschewed commitment and were promiscuous because there was no other way to categorically describe the bisexual drive that paralleled the homosexual or heterosexual drive. We have already seen how essentialists had put up a fence between homosexuality and heterosexuality. Then they froze the motion which bisexuals made from one side to the other, and began calling us fence-sitters!

In order for constructionism to reclaim some authority in lesbian/gay communities, it will have to prove that it is gay-positive and, more importantly, gay culture-positive. Its value-free position as a purely descriptive social science tool is not likely to be of help to these communities today. Essentialism, for all of its limitations, has allowed them to provide support and encouragement for gay children without wondering whether they will be gay throughout their lives. Moreover, as I said before, many of the elements which make the communities what they are would be considered essentialist. If constructionism is to be taken seriously, it must strengthen the communities as well as helping people to grow beyond them.

Essentialism, which has pointed out how well constructionism can be twisted to justify denying lesbians and gay men their rights, can also be twisted by bigots, and in a more dangerous way. Attempts to justify racism,[9] slavery and sexism have historically been based on biological essentialism; only in the last decade or so have they touched on sociology or constructionism for their justification. Theories of biology-based sexual deviancy are the more dangerous for their apparent blandness, which effectively covers a great deal of homophobia. If a "gay gene," chromosome, nutritional deficiency or other biological difference were actually found, scientists would undoubtedly receive large grants from religious leaders and politicians to engineer it out or otherwise get rid of it. The fact that one has not been found and, by constructionist reckonings, will not be, is not the end of the problem. Essentialists have to face up to the danger of essentialism, as well as constructionism, in homophobic hands.

In closing, I want to say a few things about unlearning biphobia. Homophobia education is extremely similar to biphobia education. All of us, bisexuals as well as lesbians and gay men, ask people to listen to our stories and trust that we are better equipped to describe ourselves and our lives than a bigot with an antigay or antibi agenda. Whatever the bigot's sexuality, he or she is no authority if his or her words are spoken out of hatred and fear. Our sexual differences from society at large do not imply differences in gender identity, culture, morality, or politics.[10] None of us chooses our sexual attractions, but each of us must choose how we respond to them; for any of us to hide our sexuality is tantamount to self-hatred. We ask people who hate or fear us to undergo a process of learning and acceptance. We do not think the entire world is, or ought to be bisexual; we only want the same respect that any lesbian or gay man would legitimately request. Or any other human being.

Today we may be justifiably proud of who we are and of how far we have come. Let us celebrate our sexualities in all of their rich variety. Let us think about what lies ahead, and remember that we who are lesbians and gay men and bisexual people have an important role to play in building our future. Finally, let us work together in peace; God knows the world needs the healing that we can bring.

NOTES

1. Orlando, Lisa. "Loving Whom We Choose: Bisexuality." *Gay Community News,* January 25,1984: 8.
2. Of course, there are people whose bisexuality does seem to represent a stage on the way to exclusive homosexuality or who are "experimenting." And there are bisexual people who deeply hurt their same sex lovers and whose lives appear to be about retaining heterosexual privilege. And there are gay men who abuse children and lesbians who want to be men. There is always a small minority which fits some of the stereotypes, but that minority clearly does not justify the existence of homophobia. The same is true of biphobia.
3. I am indebted to Bill Marsiglio, of the University of Florida, Gainesville, Sociology Department, for introducing me to this debate as it affects sexuality.
4. In a letter to *Gay Community News,* March 26–April 1, 1989, p. 5.
5. Here I am indebted to Paul Horowitz's article, "Beyond the Gay Nation: Where Are We Marching?" *Out/Look,* Vol. 1, No. 1, Spring 1988, pp. 7–21.
6. For the purposes of this paper, I have chosen to ignore the male-female splits in the early post-Stonewall period: divisions of thought, action, and community which were very important in many ways but which did not have an appreciable impact on this question.
7. Epstein, Steven. "Gay Politics, Ethnic Identity: The Limits of Social Constructionism." *Socialist Review* 93/4 (May–August 1987): 12.
8. I am indebted to Deb Reiner for this way of describing reification.
9. Neo-racism, which blames the victim by describing the awful ghetto conditions under which many people of color live, still avoids questions of institutional exploitation (why they live so badly), and thus is still racism. It is called "neo-racism" to distinguish it from the biological arguments for the inferiority of people of color which constituted traditional racism.

10. "Bisexuality does not exist as either a social institution or a psychological 'truth.' It exists only as a catch-all term for different erotic and social patterns whose common ground is an attempt to combine homo- and heterosexuality in a variety of ways. The term 'bisexual,' then, merely tells us that someone can or does eroticize both men and women. It does not tell us anything about the morality or politics of that person." Valverde, Marianna. *Sex, Power, and Pleasure.* Toronto: Women's Press, 1985: 119.

DISCUSSION QUESTIONS

1. Think about Udis-Kessler's views regarding how much "choice" individuals have in choosing their sexuality. What are your views on this issue? Do individuals choose heterosexuality in the same way that they choose non-hegemonic sexual identities (e.g., homosexuality, bisexuality, and asexuality)?
2. In what ways, if any, do heterosexual stereotypes of bisexuals differ from those of gays and lesbians? Does the existence of bisexuality challenge the heterosexual community in the same way it challenges the homosexual community? Why or why not?
3. Do you think the heterosexual privilege that bisexuals gain outweighs the costs of bisexuality listed by the author? Why or why not?
4. Udis-Kessler describes bisexuality as a "fluid" category of sexual existence. In your view does bisexuality constitute a single, discrete, cohesive, and meaningful category of sexuality, or does it erroneously mask the existence of multiple-sexual possibilities? Do such distinctions matter? Why or why not?

A Queer Encounter

STEVEN EPSTEIN

In the following excerpt Steven Epstein outlines the origins and key elements of Queer Theory. This perspective emerged in the late 1990s extending the constructionist critique of essentialistic categorization and the institutionalization of heterosexuality as the dominant social norm. Proposing to shift away from perspectives over-concerned with particularized sexualities, Queer Theory embraces the multi-dimensionality of human existence, arguing that the self is a patchwork of multiple identities and situational subjectivities.

In just the past few years in much of the English-speaking world, the term *queer*—formerly a word that nice people didn't use—has escaped the bounds of quotation marks. Its growing currency reflects three roughly congruent, yet uneasily related, developments: the emergence of new repertoires of political mobilization in groups such as Queer Nation, ACT UP, and (in England) Outrage; the foothold gained by new programs of lesbian and gay studies within the academy; and—partially in response to both of the above—the rise of an intellectual enterprise explicitly calling itself queer theory.

From "A Queer Encounter: Sociology and the Study of Sexuality" by Steven Epstein in *Sociological Theory* 12(2), 1994. Reprinted by permission of John Wiley and Sons, Inc.

In the 1960s and 1970s, sociologists (along with anthropologists and others) contributed significantly to a fundamental shift in the theorization of sexuality and homosexuality. Against naturalized conceptions of sexuality as a biological given, against Freudian models of the sexual drive, and against the Kinseyan obsession with the tabulation of behavior, sociologists asserted that sexual meanings, identities, and categories were inter-subjectively negotiated social and historical products—that sexuality was, in a word, *constructed.* Though sexuality never became institutionalized as a formal subfield of sociological study, the "social constructionist" perspective on sexuality drew much of its theoretical firepower from important currents within sociology at the time, particularly symbolic interactionism and labeling theory. Without seeking to minimize the importance of other disciplines, I would suggest that neither queer theory nor lesbian and gay studies in general could be imagined in their present forms without the contributions of sociological theory.[1] Yet to some recent students of sexuality working outside sociology, the concept of social construction is assumed to have sprung, like Athena, fully formed from the head of Michel Foucault; meanwhile the analyses presented by queer theorists, expressed in their own particular, often postmodern, vocabulary, confront sociologists as an alien power, unrecognizable as anything related in any way to the product of their own labor.

DECIPHERING QUEERNESS

The late 1980s marked the adoption, in various circles, of the word *queer* as a new characterization of "lesbian and gay" politics and, indeed, as a potential replacement for the very terms *lesbian* and *gay* (Bérubé and Escoffier 1991; Duggan 1992; A. Stein 1992). The term was explicitly associated with the activist group Queer Nation, which sprang up in dozens of cities around the United States; more generally, it reflected new political tendencies and cultural emphases, particularly in a younger generation of migrants to the established lesbian and gay communities. It is a term rife with connotations, some of them contradictory:

- The invocation of the "Q-word" is an act of linguistic reclamation, in which a pejorative term is appropriated by the stigmatized group so as to negate the term's power to wound. (This sometimes has the effect of reinforcing an insider/outsider division: self-styled queers can use the word freely, while sympathetic straights often do so only nervously.)
- Queerness is frequently anti-assimilationist; it stands in opposition to the inclusionary project of mainstream lesbian and gay politics, with its reliance on the discourses of civil liberties and civil rights. In this sense, queerness is often a marker of one's distance from conventional norms in all facets of life, not only the sexual.
- Similarly, queerness describes a politics of provocation, one in which the limits of liberal tolerance are constantly pushed. Yet while confrontational politics (for example, a same-sex "kiss-in" held in a bar frequented by heterosexuals) may work to affirm one's difference, it also seeks to overturn conventional norms. This transformative impulse (an "outward-looking" focus) coexists with the emphasis on anti-assimilation and self-marginalization (an "inward-looking" focus).
- Use of the term also functions as a marker of generational difference within gay/lesbian/queer communities. Younger queers may speak with resentment of feeling excluded by the established "lesbian and gay" communities, while older gays and lesbians sometimes object bitterly to the use of the term *queer,* which they consider the language of the oppressor.
- "Queer" speaks to the ideal of a more fully "co-sexual" politics, within which men and women participate on an equal footing. To

some, the use of "queer" to describe both men and women is preferable to "gay" (which includes women in much the same way as "man" used to include women), or to "gay and lesbian" (which emphasizes gender difference).

- "Queer" offers a comprehensive way of characterizing all those whose sexuality places them in opposition to the current "normalizing regime" (Warner 1991: 16). In a more mundane sense, "queer" has become convenient shorthand as various sexual minorities have claimed territory in the space once known simply, if misleadingly, as "the gay community." As stated by an editor of the defunct New York City queer magazine *Outweek* (quoted in Duggan 1992: 21), "When you're trying to describe the community, and you have to list gays, lesbians, bisexuals, drag queens, transsexuals (post-op and pre), it gets unwieldy. Queer says it all."
- The rise of queerness reflects a postmodern "decentering" of identity (A. Stein 1992). As formerly paradigmatic patterns of identity construction (such as "the lesbian feminist") lose sway, they are replaced by a loosely related hodge-podge of lifestyle choices. Collectively these offer more individual space for the construction of identity, but none provides a clear "center" for the consolidation of community.
- Queer politics are "constructionist" politics (Duggan 1992), marked by a resistance to being labeled, a suspicion of constraining sexual categories, and a greater appreciation for fluidity of sexual expression.
- At times, however, queer politics also can be "essentialist" politics: in these expressions, the new moniker is simply reified into yet another identity category understood in separatist or nationalist terms, as the name *Queer Nation* itself can imply (Duggan 1992).

Clearly, the burdens of connotation would appear to be heavier than any single word might be expected to bear. At present, "queer" has been appropriated to describe a considerable range of political projects as well as individual and collective identities. Yet the meaning of the term is complicated further by its simultaneous employment by academics. Sometimes "queer" is put forward simply as the new and concise coinage: "gay studies," or "lesbian and gay studies," or "bisexual, lesbian, and gay studies," or "multicultural, bisexual, lesbian, and gay studies" should—for convenience, if for no other reason—be named "queer studies."[2] Sometimes, however, the invocation of "queer" signals important shifts in theoretical emphasis. In this reading, said Teresa de Lauretis, one of the organizers of a "queer theory" conference at UC-Santa Cruz, "queer" is intended "to mark a certain critical distance from the . . . by now established and often convenient, formula" of "lesbian and gay" (de Lauretis 1991: iv).

Although many works are emblematic of the "queer turn" (Butler 1990, 1993; Cohen 1991; de Lauretis 1991; Dollimore 1993; Edelman 1989, 1992; Goldberg 1991; Miller 1991; A. Parker 1991; Patton 1993; Sedgwick 1993; Seidman 1993; Terry 1991; Warner 1993), Eve Kosofsky Sedgwick's *Epistemology of the Closet* (1990) is perhaps most often cited as a canonical text (even though the term *queer theory* does not appear there). Basically a critical reinterpretation of specific works of English literature, the book opens with a strong claim:

> *Epistemology of the Closet* proposes that many of the major nodes of thought and knowledge in twentieth-century Western culture as a whole are structured—indeed, fractured—by a chronic, now endemic crisis of home/heterosexual definition, indicatively male, dating from the end of the nineteenth century. (Ibid: 1)

Furthermore (as if, perhaps, that weren't bold enough for an opening paragraph):

> The book will argue that an understanding of virtually any aspect of modern Western culture must be, not merely incomplete, but damaged

> in its central substance to the degree that it does not incorporate a critical analysis of modern homo/hetero-sexual definition. (Ibid)

All too often, studies of gays and lesbians, or of other "sexual minorities," have been cast as studies of "marginal" experience. By contrast, an "epistemology of the closet" seeks to analyze how various ways of construing sexual marginality shape the self-understanding of the culture as a whole. For example, Sedgwick argues, the very notion of the "closet" (as well as the metaphor of "coming out of the closet," now somewhat widely diffused) reflects the influence of the homosexual/heterosexual dichotomy, on broader perceptions of public and private, or secrecy and disclosure (ibid: 72). In this sense, as Michael Warner (1993: xiv) suggests, "Sedgwick's work has shown that there are specifically modern forms of association and of power than can be seen properly only from the vantage of antihomophobic inquiry."

Though Sedgwick rejects many of the terms of the so-called "essentialist-constructionist debate" (Sedgwick 1990: 40–1), her work in an important sense continues the tradition of the social constructionist perspective. "Homosexuality" and "heterosexuality" do not describe transhistorical cultural forms, despite the universality of specific sexual practices. Rather, such practices come to *mean* very different things in a society which insists that each individual, just as he or she possesses a gender, also must necessarily occupy one or the other category of sexual orientation. "It was this new development," which Sedgwick and other authors (Halperin 1990) locate around the turn of the century in western societies, "that left no space in the culture exempt from the potent incoherences of homo/heterosexual definition" (Sedgwick 1990: 2).

In constructing a genealogy of the homosexual/heterosexual divide, Sedgwick's work draws on Foucault. In general, the mark of Foucault is broadly apparent in works of this kind—in their emphasis on power and "normalization," in their understanding of the constitutive role of discourse in the construction of subjectivity, in their poststructuralist critique of conceptions of coherent selfhood (Butler 1990, 1993), and in their postmodern suspicion of identity as a totalizing construct that subsumes difference (Cohen 1991). Whereas queer politics often seem divided in their approach to identity Politics—at times subverting popular notions of stable identities, at times fashioning a new queer identity with their own enforced boundaries—queer *theory* is more consistent on this point. Indeed, the terrain of queerness provides a meeting point for those who come to the critique of identity from many different directions: those who believe that identity politics mute internal differences within the group along racial, class, gender, or other lines of cleavage (Montero 1993; Mort no date; Seidman 1991); and those who believe that subjectivities are always multiple (Ferguson 1991; Seidman 1991); and those who are simply suspicious of categorization as inherently constraining. The point (at least as I read it) is not to stop studying identity formation, or even to abandon all forms of identity politics, but rather to maintain identity and difference in productive tension, and to rely on notions of identity and identity politics for their strategic utility while remaining vigilant against reification.[3]

In subject matter, queer studies emphasize literary works, texts, and artistic and cultural forms; in analytical technique, deconstructionist and psychoanalytic approaches loom large. Yet however marked these tendencies, none of them is necessarily definitive of queer theory, whereas the assertion of the centrality of marginality is the pivotal queer move. Just as queer *politics* emphasize outsiderness as a way of constructing opposition to the regime of normalization as a whole, so queer *theory* analyzes putatively marginal experience, but in order to expose the deeper contours of the whole society and the mechanisms of its functioning.

In some sense, this idea is not altogether new: a presumed goal of the sociology of deviance, for example, was to study the processes by which people become labeled deviant, so as to reveal, by contrast, the ideological construction of "the normal." In practice, however, sociologists have tended to relegate the study of "sexual minorities" to the analytical sidelines rather than treating such study as a window onto a larger world of power, meaning, and social organization. The challenge that queer theory poses to sociological investigation is precisely in the strong claim that no facet of social life is fully comprehensible without an examination of how sexual meanings intersect with it. In no way does it disqualify such a claim to recognize it as serving a certain strategic function within the intellectual "field" (Bourdieu 1988): queer theorists are seeking to situate their work as an "obligatory passage point" (Latour 1987) through which other academics must pass if they want to fully understand their own particular subject matters. In this sense as well, queer theory, like queer politics, is locating itself simultaneously both on the margins and at the center.

Perhaps the clearest analogy, as the editors of the *Lesbian and Gay Studies Reader* note (Abelove, Barale, and Halperin 1993b: xv–xvi), is with feminist theory and women's studies programs; they have sought to argue that gender is not a "separate sphere," but rather is partially constitutive of other institutions such as the economy and the state. The goal therefore should not be to restrict concerns with gender to a bounded domain called "sociology of gender," but to introduce gendered understandings into sociological scrutiny across the board. The challenge for queer studies will be to demonstrate the links concretely in the case of sexuality—to identify the precise ways in which sexual meanings, categories, and identities are woven into the fabric of society and help give shape to diverse institutions, practices, and beliefs.

NOTES

1. One can also trace a parallel, and equally influential, lineage in anthropology, beginning perhaps with the cultural relativist perspective of Margaret Mead (1935) and proceeding through more recent work on the "sex/gender system" (Rubin 1975) and on the cultural construction of gender (Ortner and Whitehead 1981) and sexuality (Blackwood 1986; Caplan 1987; Lancaster 1988; Newton 1988; R. Parker 1991).

   "Origins stories" are always problematic. My point here is neither to say that "it all started with sociology" nor to argue that sociology's early role entitles it to special respect. Rather, I seek to highlight the curious case of a discipline whose contributions have been forgotten, both within and without.
2. This move, however, remains tentative and controversial. Indeed, the editors of *The Lesbian and Gay Studies Reader* noted in their introduction to the volume (Abelove, Barale, and Halperin 1993b: xvii): "It was difficult to decide what to title this anthology. We have reluctantly choose not to speak here and in our title of 'queer studies,' despite our own attachment to the term, because we wish to acknowledge the force of current usage."
3. For different approaches to the maintenance of such a "productive tension" in identity politics, see Clarke (1991), Gamson (no date), and Seidman (1993). For the view that "the temporary totalization performed by identity categories is a necessary error," see Butler (1993: 230).

REFERENCES

Abelove, Henry, Michèle Aina Barale, and David M. Halperin, eds. 1993a. *The Lesbian and Gay Studies Reader.* New York: Routledge.

——— 1993b. "Introduction." pp. xv–xviii in *The Lesbian and Gay Studies Reader,* edited by Henry Abelove, Michèle Aina Barale, and David M. Halperin. New York: Routledge.

Bourdieu, Pierre. 1988. *Homo Academicus,* translated by Peter Collier Stanford, CA: Stanford University Press.

Butler, Judith. 1990. *Gender Trouble: Feminism and the Subversion of Identity.* New York: Routledge.

——— 1993. *Bodies That Matter: On the Discursive Limits of "Sex."* New York: Routledge.

Cohen, Ed. 1991. "Why are 'We'? Gay 'Identity' as Political (E)motion (A Theoretical Rumination)." pp. 71–92 in *Inside/Out: Lesbian Theories, Gay Theories,* edited by Diana Fuss. New York: Routledge.

de Lauretis, Teresa. 1991. "Queer Theory and Lesbian and Gay Sexualities: An Introduction." *Differences: A Journal of Feminist Cultural Studies 3*: iii–xviii.

Dollimore, Jonathan. 1983. "Different Desires: Subjectivity and Transgression in Wilde and Gide." pp. 624–641 in *The Lesbian and Gay Studies Reader,* edited by Henry Abelove, Michèle Aina Barale, and David M. Halperin. New York: Routledge.

Duggan, Lisa. 1992. "Making It Perfectly Queer." *Socialist Review* 22: 11–31.

Edelman, Lee. 1989. "The Plague of Discourse: Politics, Literary Theory, and AIDS." *South Atlantic Quarterly* 88: 301–317.

——— 1992. "Tearooms and Sympathy, or the Epistemology of the Water Closet." pp. 263–84 in *Nationalism & Sexualists,* edited by Andrew Parker, Mark Russo, Doris Summer, and Patricia Yaeger. New York: Routledge.

Ferguson, Ann. 1991. "Lesbianism, Feminism, and Empowerment in Nicaragua." *Socialist Review* 21: 75–97.

Foucault, Michel. 1980. *The History of Sexuality,* vol. 1, translated by Robert Hurley. New York: Vintage.

Goldberg, Jonathan. 1991. "Sodomy in the New World: Anthropologies Old and New." *Social Text* 29: 46–56.

Halperin, David. 1990. *One Hundred Years of Homosexuality*. New York: Routledge.

Latour, Bruno. 1987. *Science in Action: How to Follow Scientists and Engineers through Society*. Cambridge, MA: Harvard University Press.

Miller, D. A. 1991. "Anal *Rope*." pp. 119–141 in *Inside/Out: Lesbian Theories, Gay Theories,* edited by Diana Fuss. New York: Routledge.

Montero, Oscar. 1993. "Before the Parade Passes By: Latino Queers and National Identity." *Radical America* 24: 15–26.

Mort, Frank. No date. "Essentialism Revisited? Identity politics and Late Twentieth Century Discourses of Homosexuality." Unpublished manuscript.

Parker, Andrew. 1991. "Unthinking Sex: Marx, Engels and the Scene of Writing." *Social Text* 29: 28–45.

Patton, Cindy. 1993. "Tremble, Hetero Swine!" pp. 143–177 in *Fear of a Queer Planet: Queer Politics and Social Theory,* edited by Michael Warner. Minneapolis: University of Minnesota Press.

Sedgwick, Eve Kosofsky. 1990. *Epistemology of the Closet.* Berkeley: University of California Press.

——— 1993. *Tendencies*. Durham, NC: Duke University Press.

——— 1991. "Postmodern Anxiety: The Politics of Epistemology." *Sociological Theory* 9: 180–190.

——— 1993. "Identity and Politics in 'Postmodern' Gay Culture: Some Historical and Conceptual Notes." pp.105–142 in *Fear of a Queer Planet: Queer Politics and Social Theory,* edited by Michael Warner. Minneapolis: University of Minnesota Press.

——— 1992. "Sisters and Queers: The Decentering of Lesbian Feminism." *Socialist Review* 22: 33–55.

Terry, Jennifer. 1991. "Theorizing Deviant Historiography." *Differences: A Journal of Feminist Cultural Studies* 3: 55–74.

Warner, Michael. 1991. "Introduction: Fear of a Queer Planet." *Social Text* 29: 3–17.

——— 1993. "Introduction." pp. vii–xxxi in *Fear of a Queer Planet: Queer Politics and Social Theory,* edited by Michael Warner. Minneapolis: University of Minnesota Press.

DISCUSSION QUESTIONS

1. Can pejorative terms (such as queer) be "reclaimed" by marginalized groups "so as to negate the term's power to wound," or do attempts at reclamation actually reinforce inequality and harden the boundaries between those who are marginalized and those who are not? If attempts are made to reclaim pejorative words how, then, is the use of such terms by the non-marginalized to be 'read'?
2. Does the term "queer" represent a single coherent identity? Should it? Does it matter what audience you ask?
3. Is "queer" more gender-neutral than gay? Why or why not? Ask several people to describe the mental image they have of someone calling themselves queer. Do the respondents indicate that the picture they have is primarily male or female? What conclusions can you draw from this? Are there other patterns of interest that arise in their responses?
4. In what ways does the metaphor of the closet represent the cultural split between the public and private realms?

3

Conflated Constructs: Sex, Gender, and Sexuality

Beyond the Binaries: Depolarizing the Categories of Sex, Sexuality, and Gender

JUDITH LORBER

For many in contemporary society, the concepts of sex, gender, and sexuality are interchangeable statuses that are mentally fused into a single, slippery but unitary and purportedly meaningful amalgamation of personhood—despite the fact that each status is indicative of quite different phenomena. To further complicate matters, these constructs are often oversimplified and stripped down to dichotomous polar extremes (male and female, masculine and feminine, gay and straight) that mask important diversity and variation within each status. In the following article, Judith Lorber deconstructs the conflation and polarization of these concepts and discusses the implications of such methodological slippage for sociologists and other social researchers.

. . . Most research designs in sociology assume that each person has one sex, one sexuality, and one gender, which are congruent with each other and fixed for life. Sex and gender are used interchangeably, and sex sometimes means sexuality, sometimes physiology or biology, and sometimes social status. The social construction of bodies is examined only when the focus is

From "Beyond the Binaries: Depolarizing the Categories of Sex, Sexuality, and Gender," by Judith Lorber in *Sociological Inquiry* 66(2): 143–159. Reprinted by permission of Blackwell Publishing.

medicine, sports, or procreation (Butler 1993). Variations in gender displays are ignored: A woman is assumed to be a feminine female; a man a masculine male. Heterosexuality is the uninterrogated norm against which variations are deviance (Ingraham 1994). These research variables—"sex" polarized as "homosexuals" and "heterosexuals," and "gender" polarized as "women" and "men"—reflect unnuanced series that conventionalize bodies, sexuality, and social location (Young 1994). Such designs cannot include the experiences of hermaphrodites, pseudohermaphrodites, transsexuals, transvestites, bisexuals, third genders, and gender rebels as lovers, friends, parents, workers, and sports participants. Even if the research sample is restricted to putative "normals," the use of unexamined categories of sex, sexuality, and gender will miss complex combinations of status and identity, as well as differently gendered sexual continuities and discontinuities (Chodorow 1994, 1995).

Current debates over the global assumptions of only two gender categories have led to the insistence that they must be nuanced to include race and class, but they have not gone much beyond that (Collins 1990; Spelman 1988; Staples 1982). Similarly, the addition of sexual orientation has expanded gendered sexual statuses only to four: heterosexual women and men, gays, and lesbians.

Deconstructing sex, sexuality, and gender reveals many possible categories embedded in social experiences and social practices, as does the deconstruction of race and class. As queer theorists have found, multiple categories disturb the neat polarity of familiar opposites that assume one dominant and one subordinate group, one normal and one deviant identity, one hegemonic status and one "other" (Martin 1994; Namaste 1994). But in sociology, as Barrie Thorne (1993) comments in her work on children,

> The literature moves in a circle, carting in cultural assumptions about the nature of masculinity (bonded, hierarchical, competitive, "tough"), then highlighting behavior that fits those parameters and obscuring the varied styles and range of interactions among boys as a whole. (p. 100)

Behavior that is gender-appropriate is considered normal; anything else (girls insulting, threatening, and physically fighting boys and other girls) is considered "gender deviance" (Thorne 1993, pp. 101–103). The juxtaposition both assumes and reproduces seemingly clear and stable contrasts. Deconstructing those contrasts reveals that the "normal" and the "deviant" are both the product of deliberate social practices and cultural discourses. Of all the social sciences, sociology is in the best position to analyze those practices and discourses, rather than taking their outcome for granted.

But as long as sociological research uses only the conventional dichotomies of females and males, homosexuals and heterosexuals, women and men, it will take the "normal" for granted by masking the extent of subversive characteristics and behavior. Treating deviant cases as markers of the boundaries of the "normal" (e.g., heterosexuality) does not have to be explained as equally the result of processes of socialization and social control (Ingraham 1994). Such research colludes in the muffling and suppressing of behavior that may be widespread, such as heterosexual men who frequently cross-dress, which, if not bracketed off as "deviant," could subvert conventional discourses on gender and sexuality (Stein and Plummer 1994).

Our commonsense knowledge of the real world tells us that behavior is situational and that sexual and gender statuses combined with race and social class produce many identities in one individual (West and Fenstermaker 1995). This individual heterogeneity is nonetheless overridden by the major constructs (race, class, gender) that order and stratify informal groups, formal organizations, social institutions, and social interaction. By accepting these constructs as given, by not unpacking them, sociologists collude in the relations of ruling (Smith 1990a, 1990b).

As researchers, as theorists, and as activists, sociologists have to go beyond paying lip service to the diversity of bodies, sexualities, genders, and racial–ethnic and class positions. We have to think not only about how these characteristics variously intermingle in individuals and therefore in groups but what the extent of variation is *within these categories*. For example, using conventional categories, where would we place the competitive runner in woman's competitions who has XY chromosomes and normal female genitalia (Grady 1992)? Or the lesbian transsexual (Bolin 1988)? Or the woman or man who has long-term relationships with both women and men (Weinberg, Williams, and Pryor 1994)? Or the wealthy female husband in an African society and her wife (Amadiume 1987)? These are not odd cases that can be bracketed off in a footnote (Terry 1991). As did the concept of conflicting latent statuses (e.g., black woman surgeon), they call our attention to the rich data about social processes and their outcomes that lie beneath neat comparisons of male and female, heterosexual and homosexual, men and women.

DECONSTRUCTING SEX, SEXUALITY, AND GENDER

In rethinking gender categories, it is important to split what is usually conflated as sex/gender or sex/sexuality/gender into three conceptually distinct categories: sex (or biology, physiology), sexuality (desire, sexual preference, sexual orientation), and gender (a social status, sometimes with sexual identity). Each is socially constructed but in different ways. Gender is an overarching category—a major social status that organizes almost all areas of social life. Therefore bodies and sexuality are gendered; biology, physiology, and sexuality, in contrast, do not add up to gender, which is a social institution that establishes patterns of expectations for individuals, orders the social processes of everyday life, is built into the major social organizations of society, such as the economy, ideology, the family, and politics, and is also an entity in and of itself (Lorber 1994).

For an individual, the components of gender are the sex category assigned at birth on the basis of the appearance of the genitalia; gender identity; gendered sexual orientation; marital and procreative status; a gendered personality structure; gender beliefs and attitudes; gender displays; and work and family roles. All these social components are supposed to be consistent and congruent with perceived physiology. The actual combination of genes and genitalia; prenatal, adolescent, and adult hormonal input; and procreative capacity may or may not be congruous with each other and with the components of gender and sexuality, and the components may also not line up neatly on only one side of the binary divide.

Deconstructing Sex

Anne Fausto-Sterling (1993) says that "no classification scheme could more than suggest the variety of sexual anatomy encountered in clinical practice" (p. 22), or seen on a nudists' beach. Male and female genitalia develop from the same fetal tissue, and so, because of various genetic and hormonal inputs, at least 1 of 1,000 infants is born with ambiguous genitalia, and perhaps more (Fausto-Sterling 1993). The "mix" varies; there are

> . . . the so-called true hermaphrodites . . . , who possess one testis and one ovary . . . ; the male pseudohermaphrodites . . . , who have testes and some aspects of the female genitalia but no ovaries; and the female pseudohermaphrodites . . . , who have ovaries and some aspects of the male genitalia but lack testes. Each of these categories is in itself complex; the percentage of male and female characteristics . . . can vary enormously among members of the same subgroup. (Fausto-Sterling 1993, p. 21)

Because of the need for official categorization in bureaucratically organized societies, these infants must legally be labeled "boy" or "girl" soon after birth, yet they are subject to rather arbitrary sex assignment (Epstein 1990). Suzanne Kessler (1990) interviewed six medical specialists in pediatric intersexuality and found that whether an infant with XY chromosomes and anomalous genitalia was categorized as a boy or a girl depended on the size of the penis. If the penis was very small, the child was categorized as a girl, and sex-change surgery was used to make an artificial vagina.

An anomaly common enough to be found in several feminine-looking women at every major international sports competition is the existence of XY chromosomes that have not produced male anatomy or physiology because of other genetic input (Grady 1992). Now that hormones have proved unreliable, sports authorities nonetheless continue to find ways of separating "women" from "men." From the point of view of the sociological researcher, the interesting questions are why certain sports competitions are gender-neutral and others are not, how different kinds of women's and men's bodies, and how varieties of masculinities and femininities are constructed through sports competitions (Hargreaves 1986; Messner 1992; Messner and Sabo 1994).

As for hormones, recent research suggests that testosterone and other androgens are as important to normal development in females as in males, and that in both, testosterone is converted to estrogen in the brain. Paradoxically, maximum androgen levels seem to coincide with high estrogen levels and ovulation, leading one researcher to comment: "The borders between classic maleness and femaleness are much grayer than people realized. . . . We're mixed bags, all of us" (quoted in Angier 1994).

From a societal point of view, the variety of combinations of genes, genitalia, and hormonal input can be rendered invisible by the surgical and hormonal construction of maleness and femaleness (Epstein 1990). But this variety, this continuum of physiological sex cannot be ignored. Sociologists may not want to explore the varieties of biological and physiological sexes or the psychology of the hermaphrodite, pseudohermaphrodite, or transsexual, but the rationales given for the categorization of the ambiguous as either female or male shed a great deal of light on the practices that maintain the illusion of clear-cut sex differences. Without such critical exploration, sex differences are easily invoked as the "natural causes" of what is actually socially constructed.

Deconstructing Sexuality

Categories of sexuality—conventionally, homosexual and heterosexual—also mask diversity that can be crucial for generating accurate data. Sexuality is physically sexed because female and male anatomies and orgasmic experiences differ. It is gendered because sexual scripts differ for women and for men whether they are heterosexual, homosexual, bisexual, transsexual, or transvestite. Linking the experience of physical sex and gendered social prescriptions for sexual feelings, fantasies, and actions are individual bodies, desires, and patterns of sexual behavior, which coalesce into gendered sexual statuses. There are certainly more than two gendered sexual statuses: "If one uses the criteria of linguistic markers alone, it suggests that people in most English-speaking countries . . . recognize four genders: woman, lesbian (or gay female), man and gay male" (Jacobs and Roberts 1989, p. 439). But there is not the variety we might find if we looked at what is actually out there.

Studies of bisexuality have shown that the conventional sexual categories are hard to document empirically. At what point does sexual desire become sexual preference, and what turns sexual preference into a sexual identity or social status? What sexual behavior identifies a "pure" heterosexual or a "pure" homosexual? Additionally, a sexual preference involves desired and actual sexual attraction, emotions, and fantasies, not just behavior. A sexual identity involves self-identification, a lifestyle, and social recognition of the status (Klein, Sepekoff, and Wolf 1985).

Sexual identities (heterosexual, homosexual, bisexual) are responses not just to psychic constructs but also to social and cultural strictures and pressures from family and friends. Because Western culture constructs sexuality dichotomously, many people whose sexual experiences are bisexual are forced to choose between a heterosexual and homosexual identity as their "real" identity (Blumstein and Schwartz 1976a, 1976b; Garber 1995; Rust 1992, 1993, forthcoming;

Valverde 1985, pp. 109–120). Rust's research on bisexual and lesbian sexual identity found that 90 percent of the 323 self-identified lesbians who answered her questionnaire had heterosexual experiences, 43 percent after coming out as lesbians (1992, 1993). They discounted these experiences, however; what counted for these lesbians was their current relationships. The forty-two women who identified themselves as bisexual, in contrast, put more emphasis on their sexual attraction to both women and men. Assuming that all self-identified homosexual men and lesbians have exclusively same-sex partners not only renders invisible the complexities of sexuality but can also have disastrous health outcomes, as has been found in the spread of HIV and AIDS among women (Goldstein 1995).

The interplay of gender and sexuality needs to be explored as well. One study found that heterosexual men labeled sexual provocativeness toward them by gay men sexual harassment, but heterosexual women did not feel the same about lesbians' coming on to them (Giuffre and Williams 1994). The straight men felt their masculinity was threatened by the gay men's overtures; the straight women did not feel that a lesbian's interest in them impugned their heterosexuality.

Weinberg, Williams, and Pryor (1994) found five types of bisexuals among the 49 men, 44 women, and 11 transsexuals they interviewed in 1983 (pp. 46–48). In their research, gender was as salient a factor as sexuality. On the basis of sexual feelings, sexual behaviors, and romantic feelings, they estimated that only 2 percent of the self-identified bisexual men in their research and 17 percent of the self-identified bisexual women were equally sexually and romantically attracted to and involved with women and men, but about a third of both genders were around the midpoint of their scale. About 45 percent of the men and 20 percent of the women leaned toward heterosexuality, and 15 percent of each gender leaned toward homosexuality. About 10 percent of each were varied in their feelings and behavior.

Weinberg, Williams, and Pryor (1994) found that although gender was irrelevant to choice of partner among bisexuals, sexual scripting was not only gendered, but quite conventional, with both women and men saying that women partners were more emotionally attuned and men partners were more physically sexual (pp. 49–58). Paradoxically, they say,

> In a group that often sets itself against societal norms, we were surprised to discover that bisexual respondents organized their sexual preferences along the lines of traditional gender stereotypes. As with heterosexuals and homosexuals, gender is the building material from which they put together their sexuality. Unlike these groups, however, the edifice built is not restricted to one gender. (p. 57)

The meaning of gender and sexuality to self-identified homosexuals cannot be taken for granted by researchers. Eve Kosofsky Sedgwick notes that some homosexuals want to cross into the other gender's social space (e.g., gay drag queens and butch lesbians), whereas for others (e.g., macho gay men and lesbian separatists) ". . . it is instead the most natural thing in the world that people of the same gender, people grouped under the single most determinative diacritical mark of social organization, people whose economic, institutional, emotional, physical needs and knowledges may have so much in common, should bond together also on the axis of sexual desire" (1990, p. 87).

Paula Rust (forthcoming), in her research on varieties of sexuality, found that her respondents spoke of being attracted to another person because of particular personality characteristics, ways of behaving, interests, intellect, looks, style. What heterosexuals do—choose among many possible members of the opposite sex—is true of gays and lesbians for same-sex partners, and bisexuals for either sex. The physical sex, sexual orientation, masculinity, femininity, and gender markers are just the beginning set of parameters, and they might differ for a quick sexual encounter, a romantic liaison, a long-term relationship. Rather than compare on categories of gender or sexuality, researchers might want to compare on types of relationships.

Deconstructing Gender

Gendered behavior is constantly normalized by processes that minimize or counteract contradictions to the expected. Competitive women bodybuilders downplay their size, use makeup, wear their hair long and blond, and emphasize femininity in posing by using "dance, grace, and creativity"; otherwise, they don't win competitions (Mansfield and McGinn, 1993):

> In the same way as it is necessary for the extreme gender markers of the hyper-feminine to be adopted by the male cross-dressers in order to make it clear that they wish to be recognized as "women," so too is it necessary for women body builders. . . . It seems that the female muscled body is so dangerous that the proclamation of gender must be made very loudly indeed (p. 64).

Iris Marion Young (1994) argues that gender, race, and class are *series*—comparatively passive social collectives grouped by their similar tasks, ends, or social conditioning. These locations in social structures may or may not become sources of self-identification, significant action by others, or political action. When and how they do is an area for research. For example, U.S. lesbians first identified with homosexual men in their resistance to sexual discrimination, but after experiencing the same gender discrimination as did women in the civil rights and draft-resistance movements, they turned to the feminist movement, where, unhappily, they experienced hostility to their sexuality from many heterosexual women. Subsequently, some lesbian feminists have created an oppositional, woman-identified, separatist movement that identifies heterosexuality as the main source of the oppression of women (Taylor and Rupp 1993).

David Collinson and Jeff Hearn (1994) argue that men in management exhibit multiple masculinities: aggressive authoritarianism, benevolent paternalism, competitive enterpreneurialism, buddy–buddy informalism, and individualist careerism. These multiple masculinities among men managers have different effects on relationships with men colleagues, women colleagues, as well as on sponsor–protégé interactions. Collinson and Hearn call for simultaneous emphasis on unities and differences among men. Cynthia Cockburn similarly says about women, "We can be both the same as you *and* different from you, at various times and in various ways" (1991, p. 10).

Igor Kopytoff (1990), raising the question of why it seems to be easier for women in traditional societies than in Westernized societies to claim positions of political power and rule as heads of state, uses a concept of core or existential gender identities. He argues that in Africa and many traditional societies the core of womanhood (or immanent or existential being as a woman) is childbearing—but all the rest is praxis and negotiable, transferable. Because women do not have to bring up their children to be women in traditional societies, just birth them, he argues that they are free to take on other time-consuming roles. In the West, in contrast, since the nineteenth century, being a "real" woman means one must be married with children, and must bring them up personally, while also keeping an impeccable house and attractive appearance, and looking after a husband's sexual and emotional needs. "Once existentially complete, she can then turn to other occupations," but will rarely have the time to assume a position of leadership (p. 93).

> The problem of women's roles is not whether a society recognizes women as being different from men (they invariably do) but how it organizes other things around the difference (p. 91).

CHALLENGE CATEGORIES, CHALLENGE POWER

. . . The goal of sociological research should be multiple levels of analysis that include the heterogeneity of people's lives, the varied dimensions of status categories, and the power relations between and among them. As Dorothy Smith (1990a) says:

> The social scientist must work with the constraint of actuality and is not privileged to draw relations between observables arbitrarily. A theoretical account is not fixed at the outset, but evolves in the course of inquiry dialectically as the social scientist seeks to explicate the properties of organization discovered in the way people order their activities. Hence the structure of a theoretical account is constrained by the relations generated in people's practical activities (p. 48).

Research using a variety of gendered sexual statuses has already challenged long-accepted theories. Lesbian and homosexual parenting, as well as single-parent households, call into question ideas about parenting and gendered personality development based on heterogendered nuclear families. In psychoanalytic theory, having a woman as a primary parent allows girls to maintain their close bonding and identification with women, but forces boys to differentiate and separate in order to establish their masculinity. The personality structure of adult women remains more open than that of men, whose ego boundaries make them less emotional. Women in heterosexual relationships want children to bond with as substitutes for their lack of intense emotional intimacy with their men partners. But there are lesbians who have deep and intense relationships with women who also want children, as do some homosexual men (Lewin 1993). Furthermore, not all full-time mothering is emotionally intense, nor is all intensive mothering done by women. Barbara Risman (1987), in her study of fifty-five men who became single fathers because of their wives' death, desertion, or giving up custody, found that their relationships with their children were as intimate as those of single mothers and mothers in traditional marriages. And Karen Hansen's studies (1992) of nineteenth-century heterosexual men's friendships reveal a world of feeling similar to that described by Carroll Smith-Rosenberg (1975) for women.

In work organizations, position in the hierarchy does and does not override a worker's gender. The behavior of men and women doctors sometimes reflects their professional status and sometimes their gender, and it is important to look at both aspects to understand their relationships with patients (Lorber 1985). The men workers in men's occupations cannot be lumped in a minority category. The women come up against the glass ceiling that blocks their upward mobility, whereas the men are on what Christine Williams has called a "glass escalator": They are encouraged to compete for managerial and administrative positions (Williams 1989,1992).

Joey Sprague (1991) found that because material interests reflect positions in the social relations of production and reproduction, as well as more immediate community contexts, political attitudes hew more closely to class, gender role, and affiliation with social movements than to simple division of men versus women (also see Henderson-King and Stewart 1994).

There are revolutionary possibilities inherent in rethinking the categories of gender, sexuality, and physiological sex. Sociological data that challenge conventional knowledge by reframing the questions could provide legitimacy for new ways of thinking. When one term or category is defined only by its opposite, resistance reaffirms the polarity (Fuss 1991). The margin and the center, the insider and the outsider, the conformist and the deviant are two sides of the same concept. Introducing even one more term, such as bisexuality, forces a rethinking of the oppositeness of heterosexuality and homosexuality. "A critical sexual politics, in other words, struggles to move beyond the confines of an inside/outside model" (Namaste 1994, p. 230). The politics of identity are challenged, but such political stances are already split racially and by social class. Data that undermine the supposed natural dichotomies on which the social orders of most modern societies are still based could radically alter political discourses that valorize biological causes, essential heterosexuality, and traditional gender roles in families and workplaces.

REFERENCES

Amadiume, Ifi. *Male Daughters, Female Husbands: Gender and Sex in an African Society*. London: Zed Books, 1987.

Angier, Natalie. "Does Testosterone Equal Aggression? Maybe Not." *The New York Times,* June 20, 1995.

———. "Male Hormone Molds Women, in Mind and Body." *The New York Times,* May 3, 1994.

Blumstein, Philip W., and Pepper Schwartz. "Bisexuality in Women." *Archives of Sexual Behavior* 5 (1976a), pp. 171–181.

———. "Bisexuality in Men." *Urban Life* 5 (1976b), pp. 339–358.

Bolin, Anne. *In Search of Eve: Transsexual Rites of Passage*. South Hadley, MA: Bergin & Garvey, 1988.

Butler, Judith. *Bodies that Matter: On the Discursive Limits of "Sex."* New York and London: Routledge, 1993.

———. *Gender Trouble: Feminism and the Subversion of Identity*. New York and London: Routledge, 1990.

Chodorow, Nancy. "Gender as Personal and Cultural." *Signs: Journal of Women in Culture and Society* 20 (1995), pp. 516–544.

———. *Femininities, Masculinities, Sexualities: Freud and Beyond*. Lexington: University Press of Kentucky, 1994.

Cockburn, Cynthia. *In the Way of Women: Men's Resistance to Sex Equality in Organizations*. Ithaca, NY: ILR, 1991.

Collins, Patricia Hill. *Black Feminist Thought: Knowledge, Conciousness, and the Politics of Empowerment*. Boston: Unwin Hyman, 1990.

Collinson, David, and Jeff Hearn. "Naming Men as Men: Implication for Work, Organization and Management." *Gender, Work and Organization* 1 (1994), pp. 2–22.

Epstein, Julia. "Either/Or—Neither/Both: Sexual Ambiguity and the Ideology of Gender." *Genders* 7 (1990), pp. 100–142.

Fausto-Sterling, Anne. "The Five Sexes: Why Male and Female Are Not Enough." *The Sciences,* March/April 1993, pp. 20–25.

Frye, Marilyn. *Willfull Virgin: Essays in Feminism*. Freedom, CA: The crossing Press, 1992.

———. "The Possibility of Feminist Theory." In *Theoretical Perspectives on Sexual Difference,* ed. Deborah L. Rhode, New Haven, CT: Yale University Press, 1990.

Fuss, Diana (ed.). *Inside/Out: Lesbian Theories, Gay Theories*. New York and London: Routledge, 1991.

Garber, Marjorie. *Vice versa: Bisexuality and the Eroticism of Everyday Life*. New York: Simon and Schuster, 1995.

Giuffre, Patti A., and Christine L. Williams. "Boundary Lines: Labeling Sexual Harassment in Restaurants." *Gender and Society* 8 (1994), pp. 378–401.

Goldstein, Nancy. "Lesbians and the Medical Profession: HIV/AIDS and the Pursuit of Visibility." *Women's Studies: An Interdisciplinary Journal* 24, 1995, pp. 531–552.

Grady, Denise. "Sex Text of Champions: Olympic Officials Struggle to Define What Should Be Obvious: Just Who Is a Female Athlete." *Discover* 13 (June 1992), pp. 78–82.

Hansen, Karen V. " 'Our Eyes Behold Each Other': Masculinity and Intimate Friendship in Antebellum New England." In *Men's Friendships,* ed. Peter M. Nardi. Newbury Park, CA: Sage, 1992.

Hargreaves, Jennifer A. "Where's the Virtue? Where's the Grace? A Discussion of the Social Production of Gender Relations in and through Sport." *Theory, Culture, and Society* 3 (1986), pp. 109–121.

Henderson-King, Donna H., and Abigail J. Stewart. "Women or Feminists? Assessing Women's Group Consciousness." *Sex Roles* 31 (1994), pp. 505–516.

Ingraham, Chrys. "The Heterosexual Imaginary: Feminist Sociology and Theories of Gender." *Sociological Theory* 12 (1994), pp. 202–219.

Jacobs, Sue-Ellen, and Christine Roberts. "Sex, Sexuality, Gender, and Gender Variance." In *Gender and Anthropology,* ed. Sandra Morgen. Washington, DC: American Anthropological Association, 1989.

Kessler, Suzanne J. "The Medical Construction of Gender: Case Management of Intersexed Infants." *Signs* 16 (1990), pp. 3–26.

Klein, Fritz, Barry Sepekoff, and Timothy J. Wolf. "Sexual Orientation: A Multi-Variable Dynamic Process." *Journal of Homosexuality* 11, no. 1/2 (1985), pp. 35–49.

Kopytoff, Igor. "Women's Roles and Existential Identities." In *Beyond the Second Sex: New Directions in the Anthropology of Gender,* ed. Peggy Reeves Sanday and Ruth Gallagher Goodenough, Philadelphia: University of Pennsylvania Press, 1990.

Lewin, Ellen. *Lesbian Mothers: Accounts of Gender in American Culture*. Ithaca, NY: Cornell University Press, 1993.

Lorber, Judith. *Paradoxes of Gender*. New Haven, CT: Yale University Press, 1994.

———. "More Women Physicians: Will It Mean More Humane Health Care?" *Social Policy* 16 (Summer 1985), pp. 50–54.

Mansfield, Alan, and Barbara McGinn. "Pumping Irony: The Muscular and the Feminine." In *Body Matters: Essays on the Sociology of the Body,* ed. Sue Scott and David Morgan. London: Falmer, 1993.

Martin, Jane Roland. "Methodological Essentialism, False Difference, and Other Dangerous Traps." *Signs* 19 (1994), pp. 630–657.

Messner, Michael A. *Power at Play: Sports and the Problem of Masculinity*. Boston: Beacon, 1992.

Messner, Michael A., and Donald F. Sabo. *Sex, Violence, and Power in Sports: Rethinking Masculinity*. Freedom, CA: The Crossing Press, 1994.

Namaste, Ki. "The Politics of Inside/Out: Queer Theory, Poststructuralism, and a Sociological Approach to Sexuality." *Sociological Theory* 12 (1994), pp. 220–231.

Nicholson, Linda J. (ed). *Feminism/Postmodernism*. New York: Routledge, 1990.

Risman, Barbara J. "Intimate Relationships from a Microstructural Perspective: Men Who Mother." *Gender & Society* 1 (1987), pp. 6–32.

Rust, Paula C. *The Challenge of Bisexuality to Lesbian Politics: Sex, Loyalty, and Revolution*. New York: New York University Press. Forthcoming.

———. " 'Coming Out' in the Age of Social Constructionism: Sexual Identity Formation among Lesbian and Bisexual Women." *Gender & Society* 7 (1993), pp. 50–77.

———. "The Politics of Sexual Identity: Attraction and Behavior among Lesbian and Bisexual Women." *Social Problems* 39 (1992), pp. 366–386.

Sedgewick, Eve Kosofsky. *Epistemology of the Closet*. Berkeley: University of California Press, 1990.

Smith, Dorothy E. *The Conceptual Practices of Power: A Feminist Sociology of Knowledge*. Toronto: University of California Press, 1990a.

———. *Texts, Facts, and Femininity: Exploring the Relations of Ruling*. New York and London: Routledge, 1990b.

Smith-Rosenberg, Carroll. "The Female World of Love and Ritual: Relations between Women in Nineteenth-Century America." *Signs* 1 (1975), pp. 1–29.

Spelman, Elizabeth. *Inessential Woman: Problems of Exclusion in Feminist Thought*. Boston: Beacon, 1988.

Sprague, Joey. "Gender, Class and Political Thinking." In *Research in Political Sociology* 5, ed. Philo Wasburn. Greenwich, CT: JAI, 1991.

Staples, Robert. *Black Masculinity: The Black Male's Roles in American Society*. San Francisco, CA: Scholar Press, 1982.

Stein, Arlene, and Ken Plummer. " ' I Can't Even Think Straight.' 'Queer' Theory and the Missing Sexual Revolution in Sociology." *Sociological Theory* 12 (1994), pp. 178–187.

Taylor, Verta, and Leila Rupp. "Women's Culture and Lesbian Feminist Activism: A Reconsideration of Cultural Feminism." *Signs* 19 (1993), pp. 32–61.

Terry, Jennifer. "Theorizing Deviant Histography." *Differences: A Journal of Feminist Cultural Studies* 3, no. 2 (1991), pp. 55–74.

Throne, Barrie. *Gender Play: Boys and Girls in School*. New Brunswick, NJ: Rutgers University Press, 1993.

Valverde, Mariana. *Sex, Power and Pleasure*. Toronto: Women's Press, 1985.

Weinberg, Martin S., Colin J. Williams, and Douglas W. Pryor. *Dual Attraction: Understanding Bisexuality*. New York: Oxford University Press, 1994.

West, Candace, and Sarah Fenstermaker. "Doing Difference." *Gender & Society* 9 (1995), pp. 8–37.

Williams, Christine L. "The Glass Escalator: Hidden Advantages for Men in the 'Female' Professions." *Social Problems* 39 (1992), pp. 253–267.

———. *Gender Differences at Work: Women and Men in Nontraditional Occupations*. Berkeley: University of California Press, 1989.

Young, Iris Marion. "Gender as Seriality: Thinking about Women as a Social Collective." "*Signs* 19 (1994), pp. 713–738.

DISCUSSION QUESTIONS

1. Lorber indicates that sexuality is gendered. In what ways, if any, is gender sexed/sexualized (i.e., organized and affected by sexuality)?
2. Is it problematic to assign (and often surgically "correct") individuals born with ambiguous genitalia to a particular sex category (i.e., male or female) based on the length of penis/genital tissue? How "long" should this tissue be to be assigned to the category male, how "short" must it be to be a female? Is it necessary to do this? Desirable?
3. In the previous chapter, Udis-Kessler introduced the notion of reification. In what

ways does surgical "correction" reify the concept of (biological) sex?

4. Can research that uses these essentialized categories (sex, gender, sex, etc.) produce results that are meaningful?
5. Though sex, gender, and sexuality are socially constructed, many theorists point out that these constructions have real consequences for individuals and for society. Think about your own life and provide examples of ways in which you have been complicit in perpetuating the "reality" of these social constructs (consciously or unconsciously). In what ways have you challenged or resisted them?

How Men Have (A) Sex

JOHN STOLTENBERG

In this revealing classic, writer John Stoltenberg examines the links between maleness, masculinity, male dominance, and sexuality to explore how hegemonic notions about "being a man" cultivate particular (and problematic) forms of sexual behavior in men alienating them from many aspects of sexual relationships, including intimacy, joy, ecstasy, and equality.

An address to college students
In the human species, how many sexes are there?
Answer A: *There are two sexes.*
Answer B: *There are three sexes.*
Answer C: *There are four sexes.*
Answer D: *There are seven sexes.*
Answer E: *There are as many sexes as there are people.*

I'd like to take you, in an imaginary way, to look at a different world, somewhere else in the universe, a place inhabited by a life form that very much resembles us. But these creatures grow up with a peculiar knowledge. They know that they have been born in an infinite variety. They know, for instance, that in their genetic material they are born with hundreds of different chromosome formations at the point in each cell that we would say determines their "sex." These creatures don't just come in XX or XY; they also come in XXY and XYY and XXX plus a long list of "mosaic" variations in which some cells in a creature's body have one combination and other cells have another. Some of these creatures are born with chromosomes that aren't even quite X or Y because a little bit of one chromosome goes and gets joined to another. There are hundreds of different combinations, and though all are not fertile, quite a number of them are. The creatures in this world enjoy their individuality; they delight in the fact that they are not divisible, into distinct categories. So when another newborn arrives with an esoterically rare chromosomal formation, there is a little celebration: "Aha," they say, "another sign that we are each unique."

These creatures also live with the knowledge that they are born with a vast range of genital formations. Between their legs are tissue structures that vary along a continuum, from clitorises with

---

a vulva through all possible combinations and gradations to penises with a scrotal sac. These creatures live with an understanding that their genitals all developed prenatally from exactly the same little nub of embryonic tissue called a genital tubercle, which grew and developed under the influence of varying amounts of the hormone androgen. These creatures honor and respect everyone's natural-born genitalia—including what we would describe as a microphallus or a clitoris several inches long. What these creatures find amazing and precious is that because everyone's genitals stem from the same embryonic tissue, the nerves inside all their genitals got wired very much alike, so these nerves of touch just go crazy upon contact in a way that resonates completely between them. "My gosh," they think "you must feel something in your genital tubercle that intensely resembles what I'm feeling in my genital tubercle." Well, they don't exactly *think* that in so many words; they're actually quite heavy into their feelings at that point; but they do feel very connected—throughout all their wondrous variety.

I could go on. I could tell you about the variety of hormones that course through their bodies in countless different patterns and proportions, both before birth and throughout their lives—the hormones that we call "sex hormones" but that they call "individuality inducers." I could tell you how these creatures think about reproduction: For part of their lives, some of them are quite capable of gestation, delivery, and lactation; and for part of their lives, some of them are quite capable of insemination; and for part or all of their lives, some of them are not capable of any of those things—so these creatures conclude that it would be silly to lock anyone into a lifelong category based on a capability variable that may or may not be utilized and that in any case changes over each lifetime in a fairly uncertain and idiosyncratic way. These creatures are not oblivious to reproduction; but nor do they spend their lives constructing a self-definition around their variable reproductive capacities. They don't have to, because what is truly unique about these creatures is that they are capable of having a sense of personal identity without struggling to fit into a group identity based on how they were born. These creatures are quite happy, actually. They don't worry about sorting *other* creatures into categories, so they don't have to worry about whether they are measuring up to some category they themselves are supposed to belong to.

These creatures, of course, have sex. Rolling and rollicking and robust sex, and sweaty and slippery and sticky sex, and trembling and quaking and tumultuous sex, and tender and tingling and transcendent sex. They have sex fingers to fingers. They have sex belly to belly. They have sex genital tubercle to genital tubercle. They *have* sex. They do not have *a* sex. In their erotic lives, they are not required to act out their status in a category system—because there is no category system. There are no sexes to belong to, so sex between creatures is free to be between genuine individuals—not representatives of a category. They have sex. They do not have a sex. Imagine life like that.

Perhaps you have guessed the point of this science fiction: Anatomically, each creature in the imaginary world I have been describing could be an identical twin of every human being on earth. These creatures, in fact, *are* us—in every way except socially and politically. The way they are born is the way we are born. And we are not born belonging to one or the other of two sexes. We are born into a physiological continuum on which there is no discrete and definite point that you can call "male" and no discrete and definite point that you can call "female." If you look at all the variables in nature that are said to determine human "sex," you can't possibly find one that will unequivocally split the species into two. Each of the so-called criteria of sexed-ness is itself a continuum—including chromosomal variables, genital and gonadal variations, reproductive capacities, endocrinological proportions, and any other criterion you could think of. Any or all of these different variables may line up in any number of ways, and all of the variables may vary independently of one another.[1]

What does all this mean? It means, first of all, a logical dilemma: Either human "male" and human "female" actually exist in nature as fixed and discrete entities and you can credibly base an entire social and political system on those absolute natural categories, or else the variety of human sexedness is infinite. As Andrea Dworkin wrote in 1974:

> The discovery is, of course, that "man" and "woman" are fictions, caricatures, cultural constructs. As models they are reductive, totalitarian, inappropriate to human becoming. As roles they are static, demeaning to the female, dead-ended for male and female both.[2]

The conclusion is inescapable:

> We are, clearly, a multisexed species which has its sexuality spread along a vast continuum where the elements called male and female are not discrete.[3]

"*We are . . . a multisexed species.*" I first read those words a little over ten years ago—and that liberating recognition saved my life.

All the time I was growing up, I knew that there was something really problematical in my relationship to manhood. Inside, deep inside, I never believed I was fully male—I never believed I was growing up enough of a man. I believed that someplace out there, in other men, there was something that was genuine authentic all-American manhood—the real stuff—but I didn't have it: not enough of it to convince *me* anyway, even if I managed to be fairly convincing to those around me. I felt like an impostor, like a fake. I agonized a lot about not feeling male enough, and I had no idea then how much I was not alone.

Then I read those words—those words that suggested to me for the first time that the notion of manhood is a cultural delusion, a baseless belief, a false front, a house of cards. It's not true. The category I was trying so desperately to belong to, to be a member of in good standing—it doesn't exist. Poof. Now you see it, now you don't. Now you're terrified you're not really part of it, now you're free, you don't have to worry anymore. However removed you feel inside from "authentic manhood," it doesn't matter. What matters is the center inside yourself—and how you live, and how you treat people, and what you can contribute as you pass through life on this earth, and how honestly you love, and how carefully you make choices. Those are the things that really matter. Not whether you're a real man. There's no such thing.

The idea of the male sex is like the idea of an Aryan race. The Nazis believed in the idea of an Aryan race—they believed that the Aryan race really exists, physically, in nature—and they put a great deal of effort into making it real. The Nazis believed that from the blond hair and blue eyes occurring naturally in the human species, they could construe the existence of a separate *race*—distinct category of human beings that was unambiguously rooted in the natural order of things. But traits do not a race make; traits only make traits. For the idea to be real that these physical traits comprised a race, the race had to be socially constructed. The Nazis inferiorized and exterminated those they defined as "non-Aryan." With that, the notion of an Aryan race began to seem to come true. That's how there could be a political entity known as an Aryan race, and that's how there could be for some people a personal, subjective sense that they belonged to it. This happened through hate and force, through violence and victimization, through treating millions of people as things, then exterminating them. The belief system shared by people who believed they were all Aryan could not exist apart from that force and violence. The force and violence created a racial class-system, *and* it created those people's membership in the race considered "superior." The force and violence served their class interests in large part because it created and maintained the class itself. But the idea of an Aryan race could never become metaphysically true, despite all the violence unleashed to create it, because there simply *is* no Aryan race. There is only the idea of it—and the consequences of trying to make it seem real. The male sex is very like that.

Penises and ejaculate and prostate glands occur in nature, but the notion that these anatomical traits comprise a sex—a discrete class, separate and distinct, metaphysically divisible from some other sex, *the* "other sex"—is simply that: a notion, an idea. The penises exist; the male sex does not. The male sex is socially constructed. It is a political entity that flourishes only through acts of force and sexual terrorism. Apart from the global inferiorization and subordination of those who are defined as "nonmale," the idea of personal membership in the male sex class would have no recognizable meaning. It would make no sense. No one could be a member of it and no one would think they *should* be a member of it. There would be no male sex to belong to. That doesn't mean there wouldn't still be penises and ejaculate and prostate glands and such. It simply means that the center of our selfhood would not be required to reside inside an utterly fictitious category—a category that only seems real to the extent that those outside it are put down.

We live in a world divided absolutely into two sexes, even though nothing about human nature warrants that division. We are sorted into one category or another at birth based solely on a visual inspection of our groins, and the only question that's asked is whether there's enough elongated tissue around your urethra so you can pee standing up. The presence or absence of a long-enough penis is the primary criterion for separating who's to grow up female from who's to grow up female. And among all the ironies in that utterly whimsical and arbitrary selection process is the fact that *anyone* can pee both sitting down and standing up.

Male sexual identity is the conviction or belief, held by most people born with penises, that they are male and not female, that they belong to the male sex. In a society predicated on the notion that there are two "opposite" and "complementary" sexes, this idea not only makes sense, it *becomes* sense; the very idea of a male sexual identity produces sensation, produces the meaning of sensation, becomes the meaning of how one's body feels. The sense and the sensing of a male sexual identity is at once mental and physical, at once public and personal. Most people born with a penis between their legs grow up aspiring to feel and act unambiguously male, longing to belong to the sex that is male and daring not to belong to the sex that is not, and feeling this urgency for a visceral and constant verification of their male sexual identity—for a fleshy connection to manhood—as the driving force of their life. The drive does not originate in the anatomy. The sensations derive from the idea. The idea gives the feelings social meaning; the idea determines which sensations shall be sought.

People born with penises must strive to make the idea of male sexual identity personally real by doing certain deeds, actions that are valued and chosen because they produce the desired feeling of belonging to a sex that is male and not female. Male sexual identity is experienced only in sensation and action, in feeling and doing, in eroticism and ethics. The feeling of belonging to a male sex encompasses both sensations that are explicitly "sexual" and those that are not ordinarily regarded as such. And there is a tacit social value system according to which certain acts are chosen because they make an individual's sexedness feel real and certain other acts are eschewed because they numb it. That value system is the ethics of male sexual identity—and it may well be the social origin of all injustice.

Each person experiences the idea of sexual identity as more or less real, more or less certain, more or less true, depending on two very personal phenomena: one's feelings and one's acts. For many people, for instance, the act of fucking makes their sexual identity feel more real than it does at other times, and they can predict from experience that this feeling of greater certainty will last for at least a while after each time they fuck. Fucking is not the only such act, and not only so-called sex acts can result in feelings of certainty about sexual identity; but the act of fucking happens to be a very good example of the correlation between *doing* a specific act in a specific way and *sensing* the specificity of the sexual identity to which one aspires. A person can decide to do certain acts and not others just because some acts will have the payoff of a feeling of greater certainty about sexual identity and others will give the feedback of a feeling of less. The transient reality of one's sexual identity, a person can know, is always a function of what one does and how one's acts make one feel. The feeling and the act must conjoin for the idea of the sexual identity to come true. We all keep longing for surety of our sexedness that we can feel; we all keep striving through our actions to make the idea real.

In human nature, eroticism is not differentiated between "male" and "female" in any clear-cut way. There is too much of a continuum, too great a resemblance. From all that we know, the penis and the clitoris are identically "wired" to receive and retransmit sensations from throughout the body, and the congestion of blood within the lower torso during sexual excitation makes all bodies sensate in a remarkably similar manner. Simply put, we share all the nerve and blood-vessel layouts that are associated with sexual arousal. Who can say, for instance, that the penis would not experience sensations the way that a clitoris does if this were not a world in which the penis is supposed to be hell-bent on penetration? By the time most men make it through puberty, they believe that erotic sensation is supposed to *begin* in their penis; that if engorgement has not begun there, then nothing else in their body will heat up either. There is a massive interior dissociation from sensations that do not explicitly remind a man that his penis is still there. And not only there as sensate, but *functional and operational.*

So much of most men's sexuality is tied up with gender-actualizing—with feeling like a real man—that they can scarcely recall an erotic sensation that had no gender-specific cultural meaning. As most men age, they learn to cancel out and deny erotic sensations that are not specifically linked to what they think a real man is supposed to feel. An erotic sensation unintentionally experienced in a receptive, communing mode—instead of in an aggressive and controlling and violative mode, for instance—can shut down sensory systems in an instant. An erotic sensation unintentionally linked to the "wrong" sex of another person can similarly mean sudden numbness. . . .

My point is that sexuality does not *have* a gender; it *creates* a gender. It creates for those who adapt to it in narrow and specified ways the confirmation for the individual of belonging to the idea of one sex or the other. So-called male sexuality is a learned connection between specific physical sensations and the idea of a male sexual identity. To achieve this male sex identity requires that an individual *identify with* the class of males—that is, accept as one's own the values and interests of the class. A fully realized male sexual identity also requires *nonidentificaiion with* that which is perceived to be nonmale, or female. A male must not identify with females; he must not associate with females in feeling, interest, or action. His identity as a member of the sex class men absolutely depends on the extent to which he repudiates the values and interests of the sex class "women."

I think somewhere inside us all, we have always known something about the relativity of gender. Somewhere inside us all, we know that our bodies harbor deep resemblances, that we are wired inside to respond in a profound harmony to the resonance of eroticism inside the body of someone near us. Physiologically, we are far more alike than different. The tissue structures that have become labial and clitoral or scrotal and penile have not forgotten their common ancestry. Their sensations are of the same source. The nerve networks and interlock of capillaries throughout our pelvises electrify and engorge as if plugged in together and pumping as one. That's what we feel when we feel one another's feelings. That's what can happen during sex that is mutual, equal, reciprocal, profoundly communing.

So why is it that some of us with penises think it's sexy to pressure someone into having sex against their will? Some of us actually get harder the harder the person resists. Some of us with penises actually believe that some of us without penises want to be raped. And why is it that some of us with penises think it's sexy to treat other people as objects, as things to be bought and sold, impersonal bodies to be possessed and consumed for our sexual pleasure? Why is it that some of us with penises are aroused by sex tinged with rape, and sex commoditized by pornography? Why do so many of us with penises want such antisexual sex?

There's a reason, of course. We have to make a lie seem real. Its a very big lie. We each have to do our part. Otherwise the lie will look like the lie that it is. Imagine the enormity of what we each must do to keep the lie alive in each of us. Imagine the awesome challenge we face to make the lie a social fact. It's a lifetime mission for each of us born with a penis: to have sex in such a way that the male sex will seem real—and so that we'll feel like a real part of it.

We all grow up knowing exactly what kind of sex that is. It's the kind of sex you can have when you pressure or bully someone else into it. So it's a kind of sex that makes your will more important than theirs. That kind of sex helps the lie a lot. That kind of sex makes you feel like someone important and it turns the other person into someone unimportant. That kind of sex makes you feel real, not like a fake. It's a kind of sex men have in order to feel like a real man.

There's also the kind of sex you can have when you force someone and hurt someone and cause someone suffering and humiliation. Violence and hostility in sex help the lie a lot too. Real men are aggressive in sex. Real men get cruel in sex. Real men use their penises like weapons in sex. Real men leave bruises. Real men think it's a turn-on to threaten harm: A brutish push can make an erection feel really hard. That kind of sex helps the lie a lot. That kind of sex makes you feel like someone who is powerful and it turns the other person into someone powerless. That kind of sex makes you feel dangerous and in control—like you're fighting a war with an enemy and if you're mean enough you'll win but if you let up you'll lose your manhood. It's a kind of sex men have *in order to have* a manhood.

There's also the kind of sex you can have when you pay your money into a profit system that grows rich displaying and exploiting the bodies and body parts of people without penises for the sexual entertainment of people with. Pay your money and watch. Pay your money and imagine. Pay your money and get real turned on. Pay your money and jerk off. That kind of sex helps the lie a lot. It helps support an industry committed to making people with penises believe that people without are sluts who just want to be ravished and reviled. . . .

And there's one more thing: That kind of sex makes the lie indelible—burns it onto your retinas right adjacent to your brain—makes you remember it and makes your body respond to it and so it makes you believe that the lie is in fact true: You really are a real man. That slavish and submissive creature there spreading her legs is really not. You and that creature have nothing in common. That creature is an alien inanimate thing, but your penis is completely real and alive. Now you can come. Thank god almighty—you have a sex at last.

Now, I believe there are many who are sick at heart over what I have been describing. There are many who were born with penises who want to stop collaborating in the sex-class system that needs us to need these kinds of sex. . . .

When you use sex to have a sex, the sex you have is likely to make you feel crummy about yourself. But when you have sex in which you are not struggling with your partner in order to act out "real manhood" the sex you have is more likely to bring you close.

This means several specific things;

1. *Consent is absolutely essential.* If both you and your partner have not freely given your informed consent to the sex you are about to have you can be quite certain that the sex you go ahead and have will make you strangers to each other. How do you know if there's consent? You ask. You ask again if you're sensing any doubt. Consent to do one thing isn't consent to do another. So you keep communicating, in clear words. And you don't take anything for granted.

2. *Mutuality is absolutely essential.* Sex is not something you do *to* someone. Sex is not a one-way transitive verb, with a subject, you, and an object, the body you're with. Sex that is mutual is not about doing and being done to; it's about being-with and feeling-with. You have to really be there to experience what is happening between and within the two of you—between every part of you and within both your whole bodies. It's a matter of paying attention—as if you are paying attention to someone who matters.
3. *Respect is absolutely essential.* In the sex that you have, treat your partner like a real person who, like you, has real feelings—feelings that matter as much as your own. You may or may not love—but you must always respect. You must respect the integrity of your partner's body. It is not yours for the taking. It belongs to someone real. And you do not get ownership of your partner's body just because you are having sex—or just because you have had sex.

. . . I speak to you as someone who is closer to the middle of my sexual history [than to its beginning]. As I look back, I see that I made many choices that I didn't know I was making. And as I look at men who are near my age, I see that what has happened to many of them is that their sex lives are stuck in deep ruts that began as tiny fissures when they were young. So I want to conclude by identifying what I believe are three of the most important decisions about your sexuality that you can make when you are at the beginning of your sexual history. However difficult these choices may seem to you now, I promise you they will only get more difficult as you grow older. I realize that what I'm about to give is some quite unsolicited nuts-and-bolts advice. But perhaps it will spare you, later on in your lives, some of the obsessions and emptiness that have claimed the sexual histories of many men just a generation before you. Perhaps it will not help, I don't know; but I hope very much that it will.

First, you can start choosing now not to let your sexuality be manipulated by the pornography industry. I've heard many unhappy men talk about how they are so hooked on pornography and obsessed with it that they are virtually incapable of a human erotic contact. And I have heard even more men talk about how, when they do have sex with someone, the pornography gets in the way, like a mental obstacle, like a barrier preventing a full experience of what's really happening between them and their partner. The sexuality that the pornography industry needs you to have is not about communicating and caring; it's about "pornographizing" people—objectifying and conquering them, not being with them as a person. You do not have to buy into it.

Second, you can start choosing now not to let drugs and alcohol numb you through your sex life. Too many men, as they age, become incapable of having sex with a clear head. But you need your head clear—to make clear choices, to send clear messages, to read clearly what's coming in on a clear channel between you and your partner. Sex is no time for your awareness to sign off. And another thing: Beware of relying on drugs or alcohol to give you "permission" to have sex, or to trick your body into feeling something that it's not, or so you won't have to take responsibility for what you're feeling or for the sex that you're about to have. If you can't take sober responsibility for your part in a sexual encounter, you probably shouldn't be having it—and you certainly shouldn't be zonked out of your mind *in order* to have it.

Third, you can start choosing now not to fixate on fucking—especially if you'd really rather have sex in other, noncoital ways. Sometimes men have coital sex—penetration and thrusting then ejaculating inside someone—not because they particularly feel like it but because they feel they *should* feel like it: It's expected that if you're the man, you fuck. And if you don't fuck, you're not a man. The corollary of this cultural imperative is that if two people don't have intercourse, they have not had real sex. That's baloney, of course, but the message comes down hard, especially inside men's heads: Fucking is *the* sex act, the act in which you act out what sex is supposed to be—and what sex you're supposed to be.

Like others born with a penis, I was born into a sex-class system that requires my collaboration every day, even in how I have sex. Nobody told me, when I was younger, that I could have noncoital sex and that it would be fine. Actually, much better than fine. Nobody told me about an incredible range of other erotic possibilities for mutual lovemaking—including rubbing body to body, then coming body to body; including multiple, nonejaculatory orgasms; including the feeling you get when even the tiniest place where you and your partner touch becomes like a window through which great tidal storms of passion ebb and flow, back and forth. Nobody told me about the sex you can have when you stop working at having a sex. My body told me, finally. And I began to trust what my body was telling me more than the lie I was supposed to make real.

I invite you too to resist the lie. I invite you too to become an erotic traitor to male supremacy.

NOTES

1. My source for the foregoing information about so-called sex determinants in the human species is a series of interviews I conducted with the sexologist Dr. John Money in Baltimore, Maryland, in 1979 for an article I wrote called "The Multisex Theorem," which was published in a shortened version as "Future Genders" in *Omni* magazine, May 1980, pp. 67–73ff.
2. Dworkin, Andrea. *Woman Hating* (New York: Dutton, 1974), p. 174.
3. Dworkin, *Woman Hating*, p. 183.

DISCUSSION QUESTIONS

1. In what ways (if any) does this work, like Lorber's, demonstrate the process of reification? What reified concepts does Stoltenberg reveal to be social constructions?
2. Stoltenberg uses the example of the Aryan race to demonstrate how sex is socially constructed. In what ways is the social construction of *Aryan* similar to the social construction of *sex*? In what ways is it different? Can you think of similar examples?
3. How are force and violence used to enforce sex, gender, and sexuality in this culture? Provide examples.
4. Is gender actualized through sexual acts for women? If so, in what way? Are certain acts required? Certain participants? If not, what *is* the relationship between sex and gender for women in contemporary American culture?

The Myth of the Latin Woman: I Just Met a Girl Named María

JUDITH ORTIZ COFER

Poet Judith Ortiz Cofer provides a personal account of the ways that sexuality is structured and conditioned by both gender and race. Her essay also demonstrates the ways that racism and sexism combine to produce particularly pernicious (and often sexualized) assumptions about members of subordinate social groups.

On a bus trip to London from Oxford University, where I was earning some graduate credits one summer, a young man, obviously fresh from a pub, spotted me and as if struck by inspiration went down on his knees in the aisle. With both hands over his heart he broke into an Irish tenor's rendition of "María" from *West Side Story*. My politely amused fellow passengers gave his lovely voice the round of gentle applause it deserved. Though I was not quite as amused, I managed my version of an English smile: no show of teeth, no extreme contortions of the facial muscles—I was at this time of my life practicing reserve and cool. Oh, that British control, how I coveted it. But María had followed me to London, reminding me of a prime fact of my life: you can leave the Island, master the English language, and travel as far as you can, but if you are a Latina, especially one like me who so obviously belongs to Rita Moreno's gene pool, the Island travels with you.

This is sometimes a very good thing—it may win you that extra minute of someone's attention. But with some people, the same things can make *you* an island—not so much a tropical paradise as an Alcatraz, a place nobody wants to visit. As a Puerto Rican girl growing up in the United States and wanting like most children to "belong," I resented the stereotype that my Hispanic appearance called forth from many people I met.

Our family lived in a large urban center in New Jersey during the sixties, where life was designed as a microcosm of my parents' casas on the island. We spoke in Spanish, we ate Puerto Rican food bought at the bodega, and we practiced strict Catholicism complete with Saturday confession and Sunday mass at a church where our parents were accommodated into a one-hour Spanish mass slot, performed by a Chinese priest trained as a missionary for Latin America.

As a girl I was kept under strict surveillance, since virtue and modesty were, by cultural equation, the same as family honor. As a teenager I was instructed on how to behave as a proper senorita. But it was a conflicting message girls got, since the Puerto Rican mothers also encouraged their daughters to look and act like women and to dress in clothes our Anglo friends and their mothers found too "mature" for our age. It was, and is, cultural, yet I often felt humiliated when I appeared at an American friend's party wearing a dress more suitable to a semiformal than to a playroom birthday celebration. At Puerto Rican festivities, neither the music nor the colors we wore could be too loud. I still experience a vague sense of

From "The Myth of the Latin Woman: I Just Met a Girl Named Maria" by Judith Ortiz Cofer in *The Latin Deli: Prose and Poetry*, 1993. Reprinted by permission of The University of Georgia Press.

letdown when I'm invited to a "party" and it turns out to be a marathon conversation in hushed tones rather than a fiesta with salsa, laughter, and dancing—the kind of celebration I remember from my childhood.

I remember Career Day in our high school, when teachers told us to come dressed as if for a job interview. It quickly became obvious that to the barrio girls, " dressing up" sometimes meant wearing ornate jewelry and clothing that would be more appropriate (by mainstream standards) for the company Christmas party than as daily office attire. That morning I had agonized in front of my closet, trying to figure out what a "career girl" would wear because, essentially except for Marlo Thomas on TV, I had no models on which to base my decision. I knew how to dress for school: at the Catholic school I attended we all wore uniforms; I knew how to dress for Sunday mass, and I knew what dresses to wear for parties at my relatives' homes. Though I do not recall the precise details of my Career Day outfit, it must have been a composite of the above choices. But I remember a comment my friend (an Italian-American) made in later years that coalesced my impressions of that day. She said that at the business school she was attending the Puerto Rican girls always stood out for wearing "everything at once." She meant, of course, too much jewelry, too many accessories. On that day at school, we were simply made the negative models by the nuns who were themselves not credible fashion experts to any of us. But it was painfully obvious to me that to the others, in their tailored skirts and silk blouses, we must have seemed "hopeless" and "vulgar." Though I now know that most adolescents feel out of step much of the time, I also know that for the Puerto Rican girls of my generation that sense was intensified. The way our teachers and classmates looked at us that day in school was just a taste of the culture clash that awaited us in the real world, where prospective employers and men on the street would often misinterpret our tight skirts and jingling bracelets as a come-on.

Mixed cultural signals have perpetuated certain stereotypes—for example, that of the Hispanic woman as the "Hot Tamale" or sexual firebrand. It is a one-dimensional view that the media have found easy to promote. In their special vocabulary, advertisers have designated "sizzling" and "smoldering" as the adjectives of choice for describing not only the foods but also the women of Latin America. From conversations in my house I recall hearing about the harassment that Puerto Rican women endured in factories where the "boss men" talked to them as if sexual innuendo was all they understood and, worse, often gave them the choice of submitting to advances or being fired.

It is custom, however, not chromosomes, that leads us to choose scarlet over pale pink. As young girls, we were influenced in our decisions about clothes and colors by the women—older sisters and mothers who had grown up on a tropical island where the natural environment was a riot of primary colors, where showing your skin was one way to keep cool as well as to look sexy. Most important of all, on the island, women perhaps felt freer to dress and move more provocatively, since, in most cases, they were protected by the traditions, mores, and laws of a Spanish/Catholic system of morality and machismo whose main rule was: *You may look at my sister, but if you touch her I will kill you.* The extended family and church structure could provide a young woman with a circle of safety in her small pueblo on the Island; if a man "wronged" a girl, everyone would close in to save her family honor.

This is what I have gleaned from my discussions as an adult with older Puerto Rican women. They have told me about dressing in their best parry clothes on Saturday nights and going to the town's plaza to promenade with their girlfriends in front of the boys they liked. The males were thus given an opportunity to admire the women and to express their admiration in the form of piropos: erotically charged street poems they composed on the spot. I have been subjected to a few *piropos* while visiting the Island, and they can be outrageous, although custom dictates that they must never cross into obscenity. This ritual, as I understand it, also entails a show of studied indifference on the woman's part; if she is "decent," she must not acknowledge the man's impassioned words. So I do understand how things can be lost in translation. When a Puerto Rican girl dressed in her idea of what is attractive meets a man from the mainstream culture who has been trained to react to certain types of clothing as a sexual signal, a clash is likely to take place. The line I first heard based on this aspect of the myth happened when the boy who took me to my first formal dance leaned over to plant a sloppy overeager kiss painfully on my mouth, and when I didn't respond with sufficient passion said in a resentful tone: "I thought you Latin girls were supposed to mature early"—my first instance of being thought of as a fruit or vegetable—I was supposed to ripen, not just grow into womanhood like other girls.

It is surprising to some of my professional friends that some people, including those who should know better, still put others "in their place." Though rarer, these incidents are still commonplace in my life. It happened to me most recently during a stay at a very classy metropolitan hotel favored by young professional couples for their weddings. Late one evening after the theater, as I walked toward my room with my new colleague (a woman with whom I was coordinating an arts program), a middle-aged man in a tuxedo, a young girl in satin and lace on his arm, stepped directly into our path. With his champagne glass extended toward me, he exclaimed, "Evita!"

Our way blocked, my companion and I listened as the man half-recited, half-bellowed "Don't Cry for Me, Argentina." When he finished, the young girl said: "How about a round of applause for my daddy?" We complied, hoping this would bring the silly spectacle to a close. I was becoming aware that our little group was attracting the attention of the other guests. "Daddy" must have perceived this too, and he once more barred the way as we tried to walk past him. He began to shout-sing a ditty to the tune of "La Bamba"—except the lyrics were about a girl named María whose exploits all rhymed with her name and gonorrhea. The girl kept saying "Oh, Daddy" and looking at me with pleading eyes. She wanted me to laugh along with the others. My companion and I stood silently waiting for the man to end his offensive song. When he finished, I looked not at him but at his daughter. I advised her calmly never to ask her father what he had done in the army. Then I walked between them and to my room. My friend complimented me on my cool handling of the situation. I confessed to her that I really had wanted to push the jerk into the swimming pool. I knew that this same man—probably a corporate executive, well educated, even worldly by most standards—would not have been likely to regale a white woman with a dirty song in public. He would perhaps have checked his impulse by assuming that she could be somebody's wife or mother, or at least *somebody* who might take offense. But to him, I was just an Evita or a María: merely a character in his cartoon-populated universe.

Because of my education and my proficiency with the English language, I have acquired many mechanisms for dealing with the anger I experience. This was not true for my parents, nor is it true for the many Latin women working at menial jobs who must put up with stereotypes about our ethnic group such as: "They make good domestics." This is another facet of the myth of the Latin women in the United States. Its origin is simple to deduce. Work as domestics, waitressing, and factory jobs are all that's available to women with little English and few skills. The myth of the Hispanic menial has been sustained by the same media phenomenon that made "Mammy" from *Gone with the Wind* America's idea of the black woman for generations; María, the housemaid or counter girl, is now indelibly etched into the national psyche. The big and the little screens have presented us with the picture of the funny Hispanic maid, mispronouncing words and cooking up a spicy storm in a shiny California kitchen.

This media-engendered image of the Latina in the United States has been documented by feminist Hispanic scholars, who claim that such portrayals are partially responsible for the denial of opportunities for upward mobility among Latinas in the professions. I have a Chicana friend working on a Ph.D. in philosophy at a major university. She says her doctor still shakes his head in puzzled amazement at all the "big words" she uses. Since I do not wear my diplomas around my neck for all to see, I too have on occasion been sent to that "kitchen," where some think I obviously belong.

One such incident that has stayed with me, though I recognize it as a minor offense, happened on the day of my first public poetry reading. It took place in Miami in a boat-restaurant where we were having lunch before the event. I was nervous and excited as I walked in with my notebook in hand. An older woman motioned me to her table. Thinking (foolish me) that she wanted me to autograph a copy of my brand new slender volume of verse, I went over. She ordered a cup of coffee from me, assuming that I was the waitress. Easy enough to mistake my poems for menus, I suppose. I know that it wasn't an intentional act of cruelty, yet with all the good things that happened that day, I remember that scene most clearly; because it reminded me of what I had to overcome before anyone would take me seriously. In retrospect I understand that my anger gave my reading fire, that I have almost always taken doubts in my abilities as a challenge—and that the result is, most times, a feeling of satisfaction at having won a convert when I see the cold, appraising eyes warm to my words, the body language change, the smile that indicates that I have opened some avenue for communication. That day I read to that woman and her lowered eyes told me that she was embarrassed at her little faux pas, and when I willed her to look up to me, it was my victory, and she graciously allowed me to punish her with my full attention. We shook hands at the end of the reading, and I never saw her again. She has probably forgotten the whole thing but maybe not.

Yet I am one of the lucky ones. My parents made it possible for me to acquire a stronger footing in the mainstream culture by giving me the chance at an education. And books and art have saved me from the harsher forms of ethnic and racial prejudice that many of my Hispanic *compañeras* have had to endure. I travel a lot around the United States, reading from my books of poetry and my novel, and the reception I most often receive is one of positive interest by people who want to know more about my culture. There are, however, thousands of Latinas without the privilege of an education or the entrée into society that I have. For them life is a struggle against the misconceptions perpetuated by the myth of the Latina as whore, domestic, or criminal. We cannot change this by legislating the way people look at us. The transformation, as I see it, has to occur at a much more individual level. My personal goal in my public life is to try to replace the old pervasive stereotypes and myths about Latinas with a much more interesting set of realities. Every time I give a reading, I hope the stories I tell, the dreams and fears I examine in my work, can achieve some universal truth which will get my audience past the particulars of my skin color, my accent, or my clothes.

I once wrote a poem in which I called us Latinas "God's brown daughters." This poem is really a prayer of sorts, offered upward, but also, through the human-to-human channel of art, outward. It is a prayer for communication, and for respect. In it, Latin women pray "in Spanish to an Anglo God/with a Jewish heritage," and they are "fervently hoping/that if not omnipotent,/at least He be bilingual."

DISCUSSION QUESTIONS

1. In what ways is the stereotypical sexuality of Latin women similar to and different from other ethnic stereotypes of women? What, if anything, does this suggest about subordinated-female sexualities?
2. Pay attention to magazine advertisements, television sit-coms, music videos, and other mass-media formats. Do the images of Latin women that you see generally support the "hot tamale" stereotype? What other kinds of sexualized generalizations do you detect?

3. How does the sexualization of Latin women serve to reinforce their social inequality?
4. Ortiz Cofer indicates that notions of chivalry and family honor operate in the Puerto Rican community to allow "decent" women to embrace some aspects of their sexuality (i.e., dressing provocatively). Are there any parallel opportunities for Caucasian women? What does this suggest about generalized notions of female sexual agency?
5. Is Ortiz Cofer being too generous when she states that she can "understand how things can be lost in translation" when she refers to men from the mainstream culture who respond to "certain kinds of clothing" as "a sexual signal?"

The following Bonus Reading can be accessed online through the *InfoTrac College Edition Online Library* by typing in the name of the author or keywords in the title at *http://www.infotrac-college.com.*

The Five Sexes, Revisited

ANNE FAUSTO-STERLING

Few would contest the notion that as a species we are divided into two basic groups—male and female. Yet these seemingly "natural categories" are not as straightforward as we might expect. In the following article, Anne Fausto-Sterling examines the case of the ***intersexed****, individuals who have both male and female biological properties or genitalia that appears to differ from their chromosomal sex. Her analysis also reveals the extent to which medical science is complicit in the* ***reification*** *(the process of treating something socially constructed as real) of sex as an essential biological trait.*

DISCUSSION QUESTIONS

1. Why would right-wing Christians be "outraged" at the "tongue in cheek" suggestion made by Fausto-Sterling that instead of utilizing a conceptual system that categorizes people into two distinct groups (male and female), that a five category model might be more appropriate?
2. If a five group model of sex were adopted, how would this impact the utility of contemporary categorizations of sexual identity? What would "heterosexual" mean in such a system?
3. Who should make medical decisions about surgical alteration of the genitalia of intersexed newborns? When should this decision be made? What, if anything, is at stake? Is it wrong for doctors to make this decision for a family? Is it wrong for parents to make it for their child? At what age, if any, is a child adequately capable of making such a decision? Is creating social space for five or more sexes the answer? It this practical? Should practicality matter?
4. Should there be greater public knowledge about the frequency with which children are born with ambiguous genitalia? Why or why not? What, if any, interests might keeping this knowledge relatively quiet serve?
5. Imagine that your parents have just informed you that you were born an intersexed child and had undergone reconstructive surgery to create the "you" you are today. How would this

affect you? What aspects of your sense of self are most affected? How do you think your life would be different if you not been surgically altered—that is, how would it differ if you had been allowed to live your life as an intersexed individual?

The following Bonus Reading can be accessed online through the *InfoTrac College Edition Online Library* by typing in the name of the author or keywords in the title at *http://www.infotrac-college.com.*

Dennis Rodman—'Barbie Doll Gone Horribly Wrong': Marginalized Masculinity, Cross-Dressing, and the Limitations of Commodity Culture

MICHELE D. DUNBAR

In the late 1990s, National Basketball Association star Dennis Rodman garnered significant media attention as much for his bad boy countenance, extensive body art, homoerotic flirtations, and gender-defying wardrobe choices as any of his on-court athletic accomplishments. In this article, Michele D. Dunbar examines the impact of Rodman's transgressive behaviors and their subversive potential given the context of Rodman's race, sex, and immersion in commodity culture.

DISCUSSION QUESTIONS

1. Can gender violations among male stars and athletes help to challenge existing gender norms? If not, what will? If so, under what conditions are they most likely to have an effect? Do you agree that Rodman's violations "encourage audiences to resist rigid notions of gender and sexuality in their own lives"? Why or why not?
2. What kinds of factors affect whether a transgression will be considered a critique/rebellion against prevailing norms, and when will it simply be labeled as deviance?
3. Dunbar argues that Rodman's gender transgressions are limited in their ability to challenge heteronormativity and masculine gender norms. Do you agree? Why or why not? How could his activities have been more challenging? More effective? What kind of normative violations do you think would be most successful in convincing others to challenge or change their views?
4. Why are cross-dressing males more socially disturbing than cross-dressing females?
5. What kind of social privileges, if any, do individuals who cross-dress in public risk losing? What kind of social privileges, if any, do they gain? Do you think the benefits outweigh the costs? Why would anyone take such risks?

What's Love Got to Do with It? Constructions of Desire, Love, and Intimacy

The Feminization of Love

FRANCESCA M. CANCIAN

Women talk. Men do. These common stereotypes about men and women can be used to describe many facets of male and female relations. In this classic selection, Francesca M. Cancian maintains that in the realm of love, feminized and expressive styles have been privileged, while masculine and instrumental aspects of loving relationships have been neglected and repressed. Cancian discusses the hidden dangers of this cultural construction and advocates a movement toward a more androgynous model of love.

A feminized and incomplete perspective on love predominates in the United States. We identify love with emotional expression and talking about feelings, aspects of love that women prefer and in which women tend to be more skilled than men. At the same time we often ignore the instrumental and physical aspects of love that men prefer, such as providing help, sharing activities, and sex. This feminized perspective leads us to believe that women are much more capable of love than men and that the way to make relationships more loving is for men to become more like women. This paper proposes an alternative, androgynous perspective on love, one based on the premise that love is both instrumental and expressive. From this perspective, the

From "The Feminization of Love" by Francesca M. Cancian in *Signs* 11(4):692–709.

way to make relationships more loving is for women and men to reject polarized gender roles and integrate "masculine" and "feminine" styles of love.

EVIDENCE ON WOMEN'S 'SUPERIORITY' IN LOVE

A large number of studies show that women are more interested and more skilled in love than men. However, most of these studies use biased measures based on feminine styles of loving, such as verbal self-disclosure, emotional expression, and willingness to report that one has close relationships. When less biased measures are used, the differences between women and men are often small.

Women have a greater number of close relationships than men. At all stages of the life cycle, women see their relatives more often. Men and women report closer relations with their mothers than with their fathers and are generally closer to female kin. Thus an average Yale man in the 1970s talked about himself more with his mother than with his father and was more satisfied with his relationship with his mother. His most frequent grievance against his father was that his father gave too little of himself and was cold and uninvolved; his grievance against his mother was that she gave too much of herself and was alternately overprotective and punitive.

Throughout their lives, women are more likely to have a confidant—a person to whom one discloses personal experiences and feelings. Girls prefer to be with one friend or a small group, while boys usually play competitive games in large groups. Men usually get together with friends to play sports or do some other activity, while women get together explicitly to talk and to be together.

Men seem isolated given their weak ties with their families and friends. Among blue-collar couples interviewed in 1950, 64 percent of the husbands had no confidants other than their spouses, compared to 24 percent of the wives. The predominantly upper-middle-class men interviewed by Daniel Levinson in the 1970s were no less isolated. Levinson concludes that "close friendship with a man or a woman is rarely experienced by American men."[1] Apparently, most men have no loving relationships besides those with wife or lover; and given the estrangement that often occurs in marriages, many men may have no loving relationship at all.

Several psychologists have suggested that there is a natural reversal of these roles in middle age, as men become more concerned with relationships and women turn toward independence and achievement; but there seems to be no evidence showing that men's relationships become more numerous or more intimate after middle age, and some evidence to the contrary.

Women are also more skilled than men in talking about relationships. Whether working class or middle class, women value talking about feelings and relationships and disclose more than men about personal experiences. Men who deviate and talk a lot about their personal experiences are commonly defined as feminine and maladjusted. Working-class wives prefer to talk about themselves, their close relationships with family and friends, and their homes, while their husbands prefer to talk about cars, sports, work, and politics. The same gender-specific preferences are expressed by college students.

Men do talk more about one area of personal experience: their victories and achievements; but talking about success is associated with power, not intimacy. Women say more about their fears and disappointments, and it is disclosure of such weaknesses that usually is interpreted as a sign of intimacy. Women are also more accepting of the expression of intense feelings, including love, sadness, and fear, and they are more skilled in interpreting other people's emotions.

Finally, in their leisure time women are drawn to topics of love and human entanglements while men are drawn to competition among men. Women's preferences in television viewing run to daytime soap operas, or if they are more educated, the high-brow soap operas on educational channels, while most men like to watch competitive and often aggressive sports. Reading-tastes show the same pattern. Women read novels and magazine articles about love, while men's magazines feature stories about men's adventures and encounters with death.

However, this evidence on women's greater involvement and skill in love is not as strong as it appears. Part of the reason that men seem so much less loving than women is that their behavior is measured with a feminine ruler. Much of this research considers only the kinds of loving behavior that are associated with the feminine role and rarely compares women and men in terms of qualities associated with the masculine role. When less biased measures are used, the behavior of men and women is often quite similar. For example, in a careful study of kinship relations among young adults in a southern city, Bert Adams found that women were much more likely than men to say that their parents and relatives were very important to their lives (58 percent of women and 37 percent of men). In measures of actual contact with relatives, though, there were much smaller differences: 88 percent of women and 81 percent of men whose parents lived in the same city saw their parents weekly. Adams concluded that "differences between males and females in relations with parents are discernible primarily in the subjective sphere; contact frequencies are quite similar.[2]

The differences between the sexes can be small even when biased measures are used. For example, Marjorie Lowenthal and Clayton Haven reported the finding, later widely quoted, that elderly women were more likely than elderly men to have a friend with whom they could talk about their personal troubles—clearly a measure of traditionally feminine behavior. The figures revealed that 81 percent of the married women and 74 percent of the married men had confidants—not a sizable difference.[3] On the other hand, whatever the measure, virtually all such studies find that women are more involved in close relationships than men, even if the difference is small.

In sum, women are only moderately superior to men in love: they have more close relationships and care more about them, and they seem to be more skilled at love, especially those aspects of love that involve expressing feelings and being vulnerable. This does not mean that men are separate and unconcerned with close relationships, however. When national surveys ask people what is most important in their lives, women tend to put family bonds first while men put family bonds first or second, along with work. For both sexes, love is clearly very important.

EVIDENCE ON THE MASCULINE STYLE OF LOVE

Men tend to have a distinctive style of love that focuses on practical help, shared physical activities, spending time together, and sex. The major elements of the masculine style of love emerged in Margaret Reedy's study of 102 married couples in the late 1970s. She showed individuals statements describing aspects of love and asked them to rate how well the statements described their marriages. On the whole, husband and wife had similar views of their marriage, but several sex differences emerged. Practical help and spending time together were more important to men. The men were more likely to give high ratings to such statements as: "When she needs help I help her," and "She would rather spend her time with me than with anyone else." Men also described themselves more often as sexually attracted and endorsed such statements as: "I get physically excited and aroused just thinking about her." In addition, emotional security was less important to men than to women, and men were less likely to describe the relationship as secure, sage, and comforting.[4] Another study in the late 1970s showed a similar pattern among young, highly educated couples. The husbands gave greater emphasis to feeling responsible for the partner's well-being and putting the spouse's needs first, as well as to spending time together. The wives gave greater importance to emotional involvement and verbal self-disclosure but also were more concerned than the men about maintaining their separate activities and their independence.

The difference between men and women in their views of the significance of practical help was demonstrated in a study in which seven couples recorded their interactions for several days. They noted how pleasant their relations were and counted how often the spouse did a helpful chore, such as cooking a good meal or repairing a faucet, and how often the spouse expressed acceptance or affection. The social scientists doing the study used feminized definition of love. They labeled practical help as "instrumental behavior" and expressions of acceptance or affection as "affectionate behavior," thereby denying the affectionate aspect of practical help. The wives seemed to be using the same scheme; they thought their marital relations were pleasant that day if their husbands had directed a lot of affectionate behavior to them, regardless of their husbands' positive instrumental behavior. The husbands' enjoyment of their marital relations, on the other hand, depended on their wives' instrumental actions, not on their expressions of affection. The men actually saw instrumental actions as affection. One husband who was told by the researchers to increase his affectionate behavior toward his wife decided to wash her car and was surprised when neither his wife nor the researchers accepted that as an "affectionate" act.

The masculine view of instrumental help as loving behavior is clearly expressed by a husband discussing his wife's complaints about his lack of communication: "What does she want? Proof? She's got it, hasn't she? Would I be knocking myself out to get things for her—like to keep up this house—if I didn't love her? Why does a man do things like that if not because he loves his wife and kids? I swear, I can't figure what she wants. His wife, who has a feminine orientation to love, says something very different: "It is not enough that he supports us and takes care of us. I appreciate that, but I want him to share things with me. I need for him to tell me his feelings."[5] Many working-class women agree with men that a man's job is something he does out of love for his family,[6] but middle-class women and social scientists rarely recognize men's practical help as a form of love. (Indeed, among upper-middle-class men whose jobs offer a great deal of intrinsic gratification, their belief that they are "doing it for the family" may seem somewhat self-serving.)

Other differences between men's and women's styles of love involve sex. Men seem to separate sex and love while women connect them, but paradoxically, sexual intercourse seems to be the most meaningful way of giving and receiving love for many men. A twenty-nine-year old carpenter who had been married for three years said that, after sex, "I feel so close to her and the kids. We feel like a real family then. I don't talk to her very often, I guess, but somehow I feel we have really communicated after we have made love."[7]

Because sexual intimacy is the only recognized "masculine" way of expressing love, the recent trend toward viewing sex as a way for men and women to express mutual intimacy is an important challenge to the feminization of love. However, the connection between sexuality and love is undermined both by the "sexual revolution" definition of sex as a form of casual recreation and by the view of male sexuality as a weapon—as in rape—with which men dominate and punish women.

Another paradoxical feature of men's style of love is that men have a more romantic attitude toward their partners than do women. In Reedy's study, men were more likely to select statements like "we are perfect for each other." In a survey of college students, 65 percent of the men but only 24 percent of the women said that, even if a relationship had all of the other qualities they desired, they would not marry unless they were in love. The common view of this phenomenon focuses on women. The view is that women marry for money and status and so see marriage as instrumentally, rather than emotionally, desirable. This of course is at odds with women's greater concern with self-disclosure and emotional intimacy and lesser concern with instrumental help. A better way to explain men's greater romanticism might be to focus on men. One such possible explanation is that men do not feel responsible for "working on" the emotional aspects of a relationship, and therefore see love as magically and perfectly present or absent. This is consistent with men's relative lack of concern with affective interaction and greater concern with instrumental help.

In sum, there is a masculine style of love. Except for romanticism, men's style fits the popularly conceived masculine role of being the powerful provider. From the androgynous perspective, the practical help and physical activities included in this role are as much a part of love as the expression of feelings. The feminized perspective cannot account for this masculine style of love; nor can it explain why women and men are so close in the degrees to which they are loving.

NEGATIVE CONSEQUENCES OF THE FEMINIZATION OF LOVE

The division of gender roles in our society that contributes to the two separate styles of love is reinforced by the feminized perspective and leads to political and moral problems that would be mitigated with a more androgynous approach to love. The feminized perspective works against some of the key values and goals of feminists and humanists by contributing to the devaluation and exploitation of women.

It is especially striking how the differences between men's and women's styles of love reinforce men's power over women. Men's style involves giving women important resources, such as money and protection that men control and women believe they need, and ignoring the resources that women control and men need. Thus men's dependency on women remains covert and repressed, while women's dependency on men is overt and exaggerated; and it is overt dependency that creates power, according to social exchange theory. The feminized perspective on love reinforces this power differential by leading to the belief that women need love more than do men, which is implied in the association of love with the feminine role. The effect of this belief is to intensify the asymmetrical dependency of women on men. In fact, however, evidence on the high death rates of unmarried men suggests that men need love at least as much as do women.

Sexual relations also can reinforce male dominance insofar as the man takes the initiative and intercourse is defined either as his "taking" pleasure or as his being skilled at "giving" pleasure, either way giving him control. The man's power advantage is further strengthened if the couple assumes that the man's sexual needs can be filled by any attractive woman while the woman's sexual needs can be filled only by the man she loves.

On the other hand, women's preferred ways of loving seem incompatible with control. They involve admitting dependency and sharing or losing control, and being emotionally intense. Further, the intimate talk about personal troubles that appeals to women requires of a couple a mutual vulnerability, a willingness to see oneself as weak and in need of support. It is true that a woman, like a man, can gain some power by providing her partner with services, such as understanding, sex, or cooking; but this power is largely unrecognized because the man's dependency on such services is not overt. The couple may even see these services as her duty or as her response to his requests (or demands).

The identification of love with expressing feelings also contributes to the lack of recognition of women's power by obscuring the instrumental, active component of women's love just as it obscures the loving aspect of men's work. In a culture that glorifies instrumental achievement, this identification devalues both women and love. In reality, a major way by which women are loving is in the clearly instrumental activities associated with caring for others, such as preparing meals, washing clothes, and providing care during illness; but because of our focus on the expressive side of love, this caring work of women is either ignored or redefined as expressing feelings. Thus, from the feminized perspective on love, child care is a subtle communication of attitudes, not work. A wife washing her husband's shirt is seen as expressing love, even though a husband washing his wife's car is seen as doing a job.

Gilligan, in her critique of theories of human development, shows the way in which devaluing love is linked to devaluing women. Basic to most psychological theories of development is the idea that a healthy person develops from a dependent child to an autonomous, independent adult. As Gilligan comments, "Development itself comes to be identified with separation, and attachments appear to be developmental impediments."[8] Thus women, who emphasize attachment, are judged to be developmentally retarded or insufficiently individuated.

The pervasiveness of this image was documented in a well-known study of mental health professionals who were asked to describe mental health, femininity, and masculinity. They associated both mental health and masculinity with independence, rationality, and dominance. Qualities concerning attachment, such as being tactful, gentle, or aware of the feelings of others, they associated with femininity but not with mental health.[9]

Another negative consequence of a feminized perspective on love is that it legitimates impersonal, exploitive relations in the workplace and the community. The ideology of separate spheres that developed in the nineteenth century contrasted the harsh immoral marketplace with the warm and loving home and implied that this contrast is acceptable. Defining love as expressive, feminine, and divorced from productive activity maintains this ideology. If personal relationships and love are reserved for women and the home, then it is acceptable for a manager to underpay workers or for a community to ignore a needy family. Such behavior is not unloving; it is businesslike or shows a respect for privacy. The ideology of separate spheres also implies that men are properly judged by their instrumental and economic achievements and that poor or unsuccessful men are failures who may deserve a hard life. Levinson presents a conception of masculine development itself as centering on achieving an occupational dream.[10]

Finally, the feminization of love intensifies the conflicts over intimacy between women and men in close relationships. One of the most common conflicts is that the woman wants more closeness and verbal contact while the man withdraws and wants less pressure. Her need for more closeness is partly the result of the feminization of love, which encourages her to be more emotionally dependent on him. Because love is feminine, he in turn may feel controlled during intimate contact. Intimacy is her "turf," an area where she sets the rules and expectations. Talking about the relationship, as she wants, may well feel to him like taking a test that she made up and that he will fail. He is likely to react by withdrawing, causing her to intensify her efforts to get closer. The feminization of love thus can lead to a vicious cycle of conflict where neither partner feels in control or gets what she or he wants.

CONCLUSION

An androgynous perspective on love challenges the identification of women and love with being expressive, powerless, and nonproductive and the identification of men with being instrumental, powerful, and productive. It rejects the ideology of separate spheres and validates masculine as well as feminine styles of love. This viewpoint suggests that progress could be made by means of a variety of social changes, including men doing child care, relations at work becoming more personal and nurturant, and cultural conceptions of love and gender becoming more androgynous. Changes that equalize power within close relationships by equalizing the economic and emotional dependency between men and women may be especially important in moving toward androgynous love.

The validity of an androgynous definition of love cannot be "proven"; the view that informs the androgynous perspective in that both the feminine style of love (characterized by emotional closeness and verbal self-disclosure) and the masculine style of love (characterized by instrumental help and sex) represent necessary parts of a good love relationship. Who is more loving: a couple who confide most of their experiences to each other but rarely cooperate or given each other practical help, or a couple who help each other through many crises and cooperate in running a household but rarely discuss their personal experiences? Both relationships are limited. Most people would probably choose a combination: a relationship that integrates feminine and masculine styles of loving, an androgynous love.

NOTES

1. Daniel Levinson, *The Seasons of a Man's Life* (New York: Alfred A. Knopf, 1978), 335.
2. Bert Adams, *Kinship in an Urban Setting* (Chicago: Markham Publishing Co., 1968), 169.
3. Marjorie Lowenthal and Clayton Haven, "Interaction and Adaptation: Intimacy as a Critical Variable." *American Sociological Review* 22, no.4 (1968): 20–30.
4. Margaret Reedy, "Age and Sex Differences in Personal Needs and the Nature of Love." (Ph.D. diss. University of Southern California, 1977). Unlike most studies, Reedy did not find that women emphasized communication more than men. Her subjects were upper-middle-class couples who seemed to be very much in love.
5. Lillian Rubin, *Worlds of Pain* (New York: Basic Books, 1976), 147.
6. See L. Rubin, *Worlds of Pain*; also see Richard Sennett and Jonathan Cobb, *Hidden Injuries of Class* (New York: Vintage, 1973).
7. Interview by Cynthia Garlich, "Interviews of Married Couples" (University of California, Irvine, School of Social Sciences, 1982).
8. Carol Gilligan, *In a Different Voice* (Cambridge, MA: Harvard University Press,1982), 12–13.
9. Inge Broverman, Frank Clarkson, Paul Rosenkrantz, and Susan Vogel, "Sex-Role Stereotypes and Clinical Judgments of Mental Health," *Journal of Consulting Psychology* 34, no. 1 (1970): 1–7.
10. Levinson (n.1 above).

DISCUSSION QUESTIONS

1. What impact do you think women's labor-market participation has on the feminization of love?
2. Cancian's article was written in 1986. Has much changed in the gendered landscape of love during the decade and a half since this work was published? Do women today generally feel that their sexual needs can be filled "only by the man she loves"? Do men today feel that any attractive woman can fill their sexual needs?
3. Who would benefit most from the development of an androgynous style of love? Do you believe the emergence of this style of love is possible? Why or why not?
4. What impact, if any, do you think the increased participation of fathers in child rearing would have on the gendering of love? How would this development be altered in non-traditional families (e.g., gay, lesbian, and single-parent)?
5. What links, if any, do you see between capitalism and constructions/styles of love? Patriarchy?

The Sexualization of Love

STEVEN SEIDMAN

In the following passages from his book Romantic Longings, *Steven Seidman documents historical shifts in cultural definitions of love in Western society. His work indicates that the relationship between romantic love and the erotic aspects of sex have shifted over time and have resulted in the progressive sexualization of love in the modern era.*

Intimate norms and practices in twentieth-century America which have valued an expansive notion of sexual choice, diversity, and erotic pleasure are today under assault. The value placed upon sex as a domain of pleasure and self-expression is being held responsible for a variety of contemporary personal and social ills. Our expanded sexual freedom, it is argued, has eroded relational commitments; it has created unrealistic expectations for personal happiness; it has uncoupled sex from stable social bonds. The yield of this intimate culture, some critics contend, has not been sexual and intimate fulfillment but an anomic and narcissistic culture. AIDS, herpes, escalating rates of divorce, illegitimacy and teen pregnancies, loneliness, violence against women, and the impoverishment and abandonment of our children are claimed as its bitter fruit. This critique of current sexual trends often carries a nostalgia for an idealized nineteenth-century intimate culture. In this period, it is assumed, sex was firmly embedded in permanent social bonds and intimacy was not dependent on fleeting sexual pleasures but anchored in a deep spiritual kinship and a range of social responsibilities. Although this critique is especially prevalent among conservatives, it has surfaced among liberals and the left, including feminists.

I take issue with this critical perspective on contemporary American intimate culture. . . . The Victorian period cannot serve as a standard to judge contemporary intimate life. Although the myth of the "repressed Victorian" may have been put to rest, reinventing the Victorians as moderns who successfully integrated desire and affect is no less a simplification. Our contemporary intimate conventions are anchored primarily in twentieth-century developments. In the early decades of this century an intimate culture formed which framed sex as a medium of love. By sexualizing love, this culture encouraged a heightened attention to the body, sensual pleasure and sex technique. One consequence of the "sexualization of love" was that eroticism was conceived of as a source of romantic bonding. Mutual erotic fulfillment was intended to enhance intimate solidarity in a social context where other unifying forces (e.g., kinship, patriarchy, economic dependency) were losing their power to do so. A second consequence of the development of the erotic aspects of sex was that the pleasurable and expressive qualities of sex gradually acquired legitimacy apart from settings of romantic intimacy. In the post-World War II years, discourses appeared in mainstream culture that constructed sex as having multiple meanings (procreation, love, and pleasure) and diverse legitimate social contexts. Although this development has been bewailed by some critics for causing many current ills, it has, in my view, been crucial to the expansion of sexual choice and diversity. In short, there materialized in the

---

From *Romantic Longings: Love in America, 1830–1980* by Steven Seidman, p. 225–229.

twentieth-century United States an intimate culture that framed sex as a sphere of love and romantic bonding as well as a domain of self-expression and sensual happiness.

My aim, however, is not to celebrate the American intimate culture. Discourses and representations promoting the joys and social uses of eroticism have sometimes neglected the dangers of sex; they have not adequately examined the emotional and moral context of sex; they have, at times, overloaded sex with excessive expectations of self-fulfillment. Finally, the culture of eroticism has imbued sex with ambiguous, often conflicting meanings: sex is projected as a sphere of love and romance and, alternatively, as a medium of pleasure and self-expression. . . .

The meanings and purposes we invest in love are, in part, a product of diverse discourses, representations, traditions, and legal and moral customs. These cultural forces construct love as a domain about which we hold a range of beliefs and judgments. These meanings shape the way we imagine and experience love. They define what we expect or hope for from love and how it relates to self-fulfillment as well as the welfare and future of our society. These public representations change over time. . . .

THE SEXUALIZATION OF LOVE

The Eroticization of Sex

. . . Between the early part of the nineteenth century and the later part of the twentieth century, the meaning and place of sex in relation to love, and therefore the meaning of love, underwent important changes. Love changed from having an essentially spiritual meaning to being conceived in a way that made it inseparable from the erotic longings and pleasures of sex. By the early decades of the twentieth century, the desires and pleasures associated with the erotic aspects of sex were imagined as a chief motivation and sustaining source of love. The Victorian language of love as a spiritual communion was either marginalized or fused with the language of desire and joy. I chart a process of the progressive "sexualization of love" in which the erotic dimensions of sex assume an expanded role in proving and maintaining love.

Paralleling the sexualization of love is an equally momentous dynamic, namely the legitimation of the erotic aspects of sex. To the extent that erotic pleasure was viewed as a medium of love, the pursuit of sensual pleasure received public legitimation. I am suggesting, in other words, an account of the rise in the twentieth century of an intimate culture in which sex was valued for its sensually pleasurable and expressive qualities. It became legitimate to pursue sex for carnal pleasure to the extent that eroticism acquired a higher meaning as a symbol or vehicle of love.

The sexualization of love made possible the legitimation of the erotic aspects of sex. Ironically, while the eroticization of sex was initially accepted only in a context of love, by the post-World War II period the pleasurable and expressive qualities of eroticism acquired value apart from a context of love. This development introduced new possibilities and complexities into America's intimate culture. For example, if sex now carried multiple legitimate meanings, individuals had more opportunities to design their own sexual and intimate lifestyles. Yet expanded sexual choice and the pursuit of erotic pleasure raised fears about negatively impacting on the stability of intimate bonds; concerns were voiced regarding the morality of sexual objectification and the reduction of individuals to vessels of pleasure. Moreover, sex was now invested with conflicting meanings. Despite even the most enthusiastic efforts to deromanticize sex and construct it as merely a domain of pleasure and self-expression, it remains culturally entangled with emotional and moral resonances of love. As both an expression and medium of love and a domain of pleasure and self-expression, sex carried ambiguous meanings and behavioral directives. The possibilities of expanded choice and lifestyle diversity as well as the new dangers and strains presented by these developments define the ambiguous moral meaning of contemporary American intimate culture. . . .

. . . *The Victorian period, 1830–1890.* In white, middle-class, Victorian culture; a spiritual ideal of love figured prominently. Love referred to a spiritual affinity and spiritual companionship. The Victorians, however, were by no means prudes. They affirmed the power and beneficial qualities of sex. Sex was an expected, obligatory and healthy part of marriage. However, we can discern an antagonism between sex and love in Victorian culture. Love was considered an ideal basis and the essential state of marriage. Yet sex was thought of as an equally vital part of marriage. It is the nature of sex, so many Victorians thought, to incite, sensuality, which threatens to destroy the spiritual essence of marriage by engulfing it in a sea of lust. Accordingly, Victorians sought to control the place of sex in marriage. They did this by urging the desexualization of love and the desensualization of sex. At times, Victorians defended the desexualization of marriage itself, but this contradicted their belief in the omnipresence and beneficent power of the sex instinct. Curiously, the desexualization of love and the desensualization of sex made same-sex love more acceptable. Romantic friendships or love between women and, to a lesser extent, love between men was fairly typical and carried no trace of wrongdoing or shame. The antithesis today between a pure and ennobled heterosexuality and an impure and ignoble homosexuality was absent. Love between members of the same sex and between members of the opposite sex were often viewed as complementary not mutually exclusive.

The post-Victorian, 'modern' years from 1890 to 1960. The language of love now intermingles with that of sex. Sexual attraction is taken as a sign of love; the giving and receiving of sexual pleasures are viewed as demonstrations of love; sustained sexual longing and satisfaction is thought to be a condition for maintaining love. It is, moreover, the sensual side of sex that is valued. Sensuality is legitimated as a vehicle of love. Hence, the Victorian antithesis between love and sex and especially between love and sensuality disappears. Of course, love means more than sex. The Victorian ideal of spiritual companionship is transfigured into an idealized solidarity of lovers based upon personal, social and cultural companionship. True love is thought to combine sexual fulfillment and this idealized solidarity. The sexualization of love was paralleled by the exaltation of heterosexual love and the pollution of homosexual love. Under the impact of a scientific-medical discourse of homosexuality, the nineteenth-century paradigm of romantic friendship was replaced by that of homosexual or lesbian love. Heterosexuality and homosexuality were now described as mutually exclusive categories of desire, identity and love.

The contemporary period, 1960 to 1980. The sexualization of love . . . inadvertently led to legitimating the erotic aspects of sex. In the early decades of the twentieth century, sensual pleasures were imbued with a higher value and purpose. Erotic fulfillment sustained, enhanced and revitalized love. This implied a heightened focus and value placed upon erotic technique and fulfillment. Although the value placed on eroticism was legitimated initially only for the purpose of strengthening romantic bonds, by the 1960s the pleasurable and expressive qualities of sex were appealed to as a sufficient justification of sex. Discourses and representations appeared that constructed sex as a domain of pleasure and self-expression requiring no higher purpose so long as the interpersonal context was one of consent and mutuality. Eros was released from the culture of romance that gave birth to it. . . . In the post-World-War II years we can observe a third dramatic change in the meaning of same sex intimacy. The Victorian model of romantic friendship, we recall, gave way to the model of homosexual or lesbian love in the early decades of the twentieth century. In the contemporary period "gay love" emerges as a new, positive model of same sex intimacy. . . .

. . . I take issue with critics who relate current disturbances in American intimate culture to the expansion of sexual choice and diversity. Although a liberal intimate culture may occasion excesses or discontents associated with sexual objectification, vulgarity, emotional callousness, even sexual disease, these are acceptable costs for expanded choice and an intimate culture that values erotic pleasure and variety.[1]

NOTE

1. Obviously, I am not saying that AIDS is an acceptable cost for expanded sexual choice. My claim is that the widespread eradication of HIV requires "safe sex" practices, not necessarily the restriction of intimate lifestyle options. See Steven Seidman "The transfiguring sexual identity: AIDS and the construction of homosexuality," *Social Text* (Fall 1988), pp. 19–20.

DISCUSSION QUESTIONS

1. In an ideal world that you alone had the power to construct, what would be the relationship between love, sex, sensuality, and erotic pleasure? Would your world more closely resemble one of the historical periods described by Seidman (i.e., the Victorian period, the post-Victorian "modern" period, or the contemporary period), or would your world be constructed in a different way? Why do you think this is the optimal choice?
2. Is the move toward releasing Eros from the culture of romantic love equally beneficial for men and women? Why or why not?
3. What are the dangers of viewing individuals as vessels of pleasure? What are the advantages?
4. Seidman indicates that the sexualization of love has impacted the interpretation of same-sex relationships. Explain how and why this might occur.
5. Does the sexualization of love represent a more androgynous style of love? Masculinized? Explain.

Women and Heterosexual Love: Complicity, Resistance, and Change

STEVI JACKSON

In this essay, sociologist Stevi Jackson examines the social construction of love through the medium of romance fiction. Her analysis suggests that romantic love is a site of both resistance and collusion for women within patriarchal power structures. While some critics maintain that these novels are simply vehicles that brainwash women into patriarchal subservience, Jackson counters that these writings have considerable value as a resource for creating and affirming cultural constructions of romantic love.

Romantic love has been somewhat neglected by feminists,, despite the considerable attention that has recently been paid to its fictional representation. Research on women as readers and viewers of romance, however, does reveal that it has considerable emotional resonance for them. In order fully to appreciate both the appeal of such fiction and the place of

From "Women and Heterosexual Love: Complicity, Resistance and Change" in *Heterosexuality in Question* by Stevi Jackson, pp. 113–122. Sage Publications, 1990. Reprinted by permission of Sage Publications, Inc.

romantic love in women's daily lives, we need an analysis of love itself, the ways in which it is made sense of as an emotion and how it figures in women's understanding of their own and others' relationships. In particular, rather than treating romantic desires as given, we should consider the ways in which they are culturally constructed.

In this [article] I will suggest some lines of enquiry that might be pursued and indicate some of the theoretical and political questions which love and romance raise for feminists. My remarks are directed towards heterosexual love, since it is here that the political issues are brought into sharpest relief. It is in heterosexual relationships that romantic love has been institutionalized as the basis of marriage, and it is heterosexual love which dominates cultural representations of romance. Yet it is clear that contemporary ideals of romantic love, framed within the context of a heterosexual and patriarchal social and cultural order, also impinge on those who resist the constraints of compulsory heterosexuality.

There is nothing new in feminist critiques of love, which had their origin in the period of first wave feminism. For example, the Russian revolutionary, Alexandra Kollontai, was fiercely critical of the individualism, possessiveness and exclusivity of romantic love (1972). Later Simone deBeauvoir (1972) provided foundations for analyses of romantic love developed by early second wave feminists such as Comer (1974), Firestone (1972) and Greer (1970). These accounts were unambiguously critical of romantic love. It was the bait in the marriage trap; it served to justify our subordination to men and rendered us complicit in that subordination; it involved an unequal emotional exchange in which women gave more than they received; its exclusivity was taken as indicative of the emotional impoverishment of our lives; it diverted women's energies from more worthwhile pursuits. Where these writers considered romantic fiction, as in the case of Greer, it was represented simply as "dope for dupes"—a means of brainwashing women into subservience. The emphasis, then, was unequivocally on the dangers of love and romance for women.

Since that time feminists have developed new perspectives, which take women's pleasure in romantic fiction more seriously and which offer more sophisticated accounts of women's reading practices. This shift in focus from the dangers of romance to its pleasures, however, risks clouding our critical vision. Part of the problem, as I see it, is that love itself has moved out of the picture. The emotion which romantic fiction represents and which is so central to its readers' responses to it remains relatively unexplored. Subjecting love itself to analysis may serve to sharpen our critical faculties.

I want to state very firmly that retaining a critical perspective on love and romance need not be simplistic. You do not have to see romance readers as cultural dupes in order to argue that romance is implicated in maintaining a cultural definition of love which is detrimental to women. Nor need we resort to a moralistic sackcloth-and-ashes feminism which enjoins strict avoidance of cultural products and practices which are less than ideologically sound. It is not necessary to deny the pleasures of romance or the euphoria of falling in love in order to be sceptical about romantic ideals and wary of their consequences. It is possible to recognize that love is a site of women's complicity in patriarchal relations while still noting that it can also be a site of resistance.

LOVE'S CONTRADICTIONS

Romantic love hinges on the idea of "falling in love" and this "fall" as a means for establishing an intimate and deep relationship. Yet being "in love" is also seen as radically different from other forms of love—mysterious, inexplicable, irrational, uncontrollable, compelling and ecstatic. Even feminists often resort to mystical language to describe it. Haug et al., for example, see love as a means of retrieving "the buried and forgotten stirrings of the soul" (Haug et al. 1987: 278). It appears to be experienced as a dramatic, deeply felt inner transformation, as something that lifts us above the mundane everyday world—which is of course part of its appeal and has led some feminists to defend it against its critics (see, for example Baruch 1991; Person 1988). It is different in kind from lasting, longer term affection and widely recognized as more transient.

There are fundamental contradictions between passionate, romantic attraction and longer term affectionate love, yet the first is supposed to provide the basis for the second: a disruptive, tumultuous emotion is ideally supposed to be the foundation of a secure and durable relationship. Feminists from Kollontai (1972) to Firestone (1972)—as well as mainstream social theorists—have suggested that romantic love is not really about caring for another, but is self-centred and individualistic. There is a strong suggestion in literary, psychoanalytic and social scientific writings that the excitement of love thrives only when obstacles are put in its way. Again this makes it an unlikely basis for a committed relationship. So too does the oft noted tendency to romantic idealization—the other we pursue so compulsively is frequently the product of our own imagination (Baruch 1991; Wilson 1983). Hence the transformative power of love, its ability to turn frogs into princes. One of the most obvious appeals of romantic fiction is that it enables readers to relive the excitement of romantic passion without having to confront its fading and routinization. In real life we all too often discover that our prince was only a frog after all.

The passionate compulsiveness of love raises the issue of eroticized power and violence—a persistent theme both of pornography and romantic fiction. This is suggestive of an articulation between love and violence which is rarely explored, although the related linkages of sexuality with violence and love with sexuality have received considerable attention. Although the concept of love in some senses carries connotations antithetical to violence, in its passionate, romantic form it is not a gentle feeling. It is often characterized as violent, even ruthless (Bertilsson 1986). 'More than wanting to cosset the beloved we may feel we want to eat them alive' (Goodison 1983: 51–52). It can also be a pretext for violence which, if provoked by a jealous rage, can be read as proof of love—as can rape. Good reason, I think, to maintain our critical stance on the romantic construction of love, particularly since many of us are well aware of the painful experiences of women abused by those they had loved.

Although love relationships are often seen as egalitarian, the compulsiveness and insecurity of romantic passion imply a struggle for power. To be in love is to be powerless, at the mercy of the other, but it also holds out the promise of power, of enslaving the other. It thus offers women the hope of gaining power over a man: a common theme of romance narrative is the idea that women can tame the male beast by snaring him in the bonds of love (Modleski 1984; Radway 1987). Here the themes of complicity and resistance come into play—the desire for power over a man might be read as resistance. The power it delivers is, of course, illusory. It only lasts while the man is in the throes of romantic passion, after which the beast is likely to reassert himself (Langford 1992). He may continue to be dependent on a woman's nurturance and she may continue to gain a sense of power providing it—but the structural bases of power and inequality in heterosexual relationships remain untouched. What she is providing is emotional labour which, like domestic labour, may offer her a sense of self-worth while simultaneously being exploitative (Bartky 1990; Delphy and Leonard 1992).

LOVE'S DISCONTENTS

Once heterosexual love is routinized within a committed relationship, then, the asymmetry of gender may become all the more apparent. This again raises the question of women's resistance. Dissatisfaction with a lack of emotional reciprocity, with men's incapacity to give or display love, has emerged as a source of women's discontent in numerous studies of marriage and long-term heterosexual relationships since the 1960s (see, for example, Duncombe and Marsden 1993; Komarovsky 1962; Mansfield and Collard 1988; Rubin 1976, 1983). It has also been used to explain the attraction of romantic fiction for women (Radway 1987). This may be a way in which the ideal of companionate marriage based on romantic love sows the seeds of its own destruction—or at least the destruction of a specific relationship. Women appear to be more dissatisfied with the emotional than the material inequities of marriage and heterosexual relations (Duncombe and Marsden 1993; Mansfield and Collard 1988).

One potentially subversive aspect of romance fiction suggested by Radway (1987) is that it is a means by which women provide themselves with the nurturance lacking in their relationships with men. It also clear from ethnographic studies like Radway's that, when women talk about their reading and viewing preferences, this can be an occasion for discussing gender differences, highlighting men's distance from the feminine emotional world and voicing their criticisms of the men in their lives (see also Gray 1992). Rarely, however, does this lead to any explicit critique of heterosexual relationships. As Radway herself notes, the consumption of romantic fiction is an adaptation to discontent not a challenge to its source. It also sustains the ideal of romance which produced the discontent in the first place.

There is a further issue here. It is all too tempting to simply accept that men are emotional inadequates and thereby treat women's emotional desires and capacities as given, or even as a form of feminine superiority, particularly since women have for so long been undervalued because of our imputed emotionality. We should be very wary indeed of falling into such an essentialist stance for two reasons. First, what we are dealing with is not merely an imbalance of values, but a material, structural imbalance. Our nurturant capacities are closely interwoven with our location within patriarchal relations—we should be cautious of revalorizing what might be symptomatic of our subordination. More generally, we should not treat emotions as given. Hence, whether we are talking about nurturant caring love or passionate romantic love, we need an explanation of the ways in which these emotions are constructed at the level of our subjectivities.

Earlier feminist accounts recognized that women's romantic desires were not merely an expression of some innate feminine proclivity, but often underestimated how deeply rooted in our psyches these desires were. Romance was a confidence trick which, once seen through, could be avoided, but which continued to dupe and ensnare less enlightened women. More recently the cultural dupe notion has been challenged, particularly in relation to romantic fiction.

Readers of romance are of course perfectly aware that it is not a realistic representation of the social world—indeed, that is part of its attraction (Fowler 1991; Radway 1987). They know what they are reading and they know they cannot hope to achieve this fantasy in reality. It is also the case, as numerous sociological studies tell us, that romantic aspirations in choice of life-partners are tempered with realism. The point, however, is that romanticism and realism can coexist at different levels of our subjectivities. It is perfectly possible to be critical of heterosexual monogamy, dismissive of romantic fantasy and still fall passionately in love: a fact which many feminists can themselves testify to (Gill and Walker 1993; Jackson 1993a). This should not surprise us since it is now widely recognized that our subjectivities are not coherent and consistent.[1] It is the awareness of such contradictions which has inspired much feminist writing on love and romance. Gradually feminists have broken the silence which surrounded our continued experience of "unsound" desires, have been willing to "come out" as secret fans of romance (Kaplan 1986; Modleski 1991; Taylor 1989b). Romantic ideals can be deeply embedded in our subjectivities even when we are critical of them.

LOVE, ROMANCE, AND SUBJECTIVITY

Here I find myself confronting what seems to me a major gap in feminist theory—the lack of a convincing theory of subjectivity. It has become almost conventional to introduce psychoanalytic explanations at this point. Various versions of psychoanalysis have indeed been used to explain the attractions of romance reading for women. Radway's use of Chodorow's (1978) framework may provide a coherent explanation of why women wish to be nurtured and why men are incapable of providing that nurturance, but it doesn't explain why women are so attracted to tales of passionate, even violent, desire.[2] Lacanian accounts certainly tackle desire in a way which is congruent with some of the features of romantic love which I have identified. Desire is constituted through lack, an inevitable product of our entry into language and culture and is intrinsically incapable of satisfaction (Mitchell and Rose 1982). Since this is conceptualized in terms of entry into language and culture per se, not of entering a specific culture, the implication is that "desire" is an essential part of human social nature. Lacanian psychoanalysis does not admit of the possibility of emotions being structured differently in different cultural settings and thus imagines the whole world to be beset by the same desire—an assumption that anthropologists would make us wary of (Errington and Gewert 1987; Lutz and Abu-Lughod 1990; Rosaldo 1984).

I am not convinced, either, that the Lacanian account can deal with the specifics of the ways in which language structures emotional and sexual experience even within Western culture. Emotions are not simply "felt" as internal states provoked by the unconscious sense of lost infantile satisfactions—they are actively structured and understood through culturally specific discourses. These discourses differentiate between love as nurture, being "in love", lust and sexual arousal—all of which are conflated in the psychoanalytic concept of desire. Even if we were to accept that desires are shaped at an unconscious level, that this is what surfaces in our romantic imaginings, this cannot account for the specific content of our desires and fantasies. Fantasies do not emerge fully formed into our consciousness. They are actively constructed by us, in narrative form, drawing on the cultural resources to hand.

Lacanian psychoanalysis, while ostensibly an account of the cultural construction of emotion, locates "desire" as an inner state and thus precludes the possibility of linking the experience of "love" to specific cultural contexts and to the specific discourses and narratives which give shape to our emotions. Feminist accounts of the pleasures of romance reading within this type of psychoanalytic framework, for example Alison Light (1984) on *Rebecca* and Cora Kaplan (1986) on *The Thorn Birds,* seem to me to suggest that romantic fiction reflects, gives voice to or is constructed around a set of emotions which already exist. I would argue, on the contrary, that romantic narrative itself contributes to the cultural construction of love. I do not maintain, as some early critics of romance did, that it is simply a means of brainwashing women into subservience. Rather, I am suggesting that this is but one of the resources from which we create a sense of what our emotions are.

What I would suggest is that we explore further the possibility that our subjectivities—including our emotions—are shaped by the social and cultural milieu we inhabit through processes which involve our active participation. We create for ourselves a sense of what our emotions are, of what being in love is, through positioning ourselves within discourses, constructing, narratives of self, drawing on whatever cultural resources are available to us. This perspective allows us to recognize the constraints of the culture we inhabit while allowing for human agency and therefore avoiding the "cultural dupe" syndrome, of admitting the possibility of both complicity in and resistance to patriarchal relations in the sphere of love.

CONCLUSION: RESISTANCE, COMPLICITY, AND CHANGE

If, as I have suggested, emotions are culturally constructed, they are not fixed for all time. Recent accounts of love suggest that it has indeed changed its meaning over time and that this has come about in part because personal life has been the object of political, especially feminist struggle (Baruch 1991; Cancian 1990; Giddens 1992; Seidman 1991). Where these writings comment on current trends and begin to predict future changes, however, they frequently overestimate the changes which are occurring.

A common strand running through these analyses is the claim that romantic love is being undermined as a result of changing sexual mores and women's demands for more equal relationships. For Baruch (1991) romantic love might meet its end once the denial it feeds upon gives way to too easy gratification of sexual desire, but may yet be revived by the anti-permissive climate consequent upon the spread of AIDS. While Seidman (1991, 1992) espouses a more libertarian and less romantic ethic than Baruch, he shares her view that libertarianism and romanticism are antithetical to each other, and that we are now witnessing a struggle between these opposing social currents. He argues that the progressive sexualization of love during the twentieth century created the preconditions for its demise by valorizing sexual pleasure in its own right and therefore breaking the linkage between love and sexuality. Giddens (1992) sees these same trends as leading away from the romantic quest for the "only one" with whom to share one's life towards the ideal of the "pure relationship" more contingent than lifelong monogamy, lasting only as long as it is mutually satisfying. Women are leading this trend because they are refusing to continue to service men's emotional needs at the expense of their own. Similarly, Cancian (1990) detects a move away from "feminized" love, to a more androgynous form, where men take more responsibility for the emotional well-being of their partners.

A less restrictive sexual morality does not, in itself, indicate that romantic love is losing its emotional salience, although it may well mean that love is less often regarded as a precondition for physical intimacy. Romanticism and libertarianism are not as mutually exclusive as Baruch and Seidman imply. It is not only moral strictures which place barriers in the way of the gratification of our desires, and romantic love is not in any case reducible to sexual desire. A libertarian ethic may be antithetical to a prescriptive form of romanticism which enjoins lifelong monogamy on lovers, but need not preclude falling in love. Young women's increased heterosexual activity is not necessarily evidence of an absence of romantic desires, although it may indicate a higher degree of realism about the durability of relationships founded upon them. Higher divorce rates, adultery and serial monogamy may indicate a continued search for romantic fulfillment rather than the abandonment of that quest. It may be the case that women are expecting more out of heterosexual relationships and are less likely to remain in them if these expectations are not realized. This does not mean, however, that in their search for the "pure relationship" they regard their love for their partner as contingent and conditional at the outset, or that they have ceased to entertain romantic hopes. Given the lack of evidence that women's demands are currently being met, claims that a more egalitarian form of love is emerging seem absurdly over optimistic and wilfully neglectful of the continued patriarchal structuring of heterosexuality.

It is erroneous to assume too close a correspondence between changes in patterns of sexual relationships and transformations of romantic desire. What may be happening is that the contradictions of romantic love are becoming more apparent with the partial erosion of its institutional supports. Now that premarital chastity and lifelong monogamy are no longer expected of women, it becomes obvious that romantic love does not guarantee lasting conjugal happiness—but then it never has. This may lead us to modify our expectations of intimate relationships, may render them less durable, but it does not yet herald the demise of romantic desires.

Certainly the purveyors of romantic fiction are not suffering a contraction of their markets. Rather, they are adapting their plots to suit shifts in sexual mores—but their more assertive, less virginal heroines are still seeking Mr. Right. There are, moreover, new markets being created, notably through book series for young readers. If, as I have suggested, the attraction of such romances both requires and helps constitute particular emotional responses, reports of the death of romantic love are certainly exaggerated.

NOTES

1. This insight is usually attributed to psychoanalytic and poststructuralist perspectives, but I would argue that most feminists have—at least implicitly—long recognized that this is the case (see Jackson 1992b).
2. *Editorial note:* Chodorow's psychoanalytic account explains gentler difference in terms of the differential experience of mothering undergone by boys and girls. Because of the strong bonds of identification between mothers and daughters, girls do not develop a strong sense of autonomous selfhood, but rather define themselves in relation to others. This creates both a capacity to nurture and a need for nurturance. In order to become masculine a boy must distance himself from his mother, and from all that is feminine, and thus develops a sense of himself as separate and apart and in the process denies the possibility of emotional connection to another.

BIBLIOGRAPHY

Bartky, S. (1990) *Femininity and Domination.* New York: Routledge.

Baruch, E. H. (1991) *Women, Love and Power,* New York: New York University Press.

Bertilsson, M. (1986) "Love's labour lost? A sociological view," *Theory, Culture & Society,* 3 (1): 19–35.

Cancian, F. (1990) *Love in America.* Cambridge: Cambridge University Press.

Chodorow, N. (1978) *The Reproduction of Mothering.* Berkeley: University of California Press.

Comer, L. (1974) *Wedlocked Women.* Leeds: Feminist Books.

de Beauvoir, S. (1972) *The Second Sex.* Harmondsworth: Penguin.

Delphy, C. and Leonard, D. (1992) *Familiar Exploitation: A New Analysis of Marriage in Contemporary Western Societies.* Oxford: Polity.

Duncombe, J. and Marsden, D. (1993) "Love and intimacy: the gender division of emotion and 'emotion work,'" *Sociology,* 27 (2): 221–241.

Errington, F. And Gewertz, D. (1987) *Cultural Alternatives and a Feminist Anthropology.* Cambridge: Cambridge University Press.

Firestone, S. (1972) *The Dialectic of Sex.* London: Paladin.

Fowler, B. (1991) *The Alienated Reader: Women and Popular Romantic Literature in the Twentieth Century.* Hemel Hempstead: Harvester Wheatsheaf.

Giddens, A. (1992) *The Transformation of Intimacy.* Oxford: Polity.

Gill, R. and Walker, R. (1993) "Heterosexuality, feminism, contradiction: on being young, white heterosexual feminists in the 1990s," in S. Wilkinson and C. Kitzinger (eds), *Heterosexuality.* London: Sage, pp. 68–72.

Goodison, L. (1983) "Really being in love means wanting to live in a different world," in S. Cartledge and J. Ryan (eds), *Sex and Love: New Thoughts on Old Contradictions.* London: Women's Press, pp. 48–66.

Gray, A. (1992) *Video Playtime: The Gendering of a Leisure Technology.* London: Routledge.

Greer, G. (1970) *The Female Eunuch.* London: Paladin.

Griffin, C. (1982) "Cultures of femininity: romance revisited," Centre for Contemporary Cultural Studies Occasional Paper, University of Birmingham.

Haug, F. et al. (1987) *Female Sexualization.* London: Verso.

Jackson, S. (1992b) "The amazing deconstructing woman: the perils of postmodern feminism," *Trouble & Strife,* 25: 25–31.

Jackson, S. (1993a) "Even sociologists fall in love: an exploration in the sociology of emotions," in *Sociology,* 27 (2): 201–220.

Kaplan, C. (1986) *Sea Changes.* London: Verso.

Kollontai, A. (1972) *Sexual Relations and the Class Struggle.* Bristol: Falling Wall Press. (Orig. pub. 1919.)

Komarovsky, M. (1962) *Blue Collar Marriage.* New York: W.W Norton.

Lacan, J. (1977) *Écrits.* London: Tavistock.

Langford, W. (1992) "Gender, power and self-esteem: women's poverty in the economy of love," unpublished paper presented to the Women's Studies Network (UK) Conference, University of Central Lancashire.

Light, A. (1984) "'Returning to Manderley'—romance fiction, female sexuality and class," *Feminist Review*, 16: 7–25.

Lutz, C. (1990) "Engendered emotion: gender, power and the rhetoric of emotional control in American discourse," in C. Lutz and L. Abu-Lughod (eds), *Language and the Politics of Emotion*. Cambridge: Cambridge University Press, pp. 69–91.

Mansfield, P. and Collard, J. (1988) *The Beginning of the Rest of Your Life*. London: Macmillan.

Mitchell, J. and Rose, J. (eds) (1982) *Feminine Sexuality: Jacques Lacan and the École Freudienne*. London: Macmillan.

Modleski, T. (1984) *Loving with a Vengeance*. London: Methuen.

Modleski, T. (1991) *Feminism Without Women*. New York: Routledge.

Person, E. S. (1988) *Love and Fateful Encounters: The Power of Romantic Passion*. New York: W.W Norton.

Radway, J. (1987) *Reading the Romance*. London: Verso.

Rosaldo, M. (1984) "Towards an anthropology of self and feeling," in R. A. Shweder and R. A. Levine (eds), *Culture Theory*. Cambridge: Cambridge University Press, pp. 137–157.

Rubin, L. (1976) *Worlds of Pain*. New York: Basic Books.

Rubin, L. (1983) *Intimate Strangers*. New York: Harper and Row.

Seidman, S. (1991) *Romantic Longings: Love in America 1830–1980*. New York: Routledge.

Seidman, S. (1992) *Embattled Eros: Sexual Politics and Ethics in Contemporary America*. New York: Routledge.

Taylor, H. (1989b) *Scarlett's Women: Gone with the Wind and Its Female Fans*. London: Virago.

Wilson, E. (1983) "A new romanticism?," in E. Phillips (ed.), *The Left and the Erotic*, London: Lawrence and Wishart, pp. 37–52.

DISCUSSION QUESTIONS

1. Aside from romance novels, what other cultural resources are used to construct love and desire? Are these resources gendered? Raced? Classed? Etc.?
2. Is there a relationship between changes in patterns of sexual relationships and transformations of desire? Do you think that an increase in sexual activity for the sake of physical pleasure *necessarily* indicates a decrease in the role of romantic love? Why or why not?
3. What could be added to Jackson's account by including consideration of gay, lesbian, bisexual, and transgendered experiences?
4. What is the relationship of heterosexual men to love? In what ways do they comply with, resist, and attempt to change normative expectations of love? Do women have more power in this realm, or, as Jackson and others suggest, is this perceived power actually reflective of their subordination?
5. Discuss the ways in which love has been a locus of both feminine complicity in, and resistance to, male domination.

5

Constructing the Sexual Self: The Negotiation and Actualization of Sexual Identity and Behavior

Becoming 100 Percent Straight

MICHAEL A. MESSNER

Social identities are both individually forged and externally imposed. Fluid and ever-changing, they are constructed, contested, assumed, and resisted. Yet, much of the work involved in shaping and maintaining our sense of who we are may take place at, or just beneath, the surface of conscious thought and action. In this pointedly subjective analysis Michael A. Messner works to uncover much of the hidden and forgotten labor of becoming "100 percent" straight.

In 1995, as part of my job as the President of the North American Society for the Sociology of Sport, I needed to prepare an hour-long presidential address for the annual meeting of some 200 people. This presented a challenge to me: how might I say something to my colleagues that was interesting, at least somewhat original, and above all, not boring. Students may think that their professors are especially dull in the classroom but, believe me, we are usually much worse at professional meetings. For some reason, many of us who are able to speak to our classroom students in a relaxed manner, using relatively jargon-free language, seem to become robots, dryly reading our papers—packed with impressively unclear jargon—to our yawning colleagues.

Since I desperately wanted to avoid putting 200 sport studies scholars to sleep, I decided to

From "Becoming 100 Percent Straight" by Michael A. Messner in *Inside Sports,* ed. Jay Coakley and Peter Donnelly, pp. 104–110. Reprinted by permission of Routledge/Taylor & Francis Ltd.

deliver a talk which I entitled "Studying up on sex." The title, which certainly did get my colleagues' attention, was intended as a play on words, a double entendre. "Studying up" has one generally recognizable colloquial meaning, but in sociology it has another. It refers to studying "up" in the power structure. Sociologists have perhaps most often studied "down"—studying the poor, the blue- or pink-collar workers, the "nuts, sluts and perverts," the incarcerated. The idea of "studying up" rarely occurs to sociologists unless and until we live in a time when those who are "down" have organized movements that challenge the institutional privileges of elites.

Much of my research, inspired by feminism, has involved a studying up on the social construction of masculinity in sport. Studying up, in these cases, has raised some fascinating new and important questions about the workings of power in society.

In sport, just as in the larger society, we seem obsessed with asking "how do people become gay?" Imbedded in this question is the assumption that people who identify as heterosexual, or "straight," require no explanation, since they are simply acting out the "natural" or "normal" sexual orientation. We seem to be saying that the "sexual deviants" require explanation, while the experience of heterosexuals, because we are considered normal, seems to require no critical examination or discussion. I decided to challenge myself and my colleagues by arguing that although we have begun to "study up" on corporate elites in sport, on whiteness, on masculinity, it is now time to extend that by studying up on heterosexuality.

But in the absence of systematic research on this topic, where could I start? How could I explore, raise questions about, and begin to illuminate the social construction of heterosexuality for my colleagues? Fortunately, for the previous two years I had been working with a group of five men (three of whom identified as heterosexual, two as gay) mutually to explore our own biographies in terms of the earlier bodily experiences that helped to shape our gender and sexual identities. We modeled our project after that of a German group of feminist women, led by Frigga Haug, who created a research method that they call "memory work." In short, the women would mutually choose a body part, such as "hair," and each would then write a short story based on a particularly salient childhood memory that related to their hair (for example, being forced by parents to cut one's hair, deciding to straighten one's curly hair in order to look more like other girls, etc.). Then the group would read all of the stories and discuss them one by one in the hope of gaining a more general understanding of, and raising new questions about, the social construction of "femininity."

A clear understanding of the subjective aspect of social life—one's bodily feelings, emotions, and reactions to others—is an invaluable window that allows us to see and ask new sociological questions about group interaction and social structure. In short, group memory work can provide an important, productive, and fascinating insight on social reality, though not a complete (or completely reliable) picture.

As I pondered the lack of existing research on the social construction of heterosexuality in sport, I decided to draw on one of my own stories from my memory work in the men's group.

Many years ago I read some psychological studies that argued that even for self-identified heterosexuals it is a natural part of their development to have gone through "bisexual" or even "homosexual" stages of life. When I read this, it seemed theoretically reasonable, but did not ring true in my experience. I have always been, I told myself, 100 percent heterosexual! The group process of analyzing my own autobiographical stories challenged the concept I had developed of myself, and also shed light on the way in which the institutional context of sport provided a context for the development of my definition of myself as "100 percent straight." Here is one of the stories.

> When I was in the ninth grade, I played on a "D" basketball team, set up especially for the smallest of high school boys. Indeed, though I was pudgy with baby fat, I was a short 5′2″, still pre-pubescent with no facial hair and a high voice that I artificially tried to lower. The first day of practice, I was immediately attracted to a boy I'll call Timmy, because he looked like the boy who played in the *Lassie* TV show. Timmy was short, with a high voice, like me. And like me, he had no facial hair yet. Unlike me, he was very skinny. I liked Timmy right away, and soon we were together a lot. I noticed things about him that I didn't notice about other boys: he said some words a certain way, and it gave me pleasure to try to talk like him.
>
> I remember liking the way the light hit his boyish, nearly hairless body. I thought about him when we weren't together. He was in the school band, and at the football games, I'd squint to see where he was in the mass of uniforms. In short, though I wasn't conscious of it at the time, I was infatuated with Timmy—I had a crush on him. Later that basketball season, I decided—for no reason that I could really articulate then—that I hated Timmy. I aggressively rejected him, began to make fun of him around other boys. He was, we all agreed, a geek. He was a faggot.
>
> Three years later, Timmy and I were both on the varsity basketball team, but had hardly spoken a word to each other since we were freshman. Both of us now had lower voices, had grown to around 6 feet tall, and we both shaved, at least a bit. But Timmy was a skinny, somewhat stigmatized reserve on the team, while I was the team captain and starting point guard. But I wasn't so happy or secure about this. I'd always dreamed of dominating games, of being the hero. Halfway through my senior season, however, it became clear that I was not a star, and I figured I knew why. I was not aggressive enough.
>
> I had always liked the beauty of the fast break, the perfectly executed pick and roll play between two players, and especially the long 20-foot shot that touched nothing but the bottom of the net. But I hated and feared the sometimes brutal contact under the basket. In fact, I stayed away from the rough fights for rebounds and was mostly a perimeter player, relying on my long shots or my passes to more aggressive teammates under the basket. But now it became apparent to me that time was running out in my quest for greatness: I needed to change my game, and fast. I decided one day before practice that I was gonna get aggressive. While practicing one of our standard plays, I passed the ball to a teammate, and then ran to the spot at which I was to set a pick on a defender. I knew that one could sometimes get away with setting a face-up screen on a player, and then as he makes contact with you, roll your back to him and plant your elbow hard in his stomach. The beauty of this move is that your own body "roll" makes the elbow look like an accident. So I decided to try this move. I approached the defensive player, Timmy, rolled, and planted my elbow deeply into his solar plexus. Air exploded audibly from Timmy's mouth, and he crumbled to the floor momentarily.
>
> Play went on as though nothing had happened, but I felt bad about it. Rather than making me feel better, it made me feel guilty and weak. I had to admit to myself why I'd chosen Timmy as the target against whom to test out my new aggression. He was the skinniest and weakest player on the team.

At the time, I hardly thought about these incidents, other than to try to brush them off as incidents that made me feel extremely uncomfortable. Years later, I can now interrogate this as a sexual story, and as a gender story unfolding within, the context of the heterosexualized and masculinized institution of sport. Examining my story in light of research conducted by Alfred Kinsey a half-century ago, I can recognize in myself what Kinsey saw as a very common fluidity and changeability of sexual desire over the lifecourse. Put simply, Kinsey found that large numbers of adult, "heterosexual" men had previously, as adolescents and young adults, experienced sexual desire for males. A surprisingly large number of these men had experienced sexual contact to the point of orgasm with other males during adolescence or early adulthood. Similarly, my story invited me to consider what is commonly called the "Freudian theory of bisexuality." Sigmund Freud shocked the post-Victorian world by suggesting that all people go through a stage, early in life, when they are attracted to people of the same sex.[1] Adult experiences, Freud argued, eventually led most people to shift their sexual desire to what he called an appropriate "love object"—a person of the opposite sex. I also considered my experience in light of what lesbian feminist author Adrienne Rich called the institution of compulsory heterosexuality. Perhaps the extremely high levels of homophobia that are often endemic in boys' and men's organized sports led me to deny and repress my own homoerotic desire through a direct and overt rejection of Timmy, through homophobic banter with male peers, and the resultant stigmatization of the feminized Timmy. Eventually I considered my experience in the light of what radical theorist Herbert Marcuse called the sublimation of homoerotic desire into an aggressive, violent act as serving to construct a clear line of demarcation between self and other. Sublimation, according to Marcuse, involved the driving underground, into the unconscious, of sexual desires that might appear dangerous due to their socially stigmatized status. But sublimation involves more than simple repression into the unconscious. It involves a transformation of sexual desire into something else—often into aggressive and violent acting out toward others. These acts clarify the boundaries between oneself and others and therefore lessen any anxieties that might be attached to the repressed homoerotic desire.

Importantly, in our analysis of my story, the memory group went beyond simply discussing the events in psychological terms. The story did perhaps suggest some deep psychological processes at work, but it also revealed the importance of social context—in this case, the context of the athletic team. In short, my rejection of Timmy and the joining with teammates to stigmatize him in ninth grade stands as an example of what sociologist R. W. Connell calls a moment of engagement with hegemonic masculinity, where I actively took up the male group's task of constructing heterosexual/masculine identities in the context of sport. The elbow in Timmy's gut three years later can be seen as a punctuation mark that occurred precisely because of my fears that I might be failing in this goal.

It is helpful, I think, to compare my story with gay and lesbian "coming out" stories in sport. Though we have a few lesbian and bisexual coming out stories among women athletes, there are very few from gay males. Tom Waddell, who as a closeted gay man finished sixth in the decathlon in the 1968 Olympics, later came out and started the Gay Games, an athletic and cultural festival that draws tens of thousands of people every four years. When I interviewed Tom Waddell over a decade ago about his sexual identity and athletic career, he made it quite clear that for many years sports was his closet:

> When I was a kid, I was tall for my age, and was very thin and very strong. And I was usually faster than most other people. But I discovered rather early that I liked gymnastics and I liked dance. I was very interested in being a ballet dancer . . . [but] something became obvious to me right away—that male ballet dancers were effeminate, that they were what most people would call faggots. And I thought I just couldn't handle that . . . I was totally closeted and very concerned about being male. This was the fifties, a terrible time to live, and everything was stacked against me. Anyway, I realized that I had to do something to protect my image of myself as a male—because at that time homosexuals were thought of primarily as men who wanted to be women. And so I threw myself into athletics—I played football, gymnastics, track

> and field . . . I was a jock—that's how I was viewed, and I was comfortable with that.

Tom Waddell was fully conscious of entering sports and constructing a masculine/heterosexual athletic identity precisely because he feared being revealed as gay. It was clear to him, in the context of the 1950s, that being known as gay would undercut his claims to the status of manhood. Thus, though he described the athletic closet as "hot and stifling," he remained there until several years after his athletic retirement. He even knowingly played along with locker room discussions about sex and women as part of his "cover."

> I wanted to be viewed as male, otherwise I would be a dancer today. I wanted the male, macho image of an athlete. So I was protected by a very hard shell. I was clearly aware of what I was doing . . . I often felt compelled to go along with a lot of locker room garbage because I wanted that image—and I know a lot of others who did too.

Like my story, Waddell's points to the importance of the athletic institution as a context in which peers mutually construct and reconstruct narrow definitions of masculinity. Heterosexuality is considered to be a rock-solid foundation of this concept of masculinity. But unlike my story, Waddell's may invoke a dramaturgical analysis.[2] He seemed to be consciously "acting" to control and regulate others' perceptions of him by constructing a public "front stage" persona that differed radically from what he believed to be his "true" inner self. My story, in contrast, suggests a deeper, less consciously strategic repression of my homoerotic attraction. Most likely, I was aware on some level of the dangers of such feelings, and was escaping the risks, disgrace, and rejection that would likely result from being different. For Waddell, the decision to construct his identity largely within sport was to step into a fiercely heterosexual/masculine closet that would hide what he saw as his "true" identity. In contrast, I was not so much stepping into a "closet" that would hide my identity; rather, I was stepping out into an entire world of heterosexual privilege. My story also suggests how a threat to the promised privileges of hegemonic masculinity—my failure as an athlete—might trigger a momentary sexual panic that can lay bare the constructedness, indeed, the instability of the heterosexual/masculine identity.

In either case, Waddell's or mine, we can see how, as young male athletes, heterosexuality and masculinity was not something we "were," but something we were doing. It is significant, I think, that although each of us was "doing heterosexuality," neither of us was actually "having sex" with women (though one of us desperately wanted to). This underscores a point made by some recent theorists that heterosexuality should not be thought of simply as sexual acts between women and men. Rather, heterosexuality is a constructed identity, a performance, and an institution that is not necessarily linked to sexual acts. Though for one of us it was more conscious than for the other, we were both "doing heterosexuality as an ongoing practice through which we sought to do two things:

- avoid stigma, embarrassment, ostracism, or perhaps worse if we were even suspected of being gay;
- link ourselves into systems of power, status, and privilege that appear to be the birthright of "real men" (i.e., males who are able to compete successfully with other males in sport, work, and sexual relations with women).

In other words, each of us actively scripted our own sexual and gender performances, but these scripts were constructed within the constraints of a socially organized (institutionalized) system of power and pleasure.

QUESTIONS FOR FUTURE RESEARCH

As I prepared to tell this sexual story publicly to my colleagues at the sport studies conference, I felt extremely nervous. Part of the nervousness was due to the fact that I knew some of them would object to my claim that telling personal stories can be a source of sociological insights. But a larger part of the reason for my nervousness was due to the fact that I was revealing something very personal about my sexuality in such a public way. Most of us are not accustomed to doing this, especially in the context of a professional conference. But I had learned long ago, especially from feminist women scholars, and from gay and lesbian scholars, that biography is linked to history. Part of "normal" academic discourse has been to hide "the personal" (including the fact that the researchers are themselves people with values, feelings, and yes, biases) behind a carefully constructed facade of "objectivity." Rather than trying to hide or be ashamed of one's subjective experience of the world, I was challenging myself to draw on my experience of the world as a resource. Not that I should trust my experience as the final word on "reality." White, heterosexual males like me have made the mistake for centuries of calling their own experience "objectivity," and then punishing anyone who does not share their worldview by casting them as "deviant." Instead, I hope to use my experience as an example of how those of us who are in dominant sexual/racial/gender/class categories can get a new perspective on the "constructedness" of our identities by juxtaposing our subjective experiences against the recently emerging worldviews of gay men and lesbians, women, and people of color.

Finally, I want to stress that in juxtaposition neither my own nor Tom Waddell's story sheds much light on the question of why some individuals "become gay" while others "become" heterosexual or bisexual. Instead, I should like to suggest that this is a dead-end question, and that there are far more important and interesting questions to be asked:

- How has heterosexuality, as an institution and as an enforced group practice, constrained and limited all of us—gay, straight, and bi?
- How has the institution of sport been an especially salient institution for the social construction of heterosexual masculinity?
- Why is it that when men play sports they are almost always automatically granted masculine status, and thus assumed to be heterosexual, while when women play sports, questions are raised about their "femininity" and sexual orientation?

These kinds of questions aim us toward an analysis of the working of power within institutions—including the ways that these workings of power shape and constrain our identities and relationships—and point us toward imagining alternative social arrangements that are less constraining for everyone.

NOTES

1. The fluidity and changeability of sexual desire over the life course is now more obvious in evidence from prison and military populations, and single-sex boarding schools. The theory of bisexuality is evident, for example, in childhood crushes on same-sex primary schoolteachers.
2. Dramaturgical analysis, associated with Erving Goffman, uses the theater and performance to develop an analogy with everyday life.

REFERENCES

Haug, Frigga (1987) *Female Sexualization: A Collective Work of Memory,* London: Verso.

Lenskyj, Helen (1986) *Out of Bounds: Women, Sport and Sexuality,* Toronto: Women's Press.

——— (1997) "No fear? Lesbians in sport and physical education," *Women in Sport and Physical Activity Journal* 6(2): 7–22.

Messner, Michael A. (1992) *Power at Play: Sports and the Problem of Masculinity,* Boston: Beacon Press.

——— (1994) "Gay athletes and the Gay Games: in interview with Tom Waddell," in M. A. Messner and D. F. Sabo (eds) *Sex, Violence and Power in Sports: Rethinking Masculinity,* Freedom, CA: The Crossing Press, pp. 113–119.

Pronger, Brian (1990) *The Arena of Masculinity: Sports, Homosexuality, and the Meaning of Sex,* New York: St. Martin's Press.

DISCUSSION QUESTIONS

1. What other social mechanisms exist as "proving grounds" for masculinity and/or heterosexuality and are these one in the same? For women and femininity? What are the key features of these mechanisms that contribute to their ability to establish true manhood/womanhood?
2. Are there any experiences in your own life where you have sublimated homoerotic desire? Did this sublimation involve overt hostility, violence, or aggression toward others? Is the notion of sublimation sufficient to fully explain ***gay-bashing*** (violence directed toward people because they are perceived to be gay or lesbian)? Are there gendered differences in the forms that homo-erotic sublimation takes?
3. What is the basis for sexual identity, that is, how do we decide who is "straight" or "gay"? For example, should sexual identity be based on the acts people engage in (e.g., do your sexual behaviors involve people of the same sex or the opposite sex)? If so, how do we deal with individuals who have same-sex sexual experiences yet identify themselves as heterosexual? How many (or what proportion of) same-sex sexual acts are required to place someone in the homosexual or heterosexual category? What counts as bisexuality? Think back to the first article in this volume—what act or acts constitute the criteria upon which sexual identity should be based? How do we determine the sexual identity of individuals who have not engaged in any sexual acts (e.g., "virgins")?

The Impact of Multiple Marginalization

PAULA RUST

Paula Rust examines how membership in marginalized social groups can create particularly complex tensions for individuals who do not identify as heterosexual. In particular, because of homophobia within some ethnic minority communities, many members may find themselves torn between the desire to be open about their sexual identity and loyalty to their community and culture.

One's sexuality is affected not only by the sexual norms of one's culture of origin but also by the position of one's culture of origin vis-à-vis the dominant culture of the United States. For individuals who belong to marginalized racial-ethnic, religious, or socioeconomic groups, the effects are numerous. Marginalized groups sometimes adopt the attitudes of the mainstream; other times, they reject these attitudes as foreign or inapplicable. McKeon (1992) notes that both processes shape the sexual attitudes of the white working class. On the one hand, the working class absorbs the homophobic attitudes promoted by the middle- and upper-class controlled media. At the same time, working-class individuals are rarely exposed to "liberal concepts of tolerance" taught in institutions of higher education which help moderate overclass heterosexism. On the other

From "The Impact of Multiple Marginalization" by Paula Rust in *Bisexuality: The Psychology and Politics of an Invisible Minority,* ed. Beth A. Firestein. Reprinted by permission of Sage Publications, Inc.

hand, working-class sexual norms are less centered around the middle-class notion of "propriety"—a value that working-class individuals cannot as readily afford. The result is a set of sexual norms that differs in complex ways from those facing middle- and upper-class bisexuals.

In marginalized racial and ethnic groups, racism interacts with cultural monosexism and heterosexism in many ways. In general, the fact of racism strengthens ethnic communities' desires to preserve ethnic values and traditions, because ethnicity is embodied and demonstrated via the preservation of these values and traditions. Tremble, Schneider, and Appathurai (1989) wrote, "After all, one can abandon traditional values in Portugal and still be Portuguese. If they are abandoned in the New World, the result is assimilation" (p. 225). Thus, ethnic minorities might cling even more tenaciously to traditional cultures than Euro-Americans do, because any cultural change reflects not a change in ethnic culture but a loss of ethnic culture. To the extent that ethnic values and traditions restrict sexual expression to heterosexuality, ethnic minority bisexuals will be under particular pressure to deny same-sex feelings in demonstration of ethnic loyalty and pride. Attempts to challenge these values and traditions by coming out as bisexual will be interpreted as a challenge to ethnic culture and identity in general.

Because homosexuality represents assimilation, it is stigmatized as a "white disease" or, at least, a "white phenomenon." Individuals who claim a bisexual, lesbian, or gay identity are accused of buying into white culture and thereby becoming traitors to their own racial or ethnic group. Previous researchers have found the attitude that lesbian or gay identity is a white thing among African-Americans and Hispanics and the attitude that homosexuality is a "Western" behavior among Asian-Americans (Chan 1989; Espin 1987; H. 1989; Icard 1986; Matteson 1994; Morales 1989). In the current study, the association of gayness with whiteness was reported most often by African-American respondents. One African-American woman wrote that "when I came out, it was made clear to me that my being queer was in some sense a betrayal of my 'blackness.' Black women just didn't do 'these' kinds of things. I spent a lot of years thinking that I could not be me and be 'really' black too." Morales (1990) found that Hispanic men choose to identify as bisexual even if they are exclusively homosexual, because they see gay identity as representing "a white gay political movement rather than a sexual orientation or lifestyle" (p. 215). A Mexican woman in the current study wrote that she has "felt like . . . a traitor to my race when I acknowledge my love of women. I have felt like I've bought into the White 'disease' of lesbianism." A Puerto Rican woman reported that in Puerto Rico homosexuality is considered an import from the continental States. Chan (1989) found that Asian-Americans tend to deny the existence of gays within the Asian-American community, Wooden et al. (1983) reported this attitude among Japanese Americans, and Carrier et al. (1992) found denial of the existence of homosexuality among Vietnamese-Americans who considered homosexuality the result of seduction by Anglo-Americans. Tremble et al. (1989) suggested that viewing homosexuality as a white phenomenon might permit ethnic minority families to accept their LesBiGay members, while transferring guilt from themselves to the dominant society.

Ironically, whereas racism can strengthen commitment to ethnic values and traditions, it can also pressure ethnic minorities to conform to mainstream values in an effort to gain acceptance from culturally dominant groups. Because members of ethnic minorities are often perceived by Euro-Americans as representatives of their entire ethnic group, the nonconformist behavior of one individual reflects negatively on the whole ethnic group. For example, African-American respondents reported that homosexuality is considered shameful for the African-American community because it reflects badly on the whole African-American community in the eyes of Euro-Americans. A similar phenomenon exists among lesbians and gays, some of whom chastise their more flamboyant members with "How can you expect heterosexual society to accept us when you act like *that!*?" As one Black bisexual woman put it, "Homosexuality is frowned upon in the black community more than in the white community. It's as if I'm shaming the community that is trying so hard to be accepted by the white community."

The fact of ethnic oppression also interacts with particular elements of ethnic minority culture in ways that affect bisexuals. Specifically, the emphasis on the family found in many ethnic minority cultures is magnified by ethnic oppression in two ways. First, oppression reinforces the prescription to marry and have children among minorities which, for historical reasons, fear racial genocide (Greene 1994; Icard 1986). Second, the fact of racism makes the support of one's family even more important for ethnic minority individuals. As Morales (1989) put it, the "nuclear and extended family plays a key role and constitutes a symbol of their ethnic roots and the focal point of their ethnic identity" (p. 225). Ethnic minority individuals learn techniques for coping with racism and maintaining a positive ethnic identity from their families and ethnic communities; to lose the support of this family and community would mean losing an important source of strength in the face of the ethnic hostility of mainstream society (Amaguer 1993; Chan 1992; Icard 1986). Thus, ethnic minority bisexuals have more to lose if they are rejected by their families than do Euro-American bisexuals. At the same time, they have less to gain because of the racism of the predominantly Euro-American LesBiGay community. Whereas Euro-American bisexuals who lose the support of their families can count on receiving support from the LesBiGay community instead (albeit limited by the monosexism of that community), ethnic minority bisexuals cannot be assured of this alternative source of support.

Because of fear of rejection within their own racial, ethnic, or class communities, many bisexuals—like lesbians and gay men—remain in the closet among people who share their racial, ethnic, and class backgrounds. For example, an African-American–Chicana "decided to stay in the closet instead of risk isolation and alienation from my communities." Sometimes, individuals who remain closeted in their own racial-ethnic or class communities participate in the mainstream lesbian, gay, and bisexual community, which is primarily a Euro-American middle-class lesbian and gay community. Such individuals have to juggle two lives in two different communities, each of which is a valuable source of support for one aspect of their identity, but neither of which accepts them completely. Among people of their own racial, ethnic, or class background, they are not accepted and often not known as bisexuals, and among Euro-American lesbians and gays, they encounter both monosexism and class and racial prejudice or, at the least, a lack of support and understanding for the particular issues that arise for them because of their race, ethnicity, or class. Simultaneously, like other members of their racial, ethnic, or class community, they have to be familiar enough with mainstream Euro-American heterosexual culture to navigate daily life as a racial or ethnic minority; so they are, in effect, tricultural (Lukes and Land 1990; Matteson 1994; Morales 1989). This situation leads not only to a complex social life but might also promote a fractured sense of self, in which one separates one's sexual identity from one's racial identity from one's American identity and experiences these identities as being in conflict with each other, just as are the communities that support each identity. Some individuals attempt to resolve this dilemma by prioritizing allegiances to these communities (Espin 1987; Johnson 1982; Morales 1989), a response that Morales (1989, 1992) sees as a developmental stage preceding full integration of one's ethnic and sexual identities. More detailed descriptions of the antagonism between ethnic and gay communities and its effect on sexual minority individuals can be found in Gutiérrez and Dworkin (1992), Icard (1986), and Morales (1989, 1992).

Other bisexuals respond to the conflict between their racial/ethnic, class, and sexual communities by leaving their communities of origin in favor of mainstream LesBiGay communities. For most ethnic minority bisexuals, however, this does not solve the problem. For example, an Orthodox Jew who grieves her lost connection to the Jewish community wrote, "I still do not feel that my Jewish life and my queer life are fully integrated and am somewhat at a loss." This is true despite the large numbers of Jewish bisexuals, gays, and lesbians she has met, because "most Jewish people in the queer community are highly assimilated and are no help to me." The identities available for ethnic minorities in the LesBiGay community sometimes consist of racialized sexual stereotypes. Icard (1986), for example, describes the "Super Stud" and "Miss Thing" identities available for African-American men in the gay male community. Such stereotypical identities limit and distort the potential for integrated identity development among ethnic minority bisexuals.

Finally, some people from cultures that stigmatize homosexuality choose neither to closet themselves nor to leave their cultures and communities of origin but to remain within their communities as "out" bisexuals, lesbians, or gays to challenge homophobic and biphobic attitudes. In fact, some react positively to their own stigmatization with increased pride. A Korean-American immigrant woman explained that the "Asian shun of homosexuality/bisexuality . . . makes me even more defensive yet proud of my orientation." The African-American–Chicana mentioned earlier eventually decided to come out within the Latin and African-American communities and now uses her " 'outness' within [her] communities as a testimony to . . . diversity and to the strength [she has] developed from being raised Latina and African-American."

A positive integration of one's racial, ethnic, or class identity with one's sexual identity is greatly facilitated by support from others who share an individual's particular constellation of identities. For some, finding kindred spirits is made difficult by demographic and cultural realities. But as more and more people come out, there are inevitably more "out" members of racial and ethnic minorities and among these, more bisexuals. Many respondents described the leap forward in the development of their sexual identities that became possible when they finally discovered a community of bisexuals, lesbians, or gays with a similar racial or ethnic background. A Jewish-Chicana reported that she is "finding more people of my ethnic backgrounds going through the same thing. This is affirming." Similarly, a Chicano is "just now starting to integrate my sexuality and my culture by getting to know other gays/bis of color." An African-American woman reported that "it wasn't until I lived in Washington, D.C., for a number of years and met large numbers of Black lesbians that I was able to resolve this conflict for myself." Many Jewish respondents commented on the fact that there are many Jewish bisexuals, lesbians, and gays, and noted that receiving support from these peers was important in the development and maintenance of their positive sexual identities. One man, when asked to describe the effect of his racial or ethnic cultural heritage on his sexuality, said simply "I'm a Jewish-Agnostic Male-oriented Bisexual. There are lots of us." Some Jewish respondents also commented that being racially white facilitated their acceptance in the mainstream LesBiGay community and permitted them to receive support from this community that was not as available to individuals of other racial and ethnic backgrounds. Of course, it is this same assimilationist attitude that caused the Orthodox Jewish woman quoted earlier to find a lack of support among Jewish LesBiGays.

For individuals who belong to racial or ethnic minorities, the discovery that one is bisexual is a discovery that one is a double or triple minority. It is even more the case for bisexuals than for lesbians and gays, because bisexuals are a political and social minority within the lesbian and gay community. Many racial and ethnic minority individuals experience their coming out as a process of further marginalization from the mainstream that is, as an exacerbation of an already undesirable position. An African-American woman described being bi as "just one other negative thing I have to deal with. My race is one and my gender another." This can inhibit coming out for individuals who are reluctant to take on yet another stigmatized identity. For example, Morales (1990) reported that some Hispanic men limit their coming out, because they do not want to risk experiencing double discrimination in their careers and personal lives. A Black woman in the current study said that she is "unwilling to come too far 'out' as I already have so many strikes against me."

Many respondents found, however, that their experiences as racial or ethnic minorities facilitated their recognition and acceptance of their sexuality. This was most common among Jewish respondents, many of whom explained that their history as an oppressed people sensitized them to other issues of oppression. One man wrote, "The Jewish sense of being an outsider or underdog has spurred my rebelliousness; the emphasis on learning and questioning has helped to open my mind." A woman wrote, "My Jewish ethnicity taught me about oppression and the need to fight it. It gave me the tools to be able to assert that the homophobes (like the anti-Semites) are wrong." Some non-Jewish respondents also found that their experiences as ethnic minorities facilitated their coming out as bisexual. For example, a woman of Mexican, Dutch, and Norwegian descent wrote that her cultural background "has made me less afraid to be different." She was already ethnically different, so she was better prepared to recognize and accept her sexual difference. Similarly, an Irish Tsalagi Indian man found being outside the mainstream to be a liberating position; he wrote, "I have always felt alienated from the cultural norm, so I'm only affected in the sense that this alienation has allowed me the freedom to visualize myself on my own terms."

Many bisexuals of mixed race or ethnicity feel a comfortable resonance between their mixed heritage and their bisexuality. In a society where both racial-ethnic and sexual categories are highly elaborated, individuals of mixed heritage or who are bisexual find themselves straddling categories that are socially constructed as distinct from one another. The paradox presented by this position was described by a bisexual woman of Native-American, Jewish, and Celtic heritage who wrote, "Because I am of mixed ethnicity, I rotate between feeling 'left out' of every group and feeling 'secretly' qualified for several racial/cultural identities. I notice the same feeling regarding my sexual identity." Other respondents of mixed racial and ethnic backgrounds also saw connections between their ethnic heritage and their bisexuality. For example, an Asian-European woman wrote,

> Being multiracial, multicultural has always made me aware of nonbipolar thinking. I have always been outside people's categories, and so it wasn't such a big leap to come out as bi, after spending years explaining my [racial and cultural] identity rather than attaching a single label [to it].

A Puerto Rican who grew up alternately in Puerto Rico and a northeastern state explained,

> The duality of my cultural upbringing goes hand in hand with the duality of my sexuality. Having the best of both worlds (ethnically speaking—I look white but am Spanish) in my everyday life might have influenced me to seek the best of both worlds in my sexual life—relationships with both a man and a woman.

A Black Lithuanian Irish Scottish woman with light skin, freckles, red curly hair, and a "Black political identity," who is only recognized as Black by other Blacks, wrote, "As with my race, my sex is not to be defined by others or absoluted by myself. It is a spectrum."

However, individuals whose mixed heritages have produced unresolved cultural difficulties sometimes transfer these difficulties to their bisexuality. A "Latino-Anglo" who was raised to be a "regular, middle-class, all-American," and who later became acculturated to Latin culture, wrote,

> Since I am ethnically confused and pass as different from what I am, as I do in sexual orientation also, I spend a lot of time underground. . . . I think it has definitely been a major factor in the breakup of two very promising long-term relations.

Similarly, a transgendered bisexual respondent of mixed European, Native-American, and North African heritage believes that the pressures she feels as a transgenderist and a bisexual are closely related to the fact that her parents "felt it necessary to hide a large part of their ethnic and racial heritage," although she did not elaborate on the nature of these pressures.

In contrast to bisexuals from marginalized racial-ethnic, religious, or class backgrounds, middle- or upper-class Protestant Euro-Americans experience relatively few difficulties integrating their sexual identities with their cultural backgrounds and other identities. Euro-American bisexuals might have difficulty developing a positive bisexual identity in a monosexist culture, but unlike Bisexuals of Color, they have no particular problems integrating their sexual identity with their racial identity, because these identities are already integrated in the LesBiGay community. Being Euro-American gives them the luxury of not dealing with racial identity. Not surprisingly, when asked how their racial-ethnic background had affected their sexuality, most Euro-Americans did not mention their race at all. Instead, Euro-Americans tended to attribute their sexual upbringing to the peculiarities of their parents, their religion, their class, or their geographic location within the United States. One woman explained,

> I do not associate my racial-ethnic cultural background and my sexuality. Undoubtedly I would think and feel differently if I were of a different background but I'm not able to identify the effect of my background on my sexuality.

REFERENCES

Almaguer, T. (1993). Chicano men: A cartography of homosexual identity and behavior. In H. Abelove, M. A. Barale, & D. M. Halperin (Eds.), *The lesbian and gay studies reader.* New York: Routledge.

Carrier, J., Nguyen, B., & Su, S. (1992). Vietnamese American sexual behaviors and HIV infection. *Journal of Sex Research, 29*(4), 547–560.

Chan, C. S. (1989). Issues of identity development among Asian American lesbians and gay men. *Journal of Counseling and Development, 68*(1), 16–21.

Chan, C. S. (1992). Cultural considerations in counseling Asian American lesbians and gay men. In S. H. Dworkin & F Guitérrez (Eds.), *Counseling gay men and lesbians* (pp. 115–124). Alexandria, VA: American Association for Counseling and Development.

Espin, O. (1987). Issues of identity in the psychology of Latina lesbians. In Boston Lesbian Psychologies Collective (Eds.), *Lesbian psychologies: Explorations and challenges* (pp. 35–51). Urbana: University of Illinois Press.

Greene, B. (1994). Ethnic-minority lesbians and gay men: Mental health and treatment issues. *Journal of Consulting and Clinical Psychology, 62*(2), 243–251.

Guitérrez, F. J., and Dworkin, S. H. (1992). Gay, lesbian, and African American: Managing the integration of identities. In Dworkin & Guitérrez (Eds.), *Counseling gay men and lesbians* (pp. 141–155).

H., P. (1989). Asian American lesbians: An emerging voice in the Asian American community. In Asian Women United of California (Eds.), *Making waves: An anthology of writings by and about Asian American women* (pp. 282–290). Boston: Beacon.

Icard, L. (1986). Black gay men and conflicting social identities: Sexual orientation versus racial identity. *Journal of Social and Human Sexuality, 4*(1/2), 83–92.

Johnson, J. (1982). *The influence of assimilation on the psychosocial adjustment of Black homosexual men.* Unpublished dissertation, California School of Professional Psychology, Berkeley.

Lukes, C. A., and Land, H. (1990, March). Biculturality and homosexuality. *Social Work,* 155–161.

Matteson, D. R. (1994). *Bisexual behavior and AIDS risk among some Asian American men.* Unpublished manuscript.

McKeon, E. (1992). To be bisexual and underclass. In E. R. Weise (Ed.), *Closer to home: Bisexuality & feminism* (pp. 27–34), Seattle, WA: Seal.

Morales, E. S. (1989). Ethnic minority families and minority gays and lesbians. *Marriage and family Review, 14*(3/4), 217–239.

Morales, E. S. (1990). HIV infection and Hispanic gay and bisexual men. *Hispanic Journal of Behavioral Sciences, 12*(2), 212–222.

Morales, E. S. (1992). Counseling Latino gays and Latina lesbians. In Dworkin, S. H. and Gutiérrez, F. (Eds.), *Counseling gay men and lesbians: Journey to the end of the rainbow* (pp. 125–139).

Tremble, B., Schneider, M., & Appathurai, C, (1989). Growing up gay or lesbian in a multicultural context. *Journal of Homosexuality, 17*(1–4), 253–267.

Wooden, W. S., Kawasaki, H., & Mayeda, R. (1983). Lifestyles and identity maintenance among gay Japanese American males. *Alternative Lifestyles, 5*(4), 236–243.

DISCUSSION QUESTIONS

1. Why might bisexuality or homosexuality be viewed as assimilation to white culture by some racial and ethnic minorities?
2. How do you decide which facet of your identity is most central to how you define yourself to others? How central is your own sexual identity to your sense of self? What kind of factors might affect which facet an individual prioritizes?
3. Discuss the costs and benefits of gays, lesbians, and bisexuals who are members of racial or ethnic minorities fully "outing" themselves to their friends, families, and communities.
4. What impact does being Caucasian have on one's sexual identity? How does white privilege extend to issues of sexuality?

Tearoom Trade: Impersonal Sex in Public Places

LAUD HUMPHREYS

Sexual identity involves a complex nexus of factors that includes attraction, desire, emotions, behaviors, and fantasies. Often however, behavior and identity exist in conflict; that is, how individuals view themselves and/or present themselves to others may not be entirely consistent with their actual sexual behaviors. In the following excerpt of his now classic and controversial study of "tearooms," Laud Humphreys explores the social milieu of participants engaged in impersonal public sex. His work provides compelling data demonstrating the frequent disjuncture between sexual identity and sexual behavior and stands as an excellent example of earlier traditions in sexuality research that were primarily concerned with describing and explaining "deviant" sexual behavior.

At shortly after five o'clock on a weekday evening, four men enter a public restroom in the city park. One wears a well-tailored business suit; another wears tennis shoes, shorts and T-shirt; the third man is still clad in the khaki uniform of his filling station; the last, a salesman, has loosened his tie and left his sports coat in the car. What has caused these men to leave the company of other homeward-bound commuters on the freeway? What common interest brings these men, with their divergent backgrounds, to this public facility?

They have come here not for the obvious reason, but in a search for "instant sex." Many men—married and unmarried, those with heterosexual identities and those whose self-image is a homosexual one—seek such impersonal sex, shunning involvement, desiring kicks without commitment. Whatever reasons—social, physiological or psychological—might be postulated for this search, the phenomenon of impersonal sex persists as a widespread but rarely studied form of human interaction.

There are several settings for this type of deviant activity—the balconies of movie theaters, automobiles, behind bushes—but few offer the advantages for these men that public restrooms provide. "Tearooms," as these facilities are called in the language of the homosexual subculture, have several characteristics that make them attractive as locales for sexual encounters without involvement.

According to its most precise meaning in the argot, the only "true" tearoom is one that gains a reputation as a place where homosexual encounters occur. Presumably, any restroom could qualify for this distinction, but comparatively few are singled out at any one time.

Public restrooms are chosen by those who want homoerotic activity without commitment for a number of reasons. They are accessible, easily recognized by the initiate, and provide little public visibility. Tearooms thus offer the advantages of both public and private settings. They are available and recognizable enough to attract a large volume of potential sexual partners, providing an opportunity for rapid action with a variety of men. When added to the relative privacy of these settings, such features enhance the impersonality of the sheltered interaction.

In keeping with the drive-in craze of American society, the more popular facilities are those readily accessible to the roadways. The restrooms of public parks and beaches—and more recently the rest stops set at programmed intervals along superhighways—are now attracting the clientele that, in a more pedestrian age, frequented great buildings of the inner cities. [M]y research is focused on the activity that takes place in the restrooms of public parks, not only because (with some seasonal variation) they provide the most action but also because of other factors that make them suitable for sociological study.

There is a great deal of difference in the volumes of homosexual activity that these accommodations shelter. In some, one might wait for months before observing a deviant act (unless solitary masturbation is considered deviant). In others, the volume approaches orgiastic dimensions. One summer afternoon, for instance, I witnessed 20 acts of fellatio in the course of an hour while waiting out a thunderstorm in a tearoom. For one who wishes to participate in (or study) such activity, the primary consideration is finding where the action is.

I have chosen the term "purlieu" (with its ancient meaning of land severed from a royal forest by perambulation) to describe the immediate environs best suited to the tearoom trade. Drives and walks that separate a public toilet from the rest of the park are almost certain guides to deviant sex. The ideal setting for homosexual activity is a tearoom situated on an island of grass, with roads close by on every side. The getaway car is just a few steps away; children are not apt to wander over from the playground; no one can surprise the participants by walking in from the woods or from over a hill; it is not likely that straight people will stop there. According to my observations, the women's side of these buildings is seldom used.

VOLUME AND VARIETY

The availability of facilities they can recognize attracts a great number of men who wish, for whatever reason, to engage in impersonal homoerotic activity. Simple observation is enough to guide these participants, the researcher and, perhaps, the police to active tearooms.

Participants assure me that it is not uncommon in tearooms for one man to fellate as many as ten others in a day. I have personally watched a fellator take on three men in succession in a half hour of observation. One respondent, who has cooperated with the researcher in a number of taped interviews, claims to average three men each day during the busy season.

I have seen some waiting turn for this type of service. Leaving one such scene on a warm September Saturday, I remarked to a man who left close behind me: "Kind of crowded in there, isn't it?" "Hell, yes," he answered, "It's getting so you have to take a number and wait in line in these places!"

There are many who frequent the same facility repeatedly. Men will come to be known as regular, even daily, participants, stopping off at the same tearoom on the way to or from work. One physician in his late fifties was so punctual in his appearance at a particular restroom that I began to look forward to our daily chats. This robust, affable respondent said he had stopped at this tearoom every evening of the week (except Wednesday, his day off) for years "for a blowjob." Another respondent, a salesman whose schedule is flexible, may "make the scene" more than once a day—usually at his favorite men's room. At the time of our formal interview, this man claimed to have had four orgasms in the past 24 hours.

According to participants I have interviewed, those who are looking for impersonal sex in tearooms are relatively certain of finding the sort of partner they want. . . .

> You go into the tearoom. You can pick up some really nice things in there. Again, it is a matter of sex real quick; and, if you like this kind, fine—you've got it. You get one and he is done; and, before long, you've got another one.

. . . and when they want it:

> Well, I go there; and you can always find someone to suck your cock, morning, noon or night. I know lots of guys who stop by there on their way to work—and all during the day.

It is this sort of volume and variety that keeps the tearooms viable as market places of the one-night stand variety.

Of the bar crowd in gay (homosexual) society, only a small percentage would be found in park restrooms. But this more overt, gay bar clientele constitutes a minor part of those in any American city who follow a predominantly homosexual pattern. The so-called closet queens and other types of covert deviants make up the vast majority of those who engage in homosexual acts—and these are the persons most attracted to tearoom encounters.

Tearooms are popular, not because they serve as gathering places for homosexuals but because they attract a variety of men, a *minority* of whom are active in the homosexual subculture [In fact,] a large group of them have no homosexual identity. For various reasons, they do not want to be seen with those who might be identified as such or to become involved with them on a "social" basis.

PRIVACY IN PUBLIC

[T]here is another aspect of the tearoom encounters that is crucial. I refer to the silence of the interaction.

Throughout most homosexual encounters in public restrooms, nothing is spoken. One may spend many hours in these buildings and witness dozens of sexual acts without hearing a word. Of 50 encounters on which I made extensive notes, only fifteen included vocal utterances. Two were encounters in which I sought to ease the strain of legitimizing myself as lookout by saying, "You go ahead—I'll watch." Four were whispered remarks between sexual partners, such as, "Not so hard!" or "Thanks." One was an exchange of greetings between friends.

The other eight verbal exchanges were in full voice and more extensive, but they reflected an attendant circumstance that was exceptional. When a group of us were locked in a restroom and attacked by several youths, we spoke for defense and out of fear. This event ruptured the reserve among us and resulted in a series of conversations among those who shared this adventure for several days afterward. Gradually, this sudden unity subsided, and the encounters drifted back into silence.

Barring such unusual events, an occasionally whispered "thanks" at the conclusion of the act constitutes the bulk of even whispered communication. At first, I presumed that speech was avoided for fear of incrimination. The excuse that intentions have been misunderstood is much weaker when those proposals are expressed in words rather than signaled by body movements. As research progressed, however, it became evident that the privacy of silent interaction accomplishes much more than mere defense against exposure to a hostile world. Even when a careful lookout is maintaining the boundaries of an encounter against intrusion, the sexual participants tend to be silent. The mechanism of silence goes beyond satisfying the demand for privacy. Like all other characteristics of the tearoom setting, it serves to guarantee anonymity, to assure the impersonality of the sex liaison.

Tearoom sex is distinctly less personal than any other form of sexual activity, with the single exception of solitary masturbation. [W]hat I mean by "less personal" [is] simply that there is less emotional and physical involvement in restroom fellatio—less, even, than in the furtive action that takes place in autos and behind bushes. In those instances, at least, there is generally some verbal involvement. Often, in tearoom stalls, the only portions of the players' bodies that touch are the mouth of the insertee and the penis of the inserter; and the mouths of these partners seldom open for speech.

Only a public place, such as a park restroom, could provide the lack of personal involvement in sex that certain men desire. The setting fosters the necessary turnover in participants by its accessibility and visibility to the "right" men. In these public settings, too, there exists a sort of democracy that is endemic to impersonal sex. Men of all racial, social, educational and physical characteristics meet in these places for sexual union. With the lack of involvement, personal preferences tend to be minimized.

If a person is going to entangle his body with another's in bed—or allow his mind to become involved with another mind—he will have certain standards of appearance, cleanliness, personality or age that the prospective partner must meet. Age, looks and other external variables are germane to the sexual action. As the amount of anticipated contact of body and mind in the sex act decreases, so do the standards expected of the partner. As one respondent told me:

> I go to bed with gay people, too. But if I am going to bed with a gay person, I have certain standards that I prefer them to meet. And in the tearooms you don't have to worry about these things—because it is just a purely one-sided affair.

Participants may develop strong attachments to the settings of their adventures in impersonal sex. I have noted more than once that these men seem to acquire stronger sentimental attachments to the buildings in which they meet for sex than to the persons with whom they engage in it. One respondent tells the following story: We had been discussing the relative merits of various facilities, when I asked him: "Do you remember that old tearoom across from the park garage—the one they tore down last winter?"

> Do I ever! That was the greatest place in the park. Do you know what my roommate did last Christmas, after they tore the place down? He took a wreath, sprayed it with black paint, and laid it on top of the snow—right where that corner stall had stood. . . . He was really broken up!

The walls and fixtures of these public facilities are provided by society at large, but much remains for the participants to provide for themselves. Silence in these settings is the product of years of interaction. It is a normative response to the demand for privacy without involvement, a rule that has been developed and taught. Except for solitary masturbation, sex necessitates joint action; but impersonal sex requires that this interaction be as unrevealing as possible. . . .

THE PEOPLE NEXT DOOR

[Tearoom activity attracts] a large number of participants—enough to produce the majority of arrests for homosexual offenses in the United States. Now, employing data gained from both formal and informal interviews, we shall consider what these men are like away from the scenes of impersonal sex. "For some people," says [Evelyn] Hooker, an authority on male homosexuality, "the seeking of sexual contacts with other males is an activity isolated from all other aspects of their lives."[1] Such segregation is apparent with most men who engage in the homosexual activity of public restrooms; but the degree and manner in which "deviant" is isolated from "normal" behavior in their lives will be seen to vary along social dimensions.

For the man who lives next door, the tearoom participant is just another neighbor—and probably a very good one at that. He may make a little more money than the next man and work a little harder for it. It is likely that he will drive a nicer car and maintain a neater yard than do other neighbors in the block. Maybe, like some tearoom regulars, he will work with Boy Scouts in the evenings and spend much of his weekend at the church. It may be more surprising for the outsider to discover that most of these men are married.

[In fact,] 54 percent of my research subjects are married and living with their wives. From the data at hand, there is no evidence that these unions are particularly unstable; nor does it appear that any of the wives are aware of their husbands, secret sexual activity. Indeed, the husbands choose public restrooms as sexual settings partly to avoid just such exposure. I see no reason to dispute the claim of a number of tearoom respondents that their preference for a form of concerted action that is fast and impersonal is largely predicated on a desire to protect their family relationships.

Superficial analysis of the data indicates that the maintenance of exemplary marriages—at least in appearance—is very important to the subjects of this study. In answering questions such as "When it comes to making decisions in your household, who generally makes them?" the participants indicate they are more apt to defer to their mates than are those in the control sample. They also indicate that they find it more important to "get along well" with their wives. In the open-ended questions regarding marital relationships, they tend to speak of them in more glowing terms.

TOM AND MYRA

This handsome couple live in ranch-style suburbia with their two young children. Tom is in his early thirties—an aggressive, muscular, and virile-looking male. He works "about 75 hours a week" at his new job as a chemist. "I am *wild* about my job," he says. "I really love it!" Both of Tom's "really close" friends he met at work.

He is a Methodist and Myra a Roman Catholic, but each goes to his or her own church. Although he claims to have broad interests in life, they boil down to "games—sports like touch football or baseball."

When I asked him to tell me something about his family, Tom replied only in terms of their "good fortune" that things are not worse:

> We've been fortunate that a religious problem has not occurred. We're fortunate in having two healthy children. We're fortunate that we decided to leave my last job. Being married has made me more stable.

They have been married for eleven years, and Myra is the older of the two. When asked who makes what kinds of decisions in his family, he said: "She makes most decisions about the family. She keeps the books. But I make the *major,* decisions."

Myra does the household work and takes care of the children. Perceiving his main duties as those of "keeping the yard up" and "bringing home the bacon," Tom sees as his wife's only shortcoming "her lack of discipline in organization." He remarked: "She's very attractive . . . has a fair amount of poise. The best thing is that she gets along well and is able to establish close relationships with other women."

Finally, when asked how he thinks his wife feels about him and his behavior in the family, Tom replied: "She'd like to have me around more—would like for me to have a closer relationship with her and the kids." He believes it is "very important" to have the kind of sex life he needs. Reporting that he and Myra have intercourse about twice a month, he feels that his sexual needs are "adequately met" in his relationships with his wife. I also know that, from time to time, Tom has sex in the restrooms of a public park.

As an upwardly mobile man, Tom was added to the sample at a point of transition in his career as a tearoom participant. If Tom is like others who share working class origins, he may have learned of the tearoom as an economical means of achieving orgasm during his navy years. Of late, he has returned to the restrooms for occasional sexual "relief," since his wife, objecting to the use of birth control devices, has limited his conjugal outlets.

Tom still perceives his sexual needs in the symbolic terms of the class in which he was socialized: "about twice a month" is the frequency of intercourse generally reported by working class men; and, although they are reticent in reporting it, they do not perceive this frequency as adequate to meet their sexual needs, which they estimate are about the same as those felt by others of their age. My interviews indicate that such perceptions of sexual drive and satisfaction prevail among working-class respondents, whereas they are uncommon for those of the upper-middle and upper classes. Among the latter, the reported perception is of both a much higher frequency of intercourse and needs greater in their estimation than those of "most other men."

AGING CRISIS

Not only is Tom moving into a social position that may cause him to reinterpret his sexual drive, he is also approaching a point of major crisis in his career as a tearoom participant. At the time when I observed him in an act of fellatio, he played the insertor role. Still relatively young and handsome, Tom finds himself sought out as "trade."[2] Not only is that the role he expects to play in the tearoom encounters, it is the role others expect of him.

"I'm not toned up anymore," Tom complains. He is gaining weight around the middle and losing hair. As he moves past 35, Tom will face the aging crisis of the tearooms. Less and less frequently will he find himself the one sought out in these meetings. Presuming that he has been sufficiently reinforced to continue this form of sexual operation, he will be forced to seek other men. As trade he was not expected to reciprocate, but he will soon be increasingly expected to serve as insertee for those who have first taken that role for him.

In most cases, fellatio is a service performed by an older man upon a younger. In one encounter, for example, a man appearing to be around 40 was observed as insertee with a man in his twenties as insertor. A few minutes later, the man of 40 was being sucked by one in his fifties. Analyzing the estimated ages of the principal partners in 53 observed acts of fellatio, I arrived at these conclusions: the insertee was judged to be older than the insertor in 40 cases; they were approximately the same age in three; and the insertor was the older in ten instances. The age differences ranged from an insertee estimated to be 25 years older than his partner to an insertee thought to be ten years younger than his insertor.

Strong references to this crisis of aging are found in my interviews with cooperating respondents, one of whom had this to say:

> Well, I started off as the straight young thing Everyone wanted to suck my cock. I wouldn't have been caught dead with one of the things in my mouth! . . . So, here I am at 40—with grown kids—and the biggest cocksucker in [the city]!

Similar experiences were expressed, in more reserved language, by another man, some 15 years his senior:

> I suppose I was around 35—or 36—when I started giving out blow jobs. It just got so I couldn't operate any other way in the park johns. I'd still rather have a good blow job any day, but I've gotten so I like it the way it is now.

Perhaps by [there is sufficient research to] have dispelled the idea that men who engage in homosexual acts may be typed by any consistency of performance in one or another sexual role. Undoubtedly, there are preferences: few persons are so adaptable, their conditioning so undifferentiated, that they fail to exercise choice between various sexual roles and positions. Such preferences, however, are learned, and sexual repertoires tend to expand with time and experience. This study of restroom sex indicates that sexual roles within these encounters are far from stable. They are apt to change within an encounter, from one encounter to another, with age, and with the amount of exposure to influences from a sexually deviant subculture.

NOTES

1. Evelyn Hooker, "Male Homosexuals and Their Worlds," in Judd Marmor, ed *Sexual Inversion* (New York: Basic Books, 1965), p. 92.
2. i.e., those men who make themselves available for acts of fellatio but who, regarding themselves as "straight," refuse to reciprocate in the sexual act.

DISCUSSION QUESTIONS

1. How significant is the fact that 54 percent of Humphreys' research subjects were married men? What does this say about the traditional stereotypes of tearoom participants? What are you comfortable saying about the sexual identity of the married participants? Of the unmarried participants?
2. To obtain the data for his research, Humphreys played the "***watchqueen***" (lookout) role at the various tearoom venues but did not inform the participants that he was a researcher gathering data. In addition, he later disguised himself and interviewed many of the tearoom participants by adding their names to a random health survey he had been hired to administer. His methodology was quite controversial and set off a string of debate and critique within the scientific community. How much deception in research is ethical? Is the information gained valuable enough to sacrifice personal privacy? Do participants who engage in sexual activities in public places have any "right to privacy?"
3. What does the fact that there is no real female equivalent of tearooms suggest about the construction of sexuality in contemporary American culture? What do the norms of activity within the tearooms imply about the ways male sexuality is constructed?
4. Age is clearly an important social status that shapes sexual interaction in the tearooms. Do the same type of dynamics exist in larger society as well? How do you think your own sexual identity and behaviors will be affected as you age? What other social statuses can you think of that may mediate behavioral expectations in sexual encounters?

Dragon Ladies, Snow Queens, and Asian-American Dykes: Reflection on Race and Sexuality

SHARON LIM-HING

The negotiation and actualization of sexual identity is shaped and constrained by a myriad of meaningful social identities such as race, class, age, and religion. These identities merge, conflict, and interact with one another often creating personal as well as political tensions in identity construction, particularly when the identities are subordinate or non-hegemonic aspects of selfhood. In the following essay, Sharon Lim-Hing provides some insight into the complexities involved in negotiating an Asian-Lesbian identity.

I'd like to approach the subject of race and sexuality two ways: first, how my race has influenced my understanding of my sexuality, and second, how my sexuality has influenced my understanding of my race. . . .

I'm Chinese. I was born in Kingston, Jamaica. My parents moved to Miami when I was 11, and I grew up in Florida. I came to Boston about six years ago for graduate school.

RACIAL UNDERSTANDING INFORMING SEXUAL UNDERSTANDING

One day when I was about 5 years old, I was masturbating in the front yard. My mother came out, saw me, and quite sternly told me not to do that. Instead of explaining to me not to play with myself in public places, she simply told me to desist—for the rest of my life, presumably.

That was the extent of the sexual education I received from both my parents until the advent of my first period. I had been kept so ignorant that I thought I was sick, or that I had internal injuries from racing on my 10-speed bike. My mom then squeamishly completed my education by telling me the function of menstruation, and how to prevent blood from getting all over my clothes.

I've talked to other Asians, and I don't think the extreme prudishness of my childhood is characteristically Asian. However, health professionals who work with Boston Chinatown residents describe Chinese attitudes toward sex as "puritanical," citing patients' avoiding the discussion of sexual matters unless they relate directly to some malady. This cultural penchant, the silencing of sex and often of subjective, private feelings, was compounded in my parents' case by the Roman Catholic Church.

If run-of-the-mill sexuality is taboo for conversations in such a family, how would other forms of sexuality be treated? Well, quite simply, it wasn't treated at all. My family looked the other way; one of my sisters knew I was a lesbian and she was very supportive. There was some tacit

From "Dragon Ladies, Snow Queens, and Asian-American Dykes: Reflections on Race and Sexuality" by Sharon Lim-Hing in *Sojourner: The Women's Forum,* May 1990.

acceptance of my proclivities. I remember giggling and holding my first lover's hand in the back seat of the family car while my mother was driving. She was dropping us off at a theater, because we were too young to drive. My mother never so much as glanced in the rearview mirror, although she took an undue dislike to Karen. Whenever stray, unidentified panties showed up in the family wash, presumably discarded by a guest who found the raiment too encumbering, the cleaned object would mysteriously appear on my dresser, as if I knew who left them. To this day I still wonder which one of my siblings knew the owner of those small, black lace panties.

The implicit message my family gave me was not so much a condemnation as an embarrassed tolerance inextricably tied to a plea for secrecy. I complied with this request, waiting until I had moved 1500 miles away from the family homestead to begin coming out. When I did come out to my mother, she was staying with me and my lover, and could easily see only one bedroom with a single large futon.

"Mom," I said hesitantly, "you know I'm gay, don't you?" I couldn't say the word "lesbian" to my mother. She began with a remark on how as a child I didn't play with dolls—which is untrue. I had a couple of favorite G.I. Joes, but I didn't argue the point.

"I never talk about things you children don't talk about first," she then said, letting me know that she already knew. Then she said that she would always love me, and that if I was happy it was alright with her. I don't know if she told my father. Like my mother, he seems to have been able to figure things out on his own. In any case, I've never felt the need to do an official coming out with him. During a recent visit home, I sensed my father fidgeting uneasily as the news broadcast a story about "outing" closeted figures. At the end of my stay, he asked me if "they" would pick me up at Logan airport, although he knows Jacquelyn's name. My father's inability to accept my being a lesbian is related to his more traditional values: family first; make money and buy land; don't stand out.

Now that I'm more or less out to my immediate family and to some of my relatives, they've all stopped sending me invitations to weddings, on the presumption that I wouldn't want to get all dressed up—in a dress—to celebrate some heterosexual union. Of course they're right, but by not inviting me, they are trying to keep me a skeleton in their closet, in keeping with the same plea made years ago: don't tell anyone we know ("we" would include the loose network of Jamaican-Chinese spread over the Americas). This tolerance, curiously ambivalent, tells me that I'm still part of the family, but that being gay or having a gay person in the family is shameful.

What about the Asian pressure to procreate? Some Asians feel as if their parents push them to get married, so they can have lots of kids—at least one male child—to feed lots of white rice to, so they in turn can grow up to get married and have lots of kids. I never felt this pressure, but maybe that's because I'm the youngest of four children, and by the time my true tendencies had fully unfurled, my siblings were well on the way to marriage (with the appropriate sex) and procreation.

Up until the exodus of many Jamaican-Chinese in the 1970s (due to fear of the island becoming communist), there was a sizable Chinese community. If we had not left Jamaica, I would have been expected to find a husband from among the Chinese men there. Throughout my childhood I remember hearing the racist Chinese term for Black people, *black ghost.* It was sometimes preceded by *damn,* which used to convey much more venom than it does today. Much later I was surprised to find that an equivalent term exists for those of a paler shade (*white ghost,* strangely enough), though it was hardly ever used. When my mother explained this term, she said that Chinese people are arrogant, believing they are superior to others. Even though my siblings have married white individuals my parents seem quite happy—though I remember the time I came home with a black male friend, my father threw a fit.

SEXUAL UNDERSTANDING INFORMING RACIAL UNDERSTANDING

How has my sexuality affected my race? Here I feel comfortable using the amorphous term *race* because non-Asian people perceive me as belonging to this huge varied group, "Asian." In fact, the first thing many of you would think if you walked into a room and saw me is "Asian woman." Not young, old, badly or well-dressed, intellectual, punk, jock, diesel dyke, girlie girl—just 'Asian." Whites get to play all the roles, while Asians are invisible or are stuck in a few stereotypes. So pervasive is the mindset that holds white as the norm that when describing a white individual to a third party, we usually don't state that person's race, but if the person being described is not white, we do specify the race. Female Asian characters make rare appearances throughout Hollywood film history as the personable Suzy Wong prostitute, the throwaway Vietnam War prostitute, the Dragon Lady, and the Submissive Lotus Flower; male Asian characters are portrayed as asexual, arch-villains, or aberrant detectives, all fantastically inhuman.

Luckily, in real life we have more choices; we have the "model minority" stereotype. This covers those typical Asian characteristics—such as introverted, dorky, hardworking, smart like computers, especially good at math and sciences, passive, and apolitical.

About two years ago I became involved with a group called the Alliance for Massachusetts Asian Lesbians and Gay Men (AMALGM), a loosely organized group with social and political aims. We have different events, some open to everyone, some for Asians only; we publish a newsletter. Through experiences and talks with AMALGM members, I've become more aware of not only racism in the gay community but also tokenism and the lack of sufficient dialogue about race. At AMALGM we talk about Asian invisibility in what is called the "gay community."

As Asians, we go into bars, and we feel less attractive or simply undesirable. This is because we have been inculcated to appreciate and emulate white standards and types of beauty, like anyone who has ever seen a billboard, TV, magazine, or film. And how well trained we are; we even have slang to describe gay Asians who lust chiefly after white people: *snow queen* or *potato queen.* One corollary of the supremacy of white beauty is the ugliness of all those who are not white.

I'd like to underscore the paradox of Asian Americans. Not all Asians were born in Asia. Some like myself don't speak any Asian language and haven't been closer to the Pacific Rim than San Francisco. We grow up in a white culture—a culture that believes it is, and prides itself on being, primarily white. Some of us grow up thinking we are white: we believe we can get a job, make good, buy a home, and somehow avoid the war raging silently in this country. Some of us know better. Then we enter a subculture of the gay community—a community of "pariahs" and "radicals." Even there we discover that we are perceived as alien entities.

Last summer I was in a Boston gay bar. I was ordering drinks. I heard a voice behind me say, "Go back to your oriental country." I turned around and I saw two white men.

"You talkin' to me?" I asked, quickly pulling myself up to my full five feet, four inches.

"No, I'm just talking to my friend here," one man replied.

I should have said, "Oh, I thought you just made a racist, asshole comment. But what I thought I heard was so ludicrously ignorant that no one would dare say such a stupid thing. Don't you agree?" Of course, I thought of that later. At the time, all I said was, "Oh."

I was stunned. It was easier for me to think I was having a hallucination than to recognize that a gay man was making a blatantly racist remark to me in a gay bar. Only later did I realize that I was operating on the assumption that a member of an oppressed group will try to understand your oppression rather than try to oppress you.

During a recent conversation on race, a white woman who is aware of many types of oppression said to me, "I would never think of having a relationship with an Asian, but I don't think I'm racist." She reminded me of the old liberal cliché, "Some of my best friends are . . . ," a strange bundle of guilt and self-deception. I wasn't mad at her because at least she had the guts to say that. Many people go around thinking unconsciously, yet not saying, "I would never have a romantic or erotic relationship with an Asian, or with a Black person, or with a Latino/a person, but I don't think I'm racist." These people might wish to reflect on their personal definition of racism.

The private realm of desire is where the little racist in each one of us will make its last stand. I am not suggesting the policing of desire, but I bring up this aspect of racism because it is this very intangible—sexual orientation—that has driven bisexuals, lesbians, and gay men to question most givens of the dominant culture. Why not put the *sex* back into *homosexuality*?

I have no solutions to the racism we carry in our hearts, except the slow process of self-questioning and self-education.

An analogy is frequently made between racism and homophobia. Well, there is at least one fundamental difference. Most of us (not all) if we really had to could pass as straight. Yet, to walk through the Somerville, Massachusetts hinterland, I can't change my clothes, my buttons, the way I walk, to avoid being thought of or harassed as a "Chink."

I'm not saying that racism is somehow worse that homophobia. In some ways, the fact that gayness has not yet been linked to biologically determined factors makes choosing our own sexuality harder to justify to our foes, who would like us to just change our behavior and conform to their standards.

Racism and homophobia are two different forms of oppression that have similar and different sources, that function differently, and that have different effects. They need to be discussed in more than a superficial way. What makes it hard to discuss is the fact that if we are gay, lesbian, or bisexual, we are supposed to be "politically correct," making it harder to admit having racist thoughts. In spite of this, I hope we will all continue to talk about the convergence of race and sexuality, and that members of the gay community will look more closely at their own racism.

DISCUSSION QUESTIONS

1. The author suggests that sexual identity is less "visible" for ethnic minorities than it is for Caucasians. Why might this be?
2. *Should* members of oppressed groups be less discriminating than members of non-oppressed groups?
3. What parallels do you see in this reading and the Boundary Lines article (Giuffre and Williams)?
4. How are the sexual-identity struggles discussed by Lim-Hing similar to and different from those faced by other gays and lesbians? How are they similar to and different from the strategies of other ethnic minorities? What larger conclusions can we draw (if any) about the similarities of the sexual identity struggles of individuals in all non-hegemonic social groups?

More Than Manly Women: How Female Transsexuals Reject Lesbian Identities

HOLLY DEVOR

The transgender community consists of those who actively challenge, resist, subvert, or transcend the prevailing gender order. Included under this rubric are **transsexuals,** *individuals whose gender identity and biological sex are at odds. In the following essay, Holly Devor examines the lives of a group of female-to-male transsexuals to determine how they negotiate their sexual identity given the clear inconsistencies between their bodies, their behaviors, and their own sense of self.*

Sexology as a discipline first began to emerge in earnest during the end of the nineteenth and the early years of the twentieth centuries. One of the main projects of sexologists in those early years was the identification and classification of some of the many varieties of human sexuality. It was therefore during this period that characterizations of lesbians were first scientifically specified. A variety of authors, including such luminaries as Havelock Ellis and Richard von Krafft-Ebing, formulated pictures of lesbians as females in whom gender had become pathologically inverted to the point that they behaved sexually and emotionally like men and wanted to be men (Ellis 1918; Krafft-Ebing 1965). Thus, the earliest sexological diagnostic criteria for lesbianism were remarkably similar to today's diagnostic criteria for female-to-male transsexualism.[1]

As the ideas of sexologists commingled with those of members of the public, the image of lesbians as manly women became firmly embedded in the popular imagination. Indeed, the 1928 publication of Radclyffe Hall's *Well of Loneliness* was a benchmark in this regard. The book's protagonist, Stephen Gordon, who was a near-perfect exemplar of the style of lesbian described by the early sexologists, became and remained emblematic of prototypical lesbianism for close to half a century (Newton 1989).

The 1970s marked a major turning point in both clinical and popular North American conceptions of lesbianism. The combined efforts of the gay and women's liberation movements shifted definitions of lesbianism away from sin and sickness and toward images of health and happiness. The success of the public relations campaigns of these liberatory movements can be noted in two major changes, one clinical and one cultural: the December 1973 removal of homosexuality from the *Diagnostic and Statistical Manual II* (DSM-II) of the American Psychiatric Association (1980), and a noticeable shift from cultural representations of lesbians as mannish women who want to be men to images of lesbians as women-identified-women who revel in their womanhood (Radicalesbians 1970).

From "More than Manly Women: How Female Transsexuals Reject Lesbian Identities" by Holly Devor in *Gender Blending,* ed. Bonnie Bullough, Vern L. Bullough, and James Elias p. 87–102.(Amherst, NY: Prometheus Books).

Concurrent with these changes in professional and public understandings of the nature of lesbianism was the development of the concept of transsexualism, which was popularly launched by the 1966 publication of Harry Benjamin's *The Transsexual Phenomenon*. By the end of the 1970s, there were an estimated 3,000 to 6,000 post-surgical transsexuals in the United States alone, and approximately 40 clinics worldwide that provided sex reassignment surgery. (Harry Benjamin International Gender Dysphoria Association 1990). In 1980, approximately six years after homosexuality was removed from the DSM-II, female-to-male transsexualism became an officially delineated diagnosis in the next edition of the DSM (American Psychiatric Association 1980). Thus, those women-who-want-to-be-men who were rapidly becoming personae non gratae among woman-identified lesbians were repatriated back into the clinical purview as female-to-male transsexuals.

Throughout the 1970s and 1980s, the older depiction of lesbianism retained currency while the newer, more radical one gained in definitional muscle. More quietly, but inexorably, the ideas and practices of transsexualism also became public knowledge during this time. Therefore, those female persons who both wanted to be men and felt sexual attractions to women during the 1970s and 1980s had all of these categories available to them as possible explanations for their feelings. However, they were not equally accessible to most people.[2]

In the earlier part of these two decades, the older idea of lesbians as women who want to be men was more widespread than the woman-identified-woman concept, which, in turn, was more readily available than the idea of female-to-male transsexualism. Toward the later end of this timespan the mannish woman concept had lost considerable ground and the idea of transsexualism was well on its way to becoming common knowledge. However, among groups of politically oriented lesbians, the idea of lesbians as mannish women became anathema very early in the 1970s.[3] Likewise, among gender-oriented clinicians, or readers of clinical literature, the diagnostic category of female-to-male transsexualism was readily at hand by the early 1970s (Pauly 1974).

Thus, female individuals who wanted to be men and found their way to self-consciously organized groups of lesbian women in the 1970s and 1980s would have been likely to find that they no longer fit the in-house definition of lesbianism. Were such individuals to continue their search for identity in libraries or in clinicians' offices, they would have been likely to find that they did fit the template for female-to-male transsexualism. If by no other means, the fascination of the popular media with transsexualism that exploded in the late 1980s eventually would have introduced them to the idea of female-to-male transsexualism.

In this report, I recount some of the ways in which these social phenomena were played out in the lives of a group of female-to-male transsexuals. In doing so, I trace some of the ways in which they came to first think of themselves as lesbian women and, later, to reject that designation in favor of identities as female-to-male transsexuals.

SUBJECTS

A total of 46 self-defined female-to-male transsexuals were interviewed. They ranged from people who had, at the time of first contact, taken no concrete steps toward becoming men, to those who had completed their transition 18 years before their participation in this research.

RESULTS AND DISCUSSION

All but two participants (95.5 percent) reported that they had been sexually or romantically attracted to women at some time during their lives as women. Thirty-five of the 43 participants who had been attracted to women (81 percent) acted upon their inclinations to some degree, all but two of whom established relationships of approximately one year or more in duration. Twenty-five of those who were attracted to women (58 percent) thought of themselves as lesbian for at least a short period of time. Those participants who never took on the title of lesbian thought of their relationships with women as heterosexual ones.

Participants Who Did Not Act Upon Their Attractions to Women

Seventeen participants (39.5 percent) went through periods during which they felt unable to act upon their sexual feelings for women, five of whom (12 percent) never became sexually involved with women while they themselves were still living as women. For all 12 of those participants who later went on to experiment with homosexual relations, their periods of reluctance to act upon their homosexual attractions were confined to their teenage years. It became clear from the stories of most participants that the generalized homophobia of the decades during which they were children and adolescents (1950s, 1960s, early 1970s) played a significant role in discouraging them from acting upon the adolescent attractions that they felt for females. Homophobia acted to abort their lesbian activities and deflect them from lesbian identities mainly in two ways.

First, information about lesbianism was not readily available.[4] Thus, the only sexual model to which many girls had access was a heterosexual one. Therefore, when they felt sexual desire for other females, the only logical interpretation that they could place on their feelings was that they should be males in order to have such lusts. Furthermore, since they were not males, there was nothing that they could do about their feelings but contain them and bide their time until an opportunity arose to somehow transform themselves into men.

For example, Bruce remembered that she had very explicitly sexual thoughts about women when she was a teenager, but did not act on them:

> I used to get out my parents' Sears catalogue and look at the women in their underwear. I got turned on. . . . But I thought, I can't do this the way I am. I have to be a boy because girls don't like girls. . . . So, I saw men and women together. So, I thought, that's what it's supposed to be, and all the girls liked me because I was a boy. . . . But I used to put those kinds of feelings behind me, I think, because I felt that I couldn't. . . . be sexual. . . . It wasn't allowed. It wasn't right. Because, how can women be attracted to women?

The second way in which the homophobia of the times acted to deny these participants opportunities to explore their lesbian urges during their teen years was through misinformation. Those participants who had heard of lesbianism had only the most negative of perspectives on the phenomenon. They absorbed the messages that their society wanted them to believe: that lesbians were sick and dangerous people; that lesbian activity was sorely stigmatized and totally taboo. Thus, those individuals who wanted to retain some modicum of self-respect and a decent standing in their society avoided tainting themselves with the stain of lesbianism.

Peter[5] for instance, thought that her childhood and adolescent attractions to girls meant that she must be "gay," but Peter did not want to accept that label as appropriate for herself. Peter recalled: "There's always been the social stigma about being gay. And I would think that probably, for a time when I was an adolescent, that, in that sense, that really backed me off. That kept me really under wraps."

The participants who were acting under the sway of these kinds of homophobia were left few alternatives. They simply could not act because they could only see their attractions as being heterosexual in form, if not in content. Nevertheless, three participants did think of themselves as lesbian on the basis of their unactualized attractions to women.

Minor Homosexual Involvements

Ten participants (23 percent) went through extended periods during which they had only limited homosexual experiences. These plateaus occurred during the adolescences of all but one such participant. They engaged mostly in kissing and in touching of their partners' breasts in the context of short-term infatuations. Only two participants thought of themselves as lesbians on the basis of these interactions. Most of these participants also kept their attractions for females relatively in check because of their fear of social stigma. Their anxieties about possible social retribution for transgressions were sufficient to deter them from any extensive homosexual adventures. Many participants had so absorbed the messages of their society that they held opinions that could be interpreted as indicative of internalized homophobia and misogyny.

Lee's comments illustrates this type of thinking:

> I knew that queers existed. . . . Things like that weren't talked about. . . . That was almost like Mafia. They's just dirt road people or something. So, that was a bad word. You didn't want to be that. . . . You knew it wasn't accepted. . . . You knew it wasn't right in the eyes of everybody. . . . It makes me sound stupid, but I didn't just sit down and think about things like that. I just did it. You knew that it wasn't right. . . . but it's something you enjoyed.

It seems plausible that, had these participants lived in a time when information about lesbianism was both more readily available and more salutary, most of them would probably have been more homosexually active. In a climate more conducive to positive lesbian identity, many of them might well have more avidly adopted a lesbian identity and cleaved to it more persistently. Be that as it may, gender identity is a different matter from sexual orientation. Later, more extensive homosexual experience and lesbian identity did not banish, but only obscured, underlying male identities. Experimentation with lesbianism for most participants was one step in the process of clarifying that a male identity was the most suitable one for them.

Major Homosexual Relationships

Thirty-five participants (81 percent) became involved in ongoing genitally sexual relationships with other females during their pretransition years. Due to the explicitly sexual and nonfleeting nature of these unions, they could not be dismissed as merely affectionate or experimental. Thus, these liaisons forced participants to confront issues of sexual identity and, by extension, issues of gender identity.

Some of these unions were undertaken by both partners with the understanding that they were, at least in all apparent aspects, lesbian relationships. In other cases participants, but not necessarily their partners, maintained the belief that they were men, and that therefore their relationships were, de facto, heterosexual ones. A few relationships foreshadowed what was to come in that both partners agreed from the start that, in their own eyes, they were in heterosexual relationships. In other cases, particular relationships became redefined as they progressed and as participants went through stages wherein they came to have better insights into themselves and into the nature of lesbianism. As they did so, they generally moved more toward the rejection of the label of lesbian and of the womanhood implicit in that title.

When participants did think of themselves as lesbian women, they did so principally for two reasons. In the first place, they were faced with the unmistakable evidence of their own, and their lovers', bodies. They knew that the definition of lesbian therefore technically included them. Some participants were also persuaded by the popular conception that lesbians are women who want to be men. As that was precisely how participants felt, they uneasily accepted the appellation of lesbian, despite the fact that it required them to acquiesce to being women. However, only a very few participants easily accepted that their intimate relationships with women fully qualified as lesbian ones. More commonly, participants recognized a superficial similarity between their own relationships and those of lesbian women, but retained a sense of themselves as different.

As participants used their intimate relationships with women as testing grounds for their sex, gender, and sexual identities, they found that their homosexual relations did not allow them to express their identities adequately. After some initial delight at the increased tolerance for their masculinity that they found among lesbian women, participants began to encounter some limitations. They found that the social and sexual values of lesbian women did not align as well with their own as they might have wished. Eventually the disjunctures between their own self-images and the images they held of what lesbian women were like became too disquieting to them and they concluded that they were not lesbian women. When they reckoned that they were beyond the range of what constituted lesbian thoughts and deeds, they became receptive to the possibilities of transsexualism as a means of realigning themselves with their social worlds.

Avoiding Lesbian Communities

Thirteen participants (30 percent) went through periods of their lives during which they were homosexually active but stayed away from places where homosexual women congregated. During those periods of their lives, they fell in love and built relationships with women, but six of them (46 percent) resisted accepting identities as lesbian women. They conceived of their partners as being attracted to them for their manly qualities and did what they could to nurture their mutual conceptualizations of their relationships as straight ones.[6]

These participants found ways, outside of established lesbian environments, to meet women with whom they could establish sexual/romantic relationships. One result of their making contact with their lovers independently of communities of similarly disposed women was that they had only popular models of the nature of lesbianism against which to measure themselves. By the time that they were making such comparisons, lesbian-feminist definitions had begun to move definitions of lesbians away from the "mannish woman" typology popular before the 1970s and toward a "woman-loving-woman" characterization that has become more dominant since then (Faderman 1991; Sedgewick 1990). As a group, they found the latter definition less acceptable than the arguably more stigmatized former one, and so they eventually rejected the label of lesbian in favor of more obtusely acceptable identities as men.

For example, Stan feared being branded as a lesbian. Although Stan remembered always having had "feelings about [women] like guys do," as a woman she felt exceptionally guilty about these feelings even as she was having several years of otherwise satisfying relationships with women. As she tried to work through this contradiction, she became so depressed at the thought that she might be lesbian that she ended up spending several months hospitalized for mental problems. Stan also reported that she later destroyed one of her relationships with the heavy drinking and marijuana smoking that she used to help her cope with her extreme aversion to being known as a lesbian woman.

As was probably common among many homosexual women who came out in the 1970s, Stan seemed to have two different views of what it meant to be a lesbian woman. On the one hand, Stan had held a more traditional view of lesbian women as sinful and sick. On the other hand, Stan had also been exposed, through the media, to a more feminist version of lesbianism. When I asked Stan about it, he described lesbians this way:

> I knew about lesbians but it just didn't occur to me that's what it was. . . . What I knew about lesbians was that two women can be together

> and it's okay if you are a lesbian. . . . It was something they did on the coast in the big cities, more liberal people did. I just didn't consider myself that liberal, that open minded. . . .To get into being a lesbian, like, you have to march for things, and you gotta go to caucuses, you gotta hate men, you gotta dress butch, and you gotta get into all that stuff, and I didn't want to do that, I didn't want to get into all that stuff.

Stan, like others who were unable to accept lesbian identities as appropriate for themselves, later enthusiastically latched onto the normalizing potential of female-to-male transsexualism.

Moving through Lesbian Communities

Another 22 participants (51 percent) who were homosexually active as women went through stages wherein they initially threw themselves wholeheartedly into lesbianism. They became friends with other homosexual women and participated in social or political activities with them. They were thus exposed to socialization processes that taught them something of what lesbian subcultures expected from women who were lesbian.

Eighteen of these 22 participants (82 percent), at some time during their lives, accepted the label "lesbian" as descriptive of themselves, only to later reject it as inadequate to the task. They came to their conclusions after making comparisons between their senses of themselves and their visions of how they believed that lesbians thought and acted.

Aaron's story illustrated how this happened. Aaron started to think of herself as "probably gay" when she was a woman of 25 and her psychiatrist diagnosed her as homosexual. Aaron accepted that label as descriptive of herself because, at that time, in the early 1960s, "the only image I could think of was women that fell in love with women, and women that dressed and wanted to be men and acted masculine. I figured that was what a gay woman was." However, when Aaron became divorced from her husband three years after this diagnosis, she decided to remain celibate and separate from other gay women because "frankly, I wouldn't have kept my kids if I wasn't." Nonetheless, during that 15-year period Aaron recalled, "I was living primarily in male clothing . . . [people] just assumed I was a dyke."

At the end of the 1970s, two days after Aaron's youngest daughter reached legal adulthood, Aaron started having a series of affairs with lesbian women at the university where she was taking courses. A number of brief affairs and one longer relationship demonstrated to Aaron that she was not like other gay women. Aaron described two aspects of her process of discovery. On the one hand, she found herself at odds with the lesbian community in which she was situated:

> Let's face it . . . they saw the woman's body, and figured I was a gay woman, and I went along with that to the point where they expected me to be female. . . . It means sticking up for the female when you get into a discussion with a bunch of women on wife beating, or sticking up for the feminist role when you get with a bunch of women and no men around, or . . . preferring the company of women . . . I was trying to get along with these women; I was trying to love some of these women . . . but I didn't fit. And the longer I was with them, the more I realized I didn't fit.

Although Aaron was able to integrate herself into a gay community she always felt different from the women around her. Aaron recalled how it was for her then:

> When I got involved with gay women and found out how frigging different I was it was obvious. Up until that point I thought that other gay females were the same as me, they wanted to be male. And when I found out that was not true, that no matter how masculine they acted, they had female identities, I realized I don't quite fit in here, but I fit in closer here than I ever had.

On a more intimate level, Aaron further found that she did not respond to her lesbian lover in the ways that both of them believed were characteristic of lesbians. Aaron drew this picture of the issues involved:

> Basically she wanted a woman. At the nitty gritty deep level I wasn't a woman. . . . Okay, concrete example . . . our lovemaking. She would resent it when I got too masculine. . . . When I became too aggressive and too demanding, too macho, whatever, it ruined it for her. . . . Hey, I want to be on top part of the time . . . figuratively and literally. And it would turn her off more, it would slow her response down and turn her off right when mine was speeding up. We didn't match.

Aaron construed these events as evidence that she did not belong among lesbians and concluded that she was a female-to-male transsexual.

Another participant, Howie, summed up well the way in which these people deduced that they were men rather than lesbian women. At first Howie thought that she and her lover were lesbian but then,

> Later . . . upon closer investigation I realized that lesbians enjoyed their womanhood and didn't want to change their bodies surgically. They were simply women who loved women. I realized I didn't fit that mould at all. . . . A lesbian is a woman, who is glad she's a woman, who happens to relate sexually to other women. She does not wish to be male. In fact, she rejoices in her femaleness and wants to be with other females . . . I knew that wasn't for me. . . . I often wish I could have accepted myself as gay, or identified as gay, because it is infinitely easier than changing.

Ron also became embedded in a lesbian community and came to be extensively committed to lesbian-feminist activities. Like Aaron and Howie, Ron also concluded, on the basis of her knowledge of lesbian social and sexual mores, that she was neither a lesbian nor a woman, and that she was better suited to being a man. Ron remembered:

> For one thing, sexual[ity] definitely played a big role. . . . had to go through and analyze for myself whether I was just a strong female, or whether I was a male. Whether I just didn't fit into the stereotypical female sexist kind of role. . . .
>
> For instance, being with women . . . the love I got was toward the woman, the physical woman. And for me, that was a conflict sexually . . . I was not making love as a woman with a woman. From my heart, it was that I was a male. . . . It's a completely different dynamic. . . . There is a different approach from a woman to her man than the approach from one woman to another woman who are lovers. . . . There were a lot of needs that I could not express with lesbians, because the lesbians that I was having relationships with were not open to anything that had anything to do with males.

These participants entered into communities of lesbian women at times during which lesbian-feminists of the 1970s and 1980s were dedicated to a redefinition of lesbianism away from the depiction of lesbians as mannish women. Instead, lesbian-feminists promoted the idea that lesbians were "women-identified-women." Participants felt excluded by that definition and therefore were left to search in other quarters for labels that more snugly fit their self-images. When they discovered female-to-male transsexualism they embraced it as both an escape and a homecoming.

SUMMARY AND CONCLUSIONS

By far the strongest pattern that emerged from the stories offered by participants was one of participants' earnest attempts to fit themselves to the available social roles of their times. Forty-three participants (95.5 percent) had been sexually attracted to women at some point in their pretransitions on lives. Thirty-five of them (81 percent) established relationships of some duration with women during their pretransition years, two participants (5 percent) had only minor sexual involvements, and another six participants (14 percent) were attracted to women but never acted upon their emotions. Drawn as these 43 participants were to being lovers of women, they were confronted with a difficult-to-deny characterization of their love as lesbian.

More than half of these participants who were sexually attracted to women (58 percent) passed through periods during which they thought of themselves as gay or lesbian women. They were originally attracted to making such identifications because of their awareness of the common social definition of lesbians as women who want to be men or as mannish women who are sexually interested in other women. However, over time, they came to make more finely sifted distinctions.

Two major issues became important to participants in their process of moving out of lesbian identities. Both of the axes on which participants judged themselves to be men rather than lesbian women were products of a particular historical period wherein the definitions of lesbianism constituted contested territory. On the one hand, all participants who once considered themselves to be lesbian ceased doing so during the 1970s and 1980s. These were years during which the proponents of lesbian-feminism were waging campaigns to supplant the idea that lesbians are mannish women with images of lesbians as women-identified women who celebrate their womanhood with other women. On the other hand, these decades were also those during which female-to-male transsexualism was being defined as a treatable medical condition, similar to, but distinct from, lesbianism and characterized by the persistent desire of females to become males. Thus, participants searching for viable words to use to identify themselves were caught up in these shifting boundaries.

Participants who lived part of their lives as lesbian women were thus often in the position of having been drawn to lesbian identities on the basis of older definitions of lesbians as women who want to be men, only to discover that the lesbian pride movements of the 1970s and 1980s required them to reject those characterizations. When participants tried to measure themselves against the more woman-centered images promulgated by lesbian-feminists they found themselves lacking on two points. First, they were ashamed, embarrassed, or disgusted by the specifically female aspects of their bodies and therefore had little desire to join with their companions in the glorification of their womanhood. Second, they were generally not interested in having their sexual partners enjoy their femaleness or attempt to provide them with pleasures in specifically female ways. In other words, when participants compared themselves to both generalized and specific lesbian others, they were struck more by the contrasts than by the similarities. It therefore became apparent to these participants that they had more in common with straight men than with lesbian women. Eventually, their discomfort with being included in the lesbian camp was alleviated by their discovery of the increasingly socially available concept of female-to-male transsexualism, which offered them a conceptually simple, and more apt, solution to their extreme gender dysphoria. Once they knew themselves to be female-to-male transsexuals, they were eager to move beyond wishing to and into actually becoming men.

Thus, in the end, participants gradually exhausted their possibilities as women. Each probed the roles for women that were available to them. As each alternative was weighed and found wanting, the field of possibilities narrowed to that which was perhaps ultimately the most suitable but also seemingly the most unobtainable: to become men. Until participants happened upon the option called female-to-male transsexualism, they were relegated to forever feeling like bizarre misfits—even among those sexual minorities who already inhabited the fringes of society. Female-to-male transsexualism offered them a way out of their dilemmas: a path toward integration and self-actualization.

NOTES

1. I use the term "female-to-male transsexuals" rather than "transsexual men" because I do not wish to distinguish between individuals at various stages of transition. I have included in this category anyone who so designated themselves to me. I use the term "lesbian" to refer to sexual/romantic relationships between two persons who have gender identities as women regardless of their anatomical sexes. I use the term "homosexual" to refer to sexual relations between persons of the same anatomical sex regardless of their gender identities. For further discussion of my use of the language of gendered sexuality see Devor 1993.
2. Consider that Krafft-Ebing described "the extreme grade of degenerative homosexuality" as "hermaphroditism" wherein "the woman of this type possesses of the feminine qualities only the genital organs; thought, sentiment, action, even external appearance are those of the man . . . the[ir] desire to adopt the active role towards the beloved person of the same sex seems to invite the use of the priapus" (264–265). Compare Krafft-Ebing's description of mannish lesbians with the diagnostic criteria for adult Gender Identity Disorder from the DSM-IV: "a stated desire to be the other sex, frequent passing as the other sex, desire to live or be treated as the other sex, or the conviction that he or she has the typical feelings and reactions of the other sex" (1994, 537).
3. My comments on this topic are partially based upon my own recollections and partially upon my analysis of the data reported herein and elsewhere.
4. In 1921 the English Parliament attempted to introduce a law that would make lesbianism a crime. Speaking against the proposition Lord Desart said, "You are going to tell the whole world that there is such an offence, to bring it to the notice of women who have never heard of it, never thought of it, never dreamt of it. I think that is a very great mischief" (quoted in Weeks 1989, 105). Clearly, many young women were still laboring under such ignorance more than 50 years later.
5. I have tried to remain true to the gender of the persons involved in this research project. When referring to a man telling a story about when he was a girl or woman, I have used gender pronouns that reflect the gender of the subject in each time frame, e.g., "He remembered that as a girl, she was a tomboy."
6. In addition, two participants who had been involved with lesbian communities also took this tack. One other participant was involved in a more-than-20-year homosexual relationship that both parties framed as a relationship between two gay men.

REFERENCES

American Psychiatric Association. 1980. *Diagnostic and statistical manual of mental disorders.* 3d ed. Washington, D.C.: American Psychiatric Association.

———. 1994. *Diagnostic and statistical manual of mental disorders.* 4th ed. Washington, D.C.: American Psychiatric Association.

Benjamin, H. 1966. *The transsexual phenomenon.* New York: Julian Press.

Devor, H. 1993. Toward a taxonomy of gendered sexuality. *Journal of Psychology and Human Sexuality* 6: 23–55.

Ellis, H. 1918. *Studies in the psychology of sex.* Vol. 2. Sexual inversion. Philadelphia: F. A. Davis.

Faderman, L. 1918. *Odd girls and twilight lovers: A history of lesbian life in twentieth century America.* New York: Penguin.

Hall, R. [1928] 1986. *The well of loneliness.* London: Hutchinson.

Harry Benjamin International Gender Dysphoria Association. 1990. *Standards of care.* Available from The Harry Benjamin International Gender Dysphoria Association, Inc., P.O. Box 1718, Sonoma, CA 95476.

Krafft-Ebing, R. von. [1906] 1965. *Psychopathia sexualis with especial reference to the antipathic sexual instinct. A medico-forensic study.* Translated by F. S. Klaf. New York: Stein & Day.

Martin, D., and P. Lyon. 1972. *Lesbian/woman.* San Francisco: Glide.

Newton. E. 1989. The mythic mannish lesbian: Radclyffe Hall and the new woman. In *Hidden from history: Reclaiming the gay and lesbian past,* edited by M. Duberman, M. Vicinus, and G. Chauncey, Jr., 281–293. New York: NAL.

Pauly, I. 1974. Female transsexualism: Parts I and II. *Archives of Sexual Behavior* 3: 487–525.

Radicalesbians. 1970. The woman identified woman. In *Radical feminism,* edited by A. Koedt, E. Levine, and A. Rapone, 240–245. New York: Quadrangle.

Sedgewick, E. 1990. *The epistemology of the closet.* Berkeley: University of California Press.

Weeks, J. 1989. *Sex, politics and society: The regulation of sexuality since 1800.* 2d ed. London: Longman.

DISCUSSION QUESTIONS

1. Devor connects the struggles of the female-to-male transsexuals in her study to changes in the social definition of lesbianism. How do you think changes in the definition of gender and sex might affect these struggles?
2. What kind of identity work do you think the romantic partners of transsexuals go through?
3. Imagine that you woke up tomorrow as a member of the "opposite" sex. How would your sense of self change (if at all)? Would your gender and sexual identity change? If so, what kind of changes would be necessary to re-establish congruence between your biology, your gender, and your sexuality? Would you ever consider surgical reassignment surgery? Why?
4. Previous readings by Lorber and Stoltenberg indicate that division of the species into two biological categories—male and female—is a social process. They argue that, biologically speaking, we could, in fact, divide people into a number of different sex categories. If a three-, four-, or even five-category system of sex assignment were to emerge, what effect might this have on gender identity, broadly speaking? Would a transgender identity still exist? Why or why not?

The following Bonus Reading can be accessed through the *InfoTrac College Edition Online Library* by typing in the name of the author or keywords in the title online at *http://www.infotrac-college.com.*

The Ambiguity of 'Having Sex': The Subjective Experience of Virginity Loss in the United States

LAURA CARPENTER

Virginity loss is typically considered to be one of the most significant events in a person's sexual life. But precisely what acts constitute virginity loss? Further, what does the loss of virginity mean *to individuals? Carpenter's analysis of 61 in-depth interviews reveals that though the concept of virginity is very much tied to hegemonic definitions of sex as a genitally-based, penetrative, heterosexual activity—definitions of virginity and virginity loss are strongly situated by sex, sexual identity, and individual volition. In addition, the author finds dramatic differences in the subjective meaning and interpretations of virginity loss which, in turn, shape individual sexual choices and expectations.*

DISCUSSION QUESTIONS

1. Taking into account the varying definitions of virginity loss described in this article, how would *you* define virginity loss? Why do you define it this way? Is this the way you would have defined it before reading this article?
2. Are there other "firsts" in your sexual history that you feel should matter more than virginity loss? Why or why not? What acts are most significant in shaping your own sexual identity? Why do you think so much emphasis is placed on virginity loss in our culture? Do you think this emphasis will increase, decrease, or remain the same in the future? Why?
3. How do you interpret virginity? As a gift? Stigma? Process? Other? What parts of your own social history do you think have most greatly influenced your interpretation? Have you ever been less than fully truthful to others about your own virginity status (ahem . . . parents are included here)? Why?
4. What kind of privilege and social power do the labels "virgin" and "non-virgin" carry?
5. If you identify as a non-virgin, how did the experience of virginity loss affect your sense of who you are? Was it what you expected? Why or why not?
6. The article points out that sex is one of the factors that can affect how and why one interprets virginity loss. What other social statuses might affect the subjective experience of virginity loss? Why? What do these differences say about the social construction of virginity and/or sex in our culture?

PART II

Sex and Social Institutions

INTRODUCTION

Anthropologist F. A. Beach has noted that "Every society shapes, structures, and constrains the development and expression of sexuality in all of its members" (1977:116). Social institutions are the mechanisms through which much of this social masonry is accomplished. So, what is a social institution? ***Social institutions*** are the core components of societal organization that are designed to meet basic social needs. Institutions are made up of an established and integrated system of norms, values, beliefs, roles, and rules that help to determine how a society will organize itself in its attempts to satisfy these needs. Traditionally, scholars have examined five standard social institutions. These include: the state, which is concerned with the maintenance of social order; the economy, which is concerned with the production and distribution of vital goods and services; the family, which oversees procreation and socialization; education, which ensures that important skills are taught and passed on to society's members; and religion, which is charged with providing society's members with a sense of meaning and purpose. In more recent years, several new institutions have been proposed by scholars, including the media and medicine. Each of these institutions will be examined in this section of the text.

From a constructionist perspective, institutions are important to study because they are the primary means through which social constructs are publicly articulated, transmitted, and enforced. It is from social institutions that we learn what is "normal" and what is "supposed to be." In doing so, institutions shape which social features will be naturalized and rendered invisible by their presupposed normality and designation as the unspoken standard against which other categories are compared and subordinated. Essentially, by defining what is expected, social institutions shape what groups, norms, and beliefs will be legitimated, socially reinforced, and empowered, and which will be eschewed, negatively sanctioned, and marginalized.

This is not to say that institutions unilaterally *determine* how we think, feel, and behave. Though institutions are powerful features of the social structure that mold and shape human experience, we are not merely passive recipients of institutional imperatives. On a daily basis, most of us make both conscious and unconscious decisions about whether to conform, accommodate, resist, or transgress the normative boundaries set forth within institutional domains.

Though the family is the institution with the most obvious interest in defining and organizing sexuality because of its link to procreation, all social institutions are complicit in the construction and regulation of sexuality. As described above, because institutions define notions of acceptability, deviance, and "normality," some forms of sexual expression are defined as more normal or legitimate than others. These forms (e.g., heterosexuality, male sexual agency, sex as pleasure) are more socially privileged than other forms. For example, heterosexuality is more privileged in American society than being gay, lesbian, bisexual, or transgendered. Examples of how social institutions privilege heterosexuality include the fact that heterosexuals are given the right to marry and thereby obtain tax breaks and workplace benefits such as health care for their spouses. Because their unions are legally sanctioned, heterosexual spouses also enjoy automatic legal recognition giving them the right to inheritance as well the right to make life and death medical decisions for an endangered spouse. To fully understand sexuality, it is therefore necessary to examine the impact social institutions have in defining, regulating, and stratifying the opportunities individuals have to freely and equally express their sexuality.

The purpose of this section is to illustrate the specific ways that social institutions construct and organize sex (including sexual norms, values, identities, and expectations), within and around the different spheres of our lives. In addition, these chapters also include materials that document how individuals routinely challenge and subvert these same social forces. The influence of seven key social institutions will be addressed including the state, sexual politics and public policy (Chapter 6); the media and popular culture (Chapter 7); and the economy and work (Chapter 8). In addition, this section includes combined chapters on religion and education (Chapter 9), as well as the family and medicine (Chapter 10). Selections will include both macro- and micro-level analyses. Key articles include John D'Emilio's classic work, "Capitalism and Gay Identity" (Chapter 8), Mimi Shipper's analysis of the negotiation of gender and sexuality in an alternative rock subculture (Chapter 7), and Patricia Hill Collins' important contributions in "The Sexual Politics of Black Womanhood" (Chapter 6).

6

The State, Sexual Politics, and Public Policy

It Takes More Than Two: The Prostitute, the Soldier, the State, and the Entrepreneur

CYNTHIA ENLOE

Cynthia Enloe examines the role of the military brothel in servicing the needs of both soldier and state. Her analysis reveals the calculated collusion of a multitude of individual and state agents in the creation, maintenance, and regulation of houses of prostitution for use by military personnel. Her work also reveals how the brothel plays a central role in the construction of militarized femininity and masculinity.

Since U.S. occupation troops in Japan are unalterably determined to fraternize, the military authorities began helping them out last week by issuing a phrase book. Sample utility phrases: "You're very pretty" . . . "How about a date?" . . . "Where will I meet you?" And since the sweet sorrow of parting always comes, the book lists no less than 14 ways to say goodbye.

TIME, JULY 15, 1946

On a recent visit to London, I persuaded a friend to play hooky from work to go with me to Britain's famous Imperial War Museum. Actually, I was quite embarrassed. In all my trips to London, I had never visited the Imperial War Museum. But now, in the wake of the Gulf War, the time seemed ripe. Maybe the museum would help put this most recent military conflict in perspective, mark its continuities with other wars, and clarify its special human, doctrinal, and technological features. I was in for a disappointment.

Only selective British experiences of the "great" wars were deemed worthy of display. Malaya, Aden, Kenya, the Falklands—these British twentieth-century war zones didn't rate display cases. In fact, Asia, Africa, and the West Indies didn't seem much on the curators' minds at all. There were two formal portraits of turbaned Indian soldiers who had won military honors for their deeds, but there were no displays to make visible to today's visitors how much the British military had relied on men and women from its colonies to fight both world wars. I made a vow on my next trip to take the train south of London to the Gurkha Museum.

The only civilians who received much attention in the Imperial War Museum were British. Most celebrated were the "plucky" cockney Londoners who coped with the German blitz by singing in the Underground. Women were allocated one glass case showing posters calling on housewives to practice domestic frugality for the cause. There was no evidence, however, of the political furor set off when white British women began to date—and have children with—African-American GIs.

Our disappointment with the museum's portrayal of Britain's wars served to make us trade hunches about what a realistic curatorial approach might be. What would we put on display besides frontline trenches (which at least showed the rats), cockney blitz-coping lyrics, and unannotated portraits of Sikh heroes?

Brothels. In my war museum there would be a reconstruction of a military brothel. It would show rooms for officers and rooms for rank-and-file soldiers. It would display separate doors for white soldiers and black soldiers. A mannequin of the owner of the business (it might be a disco rather than a formal brothel) would be sitting watchfully in the corner—it could be a man or a woman, a local citizen or a foreigner. The women serving the soldiers might be White European, Berber, Namibian, or Puerto Rican; they might be Korean, Filipino, Japanese, Vietnamese, African-American, or Indian. Depending on the era and locale, they could be dressed in sarongs, saris, or miniskirts topped with T-shirts memorializing resort beaches, soft drinks, and aircraft carriers.

In this realistic war museum, visitors would be invited to press a button to hear the voices of the women chart the routes by which they came to work in this brothel and describe the children, siblings, and parents they were trying to support with their earnings. Several of the women might compare the sexual behavior and outlook of these foreign men with those of the local men they had been involved with. Some of the women probably would add their own analyses of how the British, U.S., French, or United Nations troops had come to be in their country.

Museum goers could step over to a neighboring tape recorder to hear the voices of soldiers who patronized brothels and discos while on duty abroad. The men might describe how they imagined these women were different from or similar to the women from their own countries. The more brazen might flaunt their sexual prowess. They might compare their strength, chivalry, or earning power with that of the local men. Some of the soldiers, however, would describe their feelings of loneliness, their uncertainty of what it means to be a man when you're a soldier, their anxieties about living up to the sexual performance expectations of their officers and buddies.

War—and militarized peace—are occasions when sexual relations take on particular meanings. A museum curator—or a journalist, novelist, or political commentator—who edits out sexuality, who leaves it on the cutting-room floor, delivers to the audience a skewed and ultimately unhelpful account of just what kinds of myths, anxieties, inequalities, and state policies are required to fight a war or to sustain a militarized form of peace.

[Feminist ethnograhies and oral histories] he1p us to make sense of militaries' dependence on—yet denial of—particular presumptions about masculinity to sustain soldiers' morale and discipline. Without sexualized rest and recreation, would the U.S. military command be able to send young men off on long, often tedious sea voyages and ground maneuvers? Without myths of Asian or Latina women's compliant sexuality, would many American men be able to sustain their own identities, their visions of themselves as manly enough to act as soldiers?

Women who have come to work as prostitutes around U.S. bases tell us that a militarized masculinity is constructed and reconstructed in smoky bars and sparsely furnished rented rooms. If we confine our curiosity only to the boot camp and the battlefield—the focus of most investigations into the formations of militarized masculinity—we will be unable to explain just how masculinity is created and sustained in the peculiar ways still imagined by officials to be necessary to sustain a modern military organization.

We will also miss just how much governmental authority is being expended to insure that a peculiar definition of masculinity is sustained. Military prostitution differs from other forms of industrialized prostitution in that there are explicit steps taken by state institutions to protect the male customers without undermining their perceptions of themselves as sexualized men.

"Close to 250,000 men a month paid three dollars for three minutes of the only intimacy most were going to find in Honolulu."[1] These figures come from records kept in Hawaii during 1941 and 1944. Historians have these precise figures because Honolulu brothel managers, most of whom were white women, had to submit reports to Hawaii's military governor. American soldiers' sexual encounters with local prostitutes were not left to chance or to the market; they were the object of official policy consideration among the military, the police, and the governor's staff. Two hundred and fifty prostitutes paid $1.00 per year to be registered merely as "entertainers" with the Honolulu Police Department because the federal government had passed the May Act in 1941, making prostitution illegal, to assuage the fears of many American civilians that mobilizing for war would corrupt the country's sexual mores.[2] Hawaii's military governor disagreed. He had police and military officers on his side. They saw a tightly regulated prostitution industry as necessary to bolster male soldiers' morale, to prevent sexually transmitted diseases, and to reassure the Hawaiian white upper class that wartime would not jeopardize their moral order. The navy and the army set up prophylaxis distribution centers along Honolulu's Hotel Street, the center of the city's burgeoning prostitution industry. The two departments collaborated with the local police to try to ensure that licensed prostitutes kept their side of the bargain: in return for the license, women servicing soldiers and sailors up and down Hotel Street had to promise to have regular medical examinations, not to buy property in Hawaii, not to own an automobile, not to go out after 10:30 at night, and not to marry members of the armed forces. The objective was to keep prostitutes quite literally in their place.

Before the war, most Hotel Street brothels had two doors, one for white male customers and one for men of color, most of whom were Asian men who worked on the island's pineapple and sugar plantations. Brothel managers believed this segregation prevented violent outbursts by white men who objected to the women they were paying for servicing men of any other race. As the wartime influx of white soldiers and sailors tilted the brothels' business ever more toward a white clientele, most managers decided that any risk of offending white male customers was bad business; they did away with the second door and turned away men of color altogether.

Opening time for the typical Honolulu brothel during the war years was 9 A.M. It operated on an efficient assembly-line principle. From prostitutes and soldiers recalling the arrangement, we learn that most of the brothels used what was called a "bull-ring" setup consisting of three rooms. "In one room a man undressed, in a second the prostitute engaged her customer, in a third a man who had finished put his clothes back on."[3] Prostitutes learned to tailor their services to the sexual sophistication of their military clients. They offered oral sex to the more nervous and inexperienced men.

Today British and Belize officials work hard together to develop a complex policy to ensure a steady but safe supply of military prostitutes for the British troops stationed in that small ex-colony perched on the edge of Latin America.[4] A new 900-man batallion arrives every six months. British soldiers have special brothels designated for their patronage, although they slip out of the carefully woven policy net to meet local women in bars and discos in Belize City. Most of the women who work in the officially approved brothels are Latinas, rather than Afro-Belize women; many have traveled across the border from war-torn Guatemala to earn money as prostitutes.

The government-to-government agreement requires that every brothel worker, with the cooperation of the owners, have a photo identification card and undergo weekly medical examinations by a Belizean doctor. Prostitutes are required to use condoms with their military customers, although it is not clear how many women may be paid extra by their customers to break the condom rule. If a soldier-patron does show symptoms of a sexually transmitted disease or tests positive for HIV, it is assumed that the prostitute is to blame.

British-born soldiers and their Nepali Gurkha comrades, both in Belize under a Belize-British defense pact, have rather different racial/sexual preferences. Whereas the former are likely to frequent both Latina and Afro-Belize women, the Gurkhas reportedly prefer Latina women, which means that the Gurkhas are more likely to stick to the government-approved prostitutes. The fact that any Gurkha troops go to prostitutes at all, however, contradicts the long-standing British portrayal of Nepali militarized masculinity: though White British men's masculinity is presumed by their officers to require a diet of local sex while overseas, Nepali men's masculinity is constructed as more disciplined, faithful when home and celibate while on assignment abroad.[5] With the end of the Cold War and the relaxation of political tensions between Belize and Gustemala, the future of the government-to-government prostitution agreement has become uncertain. But in early 1992, Britain's Chief of Defense Staff, Field Marshal Sir Richard Vincent, made it known publicly that the Conservative government of John Major was hoping that the British troop rotation in Belize could be continued. Though no longer needed to defend Belize, the British army, according to the field marshal, now finds Belize's climate and topography especially attractive for jungle warfare training.[6] Do the field marshal and his superiors back in London perhaps also find the Belize government's willingness to cooperate in the control of local women's sexuality a military attraction?

The United States fashioned a rather different policy to regulate soldiers' relationships with prostitutes around major U.S. bases such as Clark and Subic Bay in the Philippines. Like the British, the Americans supported compulsory medical examinations of women working as prostitutes. Similarly, women without the license issued with these examinations were prevented from working by the local—in this case Filipino—municipal authorities. U.S. soldiers who contracted sexually transmitted diseases (STDs) were not required to report the woman whom they believe gave them the disease. Nonetheless, it was the practice of the Angeles City and Olongapo health authorities to pass on to U.S. base officials the names of sex workers who had contracted STDs. The base commanders then ordered that the photographs of infected Filipinas be pinned upside down on the public notice board as a warning to the American men.[7]

Apparently believing that "stable" relationships with fewer local women would reduce the chances that their personnel would become infected, base commanders allowed Filipinas hired out by bar owners to stay with their military boyfriends on the base. U.S. officials occasionally sent out a "contact" card to a club owner containing the name of a Filipino employee whom the Americans suspected of having infected a particular sailor or air force man. However, they refused to contribute to the treatment of prostitutes with sexually transmitted diseases or AIDS and turned down requests that they subsidize Pap smears for early cancer detection for the estimated one hundred thousand women working in the entertainment businesses around Clark and Subic.

The closing of both Clark Air Force Base and Subic Naval Base in 1992 forced many Filipinas in precarious states of health into the ranks of the country's unemployed. Their few options included migrating to Okinawa or Guam, or even to Germany, to continue working as prostitutes for U.S. military men. They may also have been vulnerable to recruiters procuring Filipino women for Japan's entertainment industry, an industry that is increasingly dependent on young women from abroad.[8] Olongapo City's businessman mayor, with his own entertainment investments now in jeopardy, has been in the forefront of promoters urging that Subic Bay's enormous facilities be converted into private enterprises, although the Filipino military is also eager to take over at least part of the operations for its own purposes. Military base conversion is always an intensely gendered process. Even if women working the entertainment sector are not at the conversion negotiation table, they will be on many of the negotiators' minds. For instance, the above-mentioned mayor, among others, has urged not only that privatized ship maintenance be developed at Subic Bay, but also that tourism development be high on the new investment list.[9] In the coming years, the politics of prostitution in Olongapo City may take on a civilian look, but many of the tourists attracted may be slightly older American men trying to relive their earlier militarized sexual adventures with Filipino women.

There is no evidence thus far that being compelled by the forces of nature and nationalism to shut down two of their most prized overseas bases has caused U.S. military planners to rethink their prostitution policies. Shifting some of the Philippines operations to Guam or Singapore or back home to the United States does not in itself guarantee new official presumptions about the kinds of sexual relations required to sustain U.S. military power in the post-Cold War world. The governments of Singapore and the United States signed a basing agreement in Tokyo in mid-1992. But, despite popular misgivings about the implications of allowing U.S. Navy personnel to use the small island nation for repairs and training, the basing agreement itself was kept secret. Thus, Singapore citizens, as well as U.S. citizens, are left with little information about what policing formulas, public health formulas, and commercial zoning formulas have been devised by the two governments to shape the sexual relations between American and Singapore men and the women of Singapore.[10]

The women who have been generous enough to tell their stories of prostitution have revealed that sexuality is as central to the complex web of relationships between civil and military cultures as are more talked-about security doctrines and economic quid pro quo. Korean and Filipino women interviewed by Sandra Sturdevant and Brenda Stoltzfus for their oral history collection *Let the Good Times Roll* also remind us of how hard it is sometimes to map the boundaries between sexual relations and economics.[11] They found that the local and foreign men who own the brothels, bars, and discos catering to soldiers are motivated by profit. These men weigh the market value of a woman's virginity, her "cherry," as well as her age. They constantly reassess their male clients' demands. Thus, by the early 1990s, bar owners and procurers concluded that AIDS-conscious U.S. soldiers were competing to have sex with younger and younger Filipinas, and so the proprietors sought to supply them, driving down the value of the sexual services supplied by "older" women—women in their early twenties.[12]

Over the decades, U.S. Navy veterans stayed in the Philippines and set up bars and discos, both because they liked living outside the United States (often with Filipino wives) and because they could make a comfortable livelihood from sexualized entertainment. Australian men immigrated to launch their own businesses in the base towns and eventually made up a large proportion of the owners of the military-dependent entertainment industry.[13] Local military personnel, especially officers, also used their status and authority in the rural areas to take part in the industry. Some men in the Philippines military have been known to supplement their salaries by acting as procurers of young rural women for the tourist and military prostitution industries.[14] Similarly, among the investors and managers of Thailand's large prostitution industry are Thai military officers.[15] Militarized, masculinized sexual desire, by itself, isn't sufficient to sustain a full-fledged prostitution industry. It requires (depends on) rural poverty, male entrepreneurship, urban commercialized demand, police protection, and overlapping governmental economic interest to ensure its success.

Yet military prostitution is not simply an economic institution. The women who told their stories to Sturdevant and Stoltzfus were less concerned, with parsing analytical categories—what is "economic," what is "social," and what is "political"—than with giving us an authentic account of the pressures, hopes, fears, and shortages they had to juggle every day in order to ensure their physical safety, hold onto some self-respect, and make ends meet for themselves and their children.

The stories that prostitutes tell also underscore something that is overlooked repeatedly in discussions of the impact of military bases on local communities: local women working in military brothels and discos mediate between two sets of men, the foreign soldiers and the local men—some of whom are themselves soldiers, but many of whom are civilians. Outside observers rarely talk about these two sets of men in the same breath. But the women who confided in Stoltzfus and Sturdevant knew that they had to be considered simultaneously. The Korean and Filipino women detailed how their relationships with local male lovers and husbands had created the conditions that initially made them vulnerable to the appeals of the labor-needy disco owners. Unfaithfulness, violent tempers, misuse of already low earnings, neglectful fathering—any combination of these behaviors by their local lovers and husbands might have launched these women into military prostitution. Children, too, have to be talked about. Most of the women servicing foreign soldiers sexually have children, some fathered by local men and others fathered by the foreign soldiers. Prostitution and men's ideas about fathering: the two are intimately connected in these women's lives.

In deeply militarized countries such as the Philippines, South Korea, Honduras, and Afghanistan, a woman working in prostitution may have to cope with local as well as foreign soldiers who need her services to shore up their masculinity. Because it is politically less awkward to concentrate on foreign soldiers' exploitation of local women, local soldiers' militarized and sexualized masculinity is frequently swept under the analytical rug, as if it were nonexistent or harmless. And in fact the local soldiery may have more respect for local women, may have easier access to noncommercialized sex, or may have too little money to spend to become major customers of local prostitutes. But none of those circumstances should be accepted as fact without a close look.

For instance, Anne-Marie Cass, an Australian researcher who spent many months in the late 1980s both with the Philippine government's troops and with insurgent forces, found that Filipino male soldiers were prone to sexualizing their power. Cass watched as many of them flaunted their sexualized masculinity in front of their female soldier trainees, women expected from respectable families to be virgins. She also reported that many Filipino soldiers "expect to and receive rides on civilian transport, and drinks and the services of prostitutes in discos and bars without payment."[16]

This is not, of course, to argue that local men are the root of the commercialized and militarized sex that has become so rife, especially in countries allied to the United States. Without local governments willing to pay the price for the lucrative R and R business, without the U.S. military's strategies for keeping male soldiers content, without local and foreign entrepreneurs willing to make their profits off the sexuality of poor women—without each of these conditions, even an abusive, economically irresponsible husband would not have driven his wife into work as an Olongapo bar girl. Nonetheless, local men must be inserted into the political equation; the women who tell their stories make this clear. In fact, we need to widen our lens considerably if we are to fully understand militarized prostitution. Here is a list—probably an incomplete list—of the men we need to be curious about, men whose actions may contribute to the construction and maintenance of prostitution around any government's military base:

- husbands and lovers
- bar and brothel owners, local and foreign
- local public health officials
- local government zoning board members
- local police officials
- local mayors
- national finance ministry officials
- national defense officials
- male soldiers in the national forces
- local civilian male prostitution customers
- local male soldier-customers
- foreign male soldier-customers
- foreign male soldiers' buddies
- foreign base commanders
- foreign military medical officers
- foreign national defense planners
- foreign national legislators

Among these men there may be diverse forms of masculinity. Women in Okinawa, Korea, and the Philippines described to Sturdevant and Stoltzfus how they had to learn what would make American men feel manly during sex; it was not always what they had learned would make their Korean, Japanese, or Filipino sexual partners feel manly.

Sexual practice is one of the sites of masculinity's—and femininity's—daily construction. That construction is international. It has been so for generations. Tourists and explorers, missionaries, colonial officials and health authorities, novelists, development technocrats, businessmen, and soldiers have long been the internationalizers of sexualized masculinity. Today the U.S. military's "R and R" policy and the industry it has spawned function only if thousands of poor women are willing and able to learn those sexual acts that U.S. military men rely on to bolster their sense of masculinity. Thus, bar owners, military commanders, and local finance ministry bureaucrats depend on local women to be alert to the historically evolving differences between masculinities.

Korean women have been among the current historical investigators of militarized prostitution. Korean women petitioners, together with a small, supportive group of Japanese feminists and Japanese historians, recently pressed the Japanese government to admit that the Japanese military had a deliberate policy of conscripting Korean, Thai, and Burmese women into prostitution during World War II.[17] In the past, Japanese officials insisted that any Asian women pressed into servicing Japanese soldiers sexually during the war were organized and controlled by civilian businessmen. The military itself was institutionally immune. Senior officers had simply accepted the prostituted women as part of the wartime landscape. This defense is strikingly similar to that employed by U.S. officials when asked about the Pentagon's current prostitution policy. Their Japanese counterparts, however, have had to give up their long-time defense in the face of convincing bureaucratic evidence uncovered by Yoshiaki Yoshimi, a professor of history at Chuo University. In the Self-Defense Agency's library he found a document entitled "Regarding the Recruitment of Women for Military Brothels" dating from the late 1930s, when the Japanese army was moving southward into China. It ordered the military to build "facilities for sexual comfort." The official rationale was that brothels would stop Japanese soldiers from raping Chinese and other women along the route of the army's invasion. Eventually, an estimated 100,000 to 200,000 Asian women were forcibly conscripted to work as *Karayuki-san,* "comfort women," in these military brothels.[18]

Although the uncovering of the document evoked a formal apology from Prime Minister Kiichi Miyazawa, the issue is not resolved. Kim Hak Sun, one of the survivors of the "comfort women" program, and other elderly Korean women are calling on their government and the Japanese government to reach a settlement that will include monetary compensation for the hardships they suffered.[19]

Furthermore, the internationalizing dynamics which have shaped military prostitution in the past grind on. Thus, the uncovering of 1930s and 1940s Japanese policy on prostitution led to a spate of articles in the U.S. media at a time when many Americans were in search of evidence that they were morally superior to, albeit economically lagging behind, Japan. Thus the story was set in a Pearl Harbor context by many U.S. readers, even if not intentionally by its authors. It could have been quite a different story. The research by Yo-shiaki Yoshimi, Nakahara Michiko, and other Japanese historians about their country's military's prostitution policies could have been written—and read—so as to draw attention to U.S., British, French, and other militaries' past and present prostitution policies.

This possibility was what inspired Rita Nakashima Brock to write to the *New York Times* in the wake of the discovery of the Tokyo document. A researcher studying the sex industries in Southeast Asia, she is also an Asian-American woman who spent her childhood on U.S. military bases in the United States, Germany, and Okinawa. She recalls that, as a girl, "I faced the assumption that any woman who looked Asian was sexually available to soldiers. I was often called 'geisha-girl' or 'Suzy Wong' (soldiers usually couldn't tell Japanese from Chinese). Every base I ever lived on . . . had a thriving red-light district near it." When she was older, Brock began to wonder about official military policies that led to the prostitution she had witnessed as a child. "A former Navy chaplain who served in Japan during the post-World War II occupation told me that when he protested the American base commander's efforts to set up prostitution centers using Japanese women, he was reassigned stateside."[20]

Thanh-Dam Truong, a Vietnamese feminist who has investigated the political economy of Thailand's prostitution industry, also reminds us to view sexuality historically. Thai women working in prostitution, she discovered, had to learn new sexual skills in the 1980s that they hadn't needed in the 1960s because by the 1980s their male customers, now mainly local and foreign civilians, had acquired new tastes, new insecurities, and new grounds for competing with other men.[21] Similarly, around the U.S. Navy base at Subic Bay in the late 1980s, bar owners, still dependent on military customers, introduced "foxy boxing." These entrepreneurs believed that having women wrestle and box each other on stage would make the American sailors in the audience more eager for sex with the Filipino employees. Women, in turn, learned that they would be paid for their performance only if at the end of a bout they could show bruises or had drawn blood.[22] At about the same time, women in the bars were instructed by their employers to learn how to pick up coins with their vaginas. This, too, was designed as a new way to arouse the American customers.[23]

Each group of men involved in militarized sexuality is connected to other groups by the women working in the base town bars. But they also may be connected to each other quite directly. At least some Filipino male soldiers are adopting what they see as an American form of militarized masculinity. The men most prone to adopting such attitudes are those in the Scout Rangers, the elite fighting force of the Philippine Constabulary. They act as though Rambo epitomizes the attributes that make for an effective combat soldier: "a soldier in khaki or camouflage, sunglasses or headbands, open shirt, bare head, and well armed, lounging in a roofless jeep traveling down a Davao City street, gun held casually, barrel waving in the air."[24] One consequence of this form of borrowed masculinized intimidation is that local prostitutes servicing Filipino soldiers perform sexual acts that they otherwise would refuse to perform.

A woman who comes to work in a foreign military brothel or disco finds that she must negotiate among all of these male actors. She has direct contact, however, with only some of them. She never hears what advice the foreign base commander passes on to his troops regarding the alleged unhealthiness or deviousness of local women. She never hears what financial arrangements local and foreign medical officials devise to guarantee the well-being of her soldier-customers. She rarely learns what a soldier who wants to marry her and support her children is told by his military chaplain or superior officer. She is not invited into the conference room when U.S., British, or French legislators decide it is politically wise not to hold hearings on their government's military prostitution policy. The Latina woman working as a prostitute in Belize or the Filipino woman working in the Philippines or Okinawa makes her assessments using only what information she has.

Much of that information comes from the women with whom she works. The women who told their stories to Sturdevant and Stoltzfus did not romanticize the sistership between women working in the bars. The environment is not designed to encourage solidarity. Women *have* engaged in collective actions—for instance, bar workers in Olongapo protested against being forced to engage in boxing matches for the entertainment of male customers. But, despite growing efforts by local feminists to provide spaces for such solidarity, collective action remains the exception. Most women rely on a small circle of friends to accumulate the information necessary to walk the minefield laid by the intricate relationships between the various groups of men who define the military prostitution industry. The women teach each other how to fake orgasms, how to persuade men to use a condom, how to avoid deductions from their pay, how to meet soldier-customers outside their employers' supervision, and how to remain appealing to paying customers when they are older and their valued status as a "cherry girl" is long past.

Women are telling their prostitution stories at a time when the end of the Cold War and the frailty of an industrialized economy are combining to pressure governments in North America and Europe to "downsize" their military establishments.

Base closings have their own sexual consequences. [Because of the closing of the Subic Bay Navy Base] U.S. military and civilian men and their Filipino lovers had to discuss the possibility of marriage, perhaps each with quite different fears and expectations. There were reports of a number of quick marriages.[25] [During] the departure of the last U.S. ship from Subic Bay. Filipino women from Olongapo's bars cried and hugged their sailor boyfriends and customers at the gates of the base.[26] What sexual expectations would the American men take home with them? Perhaps the Filipinas' tears and hugs prompted many men to imagine that they had experienced not commercialized sex but rather relationships of genuine affection. What were women shedding tears for? Perhaps for the loss of some temporary emotional support. Or maybe for the loss of their livelihoods. How many women who have lost their jobs around Subic Bay will seek out the employment agencies that, for a fee, will send them to Kuwait to work as maids.[27]

It might be tempting to listen to Asian women's stories as if they were tales of a bygone era. That would, I think, be a mistake. Large bases still exist in South Korea and Guam. Over forty thousand American military personnel were stationed in Japan (including Okinawa) at the end of 1991; even more will be redeployed from Clark and Subic Bay. In early 1992, the U.S. government made agreements with officials in Australia, Singapore, and Malaysia to use facilities in their countries for repairs, communications, and training. Even with some cutbacks, the number of American men going through those bases on long tours and on shorter-term maneuvers will be in the thousands. Governments in Seoul, Tokyo, and Manila have made no moves to cancel the R and R agreements they have with Washington, agreements that spell out the conditions for permitting and controlling the sort of prostitution deemed most useful for the U.S. military. The no-prostitution formula adopted to fight the Gulf War—a no-prostitution formula not initiated by Washington policymakers, but rather imposed on the United States by a Saudi regime nervous about its own Islamic legitimacy—has not been adopted anywhere else. What discussions have U.S. military planners had with their counterparts in Singapore, Canberra, and Kuala Lumpur about morale, commerce, health, and masculinity?

Listening to women who work as prostitutes is as important as ever. For political analysts, listening to them can provide information necessary for creating a more realistic picture of how fathering, child-rearing, man-to-man borrowing, poverty, private enterprise, and sexual practice play vital roles in the construction of militarized femininity and masculinity. For nonfeminist anti-base campaigners, listening to these women will shake the conventional confidence that has come from relying only on economic approaches to base conversion. Marriage, parenting, male violence, and self-respect will all have to be accepted as serious political agenda items if the women now living on wages from prostitution are to become actors, and not mere symbols, in movements to transform foreign military bases into productive civilian institutions. Listening is political.

NOTES

1. Beth Bailey and David Farber, *The First Strange Place: The Alchemy of Race and Sex in World War II Hawaii* (New York: Free Press, 1992), 95.
2. Ibid., 102–103. The material that follows is based on Bailey and Farber, 95–107.
3. Ibid., 102–103.
4. The information on Belize is contained in a manuscript by Stephaine C. Kane, "Prostitution and the Military: Planning AIDS Intervention in Belize" (Department of American Studies and African-American Studies, State University of New York at Buffalo, 1991); information on the Gurkhas is form correspondence from Stephaine Kane, December 11, 1991.
5. Tamag, "Nepali Women as Military Wives."
6. "Troops Want to Stay in Belize," *Carib News* (New York), March 17, 1992.
7. The information on Subic Bay and Clark bases is derived from Anne-Marie Cass, "Sex and Military: Gender and Violence in the Philippines" (Ph.D. diss., Department of Sociology and Anthropology, University of Queensland, Brisbane, Australia, 1992), 206–209; and Saundra Sturdevant and Brenda Stoltzfus, *Let the Good Times Roll: The Sale of Women's Sexual Labor around U.S. Military Bases in the Philippines, Okinawa and the Southern Part of Korea* (New York: New Press, 1992).
8. For descriptions and analyses of the lives of Filipino women and men who have migrated to Japan, including many women who went there for exploitative work in the entertainment industry catering to male customers—see Randolf S. David, "Filipino Workers in Japan: Vulnerability and Survival," *Kasarinlan: A Philippine Quarterly of Third World Studies* (Quezon City: University of the Philippines) 6, no. 3 (1991): 9–23; Rey Ventura, *Underground in Japan,* London, Jonathan Cape, 1992.
9. Rigoberto Tiglao, "Open for Offers" *Far Eastern Economic Review,* October 15, 1992, 62–63.
10. I am grateful to Suzaina Abdul Kadii, of the University of Wisconsin political science graduate program, for her analysis of the U.S.-Singaporean basing agreement process: conversation with the author, Madison, Wisconsisn, October 29, 1992.
11. Sturdevant and Stoltzfus, *Let the Good Times Roll.*
12. Cass, "Sex and the Military," 210.
13. Ibid., 205.
14. Ibid., 215.
15. The most complete account of the Thai military's role in Thailand's prostitution industry is Thanh-Dam Truong, *Sex, Money and Morality: Prostitution and Tourism in southeast Asia* (London: Zed Press, 1990). I am also indebted to Alison Cohn for sharing her as yet unpublished research in

Thailand with me at Clark University, Worcester, MA, February–April, 1992. For an investigation of Indonesia's prostitution business, a system which is not organized around either foreign tourists or foreign soldiers but is deeply affected by Indonesia's militarized national politics, see Saraswati Sunindyo's forthcoming Ph.D. dissertation (Department of Sociology, University of Wisconsin, Madison). Saraswati Sunindyo has also written a collection of poetry, entitled *Yakin* (typescript, 1992), which describes some of her own responses to conducting research in a coastal town's government-owned hotel, which was shared by a number of Indonesian women working as prostitutes servicing Indonesian military officers, civil servants, businessmen, farmers, and schoolboys.

16. Cass, "Sex and the Military."
17. Nakahara Michiko, "Forgotten Victims: Asian and Women Workers on the Thai-Burma Railway," *AMPO: Japan-Asian quarterly* 23, no. 2 (1991): 21–25; Yoshiaki Yoshimi, "Japan Battles Its Memories" (Editorial), *New York Times,* March 11, 1992; Saner, "Japan Admits"; David E. Sanger, "History Scholar in Japan Exposes a Brutal Chapter," *New York Times,* January 27, 1992.
18. Sanger, "History Scholar."
19. Ibid. See also: George Hicks, "Ghosts Gathering: Comfort Women Issue Haunts Tokyo as Pressure Mounts," *Far Eastern Economic Review,* February 18, 1993, 32–37.
20. Rita Nakashima Brock, "Japanese Didn't Invent Military Sex Industry" (Letter to the Editor), *New York Times,* February 23, 1992.
21. Truong, "Sex, Money and Morality."
22. Cass, "Sex and the Military," 210.
23. Sturdevant and Stoltzfus, *Let the Good Times Roll.* In a slide and tape show produced by Sturdevant and Stoltzfus, Filipinas describe being ashamed at having to perform demeaning acts. "Pussy Cat III," 726 Gilman St., Berkeley, CA 94710.
24. Anne-Marie Cass, "Sexuality, Gender and Violence in the Militarized Society of the Philippines" (Paper presented at the annual conference of the Australian Sociological Association, Brisbane, December 12–16, 1990), 6.
25. Donald Goertzen, "Withdrawal Trauma," *Far Eastern Economic Review,* January 30, 1992, 10.
26. Pat Ford, "Weekend Edition," National Public Radio, March 21, 1992.
27. I am grateful to Lauran Schultz for bringing to my attention the *Philippine Journal of Public Administration* 34, no. 4 (October 1990), a special issue devoted to articles on the current conditions of Filipino women, including women as migrants. See, in particular, Bievenda M. Amarles, "Female Migrant Labor: Domestic Helpers in Singapore," 365–389; Prosperina Domingo Tapales, "Women, Migration and the Mail-Order Bride Phenomenon: Focus on Australia," 311–322.

DISCUSSION QUESTIONS

1. In a previous chapter, Messner argued that men construct heterosexuality and sublimate same-sex attraction. Do you think this type of sublimation plays a role in the demand for prostitutes by military personnel? What role, if any, do you think the primarily male homo-social organizational structure of the military plays in this?
2. Discuss the logic of military organizations participating in the creation and regulation of prostitution in the hopes of stopping soldiers from raping citizens along the route of their army's invasion. Do paying for sex with a prostitute and raping enemy women serve the same purposes? What assumptions about sexuality (particularly male sexuality) does this speak to?
3. What personal examples could the prostitutes discussed in this article provide to support the argument that sex is socially constructed?
4. What costs derive from the fact that most "official" military policies appear to deny or hide their complicity in the sex trade? Who benefits from these policies? Who is harmed?
5. Each year more and more women are serving in the U.S. military. What impact, if any, do you believe the increasing presence of women on foreign bases will have on the sexual practices of male soldiers? What kind of arrangements, if any, do you think will be made to accommodate the sexual needs of female soldiers?
6. Racism clearly plays a role in military prostitution (i.e., the prostitute is the eroticized "other"). What other social hierarchies do you see being played out in these settings? What parallels, if any, can you draw between this article and other works in this reader?

The Case of Sharon Kowalski and Karen Thompson: Ableism, Heterosexism, and Sexism

JOAN L. GRISCOM

Tragic accidents can occur to anyone at any point in their life, but in the aftermath of tragedy it is the state that has the power and responsibility to decide who will make decisions about the treatment, life, and even the ultimate death of the victims. In the case described below, Joan L. Griscom examines how normative social expectations about sex, sexuality, and physical ability are expressed and enforced through law, institutional policies and practices, and interpersonal behaviors. She demonstrates how these structures operate to deny some categories of people equal rights to self-determination and the power to care for those they love.

In November, 1983, Sharon Kowalski was in a head-on collision with a drunk driver, suffered a severe brain-stem injury, became paralyzed, and lost the ability to speak. Sharon was in a committed partnership with Karen Thompson. Serious conflict soon developed between Karen and Kowalski's parents, erupting in a series of lawsuits that lasted eight years. Karen fought to secure adequate rehabilitation for Sharon as well as access to friends and family of her choice. In 1985, acing under Minnesota guardianship laws, Sharon's father placed her in a nursing home without adequate rehabilitation services and prohibited Karen and others from visiting her. Karen continued to fight through the courts and the media. In 1989, Sharon was finally transferred to an appropriate rehabilitation facility, reunited with lover and friends, and, in 1991, finally allowed her choice to live with Karen.

In this article I tell the story of Sharon Kowalski and Karen Thompson. While the story shows violations of their human rights, it is more than a story of two individuals. The injustices they encountered were modes of oppression that operate at a social-structural level and affect many other people. These oppressions include ableism, discrimination against disabled persons; heterosexism, the structuring of our institutions to legitimate only heterosexual relationships; and sexism, discrimination against women. Their story shows the power of structural discrimination, the intertwining of both our medical and legal systems in ways that denied both of them the fullest quality of life.

A HISTORY OF THE EVENTS

By November 1983, Sharon and Karen had lived in partnership for almost four years. Karen was thirty-six, teaching physical education at St. Cloud State University, devoutly religious, conservative.

Sharon was twenty-seven, a fine athlete who had graduated from St. Cloud in physical education and just accepted a staff coaching position. She had grown up in the Iron Mine area of Minnesota, a conservative world where women are expected to marry young. Defying such expectations, she became first in her family to attend college, earning tuition working part-time in the mines. After she and Karen fell in love, they exchanged rings, bought a house together, and vowed lifetime commitment.

After the accident Sharon lay in a coma for weeks, and doctors were pessimistic about her recovery. Karen spent hours, daily, talking to her, reading the Bible, massaging and stretching her neck, shoulders, and hands. It is essential to massage and stretch brain-injured persons in comas, for their muscles tend to curl up tightly and incur permanent damage. Early in 1984, Karen saw Sharon was moving her right index finger, and found that she could indicate answers to questions by moving it. Later she began to tap her fingers, then slowly learned to write letters and words.

The Kowalski parents became suspicious of the long hours Karen was spending with her, and increasingly Karen feared they would try to exclude her from Sharon's life. After consulting a psychologist, she wrote them a letter explaining their love, in hopes they would understand her importance to Sharon. They reacted with shock, denial, and rage. As the nightmare deepened, Karen consulted a lawyer and learned she had no legal rights, unless she won guardianship. In March,1984, she therefore filed for guardianship, and Donald Kowalski counterfiled.

Guardianship was awarded to Kowalski, but Karen was granted equal access to medical and financial information and full visitation rights. She continued to participate in both physical and occupational therapy. Sharon improved slowly; Karen made her an alphabet board, and she began to spell out answers to questions. Later she began to communicate by typewriter, and in August spoke a few words. But conflicts continued. The day after the court decision Kowalski incorrectly told Karen she did not have visitation rights, and later tried to cancel her work with Sharon's therapists. When Karen and others took Sharon out on day passes, he objected, subsequently testifying in court that he did not want her out in public. In October, Sharon was moved further away, and Kowalski filed to gain full power as guardian. Karen counterfiled to remove him as guardian.

Months elapsed while the legal battles were fought. Sharon was moved several times, regressed in her skills, and became clinically depressed. The Minnesota Civil Liberties Union (MCLU) entered the case, arguing that under the First Amendment Sharon's rights of free speech and free association were being violated. A tri-county Handicap Services Program submitted testimony of Sharon's capacity to communicate, including a long conversation in which she stated she was gay and Karen was her lover. At Sharon's request, the MCLU asked to represent her and suggested she might testify for herself. The court refused both requests, finding that Sharon lacked understanding to make decisions for herself. In July, 1985, Kowalski was awarded full guardianship. Within a day, he denied visitation to Karen, other friends, the MCLU, and disability rights groups; in two days he transferred her to a nursing home near his home with only minimal rehabilitation facilities. In August, 1985, Karen saw Sharon for what would be the last time for over three years.

As this summary indicates, the medical system failed Sharon in at least three respects. First, it failed to supply rehabilitation in the years when it was vital to her recovery. Stark in the medical record is the fact that this woman who was starting to stand and to feed herself was locked away for over three years with an implanted feeding tube, left insufficiently stretched so that muscles that had been starting to work curled back on themselves again. Second, she was deprived of the bombardment of emotional and physical stimulation needed to regenerate her cognitive faculties. Once in the nursing home, for example, she was forbidden regenerative outside excursions. Third, although medical staff often recognized Sharon's unusual response to Karen, they failed to explain to her parents its importance. Despite an urgent need for counseling to assist the parents, none, except for one court-mandated session, took place.

The failure of the medical system was consistently supported by the legal system. Initially the court ruled the Sharon must have access to a young-adult rehabilitation ward. But once Kowalski won full guardianship, he was able to move her to a nursing home without such a ward. In 1985 the Office of Health Facility Complaints investigated Sharon's right to choose visitors, a right guaranteed by the Minnesota Patient Bill of Rights, and found that indeed her right was being violated. However, the appeals court held that the Patient bill of Rights was inapplicable, since the healthcare facility was not restricting the right of visitation, the guardian was.

The deficiencies of guardianship law are a central problem in this case. First, a guardian can restrict a person's rights, without legal recourse. As is often said, under present laws a guardian can lock up a person and throw away the key. This is a national problem, affecting the disabled, the elderly, anyone presumed incompetent. Second, guardians are inadequately supervised. Under Minnesota law, a guardian is required to have the ward tested annually for competence. Kowalski never did, and for over three years the courts did not require him to. In 1985, Karen first filed a motion to hold him in contempt for failure to arrange testing and for failure to heed Sharon's wishes for visitation. The courts routinely rejected such motions.

Between 1985 and 1988, Karen and the MCLU pursued repeated appeals to various Minnesota courts, all denied. Karen began to seek help from the media, also disability, gay/lesbian, women's, and church groups. She recognized that the legal precedents could be devastating for others, e.g. gay/lesbian, couples or unmarried heterosexual couples. The reserved, closeted, conservative professor was slowly transformed into a passionate public speaker in her quest to secure freedom and rehabilitation for Sharon; and slowly she gained national attention. The alternative press responded; national groups such as the National Organization for Women were supportive; the National Committee to Free Sharon Kowalski formed, with regional chapters. Finally the mainstream media began publishing concerned articles; Karen appeared on national TV programs; state and national politicians, including Jesse Jackson, spoke out. Meanwhile Sharon remained in the nursing home, cut off from friends, physically regressed, psychologically depressed.

The first break in the case came in February, 1988. In response to a new motion from Karen, requesting that Sharon be tested for competence, testing was ordered. In January, 1989, she was moved to the Miller-Dwan Medical Center for a 60-day evaluation. Kowalski unsuccessfully argued in court against both the move and the testing. Sharon immediately expressed her wish to see Karen. On February 2, 1989, Karen visited her for the first time in three and a half years, an event which made banner headlines in the alternative press across the nation. Sharon was, however, highly depressed, with numerous physical problems: for example, her feet had curled up so tightly that she was no longer able to stand. More significant was her cognitive ability; to this day, her short-term memory loss remains considerable.

The competency evaluation nevertheless demonstrated that she could communicate on an adult level and had significant potential for rehabilitation. The report recommended "her return to pre-morbid home environment," and added:

> We believe Sharon has shown areas of potential and ability to make rational choices in many areas of her life. She has consistently indicated a desire to return home . . . to live with Karen Thompson again.

Donald Kowalski subsequently resigned as guardian, for both financial and health reasons, and the parents stopped attending medical conferences. In June, 1989, Sharon was transferred to a long-term rehabilitation center for brain-injured young adults. Here she had extensive occupational, physical, and speech therapy. Again Karen spent hours with her and took her out on trips. She had surgery on her legs, feet, toes, left shoulder and arm to reverse the results of three years of inadequate care. She began to use a speech synthesizer and a motorized wheelchair.

Karen subsequently filed for guardianship. Medical staff testified unanimously that Sharon was capable of deciding for herself what relationships she wanted and where she wished to live. They testified that she was capable of living outside an institution and Karen was best qualified to care for her in a home environment. Witnesses for the Kowalskis opposed the petition. The judge appeared increasingly uncomfortable with the national publicity. While in 1990 he allowed Sharon and Karen to fly out to San Francisco where each received a Woman of Courage Award from the National Organization for Women, he refused Sharon permission to attend the first Disability Pride Day in Boston. He issued a gag order against Karen, which was overturned on appeal. Finally, in April, 1991 he denied Karen guardianship and awarded it to a supposedly "neutral third party," a former classmate of Sharon who lived near the Kowalski parents and had testified against Karen in a 1984 hearing. This decision raised the alarming possibility that Sharon might be returned to the inadequate facility. Karen appealed it.

In December, 1991, the appeals court reversed the judge's ruling and granted guardianship to Karen, on two bases: first, the medical testimony that Sharon was able to make her own choices; and second, the fact that the two women are "a family of affinity" that deserves respect. This is a major decision in U.S. legal history, setting important legal precedents both for disabled people and gay/lesbian families. Sharon and Karen now live together.

THE THREE MODES OF OPPRESSION

Sharon and Karen were denied their rights by three interacting systems of oppression: ableism, heterosexism, and sexism. Originally Karen believed that their difficulties were merely personal problems. All her life she had believed that our social institutions are basically fair, designed to support individual rights. In the book[1] she co-authored with Julie Andrzejewski, she documented her growing awareness that widespread social/political forces were involved in their supposedly personal problems and that the oppression they experienced was systemic.

Ableism was rampant throughout. Sharon's inability to speak was often construed as incompetence, and her particular kinds of communication were not recognized. Quite early Karen noticed some did not speak to Sharon, some talked loudly as if she was deaf, others spoke to her as if she were a child. One doctor discussed her in her presence as if she was not there. When Karen later asked how she felt about this, she typed out "Shitty." Probably one reason she responded to Karen more than anyone was that Karen talked extensively and read to her, played music, asked questions, and constantly consulted her wishes. Although the MCLU and the Handicap Services Program submitted transcripts of long conversations with her, the courts did not accept these as evidence of competence, relying instead on testimony from people who had much less interaction with her. A major article in the St. Paul Pioneer Press (1987) described the Kowalskis visiting the room "where their eerily silent daughter lies trapped in her twisted body." Eerily silent? This is the person who typed out "columbine" when asked her favorite flower, answered arithmetical questions correctly, and responded to numerous questions about her life, feelings, and wishes. She also communicates nonverbally in many ways: gestures, smiles, tears, and laughter.

Thanks to ableism, Sharon was often stereotyped as helpless. The presumption of helplessness "traps" her far more severely than her "twisted body." Once a person is labeled helpless, there is no need to consult her wishes, consider her written communications, hear her testimony. When Sharon arrived at Millet-Dwan for competency testing, Karen reported with joy that staff was giving her information and allowing her choices, even if her choice was to do nothing. Most seriously, if a person is seen as helpless, then there is no potential for rehabilitation. As Ellen Bilofsky[2] has written, Sharon was presumed "incompetent until proven competent." If Karen's legal motion for competency testing had not been accepted, Sharon might have remained in the nursing home indefinitely, presumed incompetent.

Finally, ableism can lead to keeping disabled persons hidden, literally out of sight. Kowalski argued against day passes, resisted Karen's efforts to take Sharon out, and testified he would not take her to a church or shopping center because he did not wish to put her "on display . . . in her condition." Although medical staff could see that outside trips provided Sharon with pleasure and stimulation, both important for cognitive rehabilitation, they cooperated with the father in denying them. According to an article in the Washington Post, he once said, "What the hell difference does it make if she's gay or lesbian or straight or anything because she's laying there in diapers? . . . let the poor kid rest in peace."

Invisible in the nursing home, cut off from lover and friends, Sharon had little chance to demonstrate competence. The wonder is that after three and a half years of loss, loneliness, and lack of care, she was able to emerge from her depression and respond to her competency examiners. To retain her capacity for response, through such an experience, suggests a strong spirit.

The second mode of oppression infusing this case is heterosexism, the structuring of our institutions so as to legitimate heterosexuality only. Glaringly apparent is the failure to recognize gay/lesbian partnerships. When Karen was first to arrive at the hospital after the accident, she was not allowed access to Sharon or even any information, because she was not "family." Seeing her anguish, a Roman Catholic priest interceded, brought information, and arranged for a doctor to speak with her. Although the two women considered themselves married, in law they were not, and therefore lacked any legal rights as a couple. If heterosexual, there would have been no denial of visitation, no long nightmare of the three-and-half year separation. While unmarried heterosexual partners might have trouble securing guardianship, married partners would not.

Because of heterosexism, Sharon's emotional need for her partner and Karen's rehabilitative effect on her were not honored. Because of Sharon's response, Karen was often included in the therapeutic work. Yet, prior to 1989, medical staff often refused to testify to this positive effect. Perhaps they feared condoning the same-sex relationship, perhaps they wished to stay out of the conflict. One neurologist, Dr. Keith Larson, did testify, although stipulating that he spoke as friend of the court, not as witness for Karen.

> The reason I'm here today is . . . to deliver an observation that I have agonized over, and thought a great deal about, and prayed a little bit . . . I cannot help but say that Sharon's friend, Karen, can get out of Sharon physical actions, attempts at vocalization, and longer periods of alertness and attention than can really any of our professional therapists.

Why was it necessary to "agonize" over this testimony? pray about it? make such a tremendous effort? Clearly, were one of the partners male, Larson would have had no difficulty. He simply would have reported that the patient responded to her partner. Some medical staff did testify positively, without effort; and after 1989, testimony from medical personnel was strong and unanimous. However, repeatedly, the courts ignored it.

Finally, heterosexism is evident in a consistent tendency to exaggerate the role of sex in same-sex relationships. Many believe that the lives of gay/lesbian people revolve around sex, though evidence from all social-psychological research is that homosexual people are no more sexually active than heterosexual people. Further, gay/lesbian sex is often perceived as sexual exploitation rather than an expression of mutual caring. The final denial of Karen's visitation rights was based on the charge that she might sexually abuse Sharon. A physician hired by the Kowalskis, Dr. William L. Wilson, leveled this charge:

> . . . Karen Thompson has been involved in bathing Sharon Kowalski behind a closed door for a prolonged period of time . . . Ms. Thompson has [also] alleged a sexual relationship with Sharon Kowalski that existed prior to the accident. Based on this knowledge and my best medical judgment. . . . I feel that visits by Karen Thompson at this time would expose Sharon Kowalski to a high risk of sexual abuse.

Accordingly, Wilson directed the nursing home staff not to let Karen visit. Even though under statutes, Karen could have continued to visit while the court decisions were under appeal, the nursing home was obliged to obey the doctor's order.

In this instance, ableism and heterosexism merge. If unmarried heterosexual partners, sexual abuse probably would not have been an issue. If married, the issue would not exist. Ableism often denies disabled persons their sexuality, though a person does not lose her sexuality simply because she becomes disabled. Also, a person who loses the capacity to speak has a special need for touching. What were Sharon's sexual rights? When she was starting to emerge from the coma, she once reached out and touched Karen's breast, and later placed Karen's hand on her breast. At the time Karen did not dare ask medical advice for fear of revealing their relationship. Even to raise such questions might have exposed her to more charges of sexual abuse.

While same-sex relationships are often called "anti-family" in our heterosexist society, actually such relationships create family, in that they create stable emotional and economic units. Family, in this sense, may be defined as a kin-like unit of two or more persons related by blood, marriage, adoption, or primary commitment, who usually share the same household. Sharon and Karen considered themselves married. Karen's long pilgrimage over almost nine years testifies to an extraordinary depth of commitment. Sharon consistently said she was gay, Karen was her lover, she wanted to live with her. While marriage has historically occurred between two sexes, history cannot determine its definition. In U.S. history, marriage between black and white persons was forbidden for centuries. In 1967, when the Supreme Court finally declared miscegenation laws unconstitutional, there were still such laws in sixteen states.

Sexism is sufficiently interfused with heterosexism that they are hard to separate. Often sexism enforces a social role on women in which they are subordinated to men. Women in the Minnesota world where Sharon grew up were expected to marry young and submit to their husbands' authority, an intrinsically sexist model. According to this model, her partnership with Karen was illegitimate. Sexism also is apparent in awarding guardianship to the father. Had Sharon been a man rather than a twenty-eight-year-old "girl," such a decision might be less possible; but in a sexist society, it is appropriate to assign an adult woman to her male parent. Finally, our society devalues friendship, especially between women. Once, very early, a doctor advised Karen to forget Sharon. The gist of his remarks was that "Sharon's parents will always be her parents. They have to deal with this, but you don't. Maybe you should go back to leading your own life." Friendship between the two women was unimportant. Ableism as well as sexism is apparent in these remarks.

This case makes clear that the modes of oppression work simultaneously. Like Audre Lorde,[3] I argue that "there is no hierarchy of oppression." Disability was not more important than sexuality in curtailing Sharon's freedoms; they worked together seamlessly, in her life as in the legal and medical systems. Admittedly, any individual's perspective on the case may reflect the issue most central to her or his life: e.g., the gay press, reporting the case, emphasized heterosexism, and the disability rights press emphasized ableism. Working in coalition on this case, some women were ill at ease with disability rights activists; and some disability rights groups were anxious about associating with gay/lesbian issues. But there are lesbians and gays in the disabled community, and disabled folks in women's groups. Karen experienced the inseparability of the issues once when invited to speak to a Presbyterian group. They asked her to speak only about ableism since they had already "done" gay/lesbian concerns. She tried, but found it nearly impossible; she had to censor her material, ignore basic facts, leave out crucial connections.

In each mode of oppression, one group of persons takes power over another, and this power is institutionalized. Disabled people, women, gay men and lesbians, and others are all to some degree denied their full personhood by the structures of our society. Their choices can be denied, their sexuality is controlled. On the basis of ableism, heterosexism, and sexism, both Karen Thompson's and Sharon Kowalski's opportunities for the fullest quality of life were taken. Sharon lost cognitive ability that might have been saved. As the Minnesota Civil Liberties Union put it, "The convicted criminal loses only his or her liberty; Sharon Kowalski has lost the right to choose whom she may see, who she may like, and who she may love." To change this picture took nearly nine years of struggle by a partner who lived out her vow of lifetime commitment and the work of many committed persons and groups.

CONCLUSION

Many national groups joined the struggle to provide rehabilitation for Sharon and bring her home, including disability rights activists, gays and lesbians, feminists and male supporters, and civil rights groups. In addition there were thousands of people drawn to this case by simple human rights. After all, any of us could be hit by a drunk driver, become disabled, and in the process lose our legal and medical rights. The Kowalski/Thompson case stands as a warning that in our deeply divided society, freedom is still a privilege and rights are fragile.

People living in nontraditional families need legal protection to secure legal and medical rights. Karen Thompson stresses the importance of making your relationships known to your family of birth, if possible, and informing them of your wishes in case of disability or death. Also, it is essential to execute a durable power of attorney, a document that stipulates a person to make medical and financial decisions for you, in case of need. Copies should be given to your physician.

While requirements vary between states and powers of attorney are not always enforceable, they may protect your rights. Information about how to execute them may be found in your public library, in consultation with a competent lawyer, or in Appendix B of the book Why Can't Sharon Kowalski Come Home?

NOTES

1. Karen Thompson and Julie Andrzejewski. Why Can't Sharon Kowalski Come Home? San Francisco: Spinster/Aunt Lute, 1988. All quotations in text are from this book.
2. Ellen Bilofsky. "The Fragile Rights of Sharon Kowalski." Health/PAC Bulletin, 1989, 19, 4–16.
3. Audre Lorde. "There Is No Hierarchy of Oppressions." Interracial Books for Children Bulletin, 1983, 14, 9.

DISCUSSION QUESTIONS

1. Griscom observes that the role of sex in same-sex relationships is often exaggerated; how much of your own sense of self is tied to your sexual identity? What does this account suggest about how sexuality is constructed for gays and lesbians? What does this account suggest about how sexuality is constructed for the disabled? About how competency is constructed for the disabled?

2. What model of "the family" is constructed and supported by the state? What interests does this definition serve? Why is it defined that way? Whose interests are harmed? Are there other ways it could be defined? What is your definition of a family? What social functions should the family serve? What kind of social changes, if any, would society see if it were to use your definition as the state's model?

3. Discuss how the various players in this tragic human drama accepted, resisted, challenged, and reproduced normative constructions of sex, sexuality, and disability. What would you have done if you were one of the individuals caught up in these circumstances (e.g., Karen, Sharon, Sharon's father, the medical authorities, or a judge hearing this case)? What would be the costs of resistance? What would be the costs of conformity?

The Sexual Politics of Black Womanhood

PATRICIA HILL COLLINS

Patricia Hill Collins' classic writings on race, feminism, and power provide important insights about structural oppression, multiple marginalization, and the lives and political consciousness of black women. In this essay, Collins examines how institutionalized forms of sexuality, such as pornography, prostitution, and sexual violence, have been used as weapons of subjugation to infuse and disrupt the potentially empowering erotic resources of African-American women.

Even I found it almost impossible to let her say what had happened to her as she perceived it . . . And why? Because once you strip away the lie that rape is pleasant, that children are not permanently damaged by sexual pain, that violence done to them is washed away by fear, silence, and time, you are left with the positive horror of the lives of thousands of children . . . who have been sexually abused and who have never been permitted their own language to tell about it.

ALICE WALKER
1988, 57

In *The Color Purple* Alice Walker creates the character of Celie, a Black adolescent girl who is sexually abused by her stepfather. By writing letters to God and forming supportive relationships with other Black women, Celie finds her own voice, and her voice enables her to transcend the fear and silence of her childhood. By creating Celie and giving her the language to tell of her sexual abuse, Walker adds Celie's voice to muted yet growing discussions of the sexual politics of Black womanhood in Black feminist thought. Black feminists have investigated how

---

From "The Sexual Politics of Black Womanhood" in *Black Feminist Thought: Knowledge, Consciousness, and the Politics of Empowerment* by Patricia Hill Collins, 1991, pp. 232–249.
Reprinted by permission of Routledge/Taylor & Francis Books, Inc.

rape as a specific form of sexual violence is embedded in a system of interlocking race, gender, and class oppression (Davis 1978, 1981, 1989; Hall 1983). Reproductive rights issues such as access to information on sexuality and birth control, the struggles for abortion rights, and patterns of forced sterilization have also garnered attention (Davis 1981). Black lesbian feminists have vigorously challenged the basic assumptions and mechanisms of control underlying compulsory heterosexuality and have investigated homophobia's impact on African-American women (Clarke 1983; Shockley 1983; Smith 1983; Lorde 1984).

But when it comes to other important issues concerning the sexual politics of Black womanhood, like Alice Walker, Black feminists have found it almost impossible to say what has happened to Black women. In the flood of scholarly and popular writing about Black heterosexual relationships, analyses of domestic violence against African-American women—especially those that link this form of sexual violence to existing gender ideology concerning Black masculinity and Black femininity-remain rare. Theoretical work explaining patterns of Black women's inclusions in the burgeoning international pornography industry has been similarly neglected. Perhaps the most curious omission has been the virtual silence of the Black feminist community concerning the participation of far too many Black women in prostitution. Ironically, while the image of African-American women, as prostitutes has been aggressively challenged, the reality of African-American women who work as prostitutes remains unexplored.

These patterns of inclusion and neglect in Black feminist thought merit investigation. Examining the links between sexuality and power in a system of interlocking race, gender, and class oppression should reveal how important controlling Black women's sexuality has been to the effective operation of domination overall. The words of Angela Davis, Audre Lorde, Barbara Smith, and Alice Walker provide a promising foundation for a comprehensive Black feminist analysis. But Black feminist analyses of sexual politics must go beyond chronicling how sexuality has been used to oppress. Equally important is the need to reconceptualize sexuality with an eye toward empowering African-American women.

A WORKING DEFINITION OF SEXUAL POLITICS

Sexual politics examines the links between sexuality and power. In defining sexuality it is important to distinguish among sexuality and the related terms, *sex* and *gender* (Vance 1984; Andersen 1988). Sex is a biological category attached to the body—humans are born female or male. In contrast, gender is socially constructed. The sex/gender system consists of marking the categories of biological sex with socially constructed gender meanings of masculinity and femininity. Just as sex/gender systems vary from relatively egalitarian systems to sex/gender hierarchies, ideologies of sexuality attached to particular sex/gender systems exhibit diversity. Sexuality is socially constructed through the sex/gender system on both the personal level of individual consciousness and interpersonal relationships and the social structural level of social institutions (Foucault 1980). This multilevel sex/gender system reflects the needs of a given historical moment such that social constructions of sexuality change in tandem with changing social conditions.

African-American women inhabit a sex/gender hierarchy in which inequalities of race and social class have been sexualized. Privileged groups define their alleged sexual practices as the mythical norm and label sexual practices and groups who diverge from this norm as deviant and threatening (Lorde 1984; Vance 1984). Maintaining the mythical norm of the financially independent, white middle-class family organized around a monogamous heterosexual couple requires stigmatizing African-American families as being deviant, and a primary source of this assumed deviancy stems from allegations about black sexuality. This sex/gender hierarchy not only operates on the social structural level but is potentially replicated within each individual. Differences in sexuality thus take on more meaning than just benign sexual variation. Each individual becomes a powerful conduit for social relations of domination whereby individual anxieties, fears, and doubts about sexuality can be annexed by larger systems of oppression (Hoch 1979; Foucault 1980, 99).

According to Cheryl Clarke, African-Americans have been profoundly affected by this sex/gender hierarchy:

> Like all Americans, black Americans live in a sexually repressive culture. And we have made all manner of compromise regarding our sexuality in order to live here. We have expended much energy trying to debunk the racist mythology which says our sexuality is depraved. Unfortunately, many of us have overcompensated and assimilated. . . . Like everyone else in America who is ambivalent in these respects, black folk have to live with the contradictions of this limited sexual system by repressing or closeting any other sexual/erotic urges, feelings, or desires. (Clarke 1983,199)

Embedded in Clarke's statement is the theme of self-censorship inherent when a hierarchy of any kind invades interpersonal relationships among individuals and the actual consciousness of individuals themselves. Sexuality and power as domination become intertwined.

In her ground-breaking essay, "Uses of the Erotic: The Erotic as Power," Black feminist poet Audre Lorde explores this fundamental link between sexuality and power:

> There are many kinds of power, used and unused, acknowledged or otherwise. The erotic is a resource within each of us that lies in a deeply female and spiritual plane, firmly rooted in the power of our unexpressed or unrecognized feeling. In order to perpetuate itself, every oppression must corrupt or distort those various sources of power within the culture of the oppressed that can provide energy for change. For women, this has meant a suppression of the erotic as a considered source of power and information in our lives. (Lorde 1984, 53)

For Lorde sexuality is a component of the larger construct of the erotic as a source of power in women. Lorde's notion is one of power as energy, as something people possess which must be annexed in order for larger systems of oppression to function).[1]

Sexuality becomes a domain of restriction and repression when this energy is tied to the larger system of race, class, and gender oppression. But Lorde's words also signal the potential for black women's empowerment by showing sexuality and the erotic to be a domain of exploration, pleasure, and human agency. From a Black feminist standpoint sexuality encompasses the both/and nature of human existence, the potential for a sexuality that simultaneously oppresses and empowers.

One key issue for Black feminist thought is the need to examine the processes by which power as domination on the social structural level—namely, institutional structures of racism, sexism, and social class privilege—annexes this basic power of the erotic on the personal level—that is, the construct of power as energy, for its own ends.

BLACK WOMEN AND THE SEX/GENDER HIERARCHY

The social construction of Black women's sexuality is embedded in this larger, overarching sex/gender hierarchy designed to harness power as energy to the exigencies of power as race, gender, and social class domination. . . .

Pornography, prostitution, and rape as a specific tool of sexual violence have . . . been key to the sexual politics of Black womanhood. Together they form three essential and interrelated components of the sex/gender hierarchy framing Black women's sexuality.

Pornography and Black Women's Bodies

For centuries the black woman has served as the primary pornographic "outlet" for white men in Europe and America. We need only think of the black women used as breeders, raped for the pleasure and profit of their owners. We need only think of the license the "master" of the slave women enjoyed. But, most telling of all, we need only study the old slave societies of the South to note the sadistic treatment—at the hands of white "gentlemen"—of "beautiful young quadroons and octoroons" who became increasingly (and were deliberately bred to become) indistinguishable from white women, and were the more highly prized as slave mistresses because of this. (Walker 1981, 42)

Alice Walker's description of the rape of enslaved African women for the "pleasure and profit of their owners" encapsulates several elements of contemporary pornography. First, Black women were used as sex objects for the pleasure of white men. This objectification of African-American women parallels the portrayal of women in pornography as sex objects whose sexuality is available for men (McNall 1983). Exploiting Black women as breeders objectified them as less than human because only animals can be bred against their will. In contemporary pornography women are objectified through being portrayed as pieces of meat, as sexual animals awaiting conquest.

Second, African-American women were raped, a form of sexual violence. Violence is typically an implicit or explicit theme in pornography. Moreover, the rape of Black women linked sexuality and violence, another characteristic feature of pornography (Eisenstein 1983). Third, rape and other forms of sexual violence act to strip victims of their will to resist and make them passive and submissive to the will of the rapist. Female passivity, the fact that women have things done to them, is a theme repeated over and over in contemporary pornography (McNall 1983). Fourth, the profitability of Black women's sexual exploitation for white "gentlemen" parallels pornography's financially lucrative benefits for pornographers (Eisenstein 1983). Finally, the actual breeding of "quadroons and octoroons" not only reinforces the themes of Black women's passivity, objectification, and malleability to male control but reveals pornography's grounding in racism and sexism. The fates of both Black and white women were intertwined in this breeding process. The ideal African-American woman as a pornographic object was indistinguishable from white women and thus approximated the images of beauty, asexuality, and chastity forced on white women. But inside was a highly sexual whore, a "slave mistress" ready to cater to her owner's pleasure.[2]

Contemporary pornography consists of a series of icons or representations that focus the viewer's attention on the relationship between the portrayed individual and the general qualities ascribed to that class of individuals. Pornographic images are iconographic in that they represent realities in a manner determined by the historical position of the observers, their relationship to their own time, and to the history of the conventions which they employ (Gilman 1985). The treatment of Black women's bodies in nineteenth-century Europe and the United States may be the foundation upon which contemporary pornography as the representation of women's objectification, domination, and control is based. Icons about the sexuality of Black women's bodies emerged in these contexts. Moreover, as race/gender-specific representations, these icons have implications for the treatment of both African-American and white women in contemporary pornography.

I suggest that African-American women were not included in pornography as an afterthought but instead form a key pillar on which contemporary pornography itself rests. As Alice Walker points out, "the more ancient roots of modern pornography are to be found in the almost always pornographic treatment of black women who, from the moment they entered slavery . . . were subjected to rape as the 'logical' convergence of sex and violence. Conquest, in short" (1381, 42).

One key feature about the treatment of Black women in the nineteenth century was how their bodies were objects of display. In the antebellum American South white men did not have to look at pornographic pictures of women because they could become voyeurs of Black women on the auction block. A chilling example of this objectification of the Black female body is provided by the exhibition, in early nineteenth-century Europe, of Sarah Bartmann, the so-called Hottentot Venus. Her display formed one of the original icons for Black female sexuality. An African woman, Sarah Bartmann was often exhibited at fashionable parties in Paris, generally wearing little clothing, to provide entertainment. To her audience she represented deviant sexuality. At the time European audiences thought that Africans had deviant sexual practices and searched for physiological differences, such as enlarged penises and malformed female genitalia, as indications of this deviant sexuality. Sarah Bartmann's exhibition stimulated these racist and sexist beliefs. After her death in 1815, she was dissected. Her genitalia and buttocks remain on display in Paris (Gilman 1985).

Sander Gilman explains the impact that Sarah Bartmann's exhibition had on Victorian audiences:

> It is important to note that Sarah Bartmann was exhibited not to show her genitalia—but rather to present another anomaly which the European audience . . . found riveting. This was the steatopygia, or protruding buttocks, the other physical characteristic of the Hottentot female which captured the eye of early European travelers . . . . The figure of Sarah Bartmann was reduced to her sexual parts. The audience which had paid to see her buttocks and had fantasized about the uniqueness of her genitalia when she was alive could, after her death and dissection, examine both. (1985, 213)

In this passage Gilman unwittingly describes how Bartmann was used as a pornographic object similar to how women are represented in contemporary pornography. She was reduced to her sexual parts, and these parts came to represent a dominant icon applied to Black women throughout the nineteenth century. Moreover, the fact that Sarah Bartmann was both African and a woman underscores the importance of gender in maintaining notions of racial purity. In this case Bartmann symbolized Blacks as a "race." Thus the creation of the icon applied to Black women demonstrates that notions of gender, race, and sexuality were linked in overarching structures of political domination and economic exploitation.

The process illustrated by the pornographic treatment of the bodies of enslaved African women and of women like Sarah Bartmann has developed into a full-scale industry encompassing all women objectified differently by racial/ethnic category. Contemporary portrayals of Black women in pornography represent the continuation of the historical treatment of their actual bodies. African-American women are usually depicted in a situation of bondage and slavery, typically in a submissive posture, and often with two white men. As Bell observes, "this setting reminds us of all the trappings of slavery: chains, whips, neck braces, wrist clasps" (1987, 59). White women and women of color have different pornographic images applied to them. The image of Black women in pornography is almost consistently one featuring them breaking from chains. The image of Asian women in pornography is almost consistently one of being tortured (Bell 1987, 161).

The pornographic treatment of Black women's bodies challenges the prevailing feminist assumption that since pornography primarily affects white women, racism has been grafted onto pornography. African-American women's experiences suggest that Black women were not added into a preexisting pornography, but rather that pornography itself must be reconceptualized as an example of the interlocking nature of race, gender, and class oppression. At the heart of both racism and sexism are notions of biological determinism claiming that people of African descent and women possess immutable biological characteristics marking their inferiority to elite white men (Gould 1981; Fausto-Sterling 1989; Halpin 1989). In pornography these racist and sexist beliefs are sexualized. Moreover, for African-American women pornography has not been timeless and universal but was tied to Black women's experiences with the European colonization of Africa and with American slavery. Pornography emerged within a specific system of social class relationships.

This linking of views of the body, social constructions of race and gender, and conceptualizations of sexuality that inform Black women's treatment as pornographic objects promises to have significant implications for how we assess contemporary pornography. Moreover, examining how pornography has been central to the race, gender, and class oppression of African-American women offers new routes for understanding the dynamics of power as domination.

Investigating racial patterns in pornography offers one route for such an analysis. Black women have often claimed that images of white women's sexuality were intertwined with the controlling image of the sexually denigrated black woman: "In the United States, the fear and fascination of female sexuality was projected onto black women; the passionless lady arose in symbiosis with the primitively sexual slave" (Hall 1983, 333). Comparable linkages exist in pornography (Gardner 1980). Alice Walker provides a fictional account of a Black man's growing awareness of the different ways that African-American and white women are objectified in pornography: "What he has refused to see—because to see it would reveal yet another area in which he is unable to protect or defend black women—is that where white women are depicted in pornography as 'objects,' black women arc depicted as animals. Where white women are depicted as human bodies if not beings, black women are depicted as shit" (Walker 1981, 52).

Walker's distinction between "objects" and "animals" is crucial in untangling gender, race, and class dynamics in pornography. Within the mind/body, culture/nature, male/female oppositional dichotomies in Western social thought, objects occupy an uncertain interim position. As objects white women become creations of culture—in this case, the mind of white men—using the materials of nature—in this case, uncontrolled female sexuality. In contrast, as animals Black women receive no such redeeming dose of culture and remain open to the type of exploitation visited on nature overall. Race becomes the distinguishing feature in determining the type of objectification women will encounter. Whiteness as symbolic of both civilization and culture is used to separate objects from animals.

The alleged superiority of men to women is not the only hierarchical relationship that has been linked to the putative superiority of the mind to the body. Certain "races" of people have been defined as being more bodylike, more animallike; and less godlike than others (Spelman 1982, 52). Race and gender oppression may both revolve around the same axis of disdain for the body; both portray the sexuality of subordinate groups as animalistic and therefore deviant. Biological notions of race and gender prevalent in the early nineteenth century which fostered the animalistic icon of Black female sexuality were joined by the appearance of a racist biology incorporating the concept of degeneracy (Foucault 1980). Africans and women were both perceived as embodied entities, and Blacks were seen as degenerate. Fear of and disdain for the body thus formed a key element in both sexist and racist thinking (Spelman 1982).

While the sexual and racial dimensions of being treated like an animal are important, the economic foundation underlying this treatment is critical. Animals can be economically exploited, worked, sold, killed, and consumed. As "mules," African-American women become susceptible to such treatment. The political economy of pornography also merits careful attention. Pornography is pivotal in mediating contradictions in changing societies (McNall 1983). It is no accident that racist biology, religious justifications for slavery and women's subordination, and other explanations for nineteenth-century racism and sexism arose during a period of profound political and economic change. Symbolic means of domination become particularly important in mediating contradictions in changing political economies. The exhibition of Sarah Bartmann and Black women on the auction block were not benign intellectual exercises—these practices defended real material and political interests. Current transformations in international capitalism require similar ideological justifications. Where does pornography fit in these current transformations? This question awaits a comprehensive Afrocentric feminist analysis.

Publicly exhibiting Black women may have been central to objectifying Black women as animals and to creating the icon of Black women as animals. Yi-Fu Tuan (1984) offers an innovative argument about similarities in efforts to control nature—especially plant life—the domestication of animals, and the domination of certain groups of humans. Tuan suggests that displaying humans alongside animals implies that such humans are more like monkeys and bears than they are like "normal" people. This same juxtaposition leads spectators to view the captive animals in a special way. Animals acquire definitions of being like humans, only more openly carnal and sexual, an aspect of animals that forms a major source of attraction for visitors to modern zoos. In discussing the popularity of monkeys in zoos, Tuan notes: "some visitors are especially attracted by the easy sexual behavior of the monkeys. Voyeurism is forbidden except when applied to subhumans" (1984, 82). Tuan's analysis suggests that the public display of Sarah Bartmann and of the countless enslaved African women on the auction blocks of the antebellum American South—especially in proximity to animals—fostered their image as animalistic.

This linking of Black women as animals is evident in nineteenth-century scientific literature. The equation of women, Blacks, and animals is revealed in the following description of an African woman published in an 1878 anthropology text:

> She had a way of pouting her lips exactly like what we have observed in the orangutan. Her movements had something abrupt and fantastical about them, reminding one of those of the ape. Her ear was like that of many apes . . . . These are animal characters. I have never seen a human head more like an ape than that of this woman (Halpin 1989, 287).

In a climate such as this, it is not surprising that one prominent European physician even stated that Black women's "animallike sexual appetite went so far as to lead black women to copulate with apes" (Gilman 1985, 212).

The treatment of all women in contemporary pornography has strong ties to the portrayal of Black women as animals. In pornography women become nonpeople and are often represented as the sum of their fragmented body parts. Scott McNall observes:

> This fragmentation of women relates to the predominance of rear-entry position photographs . . . . All of these kinds of photographs reduce the woman to her reproductive system, and, furthermore, make her open, willing, and available—not in control . . . . The other thing rear-entry position photographs tell us about women is that they are animals. They are animals because they are the same as dogs—bitches in heat who can't control themselves (McNall 1983, 197–198).

This linking of animals and white women within pornography becomes feasible when grounded in the earlier denigration of Black women as animals.

Developing a comprehensive analysis of the race, gender, and class dynamics of pornography offers possibilities for change. Those Black feminist intellectuals investigating sexual politics imply that the situation is much more complicated than that advanced by some prominent while feminists (see, e.g., Dworkin 1981) in which "men oppress women" because they are men. Such approaches implicitly assume biologically deterministic views of sex, gender, and sexuality and offer few possibilities for change. In contrast, Afrocentric feminist analyses routinely provide for human agency and its corresponding empowerment and for the responsiveness of social structures to human action. In the short story "Coming Apart," Alice Walker describes one Black man's growing realization that his enjoyment of pornography, whether of white women as "objects" or Black women as "animals," degraded him:

> He begins to feel sick. For he realizes that he has bought some of the advertisements about women, black and white. And further, inevitably, he has bought the advertisements about himself. In pornography the black man is portrayed as being capable of fucking anything . . . even a piece of shit. He is defined solely by the size, readiness and unselectivity of his cock (Walker 1981, 52).

Walker conceptualizes pornography as a race/gender system that entraps everyone. But by exploring an African-American *man's* struggle for a self-defined standpoint on pornography, Walker suggests that a changed consciousness is essential to social change. If a Black man can understand how pornography affects him, then other groups enmeshed in the same system are equally capable of similar shifts in consciousness and action.

Prostitution and the Commodification of Sexuality

In *To Be Young, Gifted and Black,* Lorraine Hansberry creates three characters: a young domestic worker, a chic, professional, middle-age woman, and a mother in her 30s. Each speaks a variant of the following:

> In these streets out there, any little white boy from Long Island or Westchester sees me and leans out of his car and yells—"Hey there, *hot chocolate!* Say there, Jezebel! Hey you—'Hundred Dollar Misunderstanding' YOU! Bet you know where there's a good time tonight . . . ." Follow me sometimes and see if I lie. I can be coming from eight hours on an assembly line or 14 hours in Mrs. Halsey's kitchen. I can be all filled up that day with 300 years of rage so that my eyes are flashing and my flesh is trembling—and the white boys in the streets, they look at me and think . . . of sex. They look at me and that's *all* they think Baby, you could be Jesus in drag—but if you're brown they're sure you're selling! (Hansberry 1969, 98)

Like the characters in Hansberry's fiction, all Black women are affected by the widespread controlling image that African-American women are sexually promiscuous, potential prostitutes. The pervasiveness of this image is vividly recounted in Black activist lawyer Pauli Murray's description of an incident she experienced while defending two women from Spanish Harlem who had been arrested as prostitutes: "The first witness, a white man from New Jersey, testified on the details of the sexual transaction and his payment of money. When asked to identify the woman with whom he had engaged in sexual intercourse, he unhesitatingly pointed directly at me, seated beside my two clients at the defense table!" (Murray 1987, 274) Murray's clients were still convicted.

The creation of Jezebel, the image of the sexually denigrated Black woman, has been vital in sustaining a system of interlocking race, gender, and class oppression. Exploring how the image of the African-American woman as prostitute has been used by each system of oppression illustrates how sexuality links the three systems. But Black women's treatment also demonstrates how manipulating sexuality has been essential to the political economy of domination within each system and across all three.

Yi-Fu Tuan (1984) suggests that power as domination involves reducing humans to animate nature in order to exploit them economically or to treat them condescendingly as pets. Domination may be either cruel and exploitative with no affection or may be exploitative yet coexist with affection. The former produces the victim—in this case, the Black woman as "mule" whose labor has been exploited. In contrast, the combination of dominance and affection produces the pet, the individual who is subordinate but whose survival depends on the whims of the more powerful. The "beautiful young quadroons and octoroons" described by Alice Walker were bred to be pets—enslaved black mistresses whose existence required that they retain the affection of their owners. The treatment afforded these women illustrates a process that affects all African-American women: their portrayal as actual or potential victims and pets of elite white males.[3]

African-American women simultaneously embody the coexistence of the victim and the pet, with survival often linked to the ability to be appropriately subordinate as victims or pets. Black women's experiences as unpaid and paid workers demonstrate the harsh lives victims are forced to lead. While the life of the victim is difficult, pets experience a distinctive form of exploitation. Zora Neale Hurston's 1943 essay, "The 'Pet' Negro System," speaks contemptuously of this ostensibly benign situation that combines domination with affection. Written in a Black oratorical style, Hurston notes, "Brother and Sisters, I take my text this morning from the Book of Dixie . . . . Now it says here, 'And every white man shall be allowed to pet himself a Negro. Yea, he shall take a Black man unto himself to pet and cherish, and this same Negro shall be perfect in his sight'" (Walker 1979a, 156). Pets are treated as exceptions and live with the constant threat that they will no longer be "perfect in his sight," that their owners will tire of them and relegate them to the unenviable role of victim.

Prostitution represents the fusion of exploitation for an economic purpose—namely, the commodification of Black women's sexuality—with the demeaning treatment afforded pets. Sex becomes commodified not merely in the sense that it can be purchased—the dimension, of economic exploitation—but also in the sense that one is dealing with a totally alienated being who is separated from and who does not control her body: the dimension of power as domination (McNall 1983). Commodified sex can then be appropriated by the powerful. When the "white boys from Long Island" look at Black women and *all* they think about is sex, they believe that they can appropriate Black women's bodies. When they yell "Bet you know where there's a good time tonight," they expect commodified sex with Black women as "animals" to be better than sex with white women as "objects." Both pornography and prostitution commodify sexuality and imply to the "white boys" that all African-American women can be bought.

Prostitution under European and American capitalism thus exists within a complex web of political and economic relationships whereby sexuality is conceptualized along intersecting axes of race and gender. Gilman's (1985) analysis of the exhibition of Sarah Bartmann as the "Hottentot Venus" suggests another intriguing connection between race, gender, and sexuality in nineteenth-century Europe—the linking of the icon of the Black woman with the icon of the white prostitute. While the Hottentot woman stood for the essence of Africans as a race, the white prostitute symbolized the sexualized woman. The prostitute represented the embodiment of sexuality and all that European society associated with it: disease as well as passion. As Gilman points out, "it is this uncleanliness, this disease, which forms the final link between two images of women, the black and the prostitute. Just as the genitalia of the Hottentot were perceived as parallel to the diseased genitalia of the prostitute, so too the power of the idea of corruption links both images" (1985, 237). These connections between the icons of Black women and white prostitutes demonstrate how race, gender, and the social class structure of the European political economy interlock.

In the American antebellum South both of these images were fused in the forced prostitution of enslaved African women. The prostitution of Black women allowed white women to be the opposite: Black "whores" make white "virgins" possible. This race/gender nexus fostered a situation whereby white men could then differentiate between the sexualized woman-as-body who is dominated and "screwed" and the asexual woman-as-pure-spirit who is idealized and brought home to mother (Hoch 1979, 70). The sexually denigrated woman, whether she was made a victim through her rape or a pet through her seduction, could be used as the yardstick against which the cult of true womanhood was measured. Moreover, this entire situation was profitable.

Rape and Sexual Violence

Force was important in creating African-American women's centrality to American images of the sexualized woman and in shaping their experiences with both pornography and prostitution. Black women did not willingly submit to their exhibition on southern auction blocks—they were forced to do so. Enslaved African women could not choose whether to work—they were beaten and often killed if they refused. Black domestics who resisted the sexual advances of their employers often found themselves looking for work where none was to be found. Both the reality and the threat of violence have acted as a form of social control for African-American women.

Rape has been one fundamental tool of sexual violence directed against African-American women. Challenging the pervasiveness of Black women's rape and sexual extortion by white men has long formed a prominent theme in Black women's writings. Autobiographies such as Maya Angelou's *I Know Why the Caged Bird Sings* (1970) and Harriet Jacobs's "The Perils of a Slave Woman's Life" (1860/1987) from *Incidents in the Life of a Slave Girl* record examples of actual and threatened sexual assault. The effects of rape on African-American women is a prominent theme in Black women's fiction. Gayl Jones's *Corregidora* (1975) and Rosa Guy's *A Measure of Time* (1983) both explore interracial rape of Black women. Toni Morrison's *The Bluest Eye* (1970), Alice Walker's *The Color Purple* (1982), and Gloria Naylor's *The Women of Brewster Place* (1980) all examine rape within African-American families and communities. Elizabeth Clark-Lewis's (1985) study of domestic workers found that mothers, aunts, and community othermothers warned young Black women about the threat of rape. One respondent in Clark-Lewis's study, an 87-year-old North Carolina Black domestic worker, remembers, "nobody was sent out before you was told to be careful of the white man or his sons" (Clark-Lewis 1985,15).

Rape and other acts of overt violence that Black women have experienced, such as physical assault during slavery, domestic abuse, incest, and sexual extortion, accompany Black women's subordination in a system of race, class, and gender oppression. These violent acts are the visible dimensions of a more generalized, routinized system of oppression. Violence against Black women tends to be legitimated and therefore condoned while the same acts visited on other groups may remain nonlegitimated and nonexcusable. Certain forms of violence may garner the backing and control of the state while others remain uncontrolled (Edwards 1987). Specific acts of sexual violence visited on African-American women reflect a broader process by which violence is socially constructed in a race- and gender-specific manner. Thus Black women, Black men, and white women experience distinctive forms of sexual violence. As Angela Davis points out, "It would be a mistake to regard the institutionalized pattern of rape during slavery as an expression of white man's sexual urges. . . . Rape was a weapon of domination, a weapon of repression, whose covert goal was to extinguish slave women's will to resist, and in the process, to demoralize their men" (1981, 23).

Angela Davis's work (1978, 1981, 1989) illustrates this effort to conceptualize sexual violence against African-American women as part of a system of interlocking race, gender, and class oppression. Davis suggests that sexual violence has been central to the economic and political subordination of African-Americans overall. But while Black men and women were both victims of sexual violence, the specific forms they encountered were gender specific.

Depicting African-American men as sexually charged beasts who desired white women created the myth of the Black rapist.[4] Lynching emerged as the specific form of sexual violence visited on Black men, with the myth of the Black rapist as its ideological justification. The significance of this myth is that it "has been, methodically conjured up when recurrent waves of violence and terror against the Black community required a convincing explanation" (Davis 1978, 25). Black women experienced a parallel form of race- and gender-specific sexual violence. Treating African-American women as pornographic objects and portraying them as sexualized animals, as prostitutes, created the controlling image of Jezebel. Rape became the specific act of sexual violence forced on Black women, with the myth of the Black prostitute as its ideological justification.

Lynching and rape, two race/gender-specific forms of sexual violence, merged with their ideological justifications of the rapist and prostitute in order to provide an effective system of social control over African-Americans. Davis asserts that the controlling image of Black men as rapists has always "strengthened its inseparable companion: the image of the Black woman as chronically promiscuous. And with good reason, for once the notion is accepted that Black men harbor irresistible, animal-like, asexual urges, the entire race is invested with bestiality" (1978, 27). A race of "animals" can be treated as such—as victims or pets. "The mythical rapist implies the mythical whore—and a race of rapists and whores deserves punishment and nothing more" (Davis 1978, 28).

Some suggestive generalizations exist concerning the connection between the social constructions of the rapist and the prostitute and the tenets of racist biology. Tuan (1984) notes that humans practice certain biological procedures on plants and animals to ensure their suitability as pets. For animals the goal of domestication is manageability and control, a state that can be accomplished through selective breeding or, for some male animals, by castration. A similar process may have affected the historical treatment of African-Americans. Since dominant groups have generally refrained from trying to breed humans in the same way that they breed animals, the pervasiveness of rape and lynching suggests that these practices may have contributed to mechanisms of population control. While not widespread, in some slave settings selective breeding and, if that failed, rape were used to produce slaves of a certain genetic heritage. In an 1858 slave narrative, James Roberts recounts the plantation of Maryland planter Calvin Smith, a man who kept 50 to 60 "head of women" for reproductive purposes. Only whites were permitted access to these women in order to ensure that 20 to 25 racially mixed children were born annually. Roberts also tells of a second planter who competed with Smith in breeding mulattos, a group that at that time brought higher prices, the "same as men strive to raise the most stock of any kind, cows, sheep, horses, etc." (Weisbord 1975, 27). For Black men, lynching was frequently accompanied by castration. Again, the parallels to techniques used to domesticate animals, or at least serve as a warning to those Black men who remained alive, is striking.

Black women continue to deal with this legacy of the sexual violence visited on African-Americans generally and with our history as collective rape victims. One effect lies in. the treatment of rape victims. Such women are twice victimized, first by the actual rape, in this case the collective rape under slavery. But they are victimized again by family members, community residents, and social institutions such as criminal justice systems which somehow believe that rape victims are responsible for their own victimization. Even though current statistics indicate that Black women are more likely to be victimized than white women, Black women are less likely to report their rapes, less likely to have their cases come to trial, less likely to have their trials result in convictions, and, most disturbing, less likely to seek counseling and other support services. Existing evidence suggests that African-American women are aware of their lack of protection and that they resist rapists more than other groups (Bart and O'Brien 1985).

Another significant effect of this legacy of sexual violence concerns Black women's absence from antirape movements. Angela Davis argues, "if black women are conspicuously absent from the ranks of the anti-rape movement today, it is, in large part, their way of protesting the movement's posture of indifference toward the frame-up rape charge as an incitement to racist aggression" (1978, 25). But this absence fosters Black women's silence concerning a troubling issue: the fact that most Black women are raped by Black men. While the historical legacy of the triad of pornography, prostitution, and the institutionalized rape of Black women may have created the larger social context within which all African-Americans reside, the unfortunate current reality is that many Black men have internalized the controlling images of the sex/gender hierarchy and condone either Black women's rape by other Black men or their own behavior as rapists. Far too many African-American women live with the untenable position of putting up with abusive Black men in defense of an elusive Black unity.

The historical legacy of Black women's treatment in pornography, prostitution, and rape forms the institutional backdrop for a range of interpersonal relationships that Black women currently have with Black men, whites, and one another. Without principled coalitions with other groups, African-American women may not be able to effect lasting change on the social structural level of social institutions. But the first step to forming such coalitions is examining exactly how these institutions harness power as energy for their own use by invading both relationships among individuals and individual consciousness itself. Thus, understanding the contemporary dynamics of the sexual politics of Black womenhood in order to empower African-American women requires investigating how social structural factors infuse the private domain of Black women's relationships.

NOTES

1. French philosopher Michel Foucault makes a similar point: "I believe that the political significance of the problem of sex is due to the fact that sex is located at the point of intersection of the discipline of the body and the control of the population" (1980, 125). The erotic is something felt, a power that is embodied. Controlling sexuality harnesses that power for the needs of larger, hierarchical systems by controlling the body and hence the population.
2. Offering a similar argument about the relationship between race and masculinity, Paul Hoch (1979) suggests that the ideal white man is a hero who upholds honor. But inside lurks a "Black beast" of violence and sexuality, traits that the white hero deflects onto men of color.
3. Any group can be made into pets. Consider Tuan's (1984) discussion of the role that young Black boys played as exotic ornaments for wealthy white women in the 1500s to the early 1800s in England. Unlike other male servants, the boys were favorite attendants of noble ladies and gained entry into their mistresses' drawing rooms, bedchambers, and theater boxes. Boys were often given fancy collars with padlocks to wear. "As they did with their pet dogs and monkeys, the ladies grew genuinely fond of their black boys" (p. 142).
4. See Hoch's (1979) discussion of the roots of the white hero, black beast myth in Eurocentric thought. Hoch contends that white masculinity is based on the interracial competition for women. To become a "man," the white, godlike hero must prove himself victorious over the dark "beast" and win possession of the "white goddess." Through numerous examples Hoch suggests that this explanatory myth underlies Western myth, poetry, and literature. One example describing how Black men were depicted during the witch hunts is revealing. Hoch notes, "the Devil was often depicted as a lascivious black male with cloven hoofs, a tail, and a huge penis capable of super-masculine exertion—an archetypal leering "black beast from below" (1979, 44).

REFERENCES

Andersen, Margaret. 1988. *Thinking about Women: Sociological Perspectives on Sex and Gender.* 2d ed. New York: Macmillan.

Angelou, Maya. 1969. *I Know Why the Caged Bird Sings.* New York: Bantam.

Bart, Pauline B., and Patricia H. O' Brien. 1985. "Ethnicity and Rape Avoidance: Jews, White Catholics and Black." *In Stopping Rape: Successful Survival Strategies,* edited by Pauline B. Bart and Patricia H. O' Brein, 70–92. New York: Pergamon Press.

Bell, Laurie, ed. 1987. *Good Girls/Bad Girls: Feminists and Sex Trade Workers Face to Face.* Toronto: Seal Press.

Clarke, Cheryl. 1983. "The Failure to Transform: Homophobia in the Black Community." In *Home Girls: A Black Feminist Anthology,* edited by Barbara Smith, 197–208. New York: Kitchen Table Press.

Clark-Lewis, Elizabeth. 1985. *"This Work Had a 'End'" The Transition from Live-In to Day Work.* Southern Women: The Intersection of Race, Class and Gender. Working Paper #2. Memphis, TN: Center for Research on Women, Memphis State University.

Davis, Angela Y. 1978. "Rape, Racism and the Capitalist Setting." *Black Scholar* 9(7): 24–30.

———. 1981. *Women, Race and Class.* New York: Random House.

———. 1989. *Women, Culture, and Politics.* New York: Random House.

Dworkin, Andrea. 1981. *Pornography: Men Possessing Women.* New York: Perigee.

Edwards, Ann. 1987. "Male Violence in Feminist Theory: An Analysis of the Changing Conceptions of Sex/Gender Violence and Male Dominance." In *Women, Violence and Social Control,* edited by Jalna Hanmer and Mary Maynard, 13–29. Atlantic Highlands, NJ: Humanities Press.

Eisenstein, Hester. 1983. *Contemporary Feminist Thought.* Boston: G. K. Hall.

Fausto-Sterling, Anne. 1989. "Life in the XY Corral." *Women's Studies International Forum* 12(3): 319–331.

Foucalt, Michel. 1980. *Power/Knowledge: Selected Interviews and Other Writings 1972–1977,* edited by Colin Gordon. New York: Pantheon.

Gardner, Tracey A. 1980. "Racism and Pornography in the Women's Movement." In *Take Back the Night: Women of Pornography,* edited by Laura Lederer, 105–114. New York: William Morrow.

Gilman, Sander L. 1985. "Black Bodies, White Bodies: Toward an Iconography of Female Sexuality in Late Nineteenth-Century Art, Medicine and Literature." *Critical Inquiry* 12(1): 205–243.

Gould, Stephen Jay. 1981. *The Mismeasure of Man.* New York: W. W. Norton.

Guy, Rosa. 1983. *A Measure of Time.* New York: Bantam.

Hall, Jacqueline Dowd. 1983. "The Mind that Burns in Each Body: Women, Rape, and Racial Violence." In *Powers of Desire: The Politics of Sexuality,* edited by Ann Snitow, Christine Stansell, and Sharon Thompson, 329–349. New York: Monthly Review Press.

Halpin, Zuleyma Tang. 1989. "Scientific Objectivity and the Concept of 'The Other'" *Women's Studies International Forum* 12(3): 285–294.

Hansberry, Lorraine. 1969. *To Be Young, Gifted and Black.* New York: Signet.

Hoch, Paul. 1979. *White Hero Black Beast: Racism, Sexism and the Mask of Masculinity.* London: Pluto Press.

Jacobs, Harriet. [1860] 1987. "The Perils of a Slave Woman's Life." In *Invented Lives Narratives of Black Women 1860–1960,* edited by Mary Helen Washington, 16–67. Garden City, NY: Anchor.

Jones, Gayl. 1975. *Corregidora.* New York: Bantam.

Lorde, Audre. 1984. Sister Outsider. Trumansberg, NY: The Crossing Press.

McNall, Scott G. 1983. "Pornography: The Structure of Domination and the Mode of Reproduction." In *Current Perspectives in Social Theory, Volume 4,* edited by Scott Mc Nall, 181–203. Greenwich, CT: JAI Press.

Morrison, Toni. 1970. *The Bluest Eye.* New York: Pocket Books.

Murray, Pauli. 1987. *Song in a Weary Throat: An American Pilgrimage.* New York: Harper & Row.

Naylor, Gloria. 1980. *The Women of Brewster Place.* New York: Penguin.

Shockley, Ann Allen. 1983. "The Black Lesbian in American Literature: An Overview." In *Home Girls: A Black Feminist Anthology,* edited by Barbara Smith, 83–93. New York: Kitchen Table Press.

Smith, Barbara. 1983. "Introduction." In *Home Girls: A Black Feminist Anthology,* edited by Barbara Smith, xix–lvi. New York: Kitchen Table Press.

Spelman, Elizabeth V. 1982. "Theories of Race and Gender: The Erasure of Black Women." *Quest* 5(4): 36–62.

Tuan, Yi-Fu. 1984. *Dominance and Affection: The Making of Pets.* New Haven, CT: Yale University Press.

Vance, Carole S. 1984. "Pleasure and Danger: Toward a Politics of Sexuality." In *Pleasure and Danger: Exploring Female Sexuality,* edited by Carole S. Vance, 1–27. Boston: Routledge & Kegan Paul.

Walker, Alice, ed. 1979. I love myself when I Am Laughing, And Then Again When I Am Looking Mean And Impressive: A Zora Neal Hurston Reader. Old Westbury, NY: Feminist Press.

Walker, Alice. 1981. "Coming Apart." In *You Can't Keep a Good Woman Down,* 41–53. New York: Harcourt Brace Jovanovich.

———.1982. *The Color Purple.* New York: Washington Square Press.

Weisbord, Robert G. 1975. *Genocide? Birth Control and the Black American.* Westport, CT: Greenwood Press.

DISCUSSION QUESTIONS

1. What does Audre Lorde mean by the erotic being "a resource within each of us that lies in a deeply female and spiritual plane . . . " How does racism affect this resource for non-whites in American culture?
2. How do social-structural factors "infuse the private domain" of Black women's relationships?
3. Think about the cultural images and characterizations of Black women in popular culture. Is there evidence of political domination and economic exploitation interwoven into any of these images? In what ways, if any, do normative assumptions about Black women's sexuality reinforce their inequality?
4. Think back to Enloe's article earlier in this chapter. In what ways are the sexual politics of Black womanhood similar to and different from the sexual politics of other minority women? Of minority men? Of White women?
5. Can Black women effectively speak out about abusive Black men without threatening Black unity or further jeopardizing the interests, status, and struggles of Black men? Why or why not?

The following Bonus Reading can be accessed through the *InfoTrac College Edition Online Library* by typing in the name of the author listed below or keywords in the title online at *http://www.infotrac-college.com.*

Disability, Sex Radicalism, and Political Agency

ABBY WILKERSON

In this article, Abby Wilkerson demonstrates how sexual autonomy is constrained and denied in subordinated social groups through cultural mechanisms such as shame and medical discourses. Wilkerson points to the political implications of such practices and posits that the denial of sexual agency is a fundamental form of oppression that perpetuates existing social hierarchies and preserves social inequality. She maintains that sexual autonomy is not only a vital aspect of personhood, but is fundamentally central for true political agency.

DISCUSSION QUESTIONS

1. What are the interests of the state regarding the sexual autonomy of its citizens? What is the *responsibility* of the state regarding the sexual behaviors of its citizens? Should any group be denied sexual autonomy? Should some groups be held to different levels of regulation than others? Why or why not? When does regulation cross the line into oppression? Should there be limits to sexual autonomy (e.g., Can/should states prohibit child pornography? Serial rape? Consensual sadism and masochism?)? Who should decide what restrictions on sexual autonomy should be enforced?
2. Provide examples to support Wilkerson's contention that "being considered other in any way almost always renders an individual or group's sexuality socially problematic" (p.5). Can you think of any examples that do not support this argument?
3. In contemporary American society, in what context(s) is sex defined as "legitimate" and "appropriate"? What/who does this leave out?
4. How important is sexuality to one's sense of personhood?
5. In what ways does the Griscom article reflect the concerns aired by Wilkerson? How might the story have differed if Sharon Kowalski had been granted sexual autonomy?
6. Wilkerson focuses on the medical establishment in order to demonstrate its ***culpability*** (degree of blame-worthiness) in the sexual repression and marginalization of social minorities. Discuss whether/how other social institutions are similarly culpable.

7

Media and Popular Culture

A Broken Trust: Canadian Priests, Brothers, Pedophilia, and the Media

IAIN A. G. BARRIE

The media and the church are both powerful social institutions. In this article, Iain A. G. Barrie examines the Canadian media's treatment of sex scandals in the Catholic Church and finds it pointedly lacking given the scope, duration, and severity of the abuses that occurred. As you read this piece, think about the power that social institutions have in shaping public attitudes and perceptions.

Alfred, Ontario, sits on the edge of an escarpment 70 kilometers east of Ottawa, Canada's capital city. This little town, with a population of slightly more than 1,000, is bisected by Highway 17, a two-lane blacktop that parallels the Ottawa River, connecting Montreal and Ottawa. The main landmark if you approach from the west is Alfred's huge, Quebec-style, silver-spired Roman Catholic Church—St. Victor. It dominates the landscape. Motorists passing through the town cannot miss it or the adjacent rectory. The rest is architecturally non-descript—a small hotel that still features nude dancing, a gas station, a hardware store, a funeral home that also provides an ambulance service, an outdoor hockey rink, the Miss Alfred Motel, and a small liquor store. The town's major attraction for travelers is the more than a dozen mom-and-pop snack bars that feature hot dogs, hamburgers, and french fries. Some people call it the french fry capital of eastern Ontario.

Down a side street, a few meters to the south of Highway 17, sits an old, unremarkable, grey, stone-block building currently being used by the Ontario Ministry of Agriculture. The building is not hidden; it is just not particularly dominant. What was hidden from public view for decades was the horribly commonplace sexual and physical abuse that went on unchecked behind the closed doors of what was then called St. Joseph's Training School.[1] This tale of emotional and physical cruelty against young men, wards of the Crown, eventually became known as the Alfred scandal.

Between 1931 and 1974, members of a religious lay order known as the Brothers of the Christian Schools ran St. Joseph's. The province took over the reform school in 1974 and ran it until it was closed in 1981.

On November 28, 1990, 16 years after the lay brothers had left Alfred, the Canadian Press (hereafter CP) ran an unbylined, 15-paragraph story on its national general news service. The lead paragraph implied this was essentially a police story: "Criminal charges could be laid against Roman Catholic lay brothers who allegedly sexually and physically abused boys at a reform school at Alfred, Ont., police said Wednesday." The story quoted an Ontario Provincial Police (OPP) detective who said despite a 7-month investigation into allegations of abuse "there is still lots of work for us to do" and no charges would be laid until after consultation with the Crown and others.[2]

An examination of CP's database on this story reveals that CP's editors did not regard the religious dimensions of the scandal as very important. Of the 41 stories analyzed, only two received a religion slug. One story, 13 paragraphs long, dealt with a 6-day conference of Roman Catholic bishops.

The other story relied extensively on copy generated for the *Catholic New Times* magazine where an Alfred abuse victim said, "The problems we suffer today are problems that began in the church and they have to get resolved in the church.[3]

CP's attitude probably originated with its major source of copy, the *Ottawa Citizen.* Sean Upton, the *Citizen's* lead reporter on the Alfred scandal, recalled, "I saw it as a police story, not a religious story. Religion is what you see on the religion pages of the *Citizen.*"[4] With this observation, Upton was doing what most reporters at a secular newspaper would do: he relegated the religious questions and implications raised by this type of scandal to a lesser status.

There were plenty of reasons to report it as a cops and courts story. The abuse of the boys led to more than 180 separate criminal charges being laid against "19 members and former members" of the order that ran St. Joseph's and a smaller juvenile detention home in Uxbridge, north of Toronto. Eventually, 25 people were charged. "Investigators, who spent more than a year on the case, believe that at least 177 former students—some as young as seven—were victims between 1941 and 1971." The stiffest sentence handed out in the trials that followed the police investigation went to Lucien Dagenais, 67. Dagenais was convicted of 15 of the 18 charges filed and was sentenced to five years in jail. "Known as 'The Hook' to former students . . . because his left hand only has a middle finger and the stump of a thumb" an apparently unrepentant Dagenais listened as Judge Hector Soubliere said: "We built walls around that school—not of brick and mortar, but of much stronger material: walls of silence and indifference. I'm even tempted to say walls of ignorance. [The judge said] Dagenais had abused the trust placed in him.[5] Reporters covering the trial practiced conventional court reporting. Upton asked to be assigned to the trials of several of the brothers to see the story through to its logical conclusion. He is fully bilingual and was versed in the case's lurid background. The *Citizen's* religion editor, Bob Harvey, said "the story was getting a lot of ink and I saw it as a cop story. Maybe it would have been worth the try to look more seriously at the religious dimensions." Harvey suggested, with the benefit of hindsight, that "the religious dimension was ignored. I would have liked to sit down with the people involved and talk with them about their struggle with conscience . . . how the abuse affected their spiritual lives."[6]

If the story had been approached from a religious angle, some of the dimensions that could have been probed include: how religious communities cope with a group of people who violate their ethic of belief, how a religious community—living a vocation with prayerful and financial support from a larger community—could countenance the horrible actions of abusers, and what this story says about the church. Jim Travers, the editor at the *Citizen,* said in an interview that "in retrospect we should go back and check [the religious aspects] of this story. The story touched a core value in society . . . it was a story about a breach of trust . . . the religion dimension was implied and ran through the story as a sub-text.[7]

Since the Alfred scandal, the Roman Catholic Church has established a $16-million fund to compensate victims of abuse who were in its care at Alfred and Uxbridge. The fund in a sense is a secular response by the church to a secular crime and has been reported as such. But the journalists who covered this story generally ignored the religious dimension, and in doing so probably failed their readers, readers who will never understand that the horror experienced by children the Roman Catholic Church was charged with looking after has a spiritual as well as social or psychological dimensions.

In the late 1980s and early 1990s, even a casual newspaper reader could not have been blamed for thinking that Catholicism was riddled with priests and brothers who practiced buggery and other severe forms of physical, mental, and even spiritual abuse. What was less apparent was why this was happening so often. If the reporting had analyzed the patriarchal structure of the Roman Catholic Church and its deep-rooted powers over people and other institutions, perhaps the media would have served the community better.

What made the crimes at Alfred so appalling was that the church, which sets itself as a moral authority for close to 12 million Canadians, was guilty of duplicity. The public had not been hearing only about Alfred and Uxbridge. Another residential school, this one in Newfoundland, Mount Cashel, had become infamous. For decades, a Roman Catholic lay order, the Irish Christian Brothers, had been charged with looking after boys and teenagers who were wards of the Newfoundland government. These youngsters—some orphans, some from broken families that could not offer proper care—became residents of the Mount Cashel orphanage. Many of the boys in the brothers' care would be treated violently, sexually abused, controlled, degraded, exploited, and wounded emotionally.

Dereck O'Brien was just one of Mount Cashel's victims. He arrived at the gloomy institution after a dreadful childhood. His earliest memories of home were of his one-legged grandfather "drunk all the time. The floor of his room was littered with urine and rum bottles that toppled over when I opened the door . . . I slept with my clothes on and never changed them . . . being dirty was just another part of life for me . . . I begged people for food and money."[8] By 1964, the 4-year-old was on the path to Mount Cashel, via a series of foster homes where the abuse and neglect continued.

Rumors of mistreatment at Mount Cashel had been circulating since the early 1970s. Very late in the story, in 1989, the Newfoundland government set up a Royal Commission, chaired by retired Ontario Supreme Court Justice Samuel H. S. Hughes, which discovered there had been a long-time cover-up involving the police, the justice system, and the Christian Brothers.[9]

As O'Brien recalled it, the lid lifted slightly in 1975 when, "something unexpected happened. It was early Sunday morning and we had just finished breakfast. Brother [Douglas] Kenny [the orphanage superintendent] came into the dining hall and told several of us that we would be going out. He didn't say where. Myself and about six other boys piled into Kenny's station wagon. Kenny explained that we would be going to the police station and that they would be asking us some questions. 'Don't say anything about me or Brother English' [the brother in charge of dormitories], he said. 'Don't mention our names.' When he said this he squeezed my hand hard."[10] But at this point, despite the stories of abuse, neither the police nor other authorities took any action.

It was later revealed that two of the boys interviewed by the police that day "identified Brother English as a violent pedophile who had tormented them during their stays in Mount

Cashel." Another boy interviewed was John Pumphrey. His father, "Ron Pumphrey, was the host of a popular talk show on radio station VOCM in St. John's." Although the topic of Mount Cashel was raised on the show by several callers, "they were cut off, in keeping with the station's policy of not permitting criminal allegations to be made on the air." At the Hughes inquiry years later, "Ron Pumphrey would . . . say that the police had downplayed the whole affair and that he had no idea of what was really happening [in 1975] at the orphanage.[11]

The Christian Brothers knew that the allegations and facts, if exposed, would create a crisis, so they quickly posted Kenny and English to other parts of the province. Michael Harris, in his book *Unholy Orders* (1990), was able to show clearly that police, senior bureaucrats in various government agencies, the Newfoundland church hierarchy, and the brothers' international administration all knew about the gravity of the suspicions—yet they chose damage control over revelation and action, a tawdry cover-up at the expense of innocent children. The church's tentacles reached as far as the media. In 1975, the *St. John's Evening Telegram* had a couple of reporters who knew about the instances of sexual abuse at the orphanage. The stories "were suppressed by the paper.[12]

The actions that the government, police, and brothers had taken held for three years. In 1978, as a result of the separate Soper Inquiry into allegations of police cover-ups, word of the abuse at Mount Cashel slowly surfaced to again be suppressed. "The first public reference to the 1975 Mount Cashel investigation . . . ever . . . made, [happened] on May 17, 1978 when *The Daily News*" reported a few sketchy details about:

> . . . the case of three Christian Brothers alleged to have sexually assaulted two Children. . . . CBC Radio [Canada's national radio service] gave the same coverage as *The Daily News,* broadcasting the bare fact of an alleged 1975 cover-up of a sex scandal at Mount Cashel. The *Evening Telegram* . . . took a different approach, perfectly in keeping with the paper's 1975 decision to suppress the story that two of its own reporters had unearthed. All references to the Brothers were expunged from the copy, an editorial decision whose impact the Christian Brothers themselves quickly realized was to their advantage. The *Telegram* story was conveniently vague.[13]

The decision to play down the story was taken by the paper's publisher, Stephen Herder, who died in 1993. Bernie Bennett, a 26-year veteran of the *Telegram's* police and court beat, who covered all the Mount Cashel trials, said, "the reason was never given but I guess it was because the clergy [in Newfoundland] were seen to be so Almighty. The publisher took the reason to his grave and the newspaper has never released the reason.'[14]

In the late 1970s and into the following decade, several instances of child sexual abuse by priests occurred in other parts of the island province. Most of these were quickly settled by the Roman Catholic Church co-opting the police and criminal justice system, playing down the problem, and reassigning the guilty priests to other church jurisdictions—often off the island. Newfoundland media reported details of these cases but the full impact of Mount Cashel and the cover-up of the scandal would not be felt until early 1989.

Several forces involving the media coincided that year to guarantee that the Mount Cashel story would blow wide open. "A television documentary about the spate of Roman Catholic priests and Brothers who had been charged with sexual offenses" led to calls to radio station "VOCM's morning talk show, *Open line.*[15] Those calls prompted certain listeners, including a judge's wife, to action. Phone calls to the province's justice department asking about Mount Cashel could not go unheeded. The investigation that had been hidden from public view for so many years was reopened. Shane Earle, 23, a victim of Mount Cashel, took his story to the media. The unravelling began in earnest.

There is some confusion as to what media outlet really broke the story first. In the spring of 1989 the floodgates were open. Eight Christian Brothers were charged, convicted, and sentenced.

O'Brien, one of the Mount Cashel boys who had signed a police statement back in 1975, said watching the media in 1989 was like "watching a feeding frenzy—they went in feet first. O'Brien felt the Mount Cashel story was about "crime and abuse with a big helping of the church in there somewheres [*sic*]." O'Brien shared the common Newfoundland attitude that the church was omniscient, omnipresent, and untouchable. He did not blame the media for his fate or for not uncovering the scandals earlier. "I can tell you sure as I'll die that the media . . . the government couldn't have stopped the abuse before it stopped. The Catholic Church could put the squeeze on the media and even the social workers couldn't do anything. How can you fight more than 100 years of good work" by the church in Newfoundland?[16]

Rex Murphy, a Newfoundlander, Rhodes scholar, and later an acerbic columnist, analyzed how Mount Cashel might affect the cultural and spiritual fabric of the province and in doing so covered an aspect of the story neglected by most reporters. He suggested Mount Cashel "posed questions that were absolutely fundamental to Newfoundland's self-image . . . we have such shocking politics and economics—I think we clung on to the idea that we had a bit of decent domestic or personal virtue and this thing just blasted that all to hell. . . . So it was yes, at the core, a story with a religious dimension, but it had an awful lot more wings . . . than just religious."[17]

Murphy said the reporting contained "a tremendous amount of relish . . . because it was [done by] the kind of people who grew up in the 60s, who fell off church more or less, and went the kind of high secular route giving them a chance to get back at what they regarded as their petty miseries in the class-rooms with the nuns and the brothers." O'Brien's view that the media were on a feeding frenzy when the story really took off was echoed by Murphy, who questioned coverage that was "so relentless and so particularized that the overly specific, continuous, unremittent reporting of who was dangling penises" turned off many news consumers. The point that Murphy was making was that the media were playing catch up with a story many realized they had missed. Murphy also says there were "certain reporters [who] . . . had a chance to take a smack at the bishop and at the church. There should have been more of a clean take here. . . . What is the equally glaring story of this era that isn't being touched now that we are so busily congratulating ourselves on moving out these fiendish priests?"[18]

In 1984, Jason Berry, a freelance writer from New Orleans, began an epic journey that revealed a frightening side of the North American Roman Catholic Church. The product of his research into sexual misconduct by priests and brothers was a 1992 book, *Lead Us Not into Temptation: Catholic Priests and the Sexual Abuse of Children.* In his foreword to the book, Father Andrew Greeley, an author, sociology professor, and parish priest, said Berry "dug into what may be the biggest scandal in the history of religion in America and perhaps the most serious crisis Catholicism has faced since the reformation." Berry discovered: "In the decade of 1982–1992, approximately 400 priests were reported to church or civil authorities for molesting youths, and the vast majority of these men had multiple victims. By 1992, the church's financial losses—in victims' settlements, legal expenses, and medical treatment of clergy—had reached an estimated $400 million.[19]

Berry met with several of the boys who had been abused at Mount Cashel and devoted a chapter to their story. Two observations Berry makes about the Canadian experience are valuable. First, televising the Hughes Commission inquiry helped to alert Canadians to the depth of the church's problem and it may have made people think about how to prevent more of the same. He alludes to comments made by Canadian Archbishop James M. Hayes saying, "first in our compassion must be those who have been sexually abused," Second, Berry said he believes the

Canadian church may be leading the way in North America in dealing with such scandals: "No such candor has emanated from the U.S. [church] hierarchy."[20]

Before his book was published, Berry tried to sell a number of articles on sexual abuse by priests to U.S. news magazines. "I would say that on the whole, newspapers have a strong secular bent and they just don't like stories like this. It was depressing material to deal with . . . people didn't want to go near it." He took advance payments from magazines who failed to publish the resulting articles because "they just didn't want to tangle with the Church. You know when you get paid $4,000 or $5,000 from the *Los Angeles Times Magazine* . . . and they sit on it and finally the assignment chief says: 'we are asking ourselves why should we publish this' . . . do I have to lobby the guy to tell him, you know, the public has a right to know?[21]

Berry's experience illustrates that mainstream media have trouble dealing with society's central institutions like the churches. There was the perception that these organizations should be treated with the same kind of deference that other elite organizations get from the media. The special religious dimension that makes this kind of scandal even more vivid is ignored. Book publishers shied away from Berry's text even though it had been rigorously checked by lawyers. He approached more than 30 publishers. "I wanted to understand how corrupt the church was. Probably I never thought of the church in those terms, but when you have bishops proclaiming the sanctity of life in the womb and being so cavalier in their dismissal of the rights of children . . . I came to see the bishops as a bunch of politicians."[22]

When considering the media's seeming preoccupation with sin, sex and sleaze, it seems unfathomable that Berry could have faced so many problems getting his stories published. The same sort of question may be asked about Mount Cashel. How did the story stay hidden for so long? As Howard Kurtz, the press critic for the *Washington Post,* notes: "Nothing in the media business spreads faster than a hot rumor. And nothing creates more headaches for editors and reporters. The line between news and third-hand gossip has become badly blurred in recent years, and our moral compass increasingly erratic. The most titillating rumors often involve sex and public officials, which only add to the growing sense of unease in the nation's newsrooms."[23]

One reason these stories were not dealt with when they first arose was alluded to earlier: the media just did not want to tangle with the ecclesiastical power structure. Moreover, the church and the media keep their distance from each other. Few reporters have ever learned about the intricacies of the Christian church—the complexities of its structure, its absolute moral stands, its internal politics, and its 20 centuries of mystique. It would be difficult to find a general assignment reporter who understands the doctrine of papal infallibility, the rule of celibacy, or the power of a bishop. As the Right Rev. Bruce Stavert, the Anglican Bishop of the Diocese of Quebec, said in an interview: "We in the church often get cynical about the lack of knowledge in the media about our business. There's an obsession with scandal. Yet, overall there seems to be very little interest in the secular media about what we do."[24]

The secular media, although providing a massive public service by reporting directly on the scandals, does a great disservice to the community of faithful when it ignores the implications of these stories. Newfoundland, with its large Roman Catholic population, will never be the same. There has been a crisis of confidence in the church and many are disillusioned with the people they trusted to help them with their most precious attribute—their faith. Bernie Bennett, the *Telegram* reporter who listened to every word spoken in court during eight trials, will never be the same, "My faith has been shaken . . . the only time I now go to church is for weddings and funerals . . . it has brought an element of doubt . . . suspicion is there. When the priest stands at the altar I wonder what's going on in his mind and I wonder what he's done. I can't help it."[25]

The Spiritual story remains untold. Dereck O'Brien sounds tired and disillusioned. "Brothers have received counselling, legal fees, and even their pensions from the church—that should be reported. I've left the church, the priests and brothers—they make me sick, sick to my stomach when I see one of the bastards. The Roman Catholic Church is so corrupt it looks after its pedophiles . . . the hell with them."[26]

Religious news receives short shrift in many newsrooms. Although "the generalization is less true now than it was 20 years ago, the journalistic community is not very comfortable with religion. Some editors view religion stories as soft, irrelevant, and non-objective."[27]

Jim Travers, the editor of the *Ottawa Citizen,* said in an interview that newspapers should be asking: "What does religion mean in people's lives? How does it affect people's lives?" He suggested that religion today should not be ghettoized like so-called "women's news" was in the past. With 75 beat reporters and one religion specialist, Travers acknowledges that the religious dimensions to stories are sometimes neglected. "We haven't made the progress we should have."[28]

CTV's (the Canadian Television network) principal news anchor, Lloyd Robertson, was critical about both newspaper and television coverage of religion. Robertson said in an interview that he would like to see "more stories of the human spirit,"[29] but he qualified his remarks by mentioning the time constraints on television journalists.

When the Jim Bakker PTL scandal broke, followed by the Jimmy Swaggart scandal, the international media were all over the stories. News is preoccupied with power and authority and it found the abuse of power and misuse of authority in both these stories. The Canadian religious scandals were different and the Canadian media were asleep at the switch for over a decade. The boys of Mount Cashel, Uxbridge, and Alfred, and the countless children who have been victims of physical and mental abuse in residential schools and orphanages were certainly overlooked. Realistically, the media may only be partially blamed. In the late 1950s and through to the 1980s, when the scandals were unfolding, the Roman Catholic Church wielded extreme power and the media understandably backed off. But the result was that these helpless children were let down by the governments and church that were supposed to protect them. The *Washington Post* ethics code has one tenet that stands out—to listen to the voiceless. It speaks volumes about the role the media should have played.

Iain A. G. Barrie is Professor of Broadcasting in the Media Design sector of Algonquin College (Ottawa, Ontario, Canada).

NOTES

1. The story of child abuse by clergy and other religious figures recently has begun being told with increasing frequency, with this essay focusing primarily on only one story. The growing body of literature includes books such as Thomas G. Plante, ed., *Bless Me Father for I Have Sinned: Perspectives on Sexual Abuse Committed by Roman Catholic Priests* (Westport, Conn.: Praeger, 1999); Stephen J. Rossetti, *A Tragic Grace: The Catholic Church and Child Sexual Abuse* (Collegeville, Minn.: Liturgical Press, 1996); Philip Jenkins, *Pedophiles and Priests: Anatomy of a Contemporary Crisis* (New York: American Philological Association, 1996); Chris Moore, *Betrayal of Trust: The Father Brendan Smyth Affair and the Catholic Church* (Dublin: Marino, 1995); Anthony Okaiye, *The Story of an Abused Priest* (Ann Arbor, Mich.: Proctor Publications, 1995); Elinor Burkett and Frank Bruni, *A Gospel of Shame: Children, Sexual Abuse, and the Catholic Church* (New York: Viking, 1993); Stephen J. Rossetti, *Slayer of the Soul: Child Sexual Abuse and the Catholic Church* (Mystic, Conn.: Twenty Third Publications,1990); and Eamonn Flanagan, *Father and Me: A Story of Sexual Abuse at the Hands of a Priest* (North Blackburn, Australia: HarperCollins,1995). Numerous similar articles have been published in theological, public policy, cultural studies and other journals. For more insight into news coverage, see James Lull and Stephen Hinerman, *Media Scandals: Morality and Desire in the Popular Culture Marketplace* (New York: Columbia University Press,1997); David Hechler, "The Source You Shouldn't Talk To," *Columbia Journalism Review,* March–April 1992, 48; and David Hechler, "Danger Ahead: Sex Abuse Cases," *Washington Journalism Review*, September 1991, 37–40.

2. "Abuse," Canadian Press (hereafter CP) database, 28 November 1990.
3. CP Database, 15 October 1990.
4. Interview with Sean Upton, November 1994.
5. CP Database, 4 June 1992; 23 January 1993; 19 March 1993.
6. Personal interview with Bob Harvey, November 1994.
7. Personal interview with Jim Travers, November 1994.
8. Dereck O'Brien, *Suffer Little Children* (St. John's, Newfoundland.: Breakwater, 1991), 19–22.
9. For further information see *The Hughes Inquiry Report* and Michael Harris, *Unholy Orders* (Markham, Ont.: Penguin Books, 1990).
10. O'Brien, *Suffer Little Children,* 143.
11. Harris, *Unholy Orders,* 95–96.
12. Harris, *Unholy Orders,* 95–96.
13. Harris, *Unholy Orders,* 180.
14. Personal interview with Bernie Bennett, September 1994.
15. Harris, *Unholy Orders,* 260.
16. Personal interview with Dereck O'Brien, November 1994.
17. Interview with Rex Murphy.
18. Interview with Murphy.
19. Jason Berry, *Lead Us Not into Temptation* (New York: Doubleday,1994), xiii; xix.
20. Berry, *Lead Us Not,* 305.
21. Personal interview with Jason Berry, August 1994.
22. Interview with Berry.
23. Howard Kurtz, *Media Circus* (New York: Random House, 1994),151.
24. Personal interview with the Right Rev. Bruce Stavert, Anglican Bishop of the Anglican Diocese of the Quebec, November 1994.
25. Interview with Bennett.
26. Interview with O'Brien.
27. Benjamin J. Hubbard, "The Importance of the Religion Angle in Reporting on Current Events," in *Reporting Religion,* ed. Benjamin J. Hubbard (Sonoma, Calif.: Pole-bridge Press, 1990), I.
28. Interview with Travers.
29. Personal interview with Lloyd Robertson, August 1994.

DISCUSSION QUESTIONS

1. Recent interviews with many adults who were victimized as children by priests reveal that their allegations of abuse often had been discounted by parents, schools, the church, and legal authorities. Why? In what way, if any, did/do social constructions of the church and the priesthood contribute to this?
2. Are there other contemporary social roles that are constructed in ways that might similarly facilitate sexual abuse (or mask its occurrence)?
3. Given the usual eagerness of the press to publicize sex scandals, why did it take so long for this case to "break" in Canada? Was it collusion between the Church and the Canadian press? Why or why not? News of the sex scandals involving U.S. priests broke even later than the stories publicized in Canada. What factors might help to explain this difference?
4. Some argue that the Catholic Church's prohibition against marriage by priests is psychologically repressive and may cause sexuality to be expressed in unhealthy ways. Do you agree or disagree? Why? Can you think of other structural features of the Catholic Church that might help to explain this tragedy?
5. What should be done with institutions that knowingly cover up the sexual victimization of children and allow abuses to continue? What, if anything, can be done to stop abuses from happening in the first place? How much control can and should organizations (both public and private) have over those in their employ? How responsible are these organizations when abuses occur?
6. There is no indication that similar abuses have been perpetrated by nuns. What, if anything, do the crimes committed by priests have to do with sex and/or masculinity? What might Stoltenberg and/or Messner argue?

Girls, Media, and the Negotiation of Sexuality: A Study of Race, Class, and Gender in Adolescent Peer Groups

MEENAKSHI GIGI DURHAM

Meenakshi Gigi Durham's examination of the effect of the media on adolescent girls' negotiation of sexuality demonstrates that the while hegemonic norms are articulated and distributed through social institutions, social filters, such as the adolescent peer group, can have important mediating effects on the adoption and internalization of normative constructs.

Adolescence for girls in the United States has been characterized as "a troubled crossing,"[1] a period marked by severe psychological and emotional stresses. Recent research indicates that the passage out of childhood for many girls means that they experience a loss of self-esteem and self-determination as cultural norms of femininity and sexuality are imposed upon them.[2]

Much attention has been paid over the last decade or more to the role of the mass media in this cultural socialization of girls:[3] clearly, the media are crucial symbolic vehicles for the construction of meaning in girls' everyday lives. The existing data paint a disturbing portrait of adolescent girls as well as of the mass media: on the whole, girls appear to be vulnerable targets of detrimental media images of femininity. In general, the literature indicates that media representations of femininity are restrictive, unrealistic, focused on physical beauty of a type that is virtually unattainable as well as questionable in terms of its characteristics, and filled with internal contradictions. At the same time, the audience analysis that has been undertaken with adolescent girls reveals that they struggle with these media representations but are ultimately ill-equipped to critically analyze or effectively resist them.

These studies are linked to the considerable body of research documenting adolescent girls' difficulties with respect to issues such as waning self-esteem,[4] academic troubles,[5] negative body image,[6] conflicts surrounding sexuality,[7] and other issues related to girls' development. Although the majority of these studies were conducted with upper-middle-class White girls, some of them take into account the impact of race, ethnicity, and class on girls' experiences of adolescence. These findings indicate that—contrary to popular belief—girls of color and girls from lower socioeconomic backgrounds are very hard hit by adolescence and have fewer available resources for help with problems like eating disorders, pregnancies, or depression.[8]

Bearing these issues and their implications for girls in mind, this study seeks to broaden and deepen our understanding of the role of the mass media in girls' socialization, with a par-

From "Girls, Media, and the Negotiation of Sexuality: A Study of Race, Class, and Gender in Adolescent Peer Groups" by Meenakshi Gigi Durham, 1999, *Journalism and Mass Communication Quarterly* 76.

ticular emphasis on the context in which this socialization takes place. New theories of child development contend that socialization is context-specific and that the peer groups of childhood and adolescence are responsible for the transmission of cultural norms as well as the modification of children's personality characteristics.[9] However, most of the research done to date on adolescence and mass media does not take into account the peer group dynamics involved in media use, nor the race and class factors that might influence these processes.

The key question in this study, then, is how peer group activity and social context affect adolescent girls' interactions with mass media, especially in terms of their dealings with issues of gender and sexuality. This study consisted of a long-term participant observation of middle-school girls combined with in-depth interviews with the girls and their teachers. A significant aspect of the study is that the girls were from sharply varying race and class backgrounds, and these factors were crucial components of the analysis.

METHOD

Background of the Study

In order to gain a deep understanding of the social processes at work in adolescent girls' peer groups, a participant observation was conducted over a five-month period at two middle schools in a midsize city in the southwestern United States. East Middle School was situated within the city limits, in an impoverished residential neighborhood that was very close to the interstate highway. Heavily trafficked streets bounded all sides of the schoolyard. The school building was a concrete block; inside, there was little natural light. East Middle School had a total student enrollment of 1,164, of which the majority were African-American (60 percent, or 704 students). The next largest student ethnic group comprised Latino students (26 percent, or 298 students). Thirteen percent (157) of the students were Anglo/White. A majority of the students (76.5 percent) were categorized by the school district as "economically disadvantaged."

By contrast, West Middle School was located many miles outside of the city in a picturesque hilly area. It, too, was in a residential neighborhood, but the houses were half-million-dollar properties with landscaped yards. The school was on its own street. The surroundings were peaceful and quiet.

West Middle School had a total student enrollment of 804. Of these, 732 students (91 percent) were Anglo/White. The minority population was minuscule: 27 students (3 percent) were Latino; 6 students (.07 percent) were African-American; and 36 students (4 percent) were Asian/Pacific Islander. Only 2.5 percent of the West Middle School students were classified as "economically disadvantaged" by the school district.

ANALYSIS

While references to mass media abounded in the peer group conversations observed at both schools, it is important to note at the outset that none of the groups made any use of the *news* media in their day-to-day peer interactions. Discussions of politics and current events did not arise during the five months of observations; rather, popular culture was the common currency among the girls, and the media with the greatest communicatory utility were television, consumer magazines, and movies. These observations corroborate recent findings of declining use of the news media by young people.[10]

Media references cropped up much more frequently in the conversations of the students at West Middle School than at East. Students at East Middle School did use the mass media, but in their peer group discussions they were much more likely to talk about their community and church activities, their friends and relatives, and the incidents in their daily lives—for instance, the *quince-anera* celebrations which many of the Latina girls were planning. At West Middle School, by contrast, media references were almost constant: talk of movies, TV shows, and pop music featured in every conversation.

At both schools, mass media were in evidence. For example, most of the girls at both schools subscribed to *YM* and *Seventeen* magazines and carried them in their backpacks. Girls at both schools watched TV shows like *The X Files, Friends, Seinfeld, Daria, Sabrina the Teenage Witch,* and *Buffy the Vampire Slayer.* Girls at West Middle School were more likely to have seen current movies than the girls at East; they were also more involved with pop music.

The pervasiveness of popular culture at both schools was tied very closely to the single most important theme that emerged from the data: the dominance of the socio-cultural norm of heterosexuality in the girls' lives. While this was addressed and negotiated in different ways depending on various contextual factors, compulsory heterosexuality functioned as the core ideology underpinning the girls' interpersonal and intragroup transactions, although it was seldom explicitly acknowledged. What was striking was that this norm of heterosexuality was central to the social worlds of girls at both schools, and it guided the girls' behaviors and beliefs regardless of their racial, ethnic, and class differences, although it manifested itself in different ways based on these cultural variances.

The girls' efforts to understand and adapt to perceived social norms of heterosexuality played out principally through their ongoing constructions of femininity. The use of the mass media was woven into those constructions and served in various ways to cement the girls' identities within their peer groups as well as to secure their relationships to the broader social world. Themes of heterosexuality criss-crossed the girls' conversations and actions in multiple ways, but certain practices occurred frequently and repeatedly enough to constitute clearly discernible modalities of mass mediated heterosexual expression. These are described and analyzed in detail, below, under the thematic headings of (1) the discipline of the body, (2) brides and mothers, (3) homophobia and sexual confusion, and (4) iconic femininity.

The Discipline of the Body: Cosmetics, Clothes, and Diets

Bartky points out that "femininity is an artifice" and that women engage in "disciplinary practices that produce a body which in gesture and appearance is recognizably. feminine."[11] My observations of the school-girls at East and West Middle Schools indicate that these disciplinary practices are acquired fairly early in life and are essential to the maintenance of adolescent girls' peer group configurations. At both schools, peer relationships hinged on these techniques for molding the female body in group-sanctioned ways.

Cosmetics and Grooming At East Middle School, the application of cosmetics was a common group occurrence and one that sometimes transpired with the aid of magazines like *YM, Seventeen,* and *Glamour.* This happened most often among the "gangsta" girls, and it usually occurred in the classroom when students were given unstructured work time. On several occasions I observed them braiding and styling each others' hair, painting each others' fingernails, and applying makeup. During the yearbook class one afternoon, for example, when some students were writing stories or working on layouts, a group of the "gangsta" girls drifted together; one pulled out a makeup bag and began to apply cosmetics to another, while several gathered around to watch and comment.

Mariana, the 14-year-old Latina girl who was doing the make-over, went about it with great concentration, first applying lip-liner, then lipstick, then powder foundation, then eyeshadows of various colors, and finally mascara to her friend Mercedes' face. The girls were quiet and rapt while this was going on, watching the process in almost reverent silence, but after it was over they began to talk.

Laura: That looks good. That looks cute.

Mariana: I saw in *YM* that if you put white eyeshadow on like that, it makes your eyes look bigger.

Mercedes (Opening Her Eyes Wide)**:** Does it work?

Mariana: I don't know. Yeah. A little bit, maybe.

Nydia: Her hair is pretty. My hair is so ugly.

Laura: Your hair is pretty.

Nydia: Naw, it's all dry and damaged.

(*Much discussion about their hair . . .*)

Nydia: I like long hair. I want long hair. What shampoo do you use?

Mercedes: Pantene Pro-V.

Mariana: I use Wella conditioner.

Mercedes: My hair is all dry and I have split ends.

Nydia pulled a *Seventeen* magazine out of her backpack and flipped though it until she found an ad for Suave conditioner. "I need this," she said. "It's mois . . . tur . . . izing," she read from the ad, stumbling over the pronunciation. " 'To replenish moisture in dry or damaged hair.' That's me."

This episode serves as an exemplar of the girls' preoccupation with the tools and techniques required to achieve physical beauty, and their use of mass media for guidance in the acquisition of those commodities. At East Middle School the girls who were more knowledgeable about beautification were also more "popular," which is to say that they were the central figures in their peer groups.

Clothes Further, at East Middle School, peer groups were defined in part by their costuming. Conversations with the girls confirmed this: the main groups of girls were the "gangstas," students who were gang-identified; the "gangsta wannabes," who dressed and acted like gang affiliates, but who were not included in gang activities; the "preps," who were the honor students and the cheerleaders; and the "dorks," the social outcasts. The dorks, according to Ariana, a member of the prep group, "didn't know how to dress." The Latina girls in the "gangsta" group had thin plucked eyebrows, wore dark lipstick and heavy eye makeup, and had chemically lightened their hair color. The girls in the prep group wore minimal makeup, but they plucked their eyebrows and sometimes experimented with hairstyles within a very conservative range of options (ponytails or sometimes curled hair). Their clothes usually reflected current trends in shopping-mall fashions. The "dorks" wore blue jeans and t-shirts and tended not to draw attention to themselves via their costuming.

At West Middle School the groups, or "cliques" as they were called by the students, were also marked by their appearance, and costuming. As Judith, a 14-year-old Jewish girl, explained,

> Some of the cliques have um like certain styles. Like the most popular people are kind of like preps, and they wear like The Gap and J. Crew, and then the other groups that kind of get stuff like that are, the skater group who are like grunge influenced . . . and then um, like there's this one group of girls that are like really into Contempo clothes, and they wear that a lot.

Consumer fashion thus was the principal means by which group identity was demarcated, although pop music was also used in the same way. Fashion traits seemed to be derived from advertising; music was related to clothing, and these connections were made from MTV as well as peer references. The skaters listened to hard rock and heavy metal music; the preps made much of knowing which bands were currently "hot" on the charts.

Despite these identifiers, many students were emphatic about not conforming to group norms. In individual conversations, they were clear about the characterizing features of the different cliques, and they all identified themselves as iconoclasts and nonconformists even when they belonged to cliques:

Judith: My style is kind of different from everyone else's. I don't really think too much about what I wear as long as I like it.

Bobbi: If I feel comfortable, if I like what I'm wearing . . . I'm not out to please anybody else.

Lila: If I find something I like, then I wear it, but if everybody starts painting their nails silver or something then you'd have to stop.

Jenny: I try to be my own person. I just buy clothes that I like and that are comfortable.

Tara: I don't conform to fashion trends or whatever.

Jenny (to Tara): For fashion, you copy me.

In an interesting paradox, the West Middle School girls were quick to point out and criticize each others' compliance with mediated and peer standards for dress and appearance, but they denied their own participation in that system. Their peer group conversations, however, belied an intense interest in fashion and the fashion media.

Diet and Weight At East Middle School the girls were extremely critical of those who showed no interest in conforming to media-driven standards of fashion and beauty, and they were also very open about their own interest in those standards.

This was particularly striking during a conversation about food and dieting that occurred at the lunch table among the "popular" girls at East. One of the girls, Ariana, had brought to the table a newspaper article about teenagers' eating habits, which the girls all looked at. The article described teenage girls' poor eating habits, stating that teen girls are "more likely to skip meals, avoid milk, eat away from home, and fret about their weight."

Brittany: It's true, this is how we eat. Milk has calcium, doesn't it? I don't drink milk.

Rachel: I only drink milk on cereal.

Brittany: I only drink water and tea.

Rosa: I drink everything except water and milk.

Brittany: I'm addicted to tea.

Marta: I'm addicted to Dr. Pepper.

Brittany: I used to starve myself.

Ariana: I did, too. It's easy after the first day. The first day is hard. But after that it's easy, you don't notice it.

Marta: We don't eat real food, we'd get food poisoning.

They were eager to reify the connections between themselves and the girls described in the article, and it was important to them to find points of similarity between themselves and the news article's mediated construction of "teenage girls," however negative that construction might be.

Despite this, these girls did not usually discuss their bodies or their weight to any significant degree in their peer group conversations; nor did the girls in the "gangsta" group. At West Middle School, however, there was more open talk about body norms and more criticism of girls whose bodies did not conform to the ideal.

Bobbi: There is a girl who will wear like really really tight pants and like stripes with flowers or something, things that don't go together at all.

Judith: Who?

Bobbi: Kathy Smith. [*author's note: name changed to protect privacy*]

Judith: That's a fashion faux pas right there. Not to say any names or anything, but she wears like tank tops with really thin straps but she doesn't really have the body for it . . .

The girls at West school were sensitized to issues of body image and eating disorders, and these topics cropped up frequently in their conversations. Eating disorders were a serious problem at this school; teachers informed me that a student had recently died of anorexia nervosa. Yet, interestingly, the girls' discussions hovered around *resistance* to dysfunctional images of body; they used discourse around these issues to find solidarity in critiquing problematic concepts of body.

Audrey: Like anorexia and all that stuff, I don't understand that. It's just so stupid, I don't get it. How can you not eat?

Jonquil: I couldn't ever be bulimic and keep throwing up.

Emma: And there are some people at school, these girls, they'll eat like a carrot, cause they're afraid the guys will see them eating. I hate that.

Audrey: At lunch, I eat like so much, I eat like three pieces of pizza every day. I don't care what guys think. People have to eat. That's just natural . . .

Jonquil: Biology.

Audrey: Yeah. And people, they're afraid they'll think they're pigs or something if they eat too much. So they'll like go to the bathroom and eat. Some people eat lunch in the bathroom.

Emma: It's so sad. You know Laurel. . . . She kind of copies things off magazines and TV and things . . . . she's into like being perfect and skinny and not eating.

Audrey: But you have to eat! I feel like saying, eat something!

Conversations with one of the teachers revealed that Jonquil in fact had had some problems with eating disorders; how many of the others had suffered from them was not ascertainable, but the issue was clearly on their minds. Here, the peer group served as a means of consolidating ideas about rejecting and resisting damaging ideals for female bodies. Individually, their engagement with their bodies may have been different, more self-critical and less defiant,[12] but the group context appeared to moderate those tendencies in more progressive directions. The conversation reflected the paradoxical nature of eating disorders and the culture of thinness in US society:[13] while the girls understood eating disorders as pathological and abnormal, they would not admit their own involvement with these problems even as they subscribed in their daily lives to mediated norms of slenderness and beauty. As Siebecker notes:

> [Eating] disorders have been regarded as bizarre psychological phenomena that affect a minority of emotionally disturbed women. The problem has thus been isolated from the experiences of other women and marginalized into a psychological category. This has in effect thrown up a smokescreen between the clinically diagnosed eating disorder sufferer and the rest of women in society: If the smokescreen came down, what women would see is that, while we do not all actually have eating disorders, we are not so different from those sufferers.[14]

The girls' conversation kept the smokescreen up, distancing the girls with obvious eating problems from the sociocultural norms of thinness and beauty that pervaded the group members' everyday lives.

Brides and Mothers Flipping through magazines in the classroom, Nydia and Maria pause at an advertisement featuring a bride in a formal white wedding dress. They examine the photograph for several minutes.

Nydia: Oh, that's a pretty dress.

Maria: She looks so beautiful.

Nydia: I want to have a long dress like that. I want a big veil and flowers.

At East Middle School mediated images of brides and motherhood were of vital interest to the girls. Again and again, they discussed TV and celebrity weddings with admiration and obsessive attention to detail.

Ariana: You know what I did ? I saw this wedding on TV, and the guys all wore Wranglers and tuxedo shirts . . . and then the bridesmaids were late, and the bride got wet and her hair got all messed up . . .

Marta: What are you talking about?

Ariana: A wedding where these guys wore Wrangler jeans and tuxedo shirts and jackets . . .

Brittany: That's how you're gonna get married.

Marta: She's gonna have horses at the reception. (Giggles.)

Rosa: I want a formal wedding. Ariana, what color were the bridesmaids' dresses?

Ariana: Pink, a really gross pink, I'm having pastel colors for my wedding.

Brittany: Did any of you watch the Waltons reunion?

Rosa: I did!

Marta: I watched "The Brady Girls Get Married." Marcia and Jan were going to have a double wedding, but they kept fighting about what kind of wedding to have . . .

Brittany: What kind did they want?

Marta: I don't know, one wanted to go all formal, and the other one wanted something modern, I think. One ended up wearing a short gown, I think they call them tea-length? The other one was long.

Rosa: I want a long gown. And lots of bridesmaids. Big weddings are nice.

Ariana: Weddings should be special. It's your day, your special day.

On another occasion, when the girls were working in the computer lab on the Internet, a group of them found Madonna's home page and zeroed in instantly on photographs of her with her baby. They were especially delighted with one image of her when she was pregnant, exclaiming aloud about how "sweet" and "adorable" it was.

This valorization of marriage and maternity appeared to be in line with the trends in their lives. Teenage pregnancy is a significant problem at East Middle School—it is the main reason that girls drop out of that school. During the five months of this observation, on four different occasions, former seventh- and eighth-graders returned for campus visits with their babies in their arms. Their appearances were not greeted with any derision or gossip from their erstwhile classmates; rather, they were feted and embraced, the babies were cooed over, and the girls spoke with some longing of the day when they too would be mothers.

Culturally, teenage pregnancy was a norm among this student population. Many of the 14- and 15-year-old girls disclosed that their mothers were in their late twenties and early thirties. Some of the girls seemed to experience some conflict about the pressure toward maternity and their knowledge of other possibilities; they talked sometimes about the issue of abortion and whether it was sinful or not. The African-American girls tended to be more in favor of abortion as an option. The Latina girls were more opposed to it, for religious reasons.

Two of the "gangsta" girls had one extended conversation about their plans for the future; both intended to go to college and were clear and emphatic about not wanting to get pregnant before then.

Maria: I am not getting married until I'm older. Like, 60. And I'm not having no babies.

Nydia: I don't want a baby messin' up my life. I'm going to wait till after I go to college.

Maria became pregnant at the end of that academic year and dropped out of school.

By contrast, teenage pregnancy was invisible at West Middle School, and marriage and maternity were never mentioned in the girls' peer group conversations over the five months of this observation. The subjects did not seem to be relevant to the white, upper-middle-class girls' lives at all.

Homophobia and Sexual Confusion On April 30, 1997, the TV show *Ellen* aired its notorious "coming-out" episode, in which the main character declared herself a lesbian. The show precipitated a discussion of homosexuality among the girls in the "prep" group at East Middle School; this conversation reiterated the refrain of homophobia that was a constant current in the students' lives at both East and West Middle Schools.

While overt discrimination based on race was rare at both schools, homophobia was an openly declared prejudice in the peer groups that were studied. Words like "fag" and "queer" were used casually as epithets; gossip about students' sexual orientations were a way of marking the social outcasts. At East Middle School the "coming-out" episode of *Ellen* served as a catalyst for a brief, impassioned exchange about the iniquity of homosexuality.

> **Rosa:** I don't think she should have done that. It's just wrong to go on TV and put that in front of everybody.
>
> **Ariana:** She should just keep it to herself.
>
> **Brittany:** I think it's a sin and it shouldn't be on TV.

None of the girls in the peer group expressed opinions that differed from these, but later that day, Nona was looking through a *People* magazine in class and came across a photograph of Ellen DeGeneres. At that point she paused and said thoughtfully to her teacher, who was sitting nearby, "It's kind of good that they showed that on TV because it lets people who are gay know that people's lives are like that."

Her comment was unusual in a milieu where gay-bashing was considered high sport. At West Middle School homophobia was similarly open and aggressive. One student, Jenny—one of the most popular girls in the eighth grade—had a notebook covered with pictures cut out of magazines. The right side of the notebook displayed pictures of celebrity figures she disliked, and the left side was adorned with photos of people she admired. Prominent on the right side of the notebook was the band Luscious Jackson, and Jenny had written "Sucks! Dikes, too!" (*sic*) across their image.

The "regular girls" were very aware of the rhetoric of compulsory heterosexuality in teen magazines and talked about it with some anger. As Audrey put it, "They make it seem like if you don't have a boyfriend, you're just nothing, which really . . . I don't think it's true." Later, she added, "All of the articles are so superficial they make you think you have to be pretty to have friends or to have a boyfriend to be cool. . . . and that's kind of stupid . . . " Interestingly, in phrasing this resistance, she acknowledge her own susceptibility to the rhetoric. In other conversations these girls expressed aversion to the concept of homosexuality and distress at the idea of other students identifying themselves as gay or lesbian.

One of these girls, Lila, was very reflective about the homophobia that was rampant in the school. . . . Lila was clearly struggling with her sexuality and was conflicted about how to cope with her peer environment.

> **Lila:** One of my friends and I, in order to help me dump my boyfriend, we pretended to be lesbians. . . . So that went on for about a week, and then we were like, oh it's just a joke, we were just trying to get rid of my boyfriend, you know, and people at school are still keeping it up. . . . There was this guy who like came up to me last week and said, "Why don't you leave our school so that we can get the scum out of here? Why don't you go to another school and stop polluting ours?" Some guy punched me six times and threw me on the ground.
>
> **Me:** Hmm.
>
> **Lila:** Although, I didn't really care, because you know, "Oh, no! He called me a dyke! I'm going to die!" and he's really mad at me because I didn't react. I mean, he calls me a lesbian and then walks away. I'm like, "Oh, no! I'm gonna die!"

Lila was derisive of the boy's bigotry, yet she was careful to couch her experience in terms of having "pretended" to be a lesbian. Students were open in their rejection of homosexuality and their need to position themselves in the heterosexual mainstream. In order to bolster this positioning, they chose role models whom they considered to epitomize femininity in terms of the heterosexual ideal.

Iconic Femininity At both schools, the girls made frequent references to media figures who served as emblems or icons of ideal femininity.

All of the women who were chosen by the girls as role models or heroines exemplified media and sociocultural ideals of beauty; and they were admired by the girls specifically for their beauty, although other characteristics were sometimes mentioned as reasons for revering them. But none of the girls professed to admire women who had not been identified in the mass media as being physically beautiful according to dominant standards.

CONCLUSIONS

The girls' overall use of the mass media to reconstruct ideals of heterosexuality with regard to physical appearance, the goals of marriage and maternity, and active homophobia reveal a fairly direct appropriation of the dominant ideology of femininity. Race and class factors impacted the ways in which the parameters of ideal femininity were defined; but in general, the peer context was one in which emergent gender identity was consolidated via constant reference to acceptable sociocultural standards of femininity and sexuality.

This conformity was not seamless; pockets of resistance occurred in peer group discussions, but when they did, their functioning was paradoxical. In the West Middle School girls' dialogue about eating disorders, or the "gangsta" girls' rejection of the prospect of early motherhood, the privileged voices in the discussion shut out some of the participants in such a way that their personal struggles with these issues could not be recognized. Jonquil's history of eating disorders, Maria's sexual activity that culminated in pregnancy a few months later, Lila's sexual ambiguity could not be given full voice. Thus, the peer group served to achieve ideological closure in terms of how issues of gender and power could be addressed.

This functioning adds a new dimension to the studies that have looked at girls' individual responses to media texts. Duke and Kreshel found that girls "were not as uniformly vulnerable to media messages concerning the feminine ideal as was expected."[15] Frazer found girls to be aware of the distinctions between magazine portrayals and real life, although she did point out that the conventions and "registers" of discourse constrained what the girls said.[16] Numerous theorists have posited a fluid and mobile relationship between mass media and the receiver; especially in the cultural studies literature, it is supposed that readers are able to reappropriate the meanings of messages according to their various life circumstances.[17] It does appear that girls *on their own* may be somewhat more able to critically examine and deconstruct media messages than in the peer group context. Therefore, the role of the adolescent peer group is a complex one: the group dynamic serves to mask and neutralize individual experiences of social and cultural processes. The group is a microcosm for the creation of social structure via the renegotiation of dominant and oppositional ideological positions.

In the peer groups in this study, surface levels of resistance cloaked some of the participants' more private and interiorized struggles with dominant codes of femininity. The group discourse provided a text that could be analyzed, but the research cited earlier in this essay, as well as information gathered by this researcher from sources outside of the peer groups, point to the existence of subtexts that tended to be suppressed by the group process. Some evidence of this kind of suppression was provided by the West Middle School girls' insistence that they were individuals when their outward behaviors indicated complete capitulation to the norms of the group. Another mark of such masking was the conspicuous absence of teen pregnancy in the peer discourse at West Middle School; it was a taboo topic in peer discussion, so if a girl at West had undergone a pregnancy, her experience would be completely invalidated by the-group's tacit doctrine of denial.

Similarly, eating disorders were not openly discussed at East Middle School, yet the girls' casual references to starving themselves in one conversation indicated that body image issues were of more concern than was openly evidenced. Thompson points out that eating disorders among Latina and African-American women tend to be severe because they are not taken seriously or diagnosed quickly.[18] The girls at East Middle School were uncritical and unreflexive about the norm of thinness to which they subscribed.

Thus, this research indicates that while race and class were differentiators of girls' socialization and concomitant media use, the differences highlighted the ways in which their different cultures functioned to uphold different aspects of dominant ideologies of femininity.

Watkins suggests that minority youth in particular have generated cultural practices of resistance that have grown out of their social marginalization.[19] Such resistance was not obviously manifested among the girls at East and West Middle Schools, yet the potential for resistance was an ever-present subcurrent. At West Middle School, for example, the peer group discourse among the "regular girls" was more resistant than that of the "preps." Because peer acceptance is of paramount importance in girls' culture,[20] a real subversion of dominant norms *could* certainly happen in a peer group where that was part of the group identity. It is possible that the peer group's social standing with respect to other groups would influence the degree of resistance expressed in the group. A larger study in which more, and more diverse, peer groups were observed would be needed to further investigate these phenomena.

It could be argued that the girls' observed tendency to accept dominant norms of femininity was related to the fact that most of the subjects were honors students—academic achievers who conformed to social expectations in every aspect of their lives. Yet an adherence to codes of what might be called "hegemonic femininity" was also evident in the behaviors of the "gangsta" group from East Middle School, who were considered to be at risk of dropping out, delinquency, and other "antisocial" behaviors. In fact, they demonstrated even more interest in costuming, makeup, beautification, and maternity than did the more "pro-social" peer groupings. It can be tentatively concluded from these data, then, that the peer group generally serves to consolidate dominant constructions of gender and sexuality.

However, race and class factors appear to intercede in the process of meaning-making, within as well as around the peer group context. The predominantly white, upper-middle-class students at West Middle School were primary targets of advertising-driven media, and they concomitantly paid significantly more attention to the mass media than did students at East Middle School. Nonetheless, media were used to shore up systems of belief held by students at both schools.

Hermes has observed that "media use and interpretation exist by grace of unruly and unpredictable, but in retrospect understandable and interesting choices and activities of readers."[21] Among the girls in this study, a key strategy for blending into the peer group involved participating in activities that marked the limits of acceptable femininity; however, these were deployed within their racial, cultural, and class environments. Deviance from normative sexuality was a means of identifying the social outcast; conformity was a way of bonding with the group, and mass media were used as instruments in the bonding process.

These findings have multiple implications. First, they establish the centrality of mass media in adolescent society and underscore the links between socialization into dominant norms of sexuality and consumer culture. The teenagers in this study were hyperaware of the need to use the media to find their foothold in the group. Their uses of the media were more than discursive: consumption of the necessary products that openly established their acceptance and understanding of sexual norms was a necessary part of peer interaction. Thus socialization into femininity was linked to the multimillion dollar fashion, beauty and diet industries that thrive on women consumers.

Second, perhaps more important, it makes clear that the peer group must be taken into account in the contemplation of interventions or counteractions against the mediated norms that play into girls' gendered behaviors. Such interventions tend to be "top-down," devised and administered by adults; yet the significance of the peer group in girls' social lives indicates that the most effective resistant practices would germinate and take root within the peer group. In this study, the peer group was shown to be a training ground where girls learned to use the mass media to acquire the skills of ideal femininity, but it was also a place where rejection of these norms could sometimes be voiced.

While girls individually have some sense of the social environment that operates to regulate their expressions of gender and sexuality, and while they may try on an individual level to resist damaging normative constructions of femininity, the peer group dynamic tends to mitigate against such resistance. Effective interventions for girls must work within the peer context to try to encourage more nuanced and less univocal conceptualizations of normative femininity. Beyond this, the peer group's relationship to broader levels of society must be taken into account. Interventions such as media literacy efforts will not be effective unless they are sensitive to issues of race, class, and culture; a recognition of institutionalized networks of power that constrain and limit girls' autonomy is necessary before strategies for resistance and emancipation can be devised.

NOTES

1. Lyn Mikel Brown and Carol Gilligan, *Meeting at the Crossroads: Women's Psychology and Girls, Development* (New York: Ballantine Books, 1992).
2. Peggy Orenstein, *Schoolgirls: Young Women, Self-Esteem, and the Confidence Gap* (New York: Doubleday, 1994); Mary Pipher, *Reviving Ophelia: Saving the Selves of Adolescent Girls* (New York: Ballantine,1994); Lori Stern, "Disavowing the Self in Female Adolescence," in *Women, Girls and Psychotherapy: Reframing Resistance,* ed. Carol Gilligan, Annie G. Rogers, and Deborah L. Tolman (New York: Haworth Press, 1991), 105–117.
3. Margaret Duffy and Micheal Gotcher, "Crucial Advice on How to Get the Guy: The Rhetorical Vision of Power and Seduction in the Teen Magazine *YM,*" *Journal of Communication Inquiry* 21 (spring 1996): 32–48; Lisa Duke and Peggy J. Kreshel, "Negotiating Femininity: Girls in Early Adolescence Read Teen Magazines," *Journals of Communication Inquiry* 22 (January 1998): 48–71; Meenakshi Gigi Durham, "Dilemmas of Desire: Representations of Adolescent Sexuality in Two Teen Magazines," *Youth and Society* 29 (March 1998): 369–389; Ellen McCracken, *Decoding Women's Magazines: From Mademoiselle to Ms.* (New York: St. Martin's Press, 1993); Angela McRobbie, "Jackie: An Ideology of Adolescent Femininity," *in Mass Communication Review year book* vol. 4, ed. Ellen Wartella, D. Charles Whitney, and Sven Windahl (Beverly Hills, CA: Sage, 1983), 251–271; Angela McRobbie, "Shut Up and Dance: Youth Culture and Changing Modes of Femininity," in *Postmodernism and Popular Culture.* ed. Angela McRobbie (London: Routledge, 1994), 155–176; Kate Pierce, "A Feminist Theoretical Perspective on the Socialization of Teenage Girls through *Seventeen* Magazine," *Sex Roles* 23 (1990): 491–500; Kate Pierce, "Socialization of Teen age Girls through Teen-Magazine Fiction: The Making of a New Woman or an Old Lady?" *Sex Roles* 29 (1993): 59–68.
4. Brown and Gilligan, *Meeting at the Crossroads;* Pipher, *Reviving Ophelia.*
5. Orenstein, *Schoolgirls:* American Association of University Women Educational Foundation, *How Schools Shortchange Girls* (Washington, DC: American Association of University Women, 1992).
6. Susan Bordo, *Unbearable Weight: Feminism, Western Culture, and the Body* (Berkeley: University of California Press, 1993); Naomi Wolf, *The Beauty Myth: How Images of Beauty Are Used Against Women* (New York: Anchor, 1991).
7. Sue Lees, *Sugar and Spice: Sexuality and Adolescent Girls* (Harmondsworth, England: Penguin, 1993); Naomi Wolf, *Promiscuities* (New York: Random House, 1997).
8. Jill McLean Taylor, Carol Gilligan, and Amy M. Sullivan, *Between Voice and Silence: Women and Girls,*

Race and Relationship (Cambridge, MA: Harvard University Press, 1995); Becky W. Thompson, *A Hunger So Wide and Deep: American Women Speak on Eating Problems* (Minneapolis: University of Minnesota Press, 1994).

9. Judith Rich Harris, "Where Is the Child's Environment? A Group Socialization Theory of Development," *Psychological Review* 102 (1995): 458–489.
10. Kevin G. Barnhurst and Ellen Wartella, "Newspapers and Citizenship: Young Adults' Subjective Experience of Newspapers," *Critical Studies in Mass Communication* 8 (June 1991): 195–209; Kevin G. Barnhurst and Ellen Wartella, "Young Citizens, American TV Newscasts, and the Collective Memory," *Critical Studies in Mass Communication* 15 (September 1998): 279–305; Leo Bogart, *Commercial Culture: The Media System and the Public Interest* (New York: Oxford University Press, 1995).
11. Sandra Lee Bartky, "Foucault, Femininity, and the Modernization of Patriarchal Power," in *Feminism and Foucault: Reflections on Resistance,* ed. Irene Diamond and Lee Quinby (Boston: Northeastern University Press, 1988), 64.
12. See Duke and Kreshel, "Negotiating Femininity."
13. Bordo, *Unbearable Weight.*
14. July Siebecker, "Women's Oppression and the Obsession with Thinness," in *Women: Images and Realities,* ed. Amy Kesselman, Lily D. McNair, and Nancy Schniedewind (Mountain View, CA: Mayfield, 1995), 107.
15. Duke and Kreshel, "Negotiating Femininity."
16. Frazer, "Teenage Girls Reading Jackie."
17. Ien Ang, *Watching "Dallas": Soap Opera and the Melodramatic Imagination* (London: Methuen, 1985); John Fiske, "British Cultural Studies and Television," in *Channels of Discourse, Reassembled,* ed. Robert C. Allen (Chapel Hill: University of North Carolina Press, 1992), 284–326; Stuart Hall, "Encoding/Decoding," in *Culture, Media, Language: Working Papers in Cultural Studies,* 1972–1979, ed. Stuart Hall, Dorothy Hobson, Andre Lowe, and Paul Willis (London: Hutchinson, 1980), 128–138; David Morley, *The Nationwide Audience: Structure and Decoding* (London: Hutchinson, 1980).
18. Thompson, *A Hunger So Wide and Deep.*
19. S. Craig Watkins, *Representing: Hip-Hop Culture and the Production of Black Cinema* (Chicago: University of Chicago Press,1998).
20. Griffiths, *Adolescent Girls and Their Friends;* Brown and Gilligan, *Meeting at the Crossroads;* Evans and Eder, "No Exit."
21. Joke Hermes, *Reading Women's Magazines: An Analysis of Everyday Media Use* (Cambridge, England: Polity Press), 25.

DISCUSSION QUESTIONS

1. What kinds of factors are most likely to promote resistance to detrimental media images of femininity in peer groups? The least likely?
2. Generally speaking, do you think the costs of resisting dominant norms increase or decrease in the peer group context? Why?
3. How relevant to adults are the findings of Durham's research?
4. In what ways do the mass media and the female adolescent peer groups in Durham's study contribute to heterosexual socialization? Are the female peer groups Durham studied a potential source of empowerment? Why or why not?
5. Comparatively, what kind of role does the male peer group play in the negotiation of sexuality for boys?

The Social Organization of Sexuality and Gender in Alternative Hard Rock: An Analysis of Intersectionality

MIMI SCHIPPERS

Utilizing queer theory and employing participant-observational methodology, Mimi Schippers assesses the construction of gender and sexuality in a Midwestern rock subculture. Her findings reveal the existence of creative subcultural norms that, at times, explicitly challenge and redefine dominant expressions of both gender and sexuality within this cultural milieu. However, her results also demonstrate that hegemonic ideas about sex and gender hegemony underlie and confine resistance practices so that they may concomitantly reify and affirm the very normative structures they are designed to challenge.

The main purpose of this study was to identify how women and men negotiate norms for gender performance in face-to-face interaction. For two and one-half years, I conducted an ethnographic study of the face-to-face negotiation of gender in a rock music subculture in Chicago. Observations were made at small rock concerts in local bars and clubs, at the homes of *active participants,* and at parties attended by active participants. By active participants, I mean a loose network of 20 to 30 women and men who regularly attended these rock shows as audience members and/or to play as musicians. While the network was rather large, there was a core group of eight informants with whom I developed a closer relationship. It was mostly with this core group of eight that I spent time outside of the clubs. However, other active participants would attend parties or show up at the homes of these eight people.

DESCRIPTION OF THE SUBCULTURE: ALTERNATIVE HARD ROCK

The subculture in Chicago is part of a larger genre of music that is most associated with the labels *grunge* or *alternative rocks* (see Clawson 1999).

Participants in this subculture shared a general rejection of the sexism of mainstream rock (Firth 1983; Groce and Cooper 1990; Weinstein 1991). Thus, there was a concerted effort to do rock music in a way that resisted or challenged hegemonic gender relations.

From "The Social Organization of Sexuality and Gender in Alternative Hard Rock: An Analysis of Intersectionality" by Mimi Schippers, *Gender & Society* 14(6): 747–764, 2000. Reprinted by permission of Sage Publications.

THE GENDER ORGANIZATION OF THE ROCK SHOW

In mixed-gender, public discourse and in the interviews with musicians, both women and men expressed a desire to do rock differently to not reproduce the sexism of mainstream rock. For instance, there was an explicit rejection of the groupie scene.

> I think in the last few years there's been a lot of great bands that have killed that stereotype, that male stereotype . . . initially when I got into [rock music] I was like, yeah, guys are dicks, rock bands are pretty stupid, and I want to be in a rock band that doesn't patronize the sexist and racist aspects of rock. Like punk in that sense. . . . They're normal people. They're not playing some kind of rock dream of dressing up and getting girls or whatever. (Kim Thayil of Soundgarden [Man])
>
> The groupie thing just doesn't happen. It's not part of the sensibility of what we're doing. And the guys that we tour with, you know the Melvins and Wool, it's not like there are groupies back here hanging out with them. It's just not part of what we do. . . . It comes out of punk roots. It's about being more enlightened, having a sense of fairness and not being sexist pigs. You know those guys in like rock bands with all the groupies. That's fine, but they're usually fuckin' pigs trying to prove something. We're just here to rock. (Donita Sparks of L7 [Woman])

These sentiments were also expressed by alternative hard rockers in Chicago. At one show, I was talking about groupies with a few people including a woman in a successful local band.

> I always worried about being called a groupie because I've always dated musicians. But you know that's who I'm around. I admire them and what they do. And I wanted to do what they were doing. I'm a grouper, not a groupie.

Collapsing the dichotomy between musician and groupie worked to situate this woman's behavior outside of the sexualized, gender norms for rock music. The groupie scene of mainstream rock was a relatively common topic of conversation among alternative hard rockers and served as a comparison with which to define their feminist departure.

Alternative hard rockers also developed and enforced norms against men actively pursuing women sexually, or against what participants call "shmoozing," within the confines of the bar. There was a common understanding among participants that the rock show, despite taking place in bars, did not serve as a place to find potential sexual partners. They defined and maintained this normative structure by chastising men who were perceived to be shmoozing or through storytelling.

> **Bryan:** Jim can be such an asshole sometimes. It's embarrassing to be with him in public. He totally shmoozes chicks and the way he does it is so fucking obnoxious.
>
> **Me:** Like what does he do?
>
> **Bryan:** Some woman will walk by and he'll like step in front of her and [Here Bryan mimics Jim's behavior by getting really close to me, holds out his hand, smiles broadly, and says, "Hi, You're cute!"].
>
> **Me:** No way.
>
> **Bryan:** All the fucking time. I just walk away and pretend I don't know him.

Obviously, I did not have access to what men said to each other or how they acted when women were not around. However, I was able to watch men interact in the clubs and, on some occasions, eavesdrop on their conversations. While I have no way of knowing if these norms held when women were not present, it is indicative of the normative gender structure of these rock shows that, for the most part, participants followed these general guidelines.

Although there were norms against sexual interaction between women and men, this does not mean that sexual desire and sexual behavior were absent.

THE SEXUAL ORGANIZATION OF THE ROCK SHOW

At the outset and throughout data collection, feminist conceptualizations of compulsory heterosexuality, social constructions of gendered sexuality, and their relationship to male dominance guided my observations. Thus, my focus, while in the field, was on the ways in which sexual interactions reinforced or challenged hegemonic masculinity and femininity and on whether there was sexual interaction among women or among men. However, as I analyzed the data, I soon realized that conceptualizing sexuality as one facet of gender was not enough. To shed some light on my analysis of the workings of sexual desire, sexual behavior, and sexual norms, I turned to queer theory.

There were two central tenets to queer theory that I brought to my understanding of how gender and sexuality were operating to challenge or maintain gender and sexual hegemony. First, I analyzed how sexualities as stable identities (gay, lesbian, straight) were constructed, understood, and distributed among participants within the subculture. The second was to explore sexuality as having multiple levels of organization, including individual identities, but also as practices; as an organizing feature of face-to-face social interaction; and as an overarching, normative structure for these rock shows. Sexuality at the level of identity was identified by participants' talk about their own and others' sexualities. Sexual practices were coded as expressions of desire and sexual behavior by individuals, such as flirting (winking, licking lips), kissing others, erotic dancing, and genital contact with others. To identify the sexual organization of social interaction, I adopted a symbolic interactionist approach (Blumer 1969; Stryker 1979) and focused on how the heterosexual-homosexual binary took shape in participants' practices and in the ongoing process of negotiated social interaction. This included the social positions constructed during interaction and the meanings attributed to those positions and the activities of people occupying those positions. Finally, I analyzed the overall normative structure for doing sexuality while doing rock music in these clubs and whether a hetero-focused binary was assumed, reproduced, and maintained by those norms.

Sexual Practices

While there were strong norms against heterosexual contact between women and men, there was a great deal of sexual contact among women and overt expressions of sexual desire by women. For instance, women would often engage in playful, sexualized interaction with each other.

> Maddie had a cigarette hanging from her lips and asked Carrie for a light. Carrie leaned in and put her mouth over Maddie's cigarette and pretended to bite it. When she pulled away, Carrie said, "Oh, I thought you said, 'Can I have a bite.' I suppose I could give you a light, but I'd rather give you a bite."

Another time Maddie and Carrie were sitting a few seats away from each other at the bar. Carrie was looking at Maddie winking and licking her lips. After 10 or 15 minutes, Carrie yelled to Maddie, "I'm leaving for a little while, will you miss me?" Maddie responded, "Of course, darling. But we're leaving anyway." Carrie exaggerated a pout and said, "Well then, there's no reason for me to come back."

Women would also express sexual desire for women musicians, as in the following discussion of Kim Gordon of Sonic Youth among Maddie, Colleen, and Bryan:

> **Maddie:** You met Kim Gordon?!
>
> **Colleen:** Yeah. We went backstage.

Maddie: I love her. I really, like, have fallen in love with her. . . . She's so cool. They're so good, and she is just so cool.

Bryan: Oh, I know.

Maddie: I love Sonic Youth. They're one of those bands that just always puts out good music. . . . I swear, I love her.

Colleen: She's totally hot!

At the concerts, it was common to see women rubbing their bodies together and gyrating against each other as they listened to the music. On rare occasions, I observed women kissing.

Interestingly, women's sexual desire was not limited to women and men but was extended to the sound of singers' voices, the sound of guitars, the syncopation of instruments, and other tonal or musical experiences. I heard women talk about wanting to "fuck" the music or about a singer's voice as "totally fuckable." When women particularly enjoyed live or recorded music, a common expression they used to convey this was, "Just fuck me now," or they would sometimes say, "I need a cigarette" or "I'm spent" after a live performance. The sexual references were not addressed to any particular person or people; instead, they were made toward and about the music as an object of desire and as sexually gratifying. As Grossberg (1988) suggests, rock music, especially live rock music, is simultaneously an auditory and bodily experience. At these shows, the music was not only heard but also felt in the body. The rhythm and syncopation between the bass, drums, and guitars gave the bodily experience a sexual valence as expressed by the women. For this reason, it is entirely possible that women's "dirty dancing" together could have been as much about sexualizing the music as it was about their sexual desire for each other (see McRobbie 1984). In other words, sexual desire, as expressed by women in talk and through their actions, was far more diffused and fluid than the hetero-focused binary would have it. This more diffused, fluid sexuality breaks down the relevance of sexual identity labels that are referents to the gender of one's sexual object. For this reason, I want to suggest that women's sexual desire cannot be simplified as bisexual or lesbian but instead can be characterized as queer because it was opened up to include more than other women or even other individuals.

At the same time, it was uncommon to see men moving their bodies to the music, except for "head banging" (vigorously bobbing the head back and forth) or playing "air guitar" (moving the body as if playing an imaginary guitar), so dancing was not something men in this subculture usually did. The men would most often keep their eyes on the stage or each other while conversing between bands, so even overtly "checking women out" was relatively uncommon. At least in the company of women at these shows, men did not usually express sexual desire for, or attraction to, women. On the rare occasion when men did appear to be expressing an overt sexual desire for women, others would invariably make fun of them to keep the counterhegemonic, normative structure intact.

Bryan: [A band] had this girl who played guitar. She was so awesome. She played like Angus Young (of AC/DC), you know on stage and stuff. She was so cool. I went up to her once after a show, and she was like "I'm married." I just wanted to talk about music.

(Maddie, Colleen, and I laugh)

Colleen: You should have put your dick back in your pants before you went up to talk to her.

Bryan: I should have taken my coat off my dick and put it back on.

Colleen: Hide it behind a newspaper or something.

(loud laughter)

By making Bryan's sexual desire explicit and then mocking it, the women reestablished the norm against men's overt sexual subjectivity. Even Bryan went along with their mocking to demonstrate his acceptance of this norm. In other words, men's sexual subjectivity was constantly monitored and checked as people interacted with each other.

The social control over men's sexual desire is significant because, as described earlier, this is a highly sexualized, public setting where women are overtly expressing and acting on their own sexual desire. The men were, at least within the confines of the bar, taken out of these sexual dynamics. Men did not enact sexual subjectivity in relation to women or to other men, which challenges the hegemonic gender order (Connell 1995). It is significant that women experienced and expressed sexual desire for each other publicly and were not subject to men's sexual desire, at least interactively. Also, despite engaging in sexual play with other women, none of the women in the subculture referred to themselves as bisexual or lesbian. As I will discuss below, alternative hard rockers constructed gay men and lesbians as people outside the subculture and therefore as marginalized others, at the same time that the women overtly engaged in intragender sexual behavior. This suggests an important play of gender and sexual hegemony and resistance within the same practices. This mixture of counterhegemonic and hegemonic sexual and gender relations became evident at other levels of social organization as well.

Constructing Sexual Identities through Talk

When alternative hard rockers talked about sexuality, it was most often to confront or challenge homophobia or heterosexism. In their politicized, antiestablishment rock world, it was uncool to be a bigot, and they included heterosexism as bigotry. This stance translated into challenging, chastising, and making ridiculous derogatory talk about gay and lesbian people. More important, these challenges to heterosexism were often used to challenge hegemonic gender relations or to establish alternative norms for femininity or masculinity. For example, one evening, Maddie approached me and angrily said, "Dan is such an asshole sometimes." I asked her what happened:

> I was standing there talking to Dan and Nancy and some other people, and I knew you were waiting, so I was trying to get them to decide what they were going to be doing. I was like, "Come on you guys, could you just make up your minds. Let's go." He was totally ignoring me. So I was like, "Get your thumb out of your ass, and let's do something." He turned to me and said, "Why do you have to be such a dyke all the time?" I said, "Thanks for the compliment. See ya . . ." and left. What a fucking asshole. Can you believe he even said that?

Within the context of this interaction, as reported to me by Maddie, what led to Dan calling Maddie a "dyke" was not that she might have had sex with women but that she was being assertive. Maddie used the opportunity to tell this story to convey her resistive stance toward uses of heterosexism to get women back in line in terms of gender. In response to Dan's explicit attempt to set and enforce gender norms and to implicitly fortify heterosexism, Maddie snapped back, "Thanks for the compliment." By telling of her quick "Thanks," she reconstructed his attempt to insult her into a compliment, suggesting that there was not only nothing wrong with but perhaps something positive about being a "dyke."

While Maddie's story was one of challenging heterosexism, it was also a story about gender resistance. The label *dyke* was redefined as a compliment because, as told by Maddie, it named and made explicit her refusal to enact femininity in a way that reproduced hegemonic gender relations. Maddie used the label *dyke* as a marker of an alternative femininity to shift the meaning of her assertiveness. Her story exposed heterosexism and thus challenged compulsory heterosexuality. At the same time, it challenged the gender order. More important, this illustrates how a positive deployment of the identity *dyke* made gender resistance possible.

However, Maddie's story also reinforced the homosexual-heterosexual binary in her implicit validation of the homosexual label *dyke*. She fully accepted the identity meaning of the label while rejecting its evaluative meaning as negative or inferior. In fact, her story relied on, and fully supported, an underlying assumption that there are "dykes" out there, and that they are admirable because they do femininity in ways that challenge the gender order. The sexual identities and the assumed gender performances attached to those identities were held in place, even though Maddie challenged both compulsory heterosexuality and gender hegemony.

Less often, but not infrequently, alternative hard rockers also challenged compulsory heterosexuality by freely talking about their gay, lesbian, or bisexual friends, roommates, sisters, brothers, mothers, work colleagues, and others. Because of this normalizing talk, there was little question in my analysis that alternative hard rockers challenged compulsory heterosexuality. However, I never heard a participant refer to himself or herself as gay, lesbian, or homosexual, and only on a few isolated occasions did anybody talk about another member of the subculture as gay or lesbian. For the most part, participants constructed gay men and lesbians as people outside the subculture yet also as deserving of all the rights, respect, and happiness of anybody else. More important, words such as *lesbian, gay,* and *dyke* seemed to be the only way in which members of this subculture could talk about their acquaintances, suggesting, as one would expect, that they did not have an elaborate queer language or a language that did not assume stable, sexual identities. That is, while challenging compulsory heterosexuality and challenging hegemonic gender relations, they would fully reinscribe the hegemonic sexual order in their construction of identities through talk and their marginalization of gay and lesbian identities.

The Sexual Organization of Face-to-Face Interaction

In addition to explicit talk about sexual identities, sexuality was also produced through negotiated social interaction. By exaggerating heterosexism or playing different "parts" in sexualized social interaction, alternative hard rockers would sometimes expose sexual bigotry without engaging in any explicit discussions of sexual identity or the politics of sexualities. For example, the first time I met Colleen, Maddie and I were passing by her outside a club before a show. Colleen feigned to whisper to the man she was talking with but said loudly so we could hear, "Did you know she's a lesbian," referring to Maddie. Colleen laughed and then said, "Not only that, but she's a bitch too." Both Maddie and Colleen laughed, hugged, and engaged in a playful banter vying for inclusion in the "lesbian club" and "bitch club." The tone of their exchange indicated that there was something positive about being a lesbian and a bitch. When Maddie and I walked away, Colleen said to me, "Now be careful, she's a lesbian." Maddie quickly responded, "She knows. She's my girlfriend. Jealous?" Colleen laughed and said, "Yeah, but I have my own girlfriend. He (referring to the man she was talking with) is really a she!"

Colleen's initial "whisper" about Maddie being a lesbian mocked the secret of being a lesbian. Colleen immediately invoked the label *bitch,* making fun of that as well. By situating *bitch* in this playful rejection of heterosexism, Colleen implicitly made gender salient. *Bitch,* like *lesbian,* is a derogatory label hurled at women who step out of the bounds of acceptable femininity and is meant to get women back in line. Like Maddie's story about Dan's deployment of *dyke,* both Colleen and Maddie quickly turned the mocked heterosexism and sexism into a verbal competition about who really is a lesbian and who really is a bitch. This shifted the meaning of those labels to positive attributes, which worked as gender resistance because the femininities associated with both challenged hegemonic femininity.

This interaction also challenged compulsory heterosexuality. Maddie and Colleen validated lesbianism through their competitive, interactive volley for the lesbian badge. Also, Colleen's whisper to her man friend could have been a test to see how Maddie would respond to the possibility of sexualizing their relationship.

By combining a gender analysis with a sexuality analysis at the level of face-to-face interaction, the data reveal that Colleen's and Maddie's banter involved, first, a queering of sexuality by manipulating gender relations. Colleen's assertion that she has a "lesbian" relationship with a man subverted the homo-hetero binary by destabilizing the meaning of *lesbian* in the context of this interaction. There is no way to see the sexuality organizing this interaction as homosexual, heterosexual, or bisexual, for the identity *woman* was detached from the social position *lesbian*.

Second, this queer sexuality set up subversive gender relations. The man in this interaction became the "girlfriend" of Colleen. That is, Colleen constructed the social roles in the interaction in a way that subverted male dominant power relations—after all, the gender order does not confer masculine power to "girlfriends." In this interaction, queering sexuality and destabilizing gender subverted both hegemonic gender and sexual relations.

However, by also analyzing sexuality as identities constructed through talk, a complicated play of resistance and hegemony is revealed. Although Maddie and Colleen challenged compulsory heterosexuality and complacent femininity at the level of interaction, they simultaneously assumed and bolstered a hetero-focused binary. Their gender resistance was enacted by deploying sexual-identity labels. Colleen actually used the word *lesbian* to tease or test Maddie. Again, this was based on a common meaning for *lesbian* as a stable sexual identity with a whole set of corresponding characteristics. Colleen's use of this label also reflected an implicit subcultural norm for heterosexuality.

The Normative Sexual Structure for Alternative Hard Rock

Moving my analysis out to the overarching normative structure of sexuality within the subculture reveals heterosexism in Colleen's pseudowhisper. Even though she might have been testing Maddie's comfort zone with sexualizing their relationship and she made lesbian sexuality a viable possibility by speaking positively of it, Colleen had to have assumed Maddie was heterosexual for the banter about lesbians and bitches to have been not only funny but also effective as gender resistance. If Maddie were indeed a lesbian, this interaction probably would have backfired on Colleen and she would have been chastised as both heterosexist and as supporting sexism. Colleen would never have risked "outing" Maddie if there were any remote possibility that Maddie was a lesbian. Colleen must have safely assumed that people are heterosexual unless proved otherwise. That is, the overarching sexual organization of this music scene was heterosexuality despite participants' rejection of compulsory heterosexuality in their talk and in their subversion of the hetero-focused binary in their practices and interactions.

This play of subversion and hegemony developed in most contexts where alternative hard rockers talked about sexuality. For example, I was at an alternative hard rocker's house with several other people including four men in a local band. The four members of this band were talking about whether it would be a good idea for them to take a gig opening for Tribe 8. Tribe 8 is a band out of San Francisco that consists of five women who are all very much out about their sexual desire for women and who make their sexual desire a central part of their performance.

> Joe (singer and guitarist): Man, I don't know. I don't know if it would be such a good idea. It's going to be a bunch of lesbians who probably would not appreciate a bunch of aggressive guys up there. [Everybody laughs]

> Shit, I don't want to get my ass kicked! There's no way they'd put up with us if they're waiting to see Tribe 8. I don't think we should.

First, it is important to point out that saying he did not "want to get [his] ass kicked" was not, in the context of this subculture, a derogatory remark about lesbians or about the women in Tribe 8. Men in this subculture often talked about women they admired as able to "kick ass." However, despite this nod of respect, while listening to Joe, I thought that he was probably exaggerating and revealing an underlying heterosexism that fueled his apprehension.

While at the Tribe 8 show, I concluded that Joe's concerns were perhaps heterosexist but at the same time well-founded. The audience consisted of mostly women, there were more overt sexual displays among women than I had seen at other shows, and many of the women were quite aggressive about keeping men out of the space in front of the stage. In other words, the social space was transformed into one where sexual desire among women and women's use of physical aggression framed social interaction and the normative structure more than at other shows.

While Joe might have negative or antagonistic feelings about lesbians, his heterosexism was also apparent in his use of the word *lesbian* to describe the crowd. I noticed that although there were many women there I did not know and who might describe themselves as lesbians or dykes, there were also many women there who consistently attended shows where Joe's band and other local bands played, and who did not refer to themselves as lesbians or dykes. In his talk, Joe only had the label *lesbian,* although it was the practices, interactions, and normative structure that differed at the Tribe 8 show, not the identities of individuals who participated.

This tension between how sexuality was enacted or practiced and the only available identity-based language meant they consistently found it difficult to talk about their practices. For instance, the following field note excerpt demonstrates how, at another concert, Courtney Love of the band Hole vacillated between identity talk and practice talk while revealing to the audience who her sexual partner was at the time.

> Courtney Love says to the audience, "Guess who I'm fucking. If you can guess, I'll tell you. Come on, try to guess." Several people in the audience (including both men and women) raise their hands. Someone from the crowd yells, "Drew!" (The guitarist was dating Drew Barrymore.) Love says, "No. He's fucking Drew (referring to the guitarist), I'm not. I'm not a lesbian. I'm only a part-time muff-muncher." She laughs. "I only munch muff part-time. She (referring to the drummer) is a full-time muff-muncher, and she (the bass player) is a virgin."

When someone in the audience suggested Love was "fucking" Drew Barrymore, she responded by saying, "I'm not a lesbian." What did that mean to Love? It meant that she is only a "part-time muff-muncher," referring to cunnilingus. Although she first used an identity label, and by doing so supported the hetero-focused binary, she quickly shifted the emphasis to what one does in practice to define sexuality. That the base player was a "virgin" in comparison to herself, a "part-time muff-muncher," and the drummer, a "full-time muff-muncher," meant that not *doing* anything defined one's sexuality. Sexuality was constructed as what you *do* as much as who you *are.* Her immediate shift to who is a "muff-muncher" and who is not transformed an identity into a set of practices. She not only demonstrated the difficulty alternative hard rockers face when talking about sexuality in nonidentity terms but also that sexuality meant something far more complex in practice than identity markers could signify.

While alternative hard rockers fairly consistently reproduced a hetero-focused binary of sexual identities in their talk and in their assumption and reproduction of a heterosexual normative sexual structure, they sometimes queered sexuality in their practices and in their interactions. More important, both the reproduction of sexual identities in their talk and their queering of sexuality in practice worked to undermine male dominance. However, although alternative hard rockers were relatively successful in creating a social setting with norms that challenged gender hegemony, at the same time, they continuously reproduced hegemonic sexuality.

Given the ways in which sexual desire manifested at these shows, I initially concluded that alternative hard rockers were relatively successful in resisting hegemony both in terms of gender and in terms of sexuality. It was only after data were collected and I was analyzing my findings that the importance of a queer theoretical framework became salient. Only by queering my sociological analysis or, to borrow the term from Kimberle Williams Crenshaw (1991) and Patricia Hill Collins (1999), by adopting a perspective of *intersectionality,* did the layers of subversion and hegemony in participants' talk, practices, and interactions become apparent.

IMPLICATIONS AND CONCLUSION

Five main themes emerged from my observations. First, these data suggest that empirical investigations of queer sexuality must be extended to people who identify as heterosexual. Alternative hard rockers' expressions of sexual desire and sexual practices blurred or in some cases rendered meaningless the lines between heterosexual and homosexual despite an overall norm for heterosexual identities. There has been little attention paid to the queer practices of people who identify as heterosexual, and further research in this area is needed.

Second, I found a dissonance between how alternative hard rockers talked about sexuality and how they enacted sexual desire. These findings suggest that focusing only on the way people talk about sexuality will leave hidden some of the most important aspects of sexuality, namely, how it is enacted and negotiated in face-to-face interaction. This is significant because it suggests that while alternative hard rockers were facile in, and willing to think about, sexualities in terms of stable homosexual and heterosexual identities, their everyday experience of sexuality, at least to some degree, undermined identity demarcations. My guess is that if asked on a survey or in an interview, alternative hard rockers would be more than willing to "buy into" identity markers and choose one and, more important, in most cases would choose a heterosexual identity. However, survey or interview questions, because they rely on language, would be framing and therefore boxing in alternative hard rockers' sexualities into an identity framework even though in their practices this sometimes is not the case. There were ways in which sexuality was enacted and constructed in face-to-face interaction that stepped out of that framework and would likely disappear empirically if alternative hard rockers were asked to talk about sexuality. This suggests that ethnographic methods or a more innovative formulation of survey and interview questions would greatly enhance our efforts to map and understand sexuality and its relationship to gender and male dominance.

Third, these findings demonstrate that queering sexuality can be an effective form of gender resistance. There is much debate among feminists about whether deconstructing identities is politically expedient (see, e.g., Alcoff 1994; Ault 1996; Butler 1990; DiStefano 1990; Fraser and Nicholson 1990; Fuss 1989). While these findings do not provide any kind of answer to those questions, they do suggest that gender resistance and the subversion of identities are not necessarily incompatible.

Fourth, these data suggest that effective strategies of gender resistance sometimes reinscribe the hegemonic sexual order. Even when participants rejected compulsory heterosexuality, hegemonic norms for sexualized masculinity and femininity, and traditional gender scripts for sexual interaction, they still upheld the hierarchical relationship between heterosexual and homosexual as stable, fixed identities. This supports queer theorists' assertion that sexuality and gender operate separately but articulate each other. This strongly suggests that sociological theorizing and empirical research on gender must take seriously the sexual order as conceptualized by queer theorists and how it intersects with gender. Just as the social organization of race and class crosscut gender relations, the sexual order also does so.

Finally, my goal was to demonstrate the importance of conceptualizing and analyzing sexuality at multiple levels of analysis. Without a multilevel analysis, I might have concluded that alternative hard rockers successfully developed a rock music subculture that reproduced neither male dominance nor heterosexual dominance. At the level of practice, identity talk, and interaction, they rejected compulsory heterosexuality and male dominance. However, the same practices and talk that undermined compulsory heterosexuality and worked as effective gender resistance depended on the hegemonic sexual order.

As many have persuasively argued in terms of the relationships between race, ethnicity, class, and gender, adopting a perspective of intersectionality to study the relationship between gender and sexuality at multiple levels of analysis is much needed and long overdue in sociological research.

REFERENCES

Alcoff, Linda. 1994. Cultural feminism versus post-structuralism: The identity crisis in feminist theory. In *Culture/power/history: A reader in contemporary social theory,* edited by N. B. Dirks, G. Eley, and S. Ortner, 96–122. Princeton, NJ: Princeton University Press.

Ault, Amber. 1996. The dilemma of identity: Bi women's negotiations. *In Queer theory/sociology,* edited by Steven Seidman, 311–330. Cambridge, MA: Blackwell.

Blumer, Herbert. 1969. *Symbolic interactionism: Perspectives and method.* Engelwood Cliffs, NJ: Prentice Hall.

Butler, Judith. 1990. *Gender trouble: Feminism and the subversion of identity.* New York: Routledge.

Clawson, Mary Ann. 1999. When women play the bass: Instrument specialization and gender interpretation in alternative rock music. *Gender & Society* 13 (2): 193–210.

Connell, R. W. 1995. *Masculinities.* Berkeley: University of California Press.

Di Stefano, C. 1990. Dilemmas of difference Feminism, modernity, and Postmodernism. In *Feminism/postmodernism,* edited by Linda J. Nicholson, 63–82. New York: Routledge.

Fraser, Nancy, and Linda J. Nicholson. 1990. Social criticism without philosophy: An encounter between feminism and postmodernism. In *Feminism/postmodernism,* edited by Linda J. Nicholson, 19–38. New York: Routledge.

Frith, Simon 1983. *Sound effects: Youth, leisure, and the politics of rock.* London: Constable.

Fuss, Diana. 1989. *Essentially speaking: Feminism, nature and difference.* New York: Routledge.

Groce, S. B., and M. Cooper. 1990. Just me and the boys? Women in local-level rock and roll. *Gender & Society* 4: 220–229.

Grossberg L 1988. Putting the pop back in postmodernism. In *Universal abandon? The politics of post-modernism,* edited by A. Ross, 167–190. Minneapolis: University of Minnesota Press.

Hill Collins, Patricia. 1999. Moving beyond gender: Intersectionality and scientific knowledge. In *Revisioning gender,* edited by Myra Marx Ferre, Judith Lorber, and Beth B. Hess, 261–284. Thousand Oaks, CA: Sage.

McRobbie, Angela. 1984. Dance and social fantasy. In *Gender and generation,* edited by Angela McRobbie and M. Nava, 130–161. London: Macmillan.

Mulvey, Laura. 1990. Visual pleasure and narrative cinema. In *Issues in feminist film criticism,* edited by Patricia Erens, 28–40. Bloomington: University of Indiana press.

Stryker, Sheldon. 1979. *Social interactionism: A social structural approach.* New York: Benjamin-Cummings.

Weinstein, Deena. 1991. *Heavy metal: A cultural sociology.* New York: Lexington Book.

Williams Crenshaw, Kimberle. 1991. Mapping the margins: Intersectionality, identity politics, and violence against women of color. *Stanford Law Review* 43: 1241–1299.

DISCUSSION QUESTIONS

1. How is the subculture in this study different from the peer groups in Durham's study? What might account for these differences?

2. Is it possible to resist hegemonic constructs without, at some level, relying on them?
3. Is it easier for women to resist norms of sex and gender than it is for men? Why or Why not?
4. How do you think the members of this subculture would define sex? How is sexuality constructed?
5. Are the resistance practices of this subculture feminist? Why or why not?

The following Bonus Reading can be accessed through the *Infotrac College Edition Online Library* by typing in the name of the author or keywords in the title online at *http://www.infotrac-college.com*.

Normal Sins: Sex Scandal Narratives as Institutional Morality Tales

JOSHUA GAMSON

A long-held media maxim is that sex sells. The high level of public interest in most sex scandals would certainly seem to bear this out. In this work Joshua Gamson trains a sociological eye on media treatment of sex scandals. His analysis of three recent scandals reveals that the media's focus is less about exposing individual wrong-doing and culpability than it is about telling a particular story about the pathologies of the larger institutional contexts in which the players of these scandals are embedded.

DISCUSSION QUESTIONS

1. Gamson asks why, given America's tendency to look to individualistic rather than structural explanations, the opposite trend appears to occur in these stories. His answer focuses on the institutional needs of the media itself. Do you agree? Why or why not? What other explanations, if any, can you offer?
2. Gamson observes that the purchase of the sexual services of prostitutes is often presented as a demonstration of masculinity. Do you agree? If so, *how* does the purchase of a prostitute's services accomplish this? In what ways does this analysis extend the discussion of militarized prostitution in Enloe's article in Chapter 6?
3. What elements of the social construction of male sexuality normalizes men's patronage of prostitutes? What elements of the social construction of female sexuality normalizes their depiction *as* prostitutes?
4. What, if any, insights does Gamson's analysis provide for understanding the media coverage of the Canadian and/or American priest sex scandals?

8

Economy and Work

Capitalism and Gay Identity

JOHN D'EMILIO

In a previous section of this book, Jonathan Katz outlined the social history of the category heterosexual. *In this reading, John D'Emilio traces the origins of* gay *identity and links its emergence to capitalism. D'Emilio's thesis holds that wage labor and commodity production have altered the traditional functions and structure of the family thereby creating the social conditions that make gay and lesbian identities a social possibility.*

For gay men and lesbians, the 1970s were years of significant achievement. Gay liberation and women's liberation changed the sexual landscape of the nation. Hundreds of thousands of gay women and men came out and openly affirmed same-sex eroticism. We won repeal of sodomy laws in half the states, a partial lifting of the exclusion of lesbians and gay men from federal employment, civil rights protection in a few dozen cities, the inclusion of gay rights in the platform of the Democratic Party, and the elimination of homosexuality from the psychiatric profession's list of mental illnesses. The gay male subculture expanded and became increasingly visible in large cities, and lesbian feminists pioneered in building alternative institutions and an alternative culture that attempted to embody a liberatory vision of the future.

In the 1980s, however, with the resurgence of an active right wing, gay men and lesbians face the future warily. Our victories appear tenuous and fragile; the relative freedom of the past few years seems too recent to be permanent. In some parts of the lesbian and gay male community, a feeling of doom is growing: analogies with McCarthy's America, when "sexual perverts" were a special target of the Right, and with Nazi Germany, where gays were shipped to concentration camps, surface with increasing frequency. Everywhere there is the sense that new strategies are in order if we want to preserve our gains and move ahead.

From "Capitalism and Gay Identity" by John D'Emilio, pp. 100–113 in *Powers of Desire: The Politics of Sexuality*, Ann Snitow, Christine Stansell, and Sharon Thompson (eds.).
New York: Monthly Review Press.

I believe that a new, more accurate theory of gay history must be part of this political enterprise. When the gay liberation movement began at the end of the 1960s, gay men and lesbians had no history that we could use to fashion our goals and strategy. In the ensuing years, in building a movement without knowledge of our history, we instead invented a mythology. This mythical history drew on personal experience, which we read backward in time. For instance, most lesbians and gay men in the 1960s first discovered their homosexual desires in isolation, unaware of others and without resources for naming and understanding what they felt. From this experience, we constructed a myth of silence, invisibility, and isolation as the essential characteristics of gay life in the past as well as the present. Moreover, because we faced so many oppressive laws, public policies, and cultural beliefs, we projected this onto an image of the abysmal past until gay liberation, lesbians, and gay men were always the victims of systematic, undifferentiated, terrible oppression.

These myths have limited our political perspective. They have contributed, for instance, to an overreliance on a strategy of coming out—if every gay man and lesbian in America came out, gay oppression would end—and have allowed us to ignore the institutionalized ways in which homophobia and heterosexism are reproduced. They have encouraged, at times, an incapacitating despair, especially at moments like the present: How can we unravel a gay oppression so pervasive and unchanging?

There is another historical myth that enjoys nearly universal acceptance in the gay movement, the myth of the "eternal homosexual." The argument runs something like this: gay men and lesbians always were and always will be. We are everywhere; not just now, but throughout history, in all societies and all periods. This myth served a positive political function in the first years of gay liberation. In the early 1970s, when we battled an ideology that either denied our existence or defined us as psychopathic individuals or freaks of nature, it was empowering to assert that "we are everywhere." But in recent years it has confined us as surely as the most homophobic medical theories, and locked our movement in place.

Here I wish to challenge this myth. I want to argue that gay men and lesbians have *not* always existed. Instead, they are a product of history, and have come into existence in a specific era. Their emergence is associated with the relations of capitalism; it has been the historical development of capitalism—more specifically, its free labor system—that has allowed large numbers of men and women in the late twentieth century to call themselves gay, to see themselves as part of a community of similar men and women, and to organize politically on the basis of that identity.[1] Finally, I want to suggest some political lessons we can draw from this view of history.

What, then, are the relationships between the free labor system of capitalism and homosexuality? First, let me review some features of capitalism. Under capitalism, workers are "free" laborers in two ways. We have the freedom to look for a job. We own our ability to work and have the freedom to sell our labor power for wages to anyone willing to buy it. We are also freed from the ownership of anything except our labor power. Most of us do not own the land or the tools that produce what we need, but rather have to work for a living in order to survive. So, if we are free to sell our labor power in the positive sense, we are also freed, in the negative sense, from any other alternative. This dialectic—the constant interplay between exploitation and some measure of autonomy—informs all of the history of those who have lived under capitalism.

As capital—money used to make more money—expands, so does this system of free labor. Capital expands in several ways. Usually it expands in the same place, transforming small firms into larger ones, but it also expands by taking over new areas of production: the weaving of cloth, for instance, or the baking of bread. Finally, capital expands geographically. In the United States, capitalism initially took root in the Northeast, at a time when slavery was the dominant system in the South and when non-capitalist Native American societies occupied the western half of the continent. During the nineteenth century, capital spread from the Atlantic to the Pacific, and in the twentieth, U.S. capital has penetrated almost every part of the world.

The expansion of capital and the spread of wage labor have effected a profound transformation in the structure and functions of the nuclear family, the ideology of family life, and the meaning of heterosexual relations. It is these changes in the family that are most directly linked to the appearance of a collective gay life.

The white colonists in seventeenth-century New England established villages structured around a household economy, composed of family units that were basically self-sufficient, independent, and patriarchal. Men, women, and children farmed land owned by the male head of household. Although there was a division of labor between men and women, the family was truly an interdependent unit of production: the survival of each member depended on the cooperation of all. The home was a workplace where women processed raw farm produces into food for daily consumption; where they made clothing, soap, and candles; and where husbands, wives, and children worked together to produce the goods they consumed.

By the nineteenth century, this system of household production was in decline. In the Northeast, as merchant capitalists invested the money accumulated through trade in the production of goods, wage labor became more common. Men and women were drawn out of the largely self-sufficient household economy of the colonial era into a capitalist system of free labor. For women in the nineteenth century, working for wages rarely lasted beyond marriage; for men, it became a permanent condition.

The family was thus no longer an independent unit of production. But although no longer independent, the family was still interdependent. Because capitalism had not expanded very far, because it had not yet taken over—or socialized—the production of consumer goods, women still performed necessary productive labor in the home. Many families no longer produced grain, but wives still baked into bread the flour they bought with their husband's wages; or, when they purchased yarn or cloth, they still made clothing for their families. By the mid-1800s, capitalism had destroyed the economic self-sufficiency of many families, but not the mutual dependence of the members.

This transition away from the household family-based economy to a fully developed capitalist free labor economy occurred very slowly, over almost two centuries. As late as 1920, 50 percent of the U.S. population lived in communities of fewer than 2,500 people. The vast majority of blacks in the early twentieth century lived outside the free labor economy, in a system of sharecropping and tenancy that rested on the family. Not only did independent farming as a way of life still exist for millions of Americans, but even in towns and small cities women continued to grow and process food, make clothing, and engage in other kinds of domestic production.

But for those people who felt the brunt of these changes, the family took on new significance as an affective unit, an institution that produced not goods but emotional satisfaction and happiness. By the 1920s among the white middle class, the ideology surrounding the family described it as the means through which men and women formed satisfying, mutually enhancing relationships and created an environment that nurtured children. The family became the setting for a "personal life," sharply distinguished and disconnected from the public world of work and production.[2]

The meaning of heterosexual relations also changed. In colonial New England the birth rate averaged over seven children per woman of childbearing age. Men and women needed the labor of children. Producing offspring was as necessary for survival as producing grain. Sex was harnessed to procreation. The Puritans did not celebrate *hetero*sexuality but rather marriage; they condemned *all* sexual expression outside the marriage bond and did not differentiate sharply between sodomy and heterosexual fornication.

By the 1970s, however, the birth rate had dropped to under two. With the exception of the post-World War II baby boom, the decline has been continuous for two centuries, paralleling the spread of capitalist relations of production. It occurred even when access to contraceptive devices and abortion was systematically curtailed. The decline has included every segment of the population—urban and rural families, blacks and whites, ethnics and WASPs, the middle class and the working class.

As wage labor spread and production became socialized, then, it became possible to release sexuality from the "imperative" to procreate. Ideologically, heterosexual expression came to be a means of establishing intimacy, promoting happiness, and experiencing pleasure. In divesting the household of its economic independence and fostering the separation of sexuality from procreation, capitalism has created conditions that allow some men and women to organize a personal life around their erotic/emotional attraction to their own sex. It has made possible the formation of urban communities of lesbians and gay men and, more recently, of a politics based on a sexual identity.

Evidence from colonial New England court records and church sermons indicates that male and female homosexual behavior existed in the seventeenth century. Homosexual *behavior,* however, is different from homosexual *identity.* There was, quite simply, no "social space" in the colonial system of production that allowed men and women to be gay. Survival was structured around participation in a nuclear family. There were certain homosexual acts—sodomy among men, "lewdness" among women—in which individuals engaged, but family was so pervasive that colonial society lacked even the category of homosexual or lesbian to describe a person. It is quite possible that some men and women experienced a stronger attraction to their own sex than to the opposite sex—in fact, some colonial court cases refer to men who persisted in their "unnatural" attractions—but one could not fashion out of that preference a way of life. Colonial Massachusetts even had laws prohibiting unmarried adults from living outside family units.[3]

By the second half of the nineteenth century, this situation was noticeably changing as the capitalist system of free labor took hold. Only when *individuals* began to make their living through wage labor, instead of as parts of an interdependent family unit, was it possible for homosexual desire to coalesce into a personal identity—an identity based on the ability to remain outside the heterosexual family and to construct a personal life based on attraction to one's own sex. By the end of the century, a class of men and women existed who recognized their erotic interest in their own sex, saw it as a trait that set them apart from the majority, and sought others like themselves. These early gay lives came from a wide social spectrum: civil servants and business executives, department store clerks and college professors, factory operatives, ministers, lawyers, cooks, domestics, hoboes, and the idle rich: men and women, black and white, immigrant and native born.

In this period, gay men and lesbians began to invent ways of meeting each other and sustaining a group life. Already, in the early twentieth century, large cities contained male homosexual bars. Gay men staked out cruising areas, such as Riverside Drive in New York City and Lafayette Park in Washington. In St. Louis and the nation's capitol, annual drag balls brought together large numbers of black gay men. Public bathhouses and YMCAs became gathering spots for male homosexuals. Lesbians formed literary societies and private social clubs. Some working-class women "passed" as men to obtain better-paying jobs and lived with other women—lesbian couples who appeared to the world as husband and wife. Among the faculties of women's colleges, in the settlement houses, and in the professional associations and clubs that women formed, one could find lifelong intimate relationships supported by a web of lesbian friends. By the 1920s and 1930s, large cities such as New York and Chicago contained lesbian bars. These patterns of living could evolve because capitalism allowed individuals to survive beyond the confines of the family.[4]

Simultaneously, ideological definitions of homosexual behavior changed. Doctors developed theories about homosexual*ity*, describing it as a condition, something that was inherent in a person, a part of his or her "nature." These theories did not represent scientific breakthroughs, elucidations of previously undiscovered areas of knowledge; rather, they were an ideological response to a new way of organizing one's personal life. The popularization of the medical model, in turn, affected the consciousness of the women and men who experienced homosexual desire so that they came to define themselves through their erotic life.[5]

These new forms of gay identity and patterns of group life also reflected the differentiation of people according to gender, race, and class that is so pervasive in capitalist societies. Among whites, for instance, gay men have traditionally been more visible than lesbians. This partly stems from the division between the public male sphere and the private female sphere. Streets, parks, and bars, especially at night, were "male space." Yet the greater visibility of white gay men also reflected their larger numbers. The Kinsey studies of the 1940s and 1950s found significantly more men than women with predominantly homosexual histories, a situation caused, I would argue, by the fact that capitalism had drawn far more men than women into the labor force, and at higher wages. Men could more easily construct a personal life independent of attachments to the opposite sex, whereas women were more likely to remain economically dependent on men. Kinsey also found a strong positive correlation between years of schooling and lesbian activity. College-educated white women, far more able than their working-class sisters to support themselves, could survive more easily without intimate relationships with men.[6]

Among working-class immigrants in the early twentieth century, closely knit kin networks and an ethic of family solidarity placed constraints on individual autonomy that made gayness a difficult option to pursue. In contrast, for reasons not altogether clear, urban black communities appeared relatively tolerant of homosexuality. The popularity in the 1920s and 1930s of songs with lesbian and gay male themes—"B.D. Woman," "Prove It on Me," "Sissy Man," "Fairey Blues"—suggests an openness about homosexual expression at odds with the mores of whites. Among men in the rural West in the 1940s, Kinsey found extensive incidence of homosexual behavior, but, in contrast with the men in large cities, little consciousness of gay identity. Thus, even as capitalism exerted a homogenizing influence by gradually transforming more individuals into wage laborers and separating them from traditional communities, different groups of people were also affected in different ways.[7]

The decisions of particular men and women to act on their erotic/emotional preference for the same sex, along with the new consciousness that this preference made them different, led to the formation of an urban subculture of gay men and lesbians. Yet at least through the 1930s this subculture remained rudimentary, unstable, and difficult to find. How, then, did the complex, well-developed gay community emerge that existed by the time the gay liberation movement exploded? The answer is to be found during World War II, a time when the cumulative changes of several decades coalesced into a qualitatively new shape.

The war severely disrupted traditional patterns of gender relations and sexuality, and temporarily created a new erotic situation conducive to homosexual expression. It plucked millions of young men and women, whose sexual identities were just forming, out of their homes, out of towns and small cities, out of the heterosexual environment of the family, and dropped them into sex-segregated situations—as GIs, as WACs and WAVEs, in same-sex rooming houses for women workers who relocated to seek employment. The war freed millions of men and women from the settings where heterosexuality was normally imposed. For men and women already gay, it provided an opportunity to meet people like themselves. Others could become gay because of the temporary freedom to explore sexuality that the war provided.[8]

Lisa Ben, for instance, came out during the war. She left the small California town where she was raised, came to Los Angeles to find work, and lived in a women's boarding house. There she met for the first time lesbians who took her to gay bars and introduced her to other gay women. Donald Vining was a young man with lots of homosexual desire and few gay experiences. He moved to New York City during the war and worked at a large YMCA. His diary reveals numerous erotic adventures with soldiers, sailors, marines, and civilians at the Y where he worked, as well as at the men's residence club where he lived, and in parks, bars, and movie theaters. Many GIs stayed in port cities like New York, at YMCAs like the one where Vining worked. In his oral histories of gay men in San Francisco, focusing on the 1940s, Allan Bérubé has found that the war years were critical in the formation of a gay male *community* in the city. Places as different as San Jose, Denver, and Kansas City had their first gay bars in the 1940s. Even severe repression could have positive side effects. Pat Bond, a lesbian from Davenport, Iowa, joined the WACs during the 1940s. Caught in a purge of hundreds of lesbians from the WACs in the Pacific, she did not return to Iowa. She stayed in San Francisco and became part of a community of lesbians. How many other women and men had comparable experiences? How many other cities saw a rapid growth of lesbian and gay male communities?[9]

The gay men and women of the 1940s were pioneers. Their decisions to act on their desires formed the underpinnings of an urban subculture of gay men and lesbians. Throughout the 1950s and 1960s, the gay subculture grew and stabilized so that people coming out then could more easily find other gay women and men than in the past. Newspapers and magazines published articles describing gay male life. Literally hundreds of novels with lesbian themes were published.[10] Psychoanalysts complained about the new ease with which their gay male patients found sexual partners. And the gay subculture was found not just in the largest cities. Lesbian and gay male bars existed in places like Worcester, Massachusetts, and Buffalo, New York; in Columbia, South Carolina, and Des Moines, Iowa. Gay life in the 1950s and 1960s became a nationwide phenomenon. By the time of the Stonewall Riots in New York City in 1969—the event that ignited the gay liberation movement—our situation was hardly one of silence, invisibility, and isolation. A massive, grass-roots liberation movement could form almost overnight precisely because communities of lesbians and gay men existed.

Although gay community was a precondition for a mass movement, the oppression of lesbians and gay men was the force that propelled the movement into existence. As the subculture expanded and grew more visible in the post-World War II era, oppression by the state intensified, becoming more systematic and inclusive. The Right scapegoated "sexual perverts" during the McCarthy era. Eisenhower imposed a total ban on the employment of gay women and men by the federal government and government contractors. Purges of lesbians and homosexuals from the military rose sharply. The FBI instituted widespread surveillance of gay meeting places and of lesbian and gay organizations, such as the Daughters of Bilitis and the Mattachine Society. The post office placed tracers on the correspondence of gay men and passed evidence of homosexual activity on to employers. Urban vice squads invaded private homes, made sweeps of lesbian and gay male bars, entrapped gay men in public places, and fomented local witch hunts. The danger involved in being gay rose even as the possibilities of being gay were enhanced. Gay liberation was a response to this contradiction.

Although lesbians and gay men won significant victories in the 1970s and opened up some safe social space in which to exist, we can hardly claim to have dealt a fatal blow to heterosexism and homophobia. One could even argue that the enforcement of gay oppression has merely changed locales, shifting somewhat from the state to the arena of extralegal violence in the form of increasingly open physical attacks on lesbians and gay men. And, as our movements have grown, they have generated a backlash that threatens to wipe out our gains. Significantly, this New Right opposition has taken shape as a "pro family" movement. How is it that capitalism, whose structure made possible the emergence of a gay identity and the creation of urban gay communities, appears unable to accept gay men and lesbians in its midst? Why do heterosexism and homophobia appear so resistant to assault?

The answers, I think, can be found in the contradictory relationship of capitalism to the family. On the one hand, as I argued earlier, capitalism has gradually undermined the material basis of the nuclear family by taking away the economic functions that cemented the ties between family members. As more adults have been drawn into the free labor system, and as capital has expanded its sphere until it produces as commodities most goods and services we need for our survival, the forces that propelled men and women into families and kept them there have weakened. On the other hand, the ideology of capitalist society has enshrined the family as the source of love, affection, and emotional security, the place where our need for stable, intimate human relationships is satisfied.

This elevation of the nuclear family to preeminence in the sphere of personal life is not accidental. Every society needs structures for reproduction and childbearing, but the possibilities are not limited to the nuclear family. Yet the privatized family fits well with capitalist relations of production. Capitalism has socialized production while maintaining that the products of socialized labor belong to the owners of private property. In many ways, child rearing has also been progressively socialized over the last two centuries, with schools, the media, peer groups, and employers taking over functions that once belonged to parents. Nevertheless, capitalist society maintains that reproduction and child rearing are private tasks, that children "belong" to parents, who exercise the rights of ownership. Ideologically, capitalism drives people into heterosexual families: each generation comes of age having internalized a heterosexist model of intimacy and personal relationships. Materially, capitalism weakens the bonds that once kept families together so that their members experience a growing instability in the place they have come to expect happiness and emotional security. Thus, while capitalism has knocked the material foundation away from family life, lesbians, gay men, and heterosexual feminists have become the scapegoats for the social instability of the system.

This analysis, if persuasive, has implications for us today. It can affect our perception of our identity, our formulation of political goals, and our decisions about strategy.

I have argued that lesbian and gay identity and communities are historically created, the result of a process of capitalist development that has spanned many generations. A corollary of this argument is that we are *not* a fixed social minority composed for all time of a certain percentage of the population. *There are more of us* than one hundred years ago, more of us than forty years ago. And there may very well be more gay men and lesbians in the future. Claims made by gays and nongays that sexual orientation is fixed at an early age, that large numbers of visible gay men and lesbians in society, the media, and the schools will have no influence on the sexual identities of the young, are wrong. Capitalism has created the material conditions for homosexual desire to express itself as a central component of some individuals' lives; now, our political movements are changing consciousness, creating the ideological conditions that make it easier for people to make that choice.

To be sure, this argument confirms the worst fears and most rabid rhetoric of our political opponents. But our response must be to challenge the underlying belief that homosexual relations are bad, a poor second choice. We must not slip into the opportunistic defense that society need not worry about tolerating us, since only homosexuals become homosexuals. At best, a minority group analysis and a civil rights strategy pertain to those of us who already are gay. It leaves today's youth—tomorrow's lesbians and gay men—to internalize heterosexist models that it can take a lifetime to expunge.

I have also argued that capitalism has led to the separation of sexuality from procreation. Human sexual desire need no longer be harnessed to reproductive imperatives, to procreation; its expression has increasingly entered the realm of choice. Lesbians and homosexuals most clearly embody the potential of this split, since our gay relationships stand entirely outside a procreative framework. The acceptance of our erotic choices ultimately depends on the degree to which society is willing to affirm sexual expression as a form of play, positive and life-enhancing. Our movement may have begun as the struggle of a "minority," but what we should now be trying to "liberate" is an aspect of the personal lives of all people—sexual expression.[11]

Finally, I have suggested that the relationship between capitalism and the family is fundamentally contradictory. On the one hand, capitalism continually weakens the material foundation of family life, making it possible for individuals to live outside the family, and for a lesbian and gay male identity to develop. On the other hand, it needs to push men and women into families, at least long enough to reproduce the next generation of workers. The elevation of the family to ideological pre-eminence guarantees that capitalist society will reproduce not just children but also heterosexism and homophobia. In the most profound sense, capitalism is the problem.[12]

How do we avoid remaining the scapegoats, the political victims of the social instability that capitalism generates? How can we take this contradictory relationship and use it to move toward liberation?

Gay men and lesbians exist on social terrain beyond the boundaries of the heterosexual nuclear family. Our communities have formed in that social space. Our survival and liberation depend on our ability to defend and expand that terrain, not just for ourselves but for everyone. That means, in part, support for issues that broaden the opportunities for living outside traditional heterosexual family units: issues like the availability of abortion and the ratification of the Equal Rights Amendment, affirmative action for people of color and for women, publicly funded day care and other essential social services, decent welfare payments, full employment, the rights of young people—in other words, programs and issues that provide a material basis for personal autonomy.

The rights of young people are especially critical. The acceptance of children as dependents, as belonging to parents, is so deeply ingrained that we can scarcely imagine what it would mean to treat them as autonomous human beings, particularly in the realm of sexual expression and choice. Yet until that happens, gay liberation will remain out of our reach.

But personal autonomy is only half the story. The instability of families and the sense of impermanence and insecurity that people are now experiencing in their personal relationships are real social problems that need to be addressed. We need political solutions for these difficulties of personal life. These solutions should not come in the form of a radical version of the profamily position, of some left-wing proposals to strengthen the family. Socialists do not generally respond to the exploitation and economic inequality of industrial capitalism by calling for a return to the family farm and handicraft production. We recognize that the vastly increased productivity that capitalism has made possible by socializing production is one of its progressive features. Similarly, we should not be trying to turn back the clock to some mythic age of the happy family.

We do need, however, structures and programs that will help to dissolve the boundaries that isolate the family, particularly those that privatize child rearing. We need community- or worker-controlled day care, housing where privacy and community coexist, neighborhood institutions—from medical clinics to performance centers—that enlarge the social unit where each of us has a secure place. As we create structures beyond the nuclear family that provide a sense of belonging, the family will wane in significance. Less and less will it seem to make or break our emotional security.

In this respect, gay men and lesbians are well situated to play a special role. Already excluded from families as most of us are, we have had to create, for our survival, networks of support that do not depend on the bonds of blood or the license of the state, but that are freely chosen and nurtured. The building of an "affectional community" must be as much a part of our political movement as are campaigns for civil rights. In this way we may prefigure the shape of personal relationships in a society grounded in equality and justice rather than exploitation and oppression, a society where autonomy and security do not preclude each other but coexist.

NOTES

1. I do not mean to suggest that no one has ever proposed that gay identity is a product of historical change. See, for instance, Mary McInstosh, "The Homosexual Role," *Social Problems* 16 (1968): 182–192; Jeffrey Weeks, *Coming Out: Homosexual Politics in Britain* (New York: Quartet Books, 1977). It is also implied in Michel Foucault, *The History of Sexuality*, vol. 1: *An Introduction,* tr. Robert Hurley (New York: Pantheon, 1978). However, this does represent a minority viewpoint, and the works cited above have not specified how it is that capitalism as a system of production has allowed for the emergence of a gay male and lesbian identity. As an example of the "eternal homosexual" thesis, see John Boswell, *Christianity, Social Tolerance, and Homosexuality* (Chicago: University of Chicago Press, 1980), where "gay people" remains an unchanging social category through fifteen centuries of Mediterranean and Western European history.
2. See Eli Zaretsky, *Capitalism, the Family, and Personal Life* (New York: Harper and Row, 1976); and Paula Fass, *The Damned and the Beautiful: American Youth in the 1920s* (New York: Oxford University Press, 1977).
3. Robert F. Oaks, " 'Things Fearful to Name': Sodomy and Buggery in Seventeenth-Century New England," *Journal of Social History* 12 (1978): 268–281; J. R. Roberts, "The Case of Sarah Norman and Mary Hammond," *Sinister Wisdom* 24 (1980): 57–62; and Jonathan Katz, *Gay American History* (New York: Crowell, 1976), pp. 16–24, 568–571.
4. For the period from 1870 to 1940 see the documents in Katz, *Gay American History,* and idem., *Gay/Lesbian Almanac* (New York: Crowell, 1983). Other sources include Allan Bérubé, "Lesbians and Gay Men in Early San Francisco: Notes Toward a Social History of Lesbians and Gay Men in America," unpublished paper, 1979; Vern Bullough and Bonnie Bullough, "Lesbianism in the 1920s and 1930s: A Newfound Study" *Signs* 2 (Summer 1977): 895–904.
5. On the medical model see Weeks, *Coming Out,* pp. 23–32. The impact of the medical model on the consciousness of men and women can be seen in Louis Hyde, ed., *Rat and the Devil: The Journal Letters of F. O. Matthiessen and Russell Cheney* (Hamden, Conn.: Archon, 1978), p. 47; and in the story of Lucille Hart in Katz, *Gay American History,* pp. 258–279. Radclyffe Hall's classic novel about lesbianism, *The Well of Loneliness,* published in 1928, was perhaps one of the most important vehicles for the popularization of the medical model.
6. See Alfred Kinsey et al., *Sexual Behavior in the Human Male* (Philadelphia: W. B. Saunders, 1948), and *Sexual Behavior in the Human Female* (Philadelphia: W. B. Saunders, 1953).
7. On black music, see "AC/DC Blues: Gay Jazz Reissues," Stash Records, ST–106(1977); and Chris Albertson, *Bessie* (New York: Stein and Day, 1974). On the persistence of kin networks in white ethnic communities see Judith Smith, "Our Own Kind: Family and Community Networks in Providence," in *A Heritage of Her Own,* ed. Nancy F. Cott and Elizabeth H. Pleck (New York: Simon and Schuster, 1979), pp. 393–411; on differences between rural and urban male homoeroticism see Kinsey et al., *Sexual Behavior in the Human Male,* pp. 455–457, 630–631.
8. The argument and the information in this and the following paragraphs come from my book *Sexual Politics, Sexual Communities: The Making of a Homosexual Minority in the United States, 1940–1970* (Chicago: University of Chicago Press, 1983). I have also developed it with reference to San Francisco in "Gay Politics, Gay Community: San Francisco's Experience," *Socialist Review* 55 (January–February 1981): 77–104.

9. Donald Vining, *A Gay Diary, 1933–1946* (New York: Pepys Press, 1979); "Pat Bond," in Nancy Adair and Casey Adair, *Word Is Out* (New York: New Glide Publications, 1978), pp. 55–65; and Allan Bérubé, "Marching to a Different Drummer: Coming Out During World War II," a slide/talk presented at the annual meeting of the American Historical Association, December 1981, Los Angeles. A shorter version of Bérubé's presentation can be found in *The Advocate,* October 15, 1981, pp. 20–24.
10. On lesbian novels see *The Ladder*, March 1958, p. 18; February 1960, pp. 14–15; April 1961, pp. 12–13; February 1962, pp. 6–11; January 1963, pp. 6–13; February 196, pp. 12–19; February 1965 pp. 19–23; March 1966, pp. 22–26; and April 1967, pp. 8–13. *The Ladder* was the magazine published by the Daughters of Bilitis.
11. This especially needs to be emphasized today. The 1980 annual conference of the National Organization for Women, for instance, passed a lesbian rights resolution that defined the issue as one of "discrimination based on affectional/sexual preference/orientation," and explicitly disassociated the issue from other questions of sexuality such as pornography, sadomasochism, public sex, and pederasty.
12. I do not meant to suggest that homophobia is "caused" by capitalism or is to be found only in capitalist societies. Severe sanctions against homoeroticism can be found in European feudal society and in contemporary socialist countries. But my focus in this essay has been the emergence of a gay identity under capitalism, and the mechanisms specific to capitalism that made this possible and that reproduce homophobia as well.

DISCUSSION QUESTIONS

1. D'Emilio argues that capitalism has created the conditions that allow for the formation of a homosexual identity. What effects, if any, do you think capitalism's substantial cultural and economic dominance has had on the sexual identities and behaviors of people living in non-capitalistic societies?
2. Today, reproductive technologies are making childbirth a possibility for many who previously might have been unable or unwilling to have children, including heterosexual couples, single mothers, and lesbian and gay partners. How might D'Emilio integrate the increased availability and use of reproductive technologies into his analysis? What are the implications of these technologies for the future of "the family"?
3. D'Emilio's analysis suggests that homosocial environments may increase the opportunity for increased same-sex sexual behaviors. Is there a link between this fact and the demand for prostitution described previously by Enloe? If so, describe this link. If not, explain why.
4. Can you link D'Emilio's arguments of contemporary trends in divorce to Seidman's analysis of love?
5. Does the traditional family provide a sense of belonging? Can the sense of belonging that individuals derive from the "traditional" family be found elsewhere?
6. Early in this article, D'Emilio argues that the gay liberation movement suffers from an over-reliance on the strategy of coming out. If every gay man and lesbian in America came out would gay oppression end? If so, why? If not, what impact would it have?
7. Can D'Emilio's analysis be expanded to explain the emergence of other non-hegemonic sexual identities (e.g. transgender, bisexual, asexual)?

Sexuality in Organizations

PATRICIA YANCEY MARTIN AND DAVID L. COLLINSON

In the past two decades, gender theorists have revealed how the structures and policies of organizations produce, recreate, and reify gender in ways that serve to perpetuate and institutionalize social inequality. In the following except from their chapter Gender and Sexuality in Organizations, *Patricia Yancey Martin and David L. Collinson demonstrate how organizations similarly institutionalize the sexuality of their workers and the workplace environment.*

SEXUALITY IN ORGANIZATIONS

. . . Theories about organizations generally assert that sexuality has no legitimate place at work. Yet many people find life partners and partners for other sexual relationships at work. Additionally, sexual attraction, flirtation, manipulation, and coercion are pervasive in organizations (e.g., Hall 1993a) and are not easily legislated out of existence. One sexuality norm in organizations is the requirement for everyone to present him- or herself as heterosexual. Frequently, especially among men, homophobia is used to "keep each other in line" (Reskin and Padavic 1994). If sexuality were irrelevant in organizations, then "out" lesbians and gays would occasion little notice or comment; however, the opposite usually occurs. The identifiable homosexual (or bisexual) faces indignities and discrimination from presumptive heterosexuals (Dunne 1997; Raeburn 1998; Schneider 1993).

At a minimum, men use sexuality to establish a hierarchy that separates them from women and keeps women "in their place" (Pyke 1996). Numerous studies reveal how men at multiple hierarchical levels sexualize women at work (Hearn and Parkin 1987; Collinson and Collinson 1989). Hollway (1984) suggests that men invest in a "male sexual drive discourse" in which they frame men's sexuality as "incontinent," "out of control," and biologically driven. They thereby construct predatory sexual discourses and workplace cultures that derogate and undermine women (Cockburn 1983; Collinson 1992).

The greatest attention to sexuality in workplaces centers on sexual harassment. Since the late 1970s, numerous studies have documented sexual harassment as structural and commonplace rather than individual and rare (Rospenda et al. 1998; Wise and Stanley 1988). Sexual harassment involves violence, power, authority, and economic discrimination (MacKinnon 1979). Women's vulnerability to sexual harassment reflects their segregated, subordinated labor market position (Hadjifotiou 1983), and service sector employment may render them especially vulnerable because of pressures to "please customers" (Folgero and Fjeldstad 1995; Hochschild 1983; Adkins 1995). Walby (1988) views sexual harassment as a patriarchal strategy that men proactively use to keep women "in their place." Men's greater power gives them more *sexual authority* at work and conveys to women the message, "You're only a woman. And at that level you're vulnerable to me and any man" (Cockburn 1991:142). DiTomaso (1989) suggests that some work contexts license men to behave offensively so that their attempts to take sexual advantage become the norm. One respondent told her, "The men are different here than on the street. It's like they have been locked up for years" (p. 80).

Collinson and Collinson's (1996; see also 1992) study of sexual harassment in the British insurance sales industry delineates men's sexuality discourses and practices at work. Men sales agents subjected women colleagues to extensive sexual harassment; in one incident, a man displayed his penis to a woman. The men managers "normalized" other men's offensive practices with statements such as "It's a fact of life" and "It's just a bit of fun." They viewed men's harassment of women as "rites de passage," a gendered test of women's ability to deal with the "pressures" of working in a male-dominated occupation. They agreed that the women were responsible for "handling" sexual harassment and when one failed to "handle it well," they cast her as cause rather than victim of the dynamic.

Generally, men sexually harass women for the benefit of other men or to impress men colleagues (Kimmel 1996). Pinups of nude women and pornographic photos in workplaces signify in-group heterosexuality, conveying messages of exclusion to gay and bisexual men and of objectification to women. Work organizations are thus sites where men proclaim their heterosexuality by using women to establish relations with each other even while they *practice homosociality* by aggressively seeking out and preferring the company of other men (Lipman-Blumen 1976). Raeburn (1998) notes that the signification of heterosexuality at work is so pervasive as to be mandatory—for instance, displaying photographs of a spouse and children, wearing a wedding band, talking about wife, husband, or family-related concerns. These practices convey to gay and lesbian members their difference, requiring them to display discordant symbols—such as photos of a same-sex partner—or to remain silent and serve as audience for the heterosexual discourse and practice that are hegemonic in such contexts, thereby (re)constructing their own marginalization.

When harassment is extensive and women are discouraged from reporting, nearly any reaction by women is ineffective (Cockburn 1991:157). Separated from one another, women find that their strategies of resisting, integrating, showing indifference, distancing, and denial fail to produce beneficial results. Collinson and Collinson (1996) found that one woman who complained was labeled by the men as a "moaning feminist troublemaker"; another who tried to integrate into the male sexist culture was derided by the men as "unfeminine and aggressive." A third who tried to ignore the men's sexual innuendos was eventually ostracized. All responses by the women were framed by (men) line managers as evidence that the women were incompetent or "unable to fit in."

Far from diminishing, evidence suggests that sexual harassment is alive and well. A recent survey of 4,501 U.S. women medical doctors, ages 30 to 70, confirms this conclusion (Frank, Brogan, and Schiffman 1998). Fully 47.7 percent of the physicians reported having been targets of gender-based harassment, and 36.9 percent reported having been sexually harassed. The authors of the survey conclude that although "some may believe that problems of harassment will disappear in time, that they are simply a function of older, sexist physicians still being in practice," the picture is less sanguine. Even though women are 42 percent of all medical students, they conclude, "We may be continuing to train physicians in an environment where harassment is common."

Advancing a few women into male-dominated workplaces may reinforce rather than challenge harassment culture and practices (Collinson and Collinson 1996). If so, equal opportunity schemes that advance women into male-dominant arenas may subject them to increased harassment and distress (Martin 1994; Marshall 1995). Collinson and Collinson (1996) conclude that creating organizations that legitimate and value women equally with men will require a fundamental reconceptualization of their purposes, goals, and methods, not merely ameliorative policies to raise awareness or to reform "insensitive men."

In sum, men's power in organizations cannot be understood apart from sexuality. Organizations are social contexts with extensive resources that men use to enact sexuality, and the conflation-of-sexuality with women means that gender issues can be rendered invisible by powerful men, even as they engage in sexualized behavior. For example, in an interview with the first author, a 55-year-old White male vice president of a telecommunications company described a "personal policy" he had followed for 30 years. He described himself as a "Christian, happily married man, and father of four" who wanted to avoid any suspicion of sexual impropriety. His policy was to invite men but not women subordinates to accompany him on trips. When his out-of-town host was a man, they went out to dinner to prepare for the upcoming day's work. He did not go to dinner with a woman host. These practices prevented the vice president from getting to know women as well as he did men, from establishing working relations with women that were as close as his relationships with men, and from gaining firsthand knowledge of women's talents, skills, and potential. He realized only recently that his behavior had "hurt the careers of women for many years." He regretted that his policy prevented women from having the same opportunities as men and was more sympathetic toward women now, but his policy remained the same. "I still do not want anyone to accuse me of sexual misconduct. . . . I'm sorry but that's just the way it is." He was not concerned that his willingness to have dinner alone with a man could be interpreted as a sign of homosexual interest, furthermore (Hearn 1993). Yet women who openly prefer women's company and advancement at work have been accused of being lesbians (Katzenstein 1998).

THOUGHTS FOR THE FUTURE

. . . In many practical respects, gender and sexuality at work are inseparable. "Doing femininity" involves "doing heterosexuality"—looking "sexy" in heterosexist ways, deferring to men in vaguely sexual ways, being partnered with a man, and so forth (Rogers forthcoming). "Doing masculinity" is, in contrast, largely "doing dominance" (see West and Zimmerman 1987). Sexual domination epitomizes gender dominance, in many ways, given that men's greater power *as* men and *as* occupants of powerful organizational positions allows them to treat women as sexual objects. Gender disadvantage combines with organizational disadvantage to make women's resistance to sexual harassment difficult. Whereas gender and sexuality intertwine in profound ways, gender and class as well as gender and race are less consequentially bound up with one another in organizations (Tomaskovic-Devey 1993). The gender/sexuality nexus is particularly consequential in organizations because sexuality is not supposed to find expression and gender is supposed to play a minimalist role. Yet sexuality and gender are extensively conflated in everyday practice in organizations, especially, but not solely, by men.

. . . [To advance] fundamental revisioning, we need studies of *masculinities* relative to other elements of power, culture, subjectivity, and relations in organizations. We need especially to understand masculinity dynamics among men. Connell (1995) claims that masculinities that are compatible with capitalism and rational-technical bureaucracy make it possible to promote masculine values and practices without acknowledging them. We need more attention to men's *affiliative* relations at work to see whether they undercut competitiveness among men or merely restrict competition to men by, for example, excluding women. In what ways and with what consequences are masculinity practices and ideology interwoven with workplace dynamics, such as bureaucratic claims that "rationality" and efficiency inform decisions or that capitalist market imperatives require competition and domination to assure success? How do masculinities that are dominant in organizations produce racism and the marginalization of people of color, including highly educated and skilled African American women and men (Collins 1997)?

We also need research on *femininities* to explore women's agency and compliance. Pyke (1996) discusses "noncompliant femininity," which conveys a false impression of women's independence, and calls for research on how the "construction of femininities reflects and (re)constructs (or resists) the gender order and inter-male hierarchies" (p. 546). Organizations are peopled by women and men who construct both femininities and masculinities, and individual women can practice "masculinities" just as individual men can practice "femininities" (Connell 1995; Kerfoot and Knights 1998). However, only men can collectively mobilize masculinities and, in doing so, exclude both women and femininities (Martin 1998). Organizations are suffused with emotionality, and affect is the basis for much action, official and unofficial (Martin and Knopoff 1997). Socializing and affiliation on and off the job are denied by hegemonic theories of organization and management. Equated with women and femininities, they are ignored in men. Organization theory has not only neglected gender, sexuality, and women but also has rendered men's emotional ties at work invisible (Hearn 1993). Men's emotions may include fear and dismay more than liking and affirmation, but theories depicting men as guided primarily by rational, logical, unemotional sentiments are, we suggest, stereotypical and incorrect.

It would be most interesting to document the results where employers have taken steps to dismantle gender and sexuality inequalities in their own ranks. Some for-profit corporations (e.g., Microsoft, Walt Disney World) and state and local governments (e.g., the state of Washington; the cities of San Jose and San Francisco, California) have established employee benefits packages that include the same-sex partners of homosexuals or that redress pay inequities for women's jobs relative to men's. Which internal policies are helpful? How do organizational cultures, practices, and structures have to be changed so that organizations can effectively implement the policies they adopt? What ingredients of organization and action make a workplace supportive of equality?

Alliances between members of organizations and members of social movements beyond organizational borders can foster change in gender and sexuality at work (Eisenstein 1995). Such alliances will not readily occur, however, and the effort required to build and sustain them is great. Yet changes have occurred and some will continue. Our hope is that gender and sexuality, as well as race, will become less and less relevant to the social organization of work and that the myth of social neutrality in traditional organization theory will become the reality of the future.

REFERENCES

Acker, Joan. 1990. "Hierarchies, Bodies, and Jobs: A Theory of Gendered Organizations." *Gender & Society* 4:139–158.

Acker, Joan and Donald Van Houten. 1974. "Differential Recruitment and Control: The Sex Structuring of Organizations." *Administrative Science Quarterly* 19:152–163.

Adkins, L. 1995. *Gendered Work: Sexuality, Family, and the Labour Market.* Buckingham: Open University Press.

Calas, Marta B. and Linda Smircich. 1993. "Dangerous Liaisons: The 'Feminine-in-Management' Meets 'Globalization.'" *Business Horizons* (March–April): 73–83.

Cockburn, Cynthia. 1983. *Brothers.* London: Pluto

——. 1991. *In the Way of Women: Men's Resistance to Sex Equality in Organizations.* London: Macmillan.

Collins, Sharon. 1997. "Black Mobility in White Corporations: Up the Corporate Ladder but Out on a Limb." *Social Problems* 44:55–67.

——. 1992. *Managing the Shopfloor: Subjectivity, Masculinity and Workplace Culture.* Berlin: Walter de Gruyter.

Collinson, David L. 1992. *Managing the Shopfloor: Subjectivity, Masculinity and Workplace Culture.* Berlin: Walter de Gruyter.

Collinson, David L. and Margaret Collinson. 1989. "Sexuality in the Workplace: The Domination of Men's Sexuality." pp. 91–109 in *The Sexuality of Organization,* edited by Jeff Hearn, Debra L. Sheppard, Peta Tancred-Sheriff, and Gibson Burrell. London: Sage.

———. 1992. "Mismanaging Sexual Harassment: Protecting the Perpetrator and Blaming the Victim." *Women in Management Review* 7:11–17.

Collinson, Margaret and David L. Collinson. 1996. "It's Only 'Dick': The Sexual Harassment of Women

Managers in Insurance." *Work, Employment and Society* 10(1):29–56.

Connell, R. W. 1995. *Masculinities*. Berkeley: University of California Press.

DiTomaso, Nancy. 1989. "Sexuality in the Workplace: Discrimination and Harassment." pp. 71–90 in *The Sexuality of Organization,* edited by Jeff Hearn, Debra L. Sheppard, Peta Tancred-Sheriff, and Gibson Burrell. London: Sage.

Dunne, Gillian A. 1997. *Lesbian Lifestyles: Women's Work and the Politics of Sexuality*. Toronto: University of Toronto Press.

Eisenstein, Hester. 1995. *Inside Agitators: The Femocrats in Australia*. Philadelphia: Temple University Press.

Folgero, I. S. and I. H. Fjeldstad. 1995. "On Duty—Off Guard: Cultural Norms and Sexual Harassment in Service Organizations." *Organization Studies* 16:299–314.

Frank, Erica, Donna J. Brogan, and Melissa Schiffman. 1998. "Prevalence and Correlates of Sexual Harassment among U.S. Women Physicians." *Archives of Internal Medicine* 158:352–358.

Hadjifotiou, N. 1983. *Women and Harassment at Work*. London: Pluto.

Hall, Elaine. 1993a. "Smiling, Deferring, and Flirting: Doing Gender by Giving 'Good Service.'" *Work and Occupations* 20:452–471.

Hearn, Jeff. 1993. "Emotive Subjects: Organizational Men, Organizational Masculinities and the (De)Construction of 'Emotions.'" pp. 142–166 in *Emotions in Organizations,* edited by Stephen Fineman. London: Sage.

Hearn, Jeff and Wendy Parkin. 1987. *"Sex" at "Work": The Power and Paradox of Organisation Sexuality*. Brighton: Wheatsheaf.

Hochschild, Arlie R. 1983. *The Managed Heart: Commercialization of Human Feeling*. Berkeley: University of California Press.

Hollway, Wendy. 1984. "Gender Difference and the Production of Subjectivity." pp. 227–263 in *Changing the Subject,* edited by J. Henriques, W. Hollway, C. Urwin, C. Venn, and V. Walkerdine. London: Methuen.

Katzenstein, Mary. 1998. *Faithful and Fearless: Moving Feminist Protest inside the Church and Military*. Princeton, NJ: Princeton University Press.

Kerfoot, Debra and David Knights. 1998. "Managing Masculinity in Contemporary Organizational Life: A 'Man' agerial Project." *Organizations* 5:7–26.

Kimmel, Michael. 1996. *Manhood in America: A Cultural History*. New York: Free Press.

Lipman-Blumen, Jean. 1976. "Toward a Homosocial Theory of Sex Roles: An Explanation of Sex Segregation in Social Institutions." *Signs* 1:15–31.

MacKinnon, Catharine A. 1979. *The Sexual Harrassment of Working Women*. New Haven, CT: Yale University Press.

Marshall, Judi. 1995. *Women Managers Moving On: Exploring Career and Life Choices*. London: Routledge.

Martin, Joanne. 1994. "The Organization of Exclusion: Institutionalization of Sex Inequality, Gendered Faculty Jobs, and Gender Knowledge in Organizational Theory and Research." *Organization* 1:401–431.

Martin, Joanne and Kathleen Knopoff. 1997. "The Gendered Implications of Apparently Gender-Neutral Theory: Re-reading Weber." In *Ruffin Lecture Series,* vol. 3, *Business Ethics and Women's Studies,* edited by E. Freeman and A. Larson. Oxford: Oxford University-Press.

Martin, Patricia Yancey. 1998. "Theorizing Men as Men in Organizations: Men's Masculinity Mobilizations from (Some) Women's Standpoint(s)." Department of Sociology, Florida State University, Tallahassee. Unpublished manuscript.

Pyke, Karen D. 1996. "Class-Based Masculinities: The Interdependence of Gender, Class, and Interpersonal Power." *Gender & Society* 10:527–549.

Raeburn, Nicole C. 1998. "The Rise of Lesbian, Gay, and Bisexual Rights in the Workplace." Ph.D. dissertation, Department of Sociology, Ohio State University, Columbus.

Reskin, Barbara F. and Irene Padavic. 1994. *Women and Men at Work*. Thousand Oaks, CA: Pine Forge.

Rogers, Mary. Forthcoming. *Our Barbies, Our Selves*. Thousand Oaks, CA: Sage.

Rospenda, Kathleen M., Judith A. Richman, and Stephanie J. Nawyn. 1998. "Doing Power: The Confluence of Gender, Race, and Class in Contrapower Sexual Harrassment." *Gender & Society* 12:40–60.

Schneider, Beth. 1993. "Put Up and Shut Up: Workplace Sexual Assaults." pp. 57–72 in *Violence against Women: The Bloody Footprints,* edited by Pauline B. Bart and Eileen G. Moran. Newbury Park, CA: Sage.

Tomaskovic-Devey, Donald. 1993. *Gender and Racial Inequality at Work: The Sources and Consequences of Job Segregation*. Ithaca, NY: ILR.

Walby, Sylvia. 1988. *Gender Segregation at Work.* Philadelphia: Open University Press.

West, Candace and Don Zimmerman. 1987. "Doing Gender." *Gender & Society* 1:125–151.

Wise, Sue and Liz Stanley. 1988. *Georgie Porgie: Sexual Harassment in Everyday Life.* London: Pandora.

DISCUSSION QUESTIONS

1. What structural features of organizations must be changed in order for resistance to oppressive normative practices to be truly effective? What kind of features would work best to facilitate change and resistance?
2. How does the naturalization of aggressive-male sexuality color the workplace environment?
3. Where does the responsibility for changing discriminatory (e.g., racist, sexist, ablist, heterosexist) and sexualizing practices lie? Is it with organizations, individuals, or society? At which of these three levels would effective change most likely take root? What tensions exist between individual agency and organizational structures in efforts to mobilize change?
4. Select a workplace (it could be your own or one you have frequented) and spend some time analyzing its features. In what ways do gender and sexuality shape and frame this environment? For example, is there a single co-ed bathroom or separate facilities for men and women? If there are separate facilities, what assumptions about sex and sexuality does this structural arrangement assume? What other features of the environment are gendered or heterosexulized?
5. How would your occupational experience differ if you held a different sexual identity? (e.g., if you are heterosexual, how would being gay or asexual change your workplace experience? If you are bisexual, how would being either gay or straight alter your experience? etc.)
6. In what ways does the notion of separate spheres (the association of the public realm with men and the private realm with women) affect the sexual organization of the workplace?

Peepshow Feminism

TAWNYA DUDASH

Is sex work inherently exploitive and oppressive, or does it offer unique opportunities for the subversion of the normative sexual order and for female sexual empowerment? In the following pages Tawnya Dudash addresses these and other issues as she documents her experiences working in a well-known San Francisco peepshow. Augmented with observations from 15 co-workers, Dudash recounts how workers at this concern resist and challenge normative ideologies about gender, beauty, power, age, race, and female sexuality. She reveals active efforts on the part of these women to challenge and confront the dehumanizing and misogynistic acts of some male customers. In addition, she describes how through this type of sex work, workers can attain empowerment through the use of their bodies and through the unfettered expression of their own sexuality.

From "Peepshow Feminism" by Tawnya Dudash in *Whores and Other Feminists,* ed. by Jill Nagle, p. 98–118. Reprinted by permission of Routledge/Taylor & Francis Books, Inc.

INTRODUCTION

I was a college student in Connecticut in 1988 when I first heard about the particular peepshow that is the focus of this article. Several women in the feminist collective to which I belonged were discussing a friend of theirs who worked as a nude dancer in San Francisco. The theatre she danced in was woman-owned and operated, and had a reputation for being more supportive, worker-friendly, and "pro-woman" than other sex entertainment establishments. In addition, the job was lucrative and flexible in terms of time. My curiosity was immediately piqued. How could it be that something so obviously degrading to women as a peepshow was gaining acceptance among my feminist peers? What kind of place was this? . . .

Several years later I relocated to San Francisco and had the chance to answer some of these questions. Soon after arriving in the city, I auditioned at the infamous Lusty Lady theater and was hired despite my obvious nervousness. Before auditioning I had never set foot in such a place, and the newness of it all was overwhelming. The walls and ceiling of the stage were mirrored, and I was not used to seeing myself naked from so many angles. The customers seemed not quite real, since all I could see were their heads and shoulders peering through little windows, which were at the level of our crotches. I hadn't expected that the customers would be masturbating, and this was rather shocking at first. I was struck most, however, by the other dancers. Instead of conforming to my stereotype of sex workers as downtrodden, victimized, or helpless, I found these women to be strong and outspoken. Many were politically active and/or writers and artists, and I was dazzled by the richness and diversity I found. I was working alongside women who were straight, lesbian, and bisexual; high school graduates and multiple degree holders; women of all races and class backgrounds. Never had I seen so many naked women of such different shapes, colors, and sizes. There was a higher amount of body ease than I had ever before experienced with a group of women; the dancers seemed completely comfortable in their bodies and with their sexualities. I also began to notice how they interacted and what they talked about backstage, how they referred to their jobs, to men, and to their relationships with the rest of society. It became apparent that these women were constructing feminist discourses in ways that seemed much more "real" than any academic feminism I had studied. My prior experiences with feminism and sex work consisted of rather abstract discussions of pornography and censorship or the legalization of prostitution. In these discussions, the opinions of sex workers had never been included. Suddenly, however, I felt immersed in the "trenches"; the dancers I met were engaging in and *living* feminism in ways that were completely new to me.

. . . While the circumstances for most other sex workers may be different, my hope is that by documenting the subversion of and resistance to oppression and traditional beliefs about female sexuality that I saw, I may call attention to the potential for similar activities to occur in other parts of the sex industry, an environment often thought of as the quintessence of female sexual oppression. . . .

RESISTANCE TO GENDER IDEOLOGY

Emily Martin's book *The Woman in the Body*[1] is useful when discussing the resistance taking place among peepshow dancers. An environment such as a peepshow is often thought of as oppressive and misogynist, and little thought has been given to sex workers as resisters. Yet Martin points out that "the criteria for what counts as resistance have been held at an unreasonably stringent level and . . . researchers have not been looking for resistance to dominant views in the right places."[2] Martin instead states that "[There] are a multitude of ways women assert an alternative view of their bodies, react against their accustomed social roles . . . and in general struggle to achieve dignity and autonomy."[3]

At the Lusty Lady, performers have devised numerous ways to do just that. On stage, dancers have several methods of asserting their autonomy. Lola, whose common greeting to customers is, "What's a man like you doing in a nice place like this?" is selective about the men she will dance for:

> I've gotten to the point where I will not dance for people who look like they hate me. I understand there are customers who think that what they're doing is wrong and . . . sinful and that women are sinful for tempting them into doing it. . . . They think they are coming in against their will, tempted by those evil women, and they do their thing and they get their enjoyment out of it but they like to blame us. If I see that in someone's eyes, and I can almost always see it, I won't dance for him.

. . . In a dancer-created annual 'zine sold to customers, dancers fabricated a "Dear Dancer" letter from a customer asking why he was ignored when he knocked on the windows, and signed, "Sincere Customer." The response, indicative of dancers' resistance to being treated disrespectfully, was:

> Dear Sincere,
> If you are truly sincere, then you would not try tapping on the glass. To be honest, we really hate it. Gesturing, yelling, waggling your tongue, pointing at our genitalia, or knocking on the glass does not, WE REPEAT, NOT turn us on. If you give us a big smile, sit back and relax, enjoy yourself masturbating, then you are sure to get a great show.[4]

. . . Several dancers view the work as political and assert that resisting misogynist views is an important part of the job. As Rosetta states, "I feel like the men that come in [who] want to take my power, or want to be in power. I see retraining them as part of my job. . . . I almost make it political work to do that." One other woman comments,

> When women come in with their boyfriends . . . I like . . . to pay attention to the women, not to the men. I think it fucks with the dynamics that the men are trying to create and I think it empowers the women. I like that we often ask the women whether or not they want to be there and that they have the right *not* to be there, because . . . if the stripper . . . says to the woman, "You don't have to be here, you have a choice," or in front of the woman tells the guy, "You can't tell me what to do, you can't tell any woman what to do'" that is making some difference.

More overt forms of resistance have occurred only recently, when politically active dancers spearheaded a successful and nationally recognized campaign to unionize Lusty Lady employees. . . . In effect, there are many ways in which peepshow dancers resist the dominant models regarding expected female sexual behavior and dancer/customer dynamics.

EMPOWERMENT THROUGH THE BODY

. . . The intricacy and complexity of women's relationships to their bodies in patriarchal society is intensified for those of us who earn our living through exhibiting our bodies in a sexualized context. Broad concepts such as gender inequalities, sexual autonomy, and power are magnified and condensed when we depend upon our physical bodies for income. We must reconcile our thoughts, feelings, and beliefs about our bodies and sexualities with this dependence.

Gender greatly affects women's orientations to their bodies. Women of every class have fewer opportunities than men of their class in our male-dominated labor market for converting physical capital into social, economic, or cultural capital. Sex workers are perhaps more aware of this than most women, for, as Rosetta states:

> I think that's a really big problem with our society, because it's one of the only places women can even out the male-to-female pay ratio. And so society gives us this thing to do, by using our bodies, and puts us in this position, which is a really powerful position, but at the same time, it's a trap. . . . As soon as you leave, you realize, oh well, I'm not really valued for my other skills as a woman in our culture. I am valued for this one thing. . . . The problem is with the culture as a whole, it's not necessarily with the [sex] industry itself.

Dita expresses a similar sentiment: "The line I use is, I don't mind being a whore, I just mind not being able to *not* be a whore. . . . The work that's available to us at all is so limited, *so* limited, that I'm actually in a position of being grateful . . . [to] work in the sex industry. That's not right."

In addition to gender, class status also affects women's orientations toward their bodies. People of different social and economic classes tend to develop very different orientations toward their bodies. Whereas the "ruling classes" treat the body as an end in itself, working-class people may acquire an *instrumental* orientation to their bodies,[5] since their income often depends upon manual labor. Keeping healthy is viewed as a means to a desired end, not as an end in itself. Regardless of our class backgrounds, dancers fall into this instrumental category because we must use our bodies to earn our livings. In this context, dancers often define certain foods, clothing, makeup, accessories, or exercise as "investments" to help us earn more income, instead of regarding them as beneficial in their own right.

. . . For many dancers at the Lusty Lady, seeing and interacting with other naked women can serve educational and empowering purposes. In some respects, these interactions resemble a 1990s version of the now-defunct consciousness-raising groups of the 1970s. . . . Vendetta describes the impact that being with other naked women in this context has had on her:

> [Prior to entering sex work] I myself [did] not see a lot of women in a sexual context except in movies. . . . The [dancers] are all great. They have had a big impact on me just personally. . . . When I don't even talk to people outside of work, seeing them on stage. . . . I feel like I have gotten to . . . understand a lot about a lot of people there and I find that it is just great learning. Just to be in a room naked with four women is just like . . . a little exploratorium or something.

. . . For some women, sex work is the vehicle through which we can gain power and control through an increased sense of connection with our bodies. One sex worker states, "We decide when, where, and with whom we'll do what . . . we are more comfortable with our bodies and sexualities than most people. We don't have 'private parts,' dismembered from the rest; they are part of the whole."[6]

In addition to increasing body awareness, being naked for hours each day causes many dancers to question and reject the dominant ideology regarding female standards of beauty. This may seem contradictory since we are being paid to *represent* this very ideology, but after working at the theater dancers begin to realize that while we do not all conform to the feminine "ideal," we are all beautiful nevertheless. The ability to look at and interact with actual naked women has a profound impact on many dancers. Often it is the first time we have had the chance to stare at female nakedness that is not part of a magazine advertisement. Dewdrop explains:

> I am not warped with the shame of my body because the only images of my body that I get are media images, print or television stuff. I have this whole sample of all these different bodies that I see up close all the time. It has really just helped my whole perception of my body. . . . Other people have problems with their bodies still because they don't see their

> bodies. . . . Before [beginning sex work] I was just really shameful and I wouldn't make love in the light and I had problems with people seeing my body.

. . . By representing the "feminine ideal" at work we begin to see how literally *constructed* gender is, which allows us to begin *deconstructing* it. Many dancers feel less inclined to get "dolled up" outside the theater, because we realize how artificial this type of "beauty" is. As one dancer comments, "If people only knew . . . that *any* woman can look like a model." Women of different body types are employed at the theater, and dancers realize that "sexiness" reaches far beyond the confines of fashion magazines and Hollywood.

. . . Some of us realize for the first time how oppressive this gender ideology can be, while for others of us who have long been aware of these structures, the process of gender deconstruction and reconstruction going on every day in the dressing room can offer us different but equally empowering insights. Damiana avows:

> I loved dressing up in wigs and costumes and putting on makeup—all things that I had for some time rejected as "sexist oppression." I was reintroduced to the feminine role as a playful thing in a way that was really good for me, which is primarily how I see it in my life today. I think I owe this bit of reclaiming the gray area of gender socialization politics to the dykes who dug their long wigs and lipstick, and who actually had a great time.

An important part of the process of resistance to misogynist ideology is not just noticing the beauty in diversity but also speaking about it. Compliments paid by dancers to one another are a boost to everyone's self-esteem.

. . . One other factor contributing to our body image and awareness is the dancing itself. The benefits of regular exercise are widely touted, and dancing in particular can be a positive and affirming form of movement. "It's a great way to express emotion and moods physically without words. . . . Each time [you dance] your body comes more alive, grows stronger, more coordinated and balanced."[7] For many dancers, one of the most positive aspects of the job is the ability to dance and perform, and dancing in an erotic or "sexy" manner can be a playful and fun way of experiencing movement. Rosetta comments that "dancing has really opened up my sexuality, because it's gotten me so much more in touch with my body. . . . So it's been a really freeing thing for me." She expands on this theme:

> When I first started there it just seemed really silly to me, and I had a hard time. . . . I've never been a "sexy" person, and never been trained to act that way, so it was pretty silly for me, dancing around and trying to act sexy. . . . But then I ended up really enjoying it, and it brought out this whole other aspect of my personality . . . the ability to be a sexual, really sensual person. . . . And to have fun with it. . . . And to be okay showing my body to people and . . . dancing around.

As I stated previously, the entire experience of being naked and interacting with other naked women can be liberating for some dancers. However, for other women this situation brings insecurities about their bodies to the surface. One dancer I spoke with told me that her bulimia returned when she began dancing, after being gone for many years. Another described how she could never be on stage completely naked; she must keep her midsection covered because of her discomfort with that part of her body. After doing poorly in the Private Pleasures booth one day, a third dancer was convinced it was because she was "scaring the customers away" by being "too fat." While I do not believe the Lusty Lady creates poor self-esteem or negative body images where none existed before, clearly it can aggravate the situation for some women with prior negative feelings about their bodies.

. . . It can be extremely disempowering for a woman to be confronted daily with other women's bodies, as well as her own in the many mirrors, and feel as though she doesn't "measure up." Women are taught to compare and compete with one another, especially for male attention, and it is hard to undo a lifetime of socialization.

. . . [E]very dancer has days at the Lusty Lady when she feels that no one is looking at her. Dewdrop explains, "It shoots you right in the self-esteem because in the market we have to be in . . . looks are everything." The situation can be even more difficult, in some ways, for women with a feminist consciousness. Instead of thinking, "I feel fat and fat is bad," these women express feelings such as, "I feel fat, and while I know intellectually that fat is not bad, I still feel bad about it, and then I feel guilty for feeling bad." Or, as one dancer passionately states:

> I have mixed feelings on it. On the one hand, I feel very good about my body. I feel like my body is beautiful. Most of the time, I don't want to be skinny. . . . On the other hand, it really has reinforced some things. . . . I have muscles, and I really like feeling strong. And usually what I tell people is "Oh, I like feeling strong." Well, that's a lie, I also like being thinner. And I don't *like* that I like it. . . . I don't want to have feeling good about my body be dependent on looking this way. . . . I just wish I cared about different things . . . [like] being healthy and capable, and being strong. That's part of it, but it's not all of it, and the other stuff is real evil and sick. And I know it's not my fault, it's not like I blame myself for it, but it's there and I hate it. . . . Standing in a mirror next to someone who—both their butt cheeks would fit into one of mine, even if we're two different body types, there's still something that tells me, "You should look like that. If you were a good person, you would look like that." And I know it's wrong.

Such shame and guilt about one's body is common among women throughout our culture, and nude dancers cannot fully escape the repercussions of this ideology. However, we are resisting on several levels, including the refusal to be ashamed of our nakedness. Many of us are asked by those outside the industry if we're embarrassed to be naked, or if we feel like there is a power imbalance because the customers are, for the most part, clothed. One reply which I have heard expressed in many ways, is that the customer is embarassed by the dancer's nakedness, not the dancer. Being naked can be, and at this theater often is, a very powerful experience. Vendetta explains, "I think some [male friends] have looked at it [as if] it lowers me down a notch [that customers see me naked]. . . . I don't know if I see it that way. It is something I am proud of and it is something I do for a job and I don't really feel like I am one down from them by it." In our own manner, therefore, we are adding our methods to the "various ways in which women have attempted to manipulate the dominant discourse and conventions, which require that women, especially, should see their 'underparts' as shameful and unspeakable."[8]

EMPOWERMENT THROUGH SEXUALITY

. . . Though sexuality is often experienced through the physicalities of our bodies, sexuality itself is socially constructed. As Carole Vance states in the introduction to *Pleasure and Danger:*

> Sex is a social construction, articulated at many points with the economic, social, and political structures of the material world. . . . Although sexuality . . . is grounded in the body, the body's structure, physiology, and functioning do not directly or simply determine the configuration or meaning of sexuality . . .[9]

As a social construction, sexuality and sexual politics are now acknowledged as far more than "secondary, even a luxury for the self-indulgent,"[10] as had been thought by previous, often Marxist-influenced, social and political activists. Indeed, part of the discourse of women's liberation that emerged in the late 1960s and is epitomized by the slogan "the personal is political" was that something once thought completely private and personal is actually a major site of both oppression and liberation.[11]

Patriarchy and capitalism, as two of the "economic, social, and political structures" Vance mentions, play a large part in the conceptions of female sexuality. Often these conceptions are restrictive and oppressive, such as the notion that for women, sexual pleasure must be pitted against sexual safety.[12] Until recently, "safe" or socially acceptable zones for women to experience sexual desire existed within heterosexual marriage only. Although in the latter part of the twentieth century the "safe" zones have expanded somewhat, many forms of sexuality, including promiscuity, lesbianism, and sadomasochistic sex, are still considered violations of the social "rules" regarding female sexuality. These departures into sexual danger zones are thought of as violations not only by those seeking to control female sexuality, however; women themselves have often become afraid of the "danger" of their own sexualities.[13]

One manner in which this internalized fear is manifested within the realm of sexual politics is through a dread of sexual variance. Whether rooted in a deeply felt need for merging, for similarity—to no longer be different, be other[14]—or in a desire on the part of feminists to maintain a semblance of group cohesion through the minimization of difference,[15] the results are debilitating.

> Feminists, like all members of the culture, find it difficult to think about sexual difference with equanimity. The concept of benign sexual variation is a relatively new one. . . . Our relative ignorance about the actual range of sexual behavior and fantasy makes us into latterday ethnocentrists. . . . The external system of sexual hierarchy is replicated within each of us, and herein lies its power. Internalized cultural norms enforce the status quo. As each of us hesitates to admit deviations from the system of sexual hierarchy, nonconformity remains hidden, invisible, and apparently rare.[16]

A major task, then, is for women to recognize themselves and one another as sexual agents and to create a space where women can honestly express feelings and experiences of sexual variance.

Dancers at the Lusty Lady have crossed into the danger zone simply by becoming sex workers. . . . Yet our venturing into the forbidden realm of "deviant" female sexuality means it no longer holds as much of a threat for us. The bonds among us as women are strengthened, and we are able to discuss sexual variations in a nonjudgmental format, thereby breaking down notions of "normal" and "abnormal" behavior.

. . . Customers, too, can positively influence dancers' perceptions of sexuality and raise awareness and sensitivity to issues of male sexuality. The phrase, "You can't judge a book by its cover," takes on a new meaning for us, as we interact daily with conservative or average-looking men who reveal an unlimited capacity for sexual variation. Observes Summer, "Regardless of how someone looks on the outside, I know that some things are very private, and I never really thought about normal-looking people having alternative ways of being sexual." Cross-dressing, S/M fantasies, foot fetishes, bondage, anal play, and exhibitionism are but a few of the myriad activities we are able to watch our customers engage in. Some dancers formulate negative views of male sexuality from these encounters, coming to view men as "jerk-off machines." Others are disturbed by what is perceived as an "unhealthy" attitude toward sexuality displayed by some customers, as in Dita's comment, "So many men don't give themselves pleasure, you know. The way that they touch themselves is with a lot of angst and guilt and rage." For many of us, however, our positive interactions with customers serve to enlighten us to new activities and issues surrounding sexuality.

. . . For women, true sexual liberation entails not just expanding our awareness and acceptance of *women's* right to sexual variance; we must also be sensitive to the ways in which men are attempting to redefine their own sexualities in creative and healthy ways. Many men who frequent sex entertainment establishments, along with many more who do not, are doing no such redefining, preferring instead to accept and maintain the status quo by remaining deeply attached to notions of sexual inequality. Rosetta states, "I do acknowledge that there are people who come in that have fucked up ideas about women, but I think those problems exist anywhere." However, for those men who *are* attempting to discover and explore positive aspects of sexual variance, theaters such as the Lusty Lady may be one of the few supportive venues in which to do so. Without exception, the dancers with whom I have spoken say they were initially surprised by the diversity of the customers, both in terms of demographics and sexual proclivities.

. . . Many dancers are also forced to confront myths and stereotypes surrounding sexuality, because perhaps for the first time we are dealing directly with groups of people considered sexually invisible by most of society. Lola continues:

> I hadn't realized until I worked at the Lusty Lady how much our culture can dehumanize older people. I think we completely desexualize the elderly and I think it's incredibly dehumanizing. . . . I think as a culture we need to examine this. . . . I have ceased to be able to look at old men as nonsexual. I *have* to recognize that old men are sexual beings because I have been masturbated at by some of them! I think that there are races that are desexualized. . . . I think we certainly desexualize the handicapped. . . . I do think that as a culture we have to look at who we treat as nonsexual, but doing sex work has forced me as an individual to do that.

All dancers value the positive interactions we have with our customers, and many of us believe that being exposed to such racial, class, age, and sexual diversity in both dancers and customers has given us "more knowledge about [ourselves] and about people, and . . . more compassion," as Rosetta states. It is precisely this knowledge and compassion that form the basis of our resistance to socially accepted and repressive ideologies surrounding sexuality.

EMPOWERMENT CARRIERS INTO THE "REAL WORLD"

For many dancers, the sex industry has allowed us to fully realize the autonomy, power, and control we have over our bodies and sexualities, especially in relation to men. Our ability to set and articulate limits regarding what we will and will not do with our bodies increases, sometimes dramatically.

. . . All women who work at the Lusty Lady get practice learning what our own limits are and verbally communicating them to customers. Rosetta professes, "I think most of the men know, and the way I deal with most of the men [is that] I'm definitely the one in power and taking control of the situation. . . . For the most part, it's a pretty powerful experience."

For many dancers, our limit-setting abilities and our power experienced in relation to men at the theater carry over into the way we relate to men outside work. Once we realize that we do not have to "take any shit" from customers we recognize that we do not need to take it outside the theater either. A dancer who also works as a massage therapist recently told me that, had she not worked in the sex industry, she would have a much more difficult time setting limits with her male massage clients, some of whom mistakenly think there is sex involved.

Many dancers also respond differently to street harassment than they did prior to entering the sex industry. On one occasion in the dressing room, several women had a conversation about their reactions to street harassment. One dancer described how a man reached out and grabbed her buttocks as she was exiting a bus. The dancer whirled around and shouted to him, "How *dare* you touch my body! You apologize to me right now!" She then held open the back door, thus holding up the bus by preventing the driver from moving. She waited until the man apologized before exiting the bus. This dancer claimed she would have lacked the confidence to do this before becoming a sex worker. Amid the cheers of support from the other women present for this story, another dancer chimed in to say that, since working in an industry that commodifies her body, she has realized that *she* owns her body; it is not public property always and automatically available to men. She is therefore much less likely to put up with harassment from men. Dewdrop reflects on her experiences with street harassment: "I meet it with much more power, power from having dealt with it all day and managed it. It is almost like, 'I have figured you out. I know what you are all about'. . . . I feel a lot more powerful and not powerless with men because I know exactly where they are coming from." Damiana, who no longer works at the theater, similarly describes the effect dancing had on her relations with strange men: "If anyone looked at me funnily on the street, I had an easy time ignoring them. I actually once told a guy 'Hey, I get paid for that.' If my body was a commodity, I was using it for my own benefit." Rosetta explains:

> Working in the sex industry has made me even more aware of just how much power I do have in my relations with men that I didn't recognize before. Like before, if a man had asked me to come over and show him something or talk to him about something . . . on the street . . . I probably would have said "Okay," and now I refuse. And I'm very firm about that sort of thing, and about my boundaries, and about what I can and cannot give to people, have or don't have to do. [Working in] the sex industry has made that very clear.

NOTES

1. Emily Martin, *The Woman in the Body* (Boston: Beacon Press, 1987).
2. Nicholas Abercrombie, Stephen Hill, and Bryan Turner, *The Dominant Ideology Thesis* (London: George Allen and Unwin Ltd. 1980), Cited in Martin, *The Woman in the Body,* p. 183.
3. Martin, *The Woman in the Body,* p. 200. See also Ruth Behar, "The Body in the Woman, the Story in the Woman: A Book Review and Personal Essay," in *The Female Body: Figures, Styles; and Speculations,* ed. Laurence Goldstein (Michigan: University of Michigan Press, 1991), pp. 267–311.
4. Playday '92: Tales from the Lusty Side, 1992, p. 5.
5. Chris Shilling, "Educating the Body: Physical Capital and the Reproduction of Inequalities," *Sociology,* 25(4) (1991): 657. Emphasis in the original.
6. Boston Women's Health Book Collective, *The New Our Bodies, Ourselves* (New York: Simon and Schuster, 1984), p. xix.
7. Peggy Morgan, "Living on the Edge," *Sex Work: Writings by Women in the Sex Industry,* ed. Frederique Delacoste and Priscilla Alexander (Pittsburgh: Cleis Press, 1987), p. 25.
8. Boston Women's Health Book Collective, *The New Our Bodies,* p. 49.
9. Shilling, "Educating the Body," p. 664.
10. Carole Vance, "Pleasure and Danger: Toward a Politics of Sexuality," In *Pleasure and Danger: Exploring Female Sexuality,* ed. Carole Vance (Boston: Routledge & Kegan Paul, 1984), pp. 1–27.
11. Ann Snitow, Christine Stansell, and Sharon Thompson, "Introduction," *Powers of Desire: Politics of Sexuality* (New York: Monthly Review Press, 1983), p. 21.
12. Snitow, Stansell, and Thompson, "Introduction," p. 24.
13. Vance, "Pleasure and Danger," p. 3.
14. Ibid., pp. 3–4.
15. Snitow, Stansell, and Thompson, "Introduction," p. 41.
16. Vance "Pleasure and Danger," 19.

DISCUSSION QUESTIONS

1. There are many varieties of feminism. Generally speaking, how would you say the women Dudash interviewed constructed feminism? What features of female existence were con-

structed as critical sites of empowerment? Oppression? How do you think their definition of feminism is similar to, and different from, other groups' construction of feminism? How does it compare with your definition of feminism?

2. How is sex work similar to other kinds of work? How is it different? Does the fact that this type of sexual activity is legal matter? Should it?
3. How important do you think the structural features of the Lucky Lady (e.g., glass partitions, female managers) were in supporting the workers' resistance to oppression?
4. What, if anything, does the diversity of workers suggest about the appeal of these workers? What, if anything, do the workers share in common?
5. Is it problematic that sexually oriented businesses (including peepshows, strip clubs, and prostitution) remain one of the most lucrative career options for women? Why or why not? Are peepshows and other forms of sex work necessarily oppressive in patriarchal societies? Can sex be a source of empowerment for women in a patriarchal society? Is sex a source of empowerment or oppression for men under patriarchy?

The following Bonus Reading can be accessed through the *InfoTrac College Edition Online Library* by typing in the name of the author or keywords in the title online at *http://www.infotrac-college.com.*

The Locker Room and the Dorm Room: Workplace Norms and the Boundaries of Sexual Harassment in Magazine Editing

KIRSTEN DELLINGER AND CHRISTINE L. WILLIAMS

Work takes place in radically different organizational contexts. Some workplace cultures expect workers to don professional attire, some require uniforms, while others allow jeans and T-shirts. While one workplace enforces rigid rules of conduct, others are much more lax. Some are highly sexualized, some less so. These are examples of just a few of the structural expectations of organizations that shape norms of interaction within the workplace environment. In this study, Kirsten Dellinger and Christine L. Williams explore how workers negotiate sexual behavior and construct definitions of sexual harassment across differing occupational cultures. Their findings suggest that though there are some consistencies across workplace boundaries as to which behaviors constitute sexual harassment, issues of sexual propriety are quite fluid and are fashioned largely within specific organizational cultural contexts.

DISCUSSION QUESTIONS

1. This article looks at how individuals come to define sexual harassment given differing organizational contexts. Legally, however, a single standard is used to determine whether an act is defined as harassment. Should organizational context be taken into account when determining whether an act legally constitutes sexual harassment, or should a single standard or definition apply across all organizational contexts? If so, what should that definition be? If not, how should sexual harassment be enforced? How important is the issue of harm? What should be more significant, the act or the context? Why are there differences between legal standards and individual standards? What, if anything, do such differences suggest about social organization and inequality?
2. Think about the organizational contexts in which you have worked. What were the formal and informal rules that governed norms of workplace sexuality? How do they compare with the formal and informal rules governing sexual norms in the school setting?
3. In this article the authors examined work sites with quite different sex ratios. Generally speaking, how does the relative composition of a workforce (in terms or race, sex, class, sexuality, and religious affiliation, for example) affect the construction of "appropriate" and "inappropriate" workplace behaviors? Do you think more differences between workers generally means more regulation? Less? Why or why not? Which of these variables do you think most affects the nature of sexualized interactions in the workplace? Why?

9

Religion and Education

The Muslim Concept of Active Female Sexuality

FATIMA MERNISSI

In this chapter of her classic work Beyond the Veil, *Fatima Mernissi links the practice of veiling and other restrictive social customs, such as seclusion in harems and constant surveillance, to Muslim constructions of female sexuality. Using the theoretical writings of Sigmund Freud and Iman Ghazali to represent competing models of Muslim sexual dynamics (i.e., passive versus active female sexuality), Mernissi demonstrates how both models serve to circumscribe and suppress female sexuality and social equality.*

THE FUNCTION OF INSTINCTS

The Christian concept of the individual as tragically torn between two poles—good and evil, flesh and spirit, instinct and reason—is very different from the Muslim concept. Islam has a more sophisticated theory of the instincts, more akin to the Freudian concept of the libido. It views the raw instincts as energy. The energy of instincts is pure in the sense that it has no connotation of good or bad. The question of good and bad arises only when the social destiny of men is considered. The individual cannot survive except within a social order. Any social order has a set of laws. The set of laws decides which uses of the instincts are good or bad. It is the use made of the instincts, not the instincts themselves, that is beneficial or harmful to the social order. Therefore, in the Muslim order it is not necessary for the individual to eradicate his instincts or to control them for the sake of control itself, but he must use them according to the demands of religious law.

> When Muhammad forbids or censures certain human activities, or urges their omission, he does not want them to be neglected altogether, nor does he want them

From "The Muslim Concept of Active Female Sexuality" in *Beyond the Veil: Male-Female Dynamics in Modern Muslim Society* by Fatima Mernissi, pp. 28–45. Reprinted by permission of Indiana University Press.

to be completely eradicated, or the powers from which they result to remain altogether unused. He wants those powers to be employed as much as possible for the right aims. Every intention should thus eventually become the right one and the direction of all human activities one and the same.[1]

Aggression and sexual desire, for example, if harnessed in the right direction, serve the purposes of the Muslim order; if suppressed or used wrongly, they can destroy that very order:

> Muhammad did not censure wrathfulness with the intention of eradicating it as a human quality. If the power of wrathfulness were no longer to exist in man, he would lose the ability to help the truth to become victorious. There would no longer be holy war or glorification of the word of God. Muhammad censured the wrathfulness that is in the service of Satan and reprehensible purposes, but the wrathfulness that is one in God and in the service of God deserves praise.[2]
>
> . . . Likewise when he censures the desires, he does not want them to be abolished altogether, for a complete abolition of concupiscence in a person would make him defective and inferior. He wants the desire to be used for permissible purposes to serve the public interests, so that man becomes an active servant of God who willingly obeys the divine commands.[3]

Imam Ghazali (1050–1111) in his book *The Revivification of Religious Sciences*[4] gives a detailed description of how Islam integrated the sexual instinct in the social order and placed it at the service of God. He starts by stressing the antagonism between sexual desire and the social order: "If the desire of the flesh dominates the individual and is not controlled by the fear of God, it leads men to commit destructive acts."[5] But used according to God's will, the desire of the flesh serves God's and the individual's interests in both worlds, enhances life on earth and in heaven. Part of God's design on earth is to ensure the perpetuity of the human race, and sexual desires serve this purpose:

> Sexual desire was created solely as a means to entice men to deliver the seed and to put the woman in a situation where she can cultivate it, bringing the two together softly in order to obtain progeny, as the hunter obtains his game, and this through copulation.[6]

He created two sexes, each equipped with a specific anatomic configuration which allows them to complement each other in the realization of God's design.

> God the Almighty created the spouses, he created the man with his penis, his testicles and his seed in his kidneys [kidneys were believed to be the semen-producing gland]. He created for it veins and channels in the testicles. He gave the woman a uterus, the receptable and depository of the seed. He burdened men and women with the weight of sexual desire. All these facts and organs manifest in an eloquent language the will of their creator, and address to every individual endowed with intelligence an unequivocal message about the intention of His design. . . . Therefore, the man who refuses to marry fails to plant the seed, destroys it and reduces to waste the instrument created by God for this purpose.[7]

Serving God's design on earth, sexual desire also serves his design in heaven.

> . . . when the individual yields to it and satisfies it, he experiences a delight which would be without match if it were lasting. It is a foretaste of the delights secured for men in Paradise, because to make a promise to men of delights they have not tasted before would be ineffective. . . . This earthly delight, imperfect because limited in time, is a powerful motivation to incite men to try and attain the perfect delight, the eternal delight and therefore urges men to adore God so as

> to reach heaven. Therefore the desire to reach the heavenly delight is so powerful that it helps men to persevere in pious activities in order to be admitted to heaven.[8]

Because of the dual nature of sexual desire (earthly and heavenly) and because of its tactical importance in God's strategy, its regulation had to be divine as well. In accordance with God's interests, the regulation of the sexual instinct was one of the key devices in Muhammad's implementation on earth of a new social order in then-pagan Arabia.

FEMALE SEXUALITY: ACTIVE OR PASSIVE?

According to George Murdock, societies fall into two groups with respect to the manner in which they regulate the sexual instinct. One group enforces respect of sexual rules by a "strong internalization of sexual prohibitions during the socialization process," the other enforces that respect by "external precautionary safeguards such as avoidance rules," because these societies fail to internalize sexual prohibitions in their members.[9] According to Murdock, Western society belongs to the first group while societies where veiling exists belong to the second.

> Our own society clearly belongs to the former category, so thoroughly do we instill our sex mores in the consciences of individuals that we feel quite safe in trusting our internalized sanctions. . . . We accord women a maximum of personal freedom, knowing that the internalized ethics of premarital chastity and postmarital fidelity will ordinarily suffice to prevent abuse of their liberty through fornication or adultery whenever a favourable opportunity presents itself. Societies of the other type . . . attempt to preserve premarital chastity by secluding their unmarried girls or providing them with duennas or other such external devices as veiling, seclusion in harems or constant surveillance.[10]

However, I think that the difference between these two kinds of societies resides not so much in their mechanisms of internalization as in their concept of female sexuality. In societies in which seclusion and surveillance of women prevail, the implicit concept of female sexuality is active; in societies in which there are no such methods of surveillance and coercion of women's behaviour, the concept of female sexuality is passive.

In his attempt to grasp the logic of the seclusion and veiling of women and the basis of sexual segregation, the Muslim feminist Qasim Amin came to the conclusion that women are better able to control their sexual impulses than men and that consequently sexual segregation is a device to protect men, not women.[11]

He started by asking who fears what in such societies. Observing that women do not appreciate seclusion very much and conform to it only because they are compelled to, he concluded that what is feared is *fitna:* disorder or chaos. (*Fitna* also means a beautiful woman—the connotation of a *femme fatale* who makes men lose their self-control. In the way Qasim Amin used it *fitna* could be translated as chaos provoked by sexual disorder and initiated by women.) He then asked who is protected by seclusion.

> If what men fear is that women might succumb to their masculine attraction, why did they not institute veils for themselves? Did men think that their ability to fight temptation was weaker than women's? Are men considered less able than women to control themselves and resist their sexual impulse? . . . Preventing women from showing themselves unveiled expresses men's fear of losing control over their minds, falling prey to *fitna* whenever they are confronted with a non-veiled woman. The implications of such an institution lead us to think that women are believed to be better equipped in this respect than men.[12]

Amin stopped his inquiry here and, probably thinking that his findings were absurd, concluded jokingly that if men are the weaker sex, they are the ones who need protection and therefore the ones who should veil themselves.

Why does Islam fear *fitna*? Why does Islam fear the power of female sexual attraction over men? Does Islam assume that the male cannot cope sexually with an uncontrolled female? Does Islam assume that women's sexual capacity is greater than men's?

Muslim society is characterized by a contradiction between what can be called "an explicit theory" and "an implicit theory" of female sexuality, and therefore a double theory of sexual dynamics. The explicit theory is the prevailing contemporary belief that men are aggressive in their interaction with women, and women are passive. The implicit theory, driven far further into the Muslim unconscious, is epitomized in Imam Ghazali's classical work.[13] He sees civilization as struggling to contain women's destructive, all-absorbing power. Women must be controlled to prevent men from being distracted from their social and religious duties. Society can survive only by creating institutions that foster male dominance through sexual segregation and polygamy for believers.

The explicit theory, with its antagonistic, machismo vision of relations between the sexes is epitomized by Abbas Mahmud al-Aqqad.[14] In *Women in the Koran* Aqqad attempted to describe male-female dynamics as they appear through the Holy Book. Aqqad opened his book with the quotation from the Koran establishing the fact of male supremacy ("the men are superior to them by a degree") and hastily concludes that "the message of the Koran, which makes men superior to women is the manifest message of human history, the history of Adam's descendants before and after civilization."[15]

What Aqqad finds in the Koran and in human civilization is a complementarity between the sexes based on their antagonistic natures. The characteristic of the male is the will to power, the will to conquer. The characteristic of the female is a negative will to power. All her energies are vested in seeking to be conquered, in wanting to be overpowered and subjugated. Therefore, "She can only expose herself and wait while the man wants and seeks."[16]

. . . The complementarity of the sexes, according to Aqqad, resides in their antagonistic wills and desires and aspirations.

> Males in all kinds of animals are given the power—embodied in their biological structure—to compel females to yield to the demands of the instinct (that is, sex). . . . There is no situation where that power to compel is given to women over men.[17]

. . . [F]or Aqqad women experience pleasure and happiness only in their subjugation, their defeat by males. The ability to experience pleasure in suffering and subjugation is the kernel of femininity, which is masochistic by its very nature. "The woman's submission to the man's conquest is one of the strongest sources of women's pleasure."[18] The machismo theory casts the man as the hunter and the woman as his prey. This vision is widely shared and deeply ingrained in both men's and women's vision of themselves.

The implicit theory of female sexuality, as seen in Imam Ghazali's interpretation of the Koran, casts the woman as the hunter and the man as the passive victim. The two theories have one component in common, the woman's *qaid* power ("the power to deceive and defeat men, not by force, but by cunning and intrigue"). But while Aqqad tries to link the female's *qaid* power to her weak constitution, the symbol of her divinely decreed inferiority, Imam Ghazali sees her power as the most destructive element in the Muslim social order, in which the feminine is regarded as synonymous with the satanic.

The whole Muslim organization of social interaction and spacial configuration can be understood in terms of women's *qaid* power. The social order then appears as an attempt to subjugate her power and neutralize its disruptive effects. The opposition between the implicit and the explicit theories in Muslim society would appear clearly if I could contrast Aqqad and Imam Ghazali. But whereas the implicit theory is brilliantly articulated in Imam Ghazali's systematic work on the institution of marriage in Islam, the explicit theory has an unfortunate advocate in Aqqad, whose work is an amateurish mixture of history, religion and his own brand of biology and anthropology. I shall therefore contrast Imam Ghazali's conception of sexual dynamics not with Aqqad's but with that of another theoretician, one who is not a Muslim but who has the advantage of possessing a machismo theory that is systematic in the elaboration of its premises—Sigmund Freud.

IMAM GHAZALI VS. FREUD: ACTIVE VS. PASSIVE

. . . Freud did not think that he was elaborating a European theory of female sexuality; he thought he was elaborating a universal explanation of the human female. But . . . (w)e can view Freud's theory as a "historically defined" product of his culture. Linton noted that anthropological data has shown that it is culture that determines the perception of biological differences and not the other way around.

> All societies prescribe different attitudes and activities to men and to women. Most of them try to rationalize these prescriptions in terms of the physiological differences between the sexes or their different roles in reproduction. However, a comparative study of the statuses ascribed to women and men in different cultures seems to show that while such factors may have served as a starting point for the development of a division, the actual prescriptions are almost entirely determined by culture. Even the psychological characteristics ascribed to men and to women in different societies vary so much that they can have little physiological basis.[19]

. . . [We can] consider Freud's theory of sexuality in general, and of female sexuality in particular, as a reflection of his society's beliefs and not as a scientific (objective and ahistorical) theory. In comparing Freud and Imam Ghazali's theories we will be comparing the two different cultures' different conceptions of sexuality, one based on a model in which the female is passive, the other on one in which the female is active. The purpose of the comparison is to highlight the particular character of the Muslim theory of male-female dynamics, and not to compare the condition of women in the Judeo-Christian West and the Muslim East.

The novelty of Freud's contribution to Western contemporary culture was his acknowledgement of sex (sublimated, of course) as the source of civilization itself. The rehabilitation of sex as the foundation of civilized creativity led him to the reexamination of sex differences. This reassessment of the differences and of the consequent contributions of the sexes to the social order yielded the concept of female sexuality in Freudian theory.

In analysing the differences between the sexes, Freud was struck by a peculiar phenomenon—bisexuality—which is rather confusing to anyone trying to assess sex differences rather than similarities:

> Science next tells you something that runs counter to your expectations and is probably calculated to confuse your feelings. It draws your attention to the fact that portions of the male sexual apparatus also appear in women's bodies, though in an atrophied state, and vice-versa in the alternative case. It regards their occurrence as indications of bisexuality as though an individual is not a man or a woman but always both—merely a certain amount more one than the other.[20]

The deduction one expects from bisexuality is that anatomy cannot be accepted as the basis for sex differences. Freud made this deduction:

> You will then be asked to make yourself familiar with the idea that the proportion in which masculine and feminine are mixed in an individual is subject to quite considerable

> fluctuations. Since, however, apart from the very rarest cases, only one kind of sexual product, ova or semen, is nevertheless present in one person, you are bound then to have doubts as to the decisive significance of those elements and must conclude that what constitutes masculinity or femininity is an unknown characteristic which anatomy cannot lay hold of.[21]

Where then did Freud get the basis for his polarization of human sexuality into a masculine and a feminine sexuality, if he affirms that anatomy cannot be the basis of such a difference? He explains this in a footnote, apparently considering it a secondary point:

> It is necessary to make clear that the conceptions "masculine" and "feminine," whose content seems so unequivocal to the ordinary meaning, belong to the most confused terms in science and can be cut up into at least three paths. One uses masculinity and femininity at times in the sense of activity and passivity, again in the biological sense and then also in the sociological sense. The first of these three meanings is the essential one and the only one utilizable in psychoanalysis.[22]

The polarization of human sexuality into two kinds, feminine and masculine, and their equation with passivity and activity in Freudian theory helps us to understand Imam Ghazali's theory, which is characterized precisely by the absence of such a polarization. It conceives of both male and female sexuality partaking of and belonging to the same kind of sexuality.

For Freud, the sex cells' functioning is symbolic of the male-female relation during intercourse. He views it as an antagonistic encounter between aggression and submission.

> The male sex cell is actively mobile and searches out the female and the latter, the ovum, is immobile and waits passively. . . . This behaviour of the elementary sexual organism is indeed a model for the conduct of sexual individuals during intercourse. The male pursues the female for the purpose of sex union, seizes hold of her and penetrates into her.[23]

For Imam Ghazali, both the male and female have an identical cell. The word sperm (*ma'*, "water drop") is used for the female as well as for the male cell. Imam Ghazali referred to the anatomic differences between the sexes when clarifying Islam's position on coitus interruptus (*'azl*), a traditional method of birth control practised in pre-Islamic times. In trying to establish the Prophet's position on *'azl,* Imam Ghazali presented the Muslim theory of procreation and the sexes' contribution to it and respective roles in it.

> The child is not created from the man's sperm alone, but from the union of a sperm from the male with a sperm from the female . . . and in any case the sperm of the female is a determinant factor in the process of coagulation.[24]

The puzzling question is not why Imam Ghazali failed to see the difference between the male and female cells, but why Freud, who was more than knowledgeable about biological facts, saw the ovum as a passive cell whose contribution to procreation was minor compared to the sperm's. In spite of their technical advancement, European theories clung for centuries to the idea that the sperm was the only determining factor in the procreation process; babies were prefabricated in the sperm[25] and the uterus was just a cozy place where they developed.

Imam Ghazali's emphasis on the identity between male and female sexuality appears clearly in his granting the female the most uncontested expression of phallic sexuality, ejaculation. This reduces the differences between the sexes to a simple difference of pattern of ejaculation, the female's being much slower than the male's.

> The difference in the pattern of ejaculation between the sexes is a source of hostility

> whenever the man reaches his ejaculation before the woman. . . .The woman's ejaculation is a much slower process and during that process her sexual desire grows stronger and to withdraw from her before she reaches her pleasure is harmful to her.[26]

Here we are very far from the bedroom scenes of Aqqad and Freud, which resemble battlefields more than shelters of pleasure. For Imam Ghazali there is neither aggressor nor victim, just two people cooperating to give each other pleasure.

The recognition of female sexuality as active is an explosive acknowledgement for the social order with far-reaching implications for its structure as a whole. But to deny that male and female sexuality are identical is also an explosive and decisive choice. For example, Freud recognizes that the clitoris is an evident phallic appendage and that the female is consequently more bisexual than the male.

. . . Instead of elaborating a theory which integrates and elaborates the richness of both sexes' particularities, however, Freud elaborates a theory of female sexuality based on reduction: the castration of the phallic features of the female. A female child, bisexual in infancy, develops into a mature female only if she succeeds in renouncing the clitoris, the phallic appendage: "The elimination of the clitorial sexuality is a necessary precondition for the development of femininity.[27] The pubertal development process brings atrophy to the female body while it enhances the phallic potential of the male's, thus creating a wide discrepancy in the sexual potential of humans, depending on their sex:

> Puberty, which brings to the boy a great advance of libido, distinguishes itself in the girl by a new wave of repression which especially concerns the clitoral sexuality. It is a part of the male sexual life that sinks into repression. The reinforcement of the inhibitions produced in the woman by the repression of puberty causes a stimulus in the libido of the man and forces it to increase its capacity; with the height of the libido, there is a rise in the overestimation of the sexual, which can be present in its full force only when the woman refuses and denies her sexuality.[28]

The female child becomes a woman when her clitoris "acts like a chip of pinewood which is utilized to set fire to the harder wood.[29] Freud adds that this process takes some time, during which the "young wife remains anesthetic."[30] This anesthesia may become permanent if the clitoris refuses to relinquish its excitability. The Freudian woman, faced with her phallic partner, is therefore predisposed to frigidity.

> The sexual frigidity of women, the frequency of which appears to confirm this disregard (the disregard of nature for the female function) is a phenomenon that is still insufficiently understood. Sometimes it is psychogenic and in that case accessible to influence; but in other cases it suggests the hypothesis of its being constitutionally determined and even of being a contributory anatomical factor.[31]

By contrast with the passive, frigid Freudian female, the sexual demands of Imam Ghazali's female appear truly overwhelming, and the necessity for the male to satisfy them becomes a compelling social duty: "The virtue of the woman is a man's duty. And the man should increase or decrease sexual intercourse with the woman according to her needs so as to secure her virtue."[32] The Ghazalian theory directly links the security of the social order to that of the woman's virtue, and thus to the satisfaction of her sexual needs. Social order is secured when the woman limits herself to her husband and does not create *fitna,* or chaos, by enticing other men to illicit intercourse. Imam Ghazali's awe of the overpowering sexual demands of the active female appears when he admits how difficult it is for a man to satisfy a woman.

> If the prerequisite amount of sexual intercourse needed by the woman in order to

> guarantee her virtue is not assessed with precision, it is because such an assessment is difficult to make and difficult to satisfy.[33]

He cautiously ventures that the man should have intercourse with the woman as often as he can, once every four nights if he has four wives. He suggests this as a limit, otherwise, the woman's sexual needs might not be met.

. . . Freud's and Ghazali's stands on foreplay are directly influenced by their visions of female sexuality. For Freud, the emphasis should be on the coital act, which is primarily "the union of the genitals,"[34] and he deemphasizes foreplay as lying between normal (genital) union and perversion, which consists ". . . in either an anatomical transgression of the bodily regions destined for sexual union or a lingering at the intermediary relations to the sexual object which should normally be rapidly passed on the way to definite sexual union."[35]

In contrast, Imam Ghazali recommends foreplay, primarily in the interest of the woman, as a duty for the believer. Since the woman's pleasure necessitates a lingering at the intermediary stages, the believer should strive to subordinate his own pleasure, which is served mainly by the genital union.

> The Prophet said, "No one among you should throw himself on his wife like beasts do. There should be, prior to coitus, a messenger between you and her." People asked him, "What sort of messenger?" The Prophet answered, "Kisses and words."[36]

The Prophet indicated that one of the weakness in a man's character would be that:

> . . . he will approach his concubine-slave or his wife and that he will have intercourse with her without having prior to that been caressing, been tender with her in words and gestures and laid down beside her for a while, so that he does not harm her, by using her for his own satisfaction, without letting her get her satisfaction from him.[37]

THE FEAR OF FEMALE SEXUALITY

The perception of female aggression is directly influenced by the theory of women's sexuality. For Freud the female's aggression, in accordance with her sexual passivity, is turned inward. She is masochistic.

> The suppression of woman's aggressiveness which is prescribed for them constitutionally and imposed on them socially favours the development of powerful masochistic impulses, which succeed, as we know, in binding erotically the destructive trends which have been diverted inwards. Thus masochism, as people say, is truly feminine. But if, as happens so often, you meet with masochism in men, what is left for you but to say that these men exhibit very plainly feminine traits.[38]

The absence of active sexuality moulds the woman into a masochistic passive being. It is therefore no surprise that in the actively sexual Muslim female aggressiveness is seen as turned outward. The nature of her aggression is precisely sexual. The Muslim woman is endowed with a fatal attraction which erodes the male's will to resist her and reduces him to a passive acquiescent role. He has no choice; he can only give in to her attraction, whence her identification with *fitna,* chaos, and with the anti-divine and anti-social forces of the universe.

> The Prophet saw a woman. He hurried to his house and had intercourse with his wife Zaynab, then left the house and said, "When the woman comes towards you, it is Satan who is approaching you. When one of you sees a woman and he feels attracted to her, he should hurry to his wife. With her, it would be the same as with the other one.[39]

Commenting on this quotation, Imam Muslim, an established voice of Muslim tradition, reports that the Prophet was referring to the

> . . . fascination, to the irresistible attraction to women God instilled in man's soul, and he was referring to the pleasure man experiences when he looks at the woman, and the pleasure he experiences with anything related to her. She resembles Satan in his irresistible power over the individual.[40]

This attraction is a natural link between the sexes. Whenever a man is faced with a woman, *fitna* might occur: "When a man and a woman are isolated in the presence of each other, Satan is bound to be their third companion.[41]

The most potentially dangerous woman is one who has experienced sexual intercourse. It is the married woman who will have more difficulties in bearing sexual frustration. The married woman whose husband is absent is a particular threat to men.

. . . In Moroccan folk culture this threat is epitomized by the belief in Aisha Kandisha, a repugnant female demon. She is repugnant precisely because she is libidinous. She has pendulous breasts and lips and her favourite pastime is to assault men in the streets and in dark places, to induce them to have sexual intercourse with her, and ultimately to penetrate their bodies and stay with them for ever.[42] They are then said to be inhabited. The fear of Aisha Kandisha is more than ever present in Morocco's daily life. Fear of the castrating female is a legacy of tradition and is seen in many forms in popular beliefs and practices and in both religious and mundane literature, particularly novels.

The Muslim order faces two threats: the infidel without and the woman within.

> The Prophet said, "After my disappearance there will be no greater source of chaos and disorder for my nation than women.[43]

The irony is that Muslim and European theories come to the same conclusion: women are destructive to the social order—for Imam Ghazali because they are active, for Freud because they are not.

Different social orders have integrated the tensions between religion and sexuality in different ways. In the Western Christian experience sexuality itself was attacked, degraded as animality and condemned as anticivilization. The individual was split into two antithetical selves: the spirit and the flesh, the ego and the id. The triumph of civilization implied the triumph of soul over flesh, of ego over id, of the controlled over the uncontrolled, of spirit over sex.

Islam took a substantially different path. What is attacked and debased is not sexuality but women, as the embodiment of destruction, the symbol of disorder. The woman is *fitna,* the epitome of the uncontrollable, a living representative of the dangers of sexuality and its rampant disruptive potential. We have seen that Muslim theory considers raw instinct as energy which is likely to be used constructively for the benefit of Allah and His society if people live according to His laws. Sexuality *per se* is not a danger. On the contrary, it has three positive, vital functions. It allows the believers to perpetuate themselves on earth, an indispensable condition if the social order is to exist at all. It serves as a "foretaste of the delights secured for men in Paradise,"[44] thus encouraging men to strive for paradise and to obey Allah's rule on earth. Finally, sexual satisfaction is necessary to intellectual effort.

The Muslim theory of sublimation is entirely different from the Western Christian tradition as represented by Freudian psychoanalytic theory. Freud viewed civilization as a war against sexuality.[45] Civilization is sexual energy "turned aside from its sexual goal and diverted towards other ends, no longer sexual and socially more valuable."[46] The Muslim theory views civilization as the outcome of satisfied sexual energy. Work is the result not of sexual frustration but of a contented and harmoniously lived sexuality.

> The soul is usually reluctant to carry out its duty because duty [work] is against its nature. If one puts pressures on the soul in order to make it do what it loathes, the soul rebels. But if the soul is allowed to relax for some

> moments by the means of some pleasures, it fortifies itself and becomes after that alert and ready for work again. And in the woman's company, this relaxation drives out sadness and pacifies the heart. It is advisable for pious souls to divert themselves by means which are religiously lawful.[47]

According to Ghazali, the most precious gift God gave humans is reason. Its best use is the search for knowledge. To know the human environment, to know the earth and galaxies, is to know God. Knowledge (science) is the best form of prayer for a Muslim believer. But to be able to devote his energies to knowledge, man has to reduce the tensions within and without his body, avoid being distracted by external elements, and avoid indulging in earthly pleasures. Women are a dangerous distraction that must be used for the specific purpose of providing the Muslim nation with offspring and quenching the tensions of the sexual instinct. But in no way should women be an object of emotional investment or the focus of attention, which should be devoted to Allah alone in the form of knowledge-seeking, meditation, and prayer.

Ghazali's conception of the individual's task on earth is illuminating in that it reveals that the Muslim message, in spite of its beauty, considers humanity to be constituted by males only. Women are considered not only outside of humanity but a threat to it as well. Muslim wariness of heterosexual involvement is embodied in sexual segregation and its corollaries: arranged marriage, the important role of the mother in the son's life, and the fragility of the marital bond (as revealed by the institutions of repudiation and polygamy). The entire Muslim social structure can be seen as an attack on, and a defence against, the disruptive power of female sexuality.

NOTES

1. Ibn Khaldun, *The Muqaddimah, An Introduction to History,* translated by Franz Rosenthal, Princeton, NJ, 1969, pp. 160–161.
2. Ibid., p. 161.
3. Ibid.
4. Abu Hamid al-Ghazali, *Ihya Ulum al-Din,* Cairo, n.d.
5. Ibid., p. 28.
6. Ibid., p. 25.
7. Ibid.
8. Ibid., p. 27.
9. George Peter Murdock, *Social Structure,* New York, 1965, p. 273.
10. Ibid.
11. Qasim Amin, *The Liberation of Women,* Cairo, 1928, p. 64.
12. Ibid., p. 65.
13. Al-Ghazali, *The Revivification of Religious Sciences,* vol. II, chapter on marriage; and Mizan al-'Amal, *Criteria for Action,* Cairo, 1964.
14. Abbas Mahmud al-Aqqad, *Women in the Koran,* Cairo, n.d.
15. Ibid., p. 7; the verse he refers to is verse 228 of sura 2, which is striking by its inconsistency. The whole verse reads as follows:

    And they [women] have rights similar to those [of men] over them in kindness, and men are a degree above them.

    I am tempted to interpret the first part of the sentence as a simple stylistic device to bring out the hierarchical content of the second part.
16. Ibid., p. 24.
17. Ibid., p. 25. The biological assumption behind Aqqad's sweeping generalizations is obviously fallacious.
18. Ibid., p. 26.
19. Ralph Linton, *The Study of Man,* London, 1936, p. 116.
20. Sigmund Freud, *New Introductory Lectures on Psychoanalysis,* College Edition, New York, 1965, p. 114.
21. Ibid.
22. Sigmund Freud, *Three Contributions to the Theory of Sex,* 2nd ed, New York, 1909, p. 77.
23. Freud, *New Introductory Lectures,* p. 114.
24. Al-Ghazali, *Revivification of Religious Sciences,* p. 51.
25. Una Stannard, "Adam's Rib or the Woman Within," *Transaction,* November–December 1970, vol. 8, special issue on American Women, pp. 24–36.
26. Al-Ghazali, *Revivification,* p. 50. Not only is the woman granted ejaculation, she is also granted the capacity to have nocturnal ejaculation and "sees what the man sees in sleep" (Ibn Saad,

Kitab al-Tabaqat al-Kubra, Beirut, 1958, vol. 8, "On Women," p. 858.)

27. Sigmund Freud, *Sexuality and the Psychology of Love,* New York, 1963, p. 190.
28. Freud, *Three Contributions,* p. 78.
29. Ibid.
30. Ibid.
31. Freud, *New Introductory Lectures,* p. 132.
32. Al-Ghazali, *Revivification,* p. 50.
33. Ibid.
34. Freud, *Three Contributions,* p. 14.
35. Ibid., p. 15.
36. Al-Ghazali, *Revivification,* p. 50.
37. Ibid.
38. Freud, *New Introductory Lectures,* p. 116.
39. Abu Issa al-Tarmidi, *Sunan al-Tarmidi,* Medina, n.d., vol. II, p. 413, B: 9, H: 1167.
40. Abu al-Hasan Muslim, *al-Jami' al-Sahih,* Beirut, n.d., vol. III, Book of Marriage, p. 130.
41. Al-Tarmidi, *Sunan al-Tarmidi,* p. 419, B: 16, H: 1181. See also al-Bukhari, *Kitab al-Jami' al-Sahih,* Leyden, Holland, 1868, vol. III, K: 67, B: 11.
42. Edward Westermark, *The Belief in Spirits in Morocco,* Abo, Finland, 1920.
43. Abu Abdallah Muhammad Ibn Ismail al-Bukhari, *Kitab al-Jami' al-Sahih,* Leyden, Holland, 1868, p. 419, K: 67, B: 18.
44. Al-Ghazali, *Revivification,* p. 28.
45. Sigmund Freud, *Civilization and Its Discontents,* New York, 1962.
46. Sigmund Freud, *A General Introduction to Psychoanalysis,* New York, 1952, p. 27.
47. Al-Ghazali, *Revivification,* p. 32.

DISCUSSION QUESTIONS

1. Compare and contrast Freud's conceptualization of female sexuality and Christian depictions of sexuality (especially as set forth in the Bible). What is your own conceptualization of female sexuality? Is female sexuality primarily active or passive? Male sexuality?
2. Do you think sexual energy is constructive or destructive? What kind of social order would follow if you were to design a society based on your answer?
3. In what ways do patriarchy and the gendered division of labor help to facilitate cultural constructions of women as sexually passive? In what ways do they facilitate constructions of women as sexually active?
4. How might you extend Mernissi's work to incorporate an analysis of conceptualizations of homosexual sexuality (both gay and lesbian)?
5. What types of sexual behaviors would you expect to see outlawed or marginalized in societies that depict females as sexually active? What would you expect in societies that depict females as socially passive? Are there commonalties? Discuss.

Slut! Growing up Female with a Bad Reputation

LEORA TANENBAUM

The word slut *is a potent epithet that is used to mark and shame women who are perceived to be overly sexual. As such, it is a powerful instrument of social control that regulates female sexuality. To explore this phenomenon, Leora Tanenbaum draws on interviews with 50 women who had been labeled as "sluts" during middle and high school. Her research reveals that despite a climate of increased sexual freedom for women, the sexual double standard is still very much alive and well in contemporary American society.*

INTRODUCTION

Julie arrives for our meeting, at a diner across the street from her college campus, dressed for comfort: faded jeans, untucked denim shirt, olive-green army-surplus jacket. She slides opposite me into the windowed booth. I offer her a menu, but she shakes it away: she already knows she wants the French fries. Julie's face is round and friendly, her manner relaxed. But as Julie unfolds her story, I learn that she wasn't always as self-assured as she is today, at nineteen. Back in junior high in New Jersey, she was easily intimidated. Her greatest ambition in life was just to fit in. But she didn't fit in. Julie was something of an outsider in her solidly Catholic, Irish-Italian neighborhood. Her family was the only one in town that didn't attend church. She hadn't gone to the same middle school her friends had attended. And she was pudgy, a bit overweight. Too eager to conform yet too different, Julie was an easy target.

One evening when she was thirteen, Julie tells me, her soft brown eyes meeting mine squarely, she and her friends were hanging out as usual at one of their houses. They were drinking beers. Julie had recently begun to drink more and more: Alcohol enabled her to push aside the fact that her parents were sleeping in separate beds and not talking to each other. Besides, some older guys had joined them this evening, and Julie figured that it made her seem cool if she drank. That night she drank until she passed out unconscious. When she regained consciousness semiclothed and with a dull pain, she realized that someone had had sex with her. Julie pulled herself together and asked one of her friends what had happened. The guy who did it, she was informed, had been drinking with the group that night—a classmate she only slightly knew. Julie understood that she had been raped.

By Monday morning a friend of the rapist, who had witnessed the event, had spread the news that his friend and Julie had had sex. In a matter of hours, Julie tells me, she was known as a slut. "They'd call out 'slut' to me in the halls. There was graffiti. I got calls in the middle of the night, at four A.M.: 'Fat slut.' Behind the junior high school there was a playground with a handball court, and people would write graffiti there in shaving cream." Everybody in school knew about her. Even today, years later, Julie's reputation

From "Slut! Growing up Female with a Bad Reputation" by Leora Tanenbaum, 1999.
Adapted by permission of Seven Stories Press.

as a slut is known to each new crop of incoming high school students.

Julie's story of being singled out as a slut is much more common than you might think. Indeed, you would be hard-pressed to find a high school in the United States in which there is no designated slut. Two out of five girls nationwide—42 percent—have had sexual rumors spread about them, according to a 1993 poll conducted for the American Association of University Women (AAUW) on sexual harassment in schools.[1] ("Sexual rumor" sometimes means speculation that a classmate is gay, but more often it is a polite way of saying "slutty reputation.") Three out of four girls have received sexual comments or looks, and one in five has had sexual messages written about her in public areas.

Slut-bashing—as I call it—is one issue that affects every single female who grows up in this country because any preteen or teenage girl can become a target. "Slut" is a pervasive insult applied to a broad spectrum of American adolescent girls, from the girl who brags about her one-night stands to the girl who has never even kissed a boy to the girl who has been raped. Some girls are made fun of because they appear to have a casual attitude about sex (even if, in reality, they are no more sexual than their peers). Many others are picked on because they stand out in some way—being an early developer, new in school, an ethnic or class minority, overweight, or just considered "weird" for whatever reason. Some are called "sluts" because other girls dislike or envy them, and spread a sexual rumor as a form of revenge. While a girl can almost instantly acquire a "slut" reputation as a result of one well-placed rumor, it takes months, if not years, for the reputation to evaporate—if it does at all.

. . . Being known as the school slut is a terrifying experience. In school, where social hierarchy counts for everything, the school "slut" is a pariah, a butt of jokes, a loser. Girls and boys both gang up on her. She endures cruel and sneering comments—*slut* is often interchangeable with *whore* and *bitch*—as she walks down the hallway. She is publicly humiliated in the classroom and cafeteria. Her body is considered public property: She is fair game for physical harassment. There is little the targeted girl can do to stop the behavior. I was surprised to learn that teachers, generally speaking, do not intervene; they consider this behavior normal for teenagers.

. . . Slut-bashing shows us that sexism is still alive and that as boys and girls grow up, different sexual expectations and identities are applied to them. Slut-bashing is evidence of a sexual double standard that should have been eliminated decades ago, back when abortions were illegal, female office assistants were called "gal Fridays," and doctors were men and nurses were women. Slut-bashing sends the message to all girls, no matter how "pure" their reputations, that men and boys are free to express themselves sexually, but women and girls are not.

INSULT OF INSULTS

. . . [W]e are routinely evaluated and punished for our sexuality. In 1991, Karen Carter, a twenty-eight-year-old single mother, lost custody of her two-year-old daughter in a chain of events that began when she called a social service hot line to ask if it's normal to feel sexual arousal while breast feeding. Carter was charged with sexual abuse in the first degree, even though her daughter showed no signs of abuse; when she revealed in court that she had had a lifetime total of eight (adult male) lovers, her own lawyer referred to her "sexual promiscuity."[2] In 1993, when New Mexico reporter Tamar Stieber filed a sex discrimination lawsuit against the newspaper where she worked because she was earning substantially less than men in similar positions, defense attorneys deposed her former lover to ask him how often they'd had sex.[3] In the 1997 sexual-harassment lawsuits against Mitsubishi Motor Manufacturing, a company lawyer asked for the gynecological records of twenty-nine women employees charging harassment, and wanted the right to distribute them to company executives.[4] And in 1997 a North Carolina woman sued her husband's secretary for breaking up their nineteen-year-marriage and was awarded $1 million in damages by a jury. During the seven-day trial the secretary was described as a "matronly" woman who deliberately began wearing heavy makeup and short skirts in order to entice the husband into an affair.[5]

It's amazing but true: Even today a common way to damage a woman's credibility is to call her a slut.

. . . Today a teenage girl can explore her sexuality without getting married, and most do. By age eighteen over half of all girls and nearly three quarters of all boys have had intercourse at least once.[6] Yet at the same time, a fifties-era attitude lingers: Teens today are fairly conservative about sex. A 1998 *New York Times*/CBS News poll of a thousand teens found that 53 percent of girls believe that sex before marriage is "always wrong," while 41 percent of boys agree.[7] Teens may be having sex, but they also look down on others, especially girls, who are sexually active. Despite the sexual revolution, despite three decades of feminism, despite the Pill, and despite legalized abortion, teenage girls today continue to be defined by their sexuality. The sexual double standard—and the division between "good" girls and "bad" or "slutty" ones—is alive and well. Some of the rules have changed, but the playing field is startlingly similar to that of the 1950s.

. . . Lillian Rubin surveyed six hundred students in eight colleges around the country in the late 1980s and found that 40 percent of the sexually active women said that they routinely understate their sexual experience because "my boyfriend wouldn't like it if he knew," "people wouldn't understand," and "I don't want him to think I'm a slut." Indeed, these women had reason to be concerned. When Rubin queried the men about what they expected of the women they might marry, over half said that they would not want to marry a woman who had been "around the block too many times," that they were looking for someone who didn't "sleep around," and that a woman who did was a "slut."[8]

Teenage girls who are called sluts today experience slut-bashing at its worst. Caught between the conflicting pressures to have sex and maintain a "good" reputation, they are damned when they do and damned when they don't. Boys and girls both are encouraged to have sex in the teen years—by their friends, magazines, and rock and rap lyrics—yet boys alone can get away with it. "There's no way that anyone who talks to girls thinks that there's a new sexual revolution out there for teenagers," sums up Deborah Tolman, a developmental psychologist at the Wellesley College Center for Research on Women. "It's the old system very much in place." It *is* the old system, but with a twist: Today's teenage girls have grown up after the feminist movement of the late 1960s and 1970s. They have been told their whole lives that they can, and should, do anything that boys do. But soon enough they discover that sexual equality has not arrived. Certain things continue to be the privilege of boys alone.

With this power imbalance, it's no wonder high school girls report feeling less comfortable with their sexual experiences than their male counterparts do. While 81 percent of adolescent boys say that "sex is a pleasurable experience," only 59 percent of girls feel the same way.[9] The statistical difference speaks volumes. Boys and girls both succumb to early sex due to peer and media pressures, but boys still get away with it while girls don't.

. . . Most people who meet me for the first time are surprised by two things: that I am the type of person who would ever write a book with a title like *Slut!*, and that I was once known as a "slut." I have been described in print as "demure" and as "a petite brunette with wire-rimmed glasses"—code words for nice, shy, bookish. On the Oprah Winfrey show, I was presented as a nice, middle-class woman married to a nice, middle-class man. Over and over I am told, "But you're so clean-cut—you don't seem like a slut at all."

My point exactly: Any girl or woman can be labeled a "slut." Looks and attitude often have nothing to do with it.

Yet the word continues to evoke for most people an image of someone trampy and pathetic—the kind of girl or woman who wears short, tight, cleavage-enhancing clothes, always makes a beeline for the guy who enters the room, and can't string two sentences together without making a non sequitur. In short, she deserves to be called a "slut."

. . . Below are some of the comments I've received from men and women in bookstores and radio call-in shows, and from television and radio show hosts. It's clear that most people are far more concerned with the sexuality of girls than with that of boys. My responses point out that females as well as males should be entitled to express their sexual desires. Hardly a radical concept, but it can stir up a lot of hostility.

Slut-bashing is a terrible thing, but let's face it: It affects only a small number of girls. Why write a whole book about it?

A reputation acquired in adolescence can damage a young woman's self-perception for years. She may become a target for other forms of harassment and even rape, since her peers see her as "easy" and therefore not entitled to say "no." She may become sexually active with a large number of partners (even if she had not been sexually active before her reputation). Or she may shut down her sexual side completely, wearing baggy clothes and being unable to allow a boyfriend to even kiss her.

It's true that most girls escape adolescence unscathed by slut-bashing. Nevertheless, just about every girl is affected by it. Every girl internalizes the message that sex is bad—because it can earn you a reputation. The result is that even years later, when she is safely out of adolescence, a woman may suffer from a serious hangup about sex and intimacy—even if she was not herself called a "slut." Second, the fear of being called a "slut" makes many girls unlikely to carry or use contraceptives, leading of course to the risk of pregnancy or disease.

Slut-bashing also affects boys. It fosters a culture of sexual entitlement that says that "easy" girls are expendable while only "good" girls deserve to be treated well. And that means that only some girls are treated with the respect that they all deserve.

You make it seem as if we're living in the 1950s. But this is the twenty-first century. Lots of girls are having sex; most of them are not called "sluts."

Of course, a girl today has many freedoms that her 1950s counterpart did not possess, including the license to sexually experiment before marriage. But even today, the prevailing attitude is that there is something wrong with the girl who behaves just as a boy does.

. . . [I]n the summer of 1999 a dozen Virginia junior high school girls were discovered to have engaged in oral sex throughout the school year during parties and at local parks. *The Washington Post* broke the story, which was subsequently picked up by the Associated Press and reprinted in newspapers across the country. Parents, health educators, and guidance counselors weighed in with a loud chorus of condemnation. Certainly it was disturbing that kids so young were engaging in meaningless sexual encounters. But much more disturbing was that, first, the girls had been nothing more than sexual servicers to the boys; and second, that all of the censure was directed to the girls. It turns out that the school principal had called the parents of the girls to a special meeting to discuss the matter—but none of the boys or their parents was approached. Boys will be boys—but girls will be "sluts."

Girls today dress so provocatively, even to school, in skimpy outfits that expose a lot of flesh. They practically invite people to call them "sluts" and other names.

I have to admit that I am often appalled by some of the outfits I see young girls wearing these days: It's one thing for an adult woman to showcase her sexual appeal and a different thing entirely for an eighth grader to do likewise. But I don't blame the girls. On the contrary, I am sympathetic to them. These girls believe that if they attract a boyfriend and fall in love, their lives would be better and they would be happier. Sadly, many of these girls believe that their sexuality is the only power or appeal they have, and so they play it up to the hilt. They also feel competitive with other girls in a battle for the most desirable guys, so they feel the need to out-dress their peers. Dressing in sexy outfits, then, is both a strategy to obtain romance and a competition with other girls. But just because a girl dresses in a sexually provocative way doesn't mean that she is sexually promiscuous. In reality, she may not be any more sexually active than the prissy girl in tailored pants, loafers, and sweater set.

Some of your interviewees were called "sluts" even though they weren't sexually active at all. They were innocent victims. But the girls who were sexually promiscuous are a different story: They deserved what they got.

I don't believe that there should be a distinction between those who deserve a bad reputation and those who don't. Because frankly, I don't think that any girl deserves to be called a "slut." After all, boys who are sexually active are congratulated as studs.

Dividing "sluts" into the innocent and the guilty merely reinforces the sexual double standard. This is why, when I have been asked about my own sexual history—believe it or not, radio show hosts, aping Howard Stern, have felt perfectly comfortable quizzing me about the details of my sex life—I have refused to respond. Besides the fact that the answers are no one's business, they would serve only to buttonhole me as either "innocent" or "guilty," and I reject both categories. . . .

Our culture is too open about sex.

It's true that anyone with eyes is bombarded by sexual images. But while in many ways Americans are very open about sexual images, they are, at the same time, profoundly uncomfortable discussing the realities of sexuality. Last year, a popular sexuality course was nearly eliminated from the curriculum of a New York community college because of Catholic opposition. The course, which was not required, addressed abortion, birth control, and homosexuality—all standard fare for a college-level human sexuality course. The college ultimately won the battle because a federal district court judge found in its favor, but it is amazing that the opponents, who were represented by the American Catholic Lawyers Association, got as far as they did.[10] Yes, mostly naked women appear everywhere from bus shelters to *New York Times* lingerie ads, but at the same time we live in a culture that accepts and promotes moral and sexual policing. When it comes to sex, this is very much a divided culture.

If females practiced an ethic of sexual modesty, males would be more likely to treat them with respect.

Ideally, I think sex should be harnessed within a romantic relationship, but that ideal isn't possible or desirable for everybody. There are young women who perhaps would like to wait and initiate their first sexual encounter in a loving relationship, but for whatever reason, they have desire and want to act on it before they've met the "right" person—or they may never meet the "right" person. I worry that these young women are going to feel guilty and ashamed of their own sexual desire. I'm also concerned that they are going to make bad choices about who their mate is going to be, perhaps marrying too young. A sex drive is a natural appetite for males and females. If you say that females are innately modest, then you're also saying that a girl or woman who isn't modest is doing something unfemale and wrong.

. . . School sexual harassment lawsuits are getting out of hand. How can a school be monetarily responsible for sexual harassment? These lawsuits hurt everyone, since they take money away from education.

The Supreme Court ruled in May 1999 (*Davis v. Monroe County Board Education*) that school districts receiving federal money can be liable for monetary damages if they fail to prevent severe, persistent sexual harassment among students. The ruling was in response to a case brought by the mother of a fifth-grade girl in rural Georgia. The girl, LaShonda Davis, was harassed by a male classmate who made repeated unwanted sexual advances over the course of five months. At least two teachers, as well as the principal, were aware of the incidents, but no disiciplinary action was taken against the boy. Meanwhile, Davis's grades dropped and her father discovered that she had written a suicide note.

The ruling is important because it sends the message that schools must be vigilant in halting sexual harassment—which includes slut-bashing, a verbal form of sexual harassment. I agree that a school should be liable if the sexual harassment is severe and persistent and if the school is aware of the behavior but does not take steps to halt it. If, on the other hand, the school makes a good-faith effort to stop the behavior, then I don't believe it should be liable.

It's unfortunate that a ruling against a school results in a monetary loss, but it's also unfortunate that the threat of monetary payment is the most effective wake-up call to school administrators. As for the argument that these payments take money away from education, sexual harassment also impedes the ability of teachers to effectively educate and the ability of students to effectively learn.

The average age when teens begin to have sexual intercourse is fifteen. That is much too young.

I think all teenagers, boys and girls alike, should be encouraged to wait until they are in a serious relationship, and they should wait until they are emotionally ready. By and large, fifteen-year-olds do not meet both those criteria. In schools, sex education is often lacking. All teenagers should be exposed to comprehensive sex education, which encourages teens to wait until they initiate sexual activity but offers them information about birth control and abortion so that they are prepared whenever they do become sexually involved.

Yet sex education in public schools is increasingly focused on abstinence, with more than a third of districts using an abstinence-only curriculum that permits discussion of contraception only in the context of its unreliability. Fewer than half of public schools offer information on where to get birth control (45 percent). Only slightly more than a third mention abortion (37 percent) or sexual orientation (36 percent) as part of the curriculum.[11]

. . . We should learn from the example of our European counterparts, who tend to be far less inhibited about adolescent sexuality and who openly discuss contraceptives and responsible decision making with their teens. In those countries, teens initiate sexual activity at a later age. For instance, the age of first act of intercourse in the Netherlands is seventeen—two years older than in the United States.[12]

Admittedly, many parents find it awkward and difficult to raise these issues. One great bit of advice that sex educators offer is to talk about sensitive issues while riding in a car, since you have a captive audience but you can avoid potentially embarrassing eye contact.

Why are girls often worse than boys when it comes to slut-bashing?

All of us yearn for one arena in which we can wield power. For girls, this desire is often thwarted. After all, girls may get better grades—but boys, especially athletes, by and large receive more attention and congratulatory pats on the back from school administrators and teachers. Boys call out more in class and get away with it. They rule the playground. Many feel a sense of entitlement to grope girls' bodies. With these depressing realities, it's no wonder that many girls develop a sense of self-hatred. Sensing that femininity is devalued, they may feel, at some level, uncomfortable with being a girl, and therefore are reluctant to bond with other girls. Instead, they latch on to one small sphere of power they can call their own: the power to make or break reputations. Slut-bashing is a cheap and easy way to feel powerful. If you feel insecure or ashamed about your own sexual desires, all you have to do is call a girl a "slut" and suddenly you're the one who is "good" and on top of the social pecking order.

What can we do to stop slut-bashing?

Parents should be open about sexuality with their kids—and that means being open about female sexuality as well as male sexuality. They should teach their daughters and sons that girls as well as boys have sexual feelings, and that sexual feelings are entirely normal. That way they won't have to pin their sexual anxieties on a scapegoat and then distance themselves from her.

Teachers must recognize that slut-bashing is a serious problem. Too often, they dismiss it as part of the normal fabric of adolescent life. But slut-bashing is a form of sexual harassment, and it is illegal under Title IX, which entitles students to a harassment-free education. If a teacher witnesses slut-bashing, she must make sure that it stops. She must confront the ringleader and other name-callers. Of course, teachers and school administrators shouldn't wait for slut-bashing to occur. They must create and publicize awareness through sexual harassment policies for their schools.

. . . But the most important thing that all of us need to work on is this: to stop calling or thinking of women as "sluts." Face it: At one time or another, many of us have called a woman a "slut." We see a woman who's getting away with something we wish we could get away with. What do we call her? A "slut."

We see a woman who dresses provocatively, and maybe we wish we had the guts to dress that way ourselves. What do we call her? A "slut."

If we think of a woman or girl as a "slut," it's like she's not one of us. She's one of *them*. She is other. "Slut," like any other derogatory label, is a shorthand for one who is different, strange—and not worth knowing or caring about. Unlike other insults, however, it carries a unique string: the stigma of the out-of-control, trampy female. Most of us recognize that this stigma is unjust and unwarranted. Yet we have used the "slut" insult anyway: Our social conditioning runs too deep. We must will ourselves to be aware of the sexual double standard and of how we lapse into slut-bashing on an everyday level. If we become aware of our behavior, then we have the power to stop.

And never again be slut-bashers or self-bashers.

NOTES

1. Felicity Barringer, "School Hallways as Gantlets of Sexual Taunts," *The New York Times,* June 2, 1993, p. B7. The poll, conducted by Louis Harris & Associates for the American Association of University Women Educational Foundation, surveyed 1,632 students in grades eight through eleven in seventy-nine schools across the country.
2. Lauri Umansky, "Breastfeeding in the 1990s: The Karen Carter Case and the Politics of Maternal Sexuality" in Molly Ladd-Taylor and Lauri Umansky, eds., *"Bad" Mothers: The Politics of Blame in Twentieth-Century America* (New York: New York University Press, 1998), pp. 299–309. Karen Carter is a pseudonym.
3. Tamar Stieber, "Viewpoint," *Glamour,* August 1996, p. 138.
4. Stieber, p. 138.
5. Jon Jeter, "Woman Who Sued Ex-Husband's Mistress Is Awarded $1 Million," *The Washington Post,* August 7, 1997, p. A3.
6. *Sex and America's Teenagers* (New York and Washington: The Alan Guttmacher Institute, 1994), p. 20.
7. Laurie Goodstein with Marjorie Connelly, "Teen-Age Poll Finds a Turn to the Traditional," *The New York Times,* April 30, 1998, p. A20. The poll, of 1,048 teenagers ages thirteen to seventeen, was conducted by telephone in April 1998. The poll also found that only 18 percent of thirteen- to fifteen-year-olds said they had ever had sex, as against 38 percent of sixteen- and seventeen-year-olds.
8. Lillian B. Rubin, *Erotic Wars: What Happened to the Sexual Revolution?* (New York: HarperPerennial, 1991), p. 119.
9. Tamar Lewin, "Boys Are More Comfortable With Sex Than Girls Are, Survey Finds," *The New York Times,* May 18, 1994.
10. "Academic Freedom Survives Court Battle," *Censorship News,* a newsletter of the National Coalition Against Censorship (New York, NY), Spring 1999, p. 1.

11. Jodi Wilgoren, "Abstinence Is Focus of U.S. Sex Education," *The New York Times*, December 15, 1999, p. A18.
12. Judy Mann, "Wanted: A Realistic Attitude Toward Teen Sex, *The Washington Post*, January 27, 1999, p. C14.

DISCUSSION QUESTIONS

1. How does the sexual double standard reinforce male power and female subordination? Do you think the standard has changed at all in the past few years? If so, how? If not, why not? What evidence do you have to support your position? What will need to change for the double standard to be eliminated entirely?
2. How does the inclusion of race, class, age, or nationality further condition the dynamics of slut-bashing?
3. Have the rising rates of AIDS and other sexually transmitted diseases affected the sexual double standard for men and women? Should it? Why or why not? Do *you* judge men or women who have sexually transmitted diseases any differently? Why or why not?
4. In what ways, if any, are the epithets "fag" and "slut" similar? How do they differ? What interests do their shared similarities serve? Their differences? How do these terms serve to reinforce existing social constructions of male and female sexuality?
5. Make a list of other terms that are commonly used to describe male and female sexual actors. Do you see any patterns? In what ways, if any, do these terms reinforce contemporary constructions of sexuality?
6. What insights from Wilkerson's work in Chapter 6 apply to the case of sluts?

Gay and Lesbian Catholic Elementary School Teachers

EDMUNDO F. LITTON

Though public opinion polls indicate ever-increasing levels of acceptance for gays and lesbians (particularly among younger generations), much of this support erodes when people are asked about specific issues such as gay marriage and allowing openly gay, lesbian, bisexual, or transgendered individuals to teach in elementary schools. Further, in many settings (including parochial schools), gays and lesbians are barred from teaching. Edmundo F. Litton's interviews with gay and lesbian Catholic elementary school teachers reveal the oppression and tension many of these teachers face as well as the compromises, sacrifices, and survival strategies they employ to negotiate the conflicting demands of their religion, vocation, and sexual identity.

From "Voices of Courage and Hope: Gay and Lesbian Catholic Elementary School Teachers" by Edmundo F. Litton, *International Journal of Sexuality and Gender Studies* 6(3): 193–205, 2001. Reprinted by permission of the author and Kluwer Academic/Plenum Publishers.

INTRODUCTION

"What does a gay person look like? Please tell me because I have never really seen one." This was a question raised by David, a third grade boy. He asked a male teacher whom I observed while he taught a lesson on AIDS at a Catholic elementary school in Northern California. I still remember the fear that became visible on the teacher's face as he stood in front of the class, trying hard to think of a quick answer. The entire class seemed unusually attentive for an answer. Was the teacher going to ignore the question and tell the students that the question will be answered in the fourth grade? Does he tell the class that they know exactly what a gay person looks like because they have been looking at one for the past nine months? In the end, this teacher gave David the usual answer: gay people looked just like everyone else.

The incident described above served as a catalyst to my exploration of the lives of gay and lesbian teachers who teach in Catholic elementary schools. When I met with the teacher in the event described above, we spent a great deal of time discussing the conflict he experienced. He confided that he was gay and that he very much wanted to answer David's question, but he was afraid to present an answer that would make others suspect that he might be gay.

. . . The main purpose of this essay is to give voice to the stories of gay and lesbian Catholic elementary school teachers. Through interviews, these teachers are able to share their fears, their hopes, and the strategies that they use in order to be able to live truly (and not just survive) in the context of the Catholic elementary school. What role does their Catholic faith play in their decision to teach in a Catholic school? How do these gay and lesbian teachers discuss homosexuality when the question arises in their classroom? Why do these teachers continue to work for the Catholic Church despite the Church hierarchy's view on homosexuality?

Catholic school teachers do not have the same kind of protection that their public school counterparts receive under state laws. . . . While laws cannot guarantee a safe environment for any gay or lesbian teacher, the lack of legislative protection is one aspect that makes gay and lesbian Catholic teachers unique. In most Catholic elementary school systems, tenure does not exist. Teachers are on a yearly contract, and a teacher can face dismissal at the end of the time stipulated in a contract, which directs teachers to live their lives according to the moral teachings of the Church. Despite this lack of job security, many gay and lesbian teachers continue to choose to work in Catholic schools. Thus, it is important to understand the reasons behind their choice to stay in what could be an oppressive system.

BACKGROUND

In recent years, much has been written about the lives of gay and lesbian teachers and the impact of their presence on academic settings (Harbeck 1997; Jennings 1994; Kissen 1996; Woog 1995). These authors reveal that gay and lesbian teachers are dedicated, courageous, and professional people. Many of these teachers continue to serve society despite the fear that all their contributions will be forgotten as soon as their sexual orientation is discovered. While there are some teachers who "come out" and are still accepted by their communities, the majority of gay and lesbian teachers must remain in the closet if they wish to continue working in schools. Prejudice and discrimination against homosexuals in schools is often a reflection of the prejudice that exists in society as a whole . . . The burden of hiding one's sexual orientation is even heavier for gay and lesbian teachers who work in Catholic schools. This essay, however, is not the forum to discuss whether the teachings of the Church are right or wrong. The points presented here are intended to provide an explanation of the conditions that gay and lesbian Catholic elementary school teachers face in their professional setting.

Church doctrine on homosexuality can have a tremendous influence on homosexual teachers. Gay and lesbian Catholic school teachers are employees of the Catholic Church. These teachers are required to teach using church doctrine as a guideline.

. . . Despite a call for increased tolerance and acceptance, the Church has not shifted from its position of viewing homosexual acts as immoral.

. . . The Church's view on homosexuality creates tension between homosexuals and the Church. Yip (1997) conducted a study on gay and lesbian Catholics in Britain. His data came from a survey of 121 gay and lesbian Catholics who were members of Quest, a gay and lesbian organization for gay and lesbian Catholics in Britain. The results showed "there exists a discrepancy between the respondents' personal faith and the Church's official teachings on a host of controversial issues such as homosexual acts, artificial contraception and compulsory celibacy for clergy" (Yip 1997, p. 178). Despite this discrepancy, the participants in the study did not distance themselves from the Church. Yip's study shows that it is possible for gay and lesbian Catholics to reach "a point in their journey of sexuality where they have successfully developed a re-negotiated identity that incorporates their sexuality and religious beliefs" (Yip 1997, p. 178).

Homophobia in the Church and in society has a negative impact on all people, not just homosexuals since oppression of any kind hurts everyone. Blumenfeld (1992) discusses several reasons why homophobia hurts everyone. Homophobia forces people to compartmentalize their gender roles. They feel forced to act in ways that are based on socially-constructed concepts of "masculinity" and "femininity." Homophobia prevents people from being who they are and inhibits their creativity and self-expression. Homophobia also pressures heterosexuals "to treat others badly, action contrary to their basic humanity" (Blumenfeld, 1992, p. 9). In addition, homophobia also prevents heterosexuals from accepting the gifts of sexual minorities. Furthermore, homophobia discourages close relationships between people of the same sex since any close relationship between two people of the same sex may be thought to be homosexual. Most importantly, though, homophobia leads to silence on issues that are important for both the heterosexual and homosexual communities. This silence has prevented society from responding more constructively to the AIDS crisis (Blumenfeld 1992). Zapulla (1997) shows in her study that schools even find it more difficult to respond to AIDS because of the moral mission that schools are given to carry out. The silence has led to ignorance and invisibility of homosexual awareness in the school curriculum. Ignorance has also led to the high suicide rate among homosexual teenagers. Thus, "the price of our ignorance can be fatal" (Goodman, 1996, p. 9).

. . . Gay and lesbian teachers often face a conflict between wanting to be authentic role models (who can break the silence of ignorance) and their own need to feel safe. "To be a lesbian or gay teacher, in most schools, is to walk a constant line between safety and honesty" (Kissen, 1996, p. 16). Despite the desire to become less invisible, gay and lesbian teachers still do not feel that they can come out safely. Homosexual teachers who are open about their sexual identity are closely monitored by administrators wanting to be certain that these teachers do not sexually harass their students (Goodman, 1996, p. 13). Ironically, Goodman (1996) states that "almost all reported incidents of sexual harassment of students have been perpetrated by heterosexual males" (p. 13). Homosexual teachers also fear becoming targets of physical or verbal harassment, and they are often accused of wanting to influence their students to become homosexuals.

. . . Homosexual teachers often pressure themselves to work hard or to "fit in" in order to survive in educational settings. Hiding is a common survival strategy. Homosexual teachers also feel the need to work harder than their heterosexual colleagues. Kissen (1996) states that many homosexual teachers "feel they need to be outstanding in order to be seen as equal to their heterosexual colleagues" (p. 42). Homosexual teachers may also monitor their actions or manner of dressing more than heterosexual teachers because they worry about being identified as homosexuals (Kissen 1996). These self-protection strategies can lead to even more pressure because "the energy that teachers spend on hiding is more than a drain on their time, it is a drain on their minds and bodies as well" (Kissen 1996, p. 53). Whether a homosexual teacher is or is not out of the closet, Pekman's (1997) study shows that most homosexual teachers are uncomfortable about hiding their sexuality because "gay and lesbian teachers who hide do not present good role models for gay and lesbian students" (p. 194).

Efforts have been made in the last few years to make homosexuals more visible in schools by including homosexual perspectives in the school curriculum. Despite efforts by detractors, picture books portraying families with two fathers or two mothers are being used in some schools. Books such as *Heather Has Two Mommies* by Lesléa Newman (1991) and Michael Willhoite's (1991) *Daddy's Roommate* allow children to see diversify in family structures (Stewig 1996). Students are also being introduced to gay and lesbian writers in English literature classes. Making the curriculum more inclusive for homosexuals, however, remains a challenge. Prince (1996) notes that most administrators see homosexual issues as best taught in health or sex-education classes. Such a view perpetuates the stereotype that sexual activity and the ramifications of such activity should be the prime focus of discussions on homosexual issues.

While society has become somewhat more accepting and tolerant of homosexuality, schools continue to remain one of the last places where homosexuals are not safe. Homosexuals, whether they are teachers or students, are harassed and are forced to hide their true selves. The work of creating more inclusive schools is not yet complete.

METHOD

All five of the participants in this study have some connection to each other. Some of the participants have socialized with each other on a regular basis. One of the common characteristics of all the participants is their dedication to Catholic education. All the participants have taught only in Catholic schools for their entire teaching career. Two of the participants have been teaching for only three to five years, while the other three participants have been teaching in Catholic schools for more than fifteen years. All of the participants have identified themselves as either "gay" or "lesbian." Two of the participants are men while the remaining three are women. To protect the identity of the participants, all names used in this study are pseudonyms.

The data was collected using individual structured interviews for each participant. More data was gathered through dialogues that the participants had with one another at the quarterly social gatherings over a period of two years. Three of the five participants in this study are members of the group.

FINDINGS AND DISCUSSION

. . . Homosexual educators who choose to teach in Catholic schools have a unique challenge to be true to themselves while upholding the teachings of the Church. An analysis of the data shows that the Catholic school teachers involved in this study have come to live with the traditional Catholic teaching on homosexuality, and they are working with the system, as opposed to against the system, to create a more inclusive school environment.

Catholicism and Being a Gay or Lesbian Teacher

. . . Catholicism is an important factor that influenced the choice of the participants in this study to work in Catholic schools. All the participants were brought up as Catholics, and they had attended Catholic schools as children. Steve, who has been teaching for two years, says that "I was raised Catholic. It's kind of in me. I can't imagine not being Catholic." At one point in his life, Steve started studying to become a priest. Rose, a participant who has been teaching for more than fifteen years, entered religious life when she was very young. . . . She chose Catholic schools because . . . "I knew what I wanted to do, and all my experience had been with Catholic schools and I didn't really want to teach in public schools." Brett, who has been in Catholic education since 1968, states that his love for the Church is the primary reason for his desire to stay in Catholic schools. Brett clarifies, however: "I know that we do have some differences especially over the issue of homosexuality, but that does not stop me from being a loyal follower of the Church." Liz, who has been in Catholic schools for twenty-three years, says that Catholicism is important to her. She says, "I love teaching religion. I know I am a good role model for my students. I have a healthy attitude towards religion." Rose feels that she does not "have to reject the entire Church because of their position on homosexuality. I don't have a problem disagreeing with certain tenets of the Church and still staying with it. I guess my devotion is more to Christ and the Church is just a human instrument, and a human element that is certainly fallible."

Catholicism has also had a negative effect on one participant. Steve notes that the teachings of the Church made him feel like he was a bad person. Steve says, "I had internalized that I was a bad person and that was why I was afraid to deal with it [being gay]." This negative self-image led to two suicide attempts.

. . . Despite their strong connections to the Catholic Church, the participants in this study acknowledge that they continually experience a conflict between their Church and their homosexuality. Liz acknowledges, "the Catholic Church is still very oppressive with the gay and lesbian folks."

. . . To minimize the conflicts these participants feel with the issues of homosexuality and the Church, they stated that they focus their teaching of religion on the gospel message of loving one another. Liz says that when talking about the issue of homosexuality in a religion class, she tells her students that "God made people this way [gay or lesbian], and we should not make any judgments about it." When Steve teaches religion, he says, "I try to make my primary message that God loves everyone no matter who you are and no matter what you do. Our religion is not about condemning one another. I know I have gay students so I try to protect them as well."

The participants in the study strive for a balance between their acceptance of the teaching of the Church and their homosexuality. Instead of condemning everything about the Church, the participants pull out the positive aspects of the religion so that they are able to encourage their students to be more tolerant of differences.

The Conflict Surrounding Coming Out and Compromising

The gay and lesbian teachers in this study are not comfortable revealing their sexual identity to students, parents, or co-workers. These teachers acknowledge that coming out would mean losing their jobs. They live in constant fear that someone who is not sympathetic or understanding of homosexual issues will discover their secret. Steve says, "My pastor is so by the book. You know, if he's going to fire someone for being pregnant out of wedlock, I know he'll fire me for being gay." Liz shares the same sentiments when she says, "You know, after 23 years, teaching in Catholic schools is a part of my life. But they could take it all away from me if they find out I am a lesbian." Rose acknowledges that she is an employee of the Church and knows that "in a sense, I would be breaking my contract by living the homosexual lifestyle.

. . . The teachers in this study do not feel it is safe to come out to students. When asked whether they would come out to the students, all five participants said they would not. Liz says,

"I wouldn't come out to my students. Someone would have to out me. I wouldn't feel comfortable with that, just because they're at such a [impressionable] stage of their own development. I think it's very confusing. Probably, if I were in a high school, it would not be a big issue. But I think in elementary school and junior high, it is an issue." The teachers in this study, however, acknowledge that there would be some benefit to coming out. Rose believes the primary benefit of coming out to her students would be "letting the students be aware, letting them know someone who is gay, someone who holds the values they hold and worships as they do."

Coming out to the parents of the children can be equally scary for the teachers in this study. The teachers feel that the parents would pull their children out of their classrooms if the parents were uncomfortable having a homosexual teacher work with their child. Many of these teachers acknowledge that the parents of the children in their class probably suspect that they may be homosexual. The teachers who have been teaching for more than ten years are more comfortable with the parents' suspicions about their homosexuality. Liz says that the parents "know pretty much who I am. I think I'm well liked by the parents; I think overall, if someone outed me, the parents would be very supportive of me because they know me as a person, and I am a person of integrity." Teachers who have been teaching for only a few years, however, do not share the sentiments expressed by Liz. Steve says, "I know that some parents know [that I am gay], and the fact that I am not married raises eyebrows."

The teachers in the study do not feel that they would be supported by the administration of their school if they came out. They acknowledge that their principals also have some knowledge of their homosexuality, and that the principals are tolerant. The teachers feel, however, that their principals are tolerant of their homosexuality as long as they do not explicitly come out. Steve says, "My principal probably knows, but she knows she can't get too bent out of shape, or she'll end up not hiring anyone. There's no one that can really live up to the ideal Catholic person that the Church wants most people to be."

. . . Part of the decision to teach in Catholic schools involves compromise. The teachers in the study wish they did not have to hide their sexual identity. The teachers believe, however, that they have a purpose for teaching in Catholic schools. Mary chooses to stay in Catholic schools, because "from the students point of view, I'm providing them with a window that they would not otherwise have." Brett believes that his presence also provides a way for students to understand homosexuality even if this understanding may not come about until many years after a student has left school. Brett says that one of his former students returned to his school many years after she graduated. In their conversation, Brett informed her that he is gay, and she told him, "You are the first gay person I have ever met, at least the first person who has ever told me that he is gay." The student continued: "If being gay is being like you, then I don't understand what the problem is; you are a good teacher, you have taught us a lot, and you have affected my life."

Survival Strategies

. . . Being "Implicitly Out" (as defined by Griffin, 1992) is a major strategy that the teachers identified in the study as helpful to their survival in Catholic education. Throughout the interviews, the participants constantly mentioned that they feel that most people are aware of their homosexuality. A key component in this strategy is that these teachers have at least one confidant in the school who knows explicitly that they are gay or lesbian. Steve had an ally who left the school after the end of the academic year. Steve says: "I needed to let someone else at work know who I really am." Brett believes that "most of my staff knows I am gay. No one asks me about the absence of women in my romantic life. As long as I am not shouting from the rooftops about being gay, no one says anything, and I don't say anything, and that is just how it goes from day to day."

Like the homosexual teachers cited in the review of literature, the teachers in this study also acknowledge that they need to work harder than their heterosexual colleagues. There is a need for them to be model teachers. They make sure that everything is perfectly done. This strategy makes it more difficult for their administration to dismiss them. Liz says: "Oh, they won't get rid of me for being a lesbian. It will be because I missed doing my lesson plan." Brett believes that "gay educators in the Church try harder to prove themselves and to prove their worth more than a person who doesn't have to overcome the discrimination associated with homosexuality."

Creating or finding a supportive community is an essential survival strategy, according to the teachers in this study. A supportive community is one that encourages homosexual teachers to discuss their concerns and validates their decision to teach in a Catholic school. Rose says that creating a support network is important. Brett believes that "we must have one another to support." Finding the support group, however, can be difficult. The participants, who were part of the support group described earlier in this study, found it difficult to tell other gay and lesbian teachers of the presence of the group. Its members perceived the group itself as a covert operation. In some way, despite their acceptance of their sexual identity, many of the participants continued to live in fear.

CONCLUSIONS

Despite all the problems that these teachers face when they work in Catholic schools, they all choose to stay. These gay and lesbian educators hope to create change within the institution by staying in the institution. Brett says: "You can never initiate change within the organization unless you are part of the organization. You know, if I stand outside of the organization and throw stones, I end up just putting up more barriers. But by being within the institution, you can create more change."

. . . The results from this study show that there is a marginalized community within the larger Catholic school community. Catholic schools have always been credited for creating strong community ties. But, are we truly creating a just community if some people feel excluded because they are homosexual? More research also needs to be conducted on homosexual men and women who serve as priests, brothers, or nuns in Catholic schools. The present study limited itself to including the perceptions of only lay homosexual teachers. Furthermore, there are teachers in Catholic schools, especially secondary schools, who are open about their sexual identity. It would be interesting to investigate how these teachers are able to be sincere about their sexual identity in a Catholic school.

. . . In time, society will acknowledge that homosexual teachers have contributed and will continue to contribute to the educational mission of the Church. When that day comes, homosexual teachers will be able to live the hope that Mary, a participant in the study, expresses: "I just wish I didn't have to be cautious about my being a lesbian. I just wish I didn't need to lie about it. I wish I could participate in the full life of the parish and just be who I am." In the future, I look forward to the day that no one will ask the same question that the young student asked in the beginning of this paper. Or, if a homosexual teacher were asked, "What does a gay person look like? Please tell me because I have never really seen one," he or she will not experience fear in fully answering the question.

REFERENCES

Blumenfeld, W.J. (Ed.). (1992). *Homophobia: How we all pay the price*. Boston: Beacon Press.

Brown R. M. (1993). *Liberation theology: An introductory guide*. Louisville, KY: Westminster/John Knox Press.

Coleman, G. D. (1997). The teacher and the gay and lesbian student. *Momentum,* April/May, 46–48.

Congregation of the Doctrine of the Faith (1986). *Letter to the bishops of the Catholic Church on the pastoral care of homosexual persons.* World Wide Web: http://www.hli.org/issues/homosexuality/church/doc1.html.

Goodman, J. M. (1996). Lesbian, gay, and bisexual issues in education: A personal view. In D. R. Walling (Ed.). *Open lives, safe schools: Addressing gay and lesbian issues in education* (pp. 9–16). Bloomington, IN: Phi Delta Kappa.

Griffin, P. (1992). From hiding out to coming out: Empowering lesbian and gay educators. In K. M. Harbeck (Ed.). *Coming out of the classroom closet: Gay and lesbian students, teachers, and curricula* (pp. 167–197). Binghamton, NY: Harrington Park Press.

Harbeck, K. M. (1997). *Gay and lesbian educators: Personal freedoms, public constraints.* Malden, MA: Amethyst Press and Productions.

Jennings, K. (Ed.) (1994). *One teacher in 10: Gay and lesbian educators tell their stories.* Boston: Alyson Publications, Inc.

Kissen, R. M. (1996). *The last closet: The real lives of lesbian and gay teachers.* Portsmouth, NH: Heinemann.

National Conference of Catholic Bishops (1997). *Always our children: A pastoral message to parents of homosexual children and suggestions for pastoral ministers.* Washington, DC: United States Catholic Conference.

Newman, L. (author), & Souza, D. (illustrator) (1991). *Heather has two mommies.* Los Angeles: Alyson Wonderland Books.

Pekman, J. H. (1997). *Gay and lesbian educators' reflections on their experiences of oppression in the San Francisco Bay Area schools: A participatory research.* Unpublished doctoral dissertation, University of San Francisco.

Prince, T. (1996). The power of openness and inclusion in countering homophobia in schools. In D. R. Walling (Ed.). *Open lives, safe schools: Addressing gay and lesbian issues in education* (pp. 29–34). Bloomington, IN: Phi Delta Kappa.

Rowe, R. N. (1993). Are we educators who are homosexual recruiting youth? *Education, 113*(3), 508–511.

Stewig, J. W. (1996). Self-censorship of picture books about gay and lesbian families. In D. R. Walling (Ed.). *Open lives, safe schools: Addressing gay and lesbian issues in education* (pp. 71–80). Bloomington, IN: Phi Delta Kappa.

Willhoite, M. (1991). *Daddy's roommate.* Los Angeles: Alyson Wonderland Books.

Woog, D. (1995). *School's out: The impact of gay and lesbian issues on America's schools.* Boston: Alyson Publications, Inc.

Yip, A. K. T. (1997). Dare to differ: Gay and lesbian Catholics' assessment of official Catholic positions on sexuality. *Sociology of Religion 58*(2), 165–180.

Zapulla, C. (1997). *Suffering in silence: Teachers with AIDS and the moral school community.* New York: Peter Lang Publishing.

DISCUSSION QUESTIONS

1. Do you believe a day will come when gay or lesbian teachers will not be apprehensive about responding to a student who asks what a gay teacher looks like? Why or why not?
2. What are the dangers and costs of "passing" (acting heterosexual when you are not)? What are the benefits? What are the dangers and costs of "coming out" (letting others know your sexual preference/orientation)? What are the benefits? How would you handle a conflict like this (whether or not to "come out") if you were in this position (a practicing Catholic teacher working for a Catholic school)?
3. What kinds and forms of sexuality and sexual behavior do most religious institutions privilege? What similarities and differences do you see between various religions and religious denominations? What kinds of privilege derive from their shared norms and expectations; whose interests are privileged?
4. Is it harmful or beneficial for children to know an adult authority figure's sexuality? How much legitimacy do you give to the argument that gay and lesbian teachers can "turn" children gay? Why? How important is it for youth with alternate sexualities to have positive gay, lesbian, bisexual, or transgendered role models such as teachers? Why is it that the private sexual activities of heterosexuals are not similarly relevant, or are they? Are your feelings on these issues the same or different if the teachers in question teach junior high students? High school students? College students? Why?

The following Bonus Reading can be accessed through the *InfoTrac College Edition Online Library* by typing in the name of the author or keywords in the title online at *http://www.infotrac-college.com.*

Identity Puzzles: Talking Sex in Education

MARGOT FRANCIS

Drawing on historical and cross-cultural examples, Margot Francis offers a concrete example of the social construction of sexual behavior and identity. She examines the dangers of relying on dichotomous indicators of sexual identity and calls on educators to challenge the heterosexual/homosexual dichotomy in the classroom.

DISCUSSION QUESTIONS

1. In what ways are the terms we use for sexual identity insufficient to capture the differences in sexual expression?
2. Is it important for educators to include examples, such as those provided by the author, that point to different historical and cultural constructions of sexual identity? Why or why not?
3. Why is so much energy spent on the issue of whether or not sexual behavior is a choice? Why is so much of the energy spent on this issue focused on homosexual rather than heterosexual behavior?
4. What kind of sex education is best for children? Who should teach children about sexuality? When should they be taught? What should be taught? What are the dangers of abstinence-only education? What are its benefits? What is the responsibility of schools regarding sex education? What are the limits of this responsibility?

1 0

Family and Medicine

Dubious Conceptions: The Controversy over Teen Pregnancy

KRISTIN LUKER

Teen pregnancy emerged as a significant social issue during the decade of the 1970s. Kristen Luker contends that the impetus for public concern did not derive from increased teen births, but rather from a dramatic increase in the number of adolescent mothers (especially white *mothers) who chose to forgo marriage. Her analysis offers fresh insights into the relationship between teen pregnancy, poverty, and public assistance.*

The conventional wisdom has it that an epidemic of teen pregnancy is today ruining the lives of young women and their children and perpetuating poverty in America. In polite circles, people speak regretfully of "babies having babies." Other Americans are more blunt. "I don't mind paying to help people in need," one angry radio talk show host told Michael Katz, a historian of poverty, "but I don't want my tax dollars to pay for the sexual pleasure of adolescents who won't use birth control."

By framing the issue in these terms, Americans have imagined that the persistence of poverty and other social problems can be traced to youngsters who are too impulsive or too ignorant to postpone sexual activity, to use contraception, to seek an abortion, or failing all that, especially if they are white, to give their babies up for adoption to "better" parents. Defining the problem this way, many Americans, including those in a position to influence public policy, have come to believe that one attractive avenue to reducing poverty and other social ills is to reduce teen birth rates. Their remedy is to persuade teenagers to postpone childbearing, either by convincing them of the virtues of chastity (a strategy conservatives prefer) or by making abortion, sex education, and contraception more freely available (the strategy liberals prefer).

Reducing teen pregnancy would almost certainly be a good thing. After all, the rate of teen childbearing in the United States is more similar

From "Dubious Conceptions: The Controversy over Teen Pregnancy" by Kristin Luker.
Reprinted with permission from *The American Prospect,* Number 5: Spring 1991. The American Prospect, 5 Broad Street, Boston, MA 02109.

to the rates prevailing in the poor countries of the world than in the modern, industrial nations we think of as our peers. However, neither the problem of teen pregnancy nor the remedies for it are as simple as most people think.

In particular, the link between poverty and teen pregnancy is a complicated one. We do know that teen mothers are poorer than women who wait past their twentieth birthday to have a child. But stereotypes to the contrary, it is not clear whether early motherhood causes poverty or the reverse. Worse yet, even if teen pregnancy does have some independent force in making teen parents poorer than they would otherwise be, it remains to be seen whether any policies in effect or under discussion can do much to reduce teen birth rates.

These uncertainties raise questions about our political culture as well as our public choices. How did Americans become convinced that teen pregnancy is a major cause of poverty and that reducing one would reduce the other? The answer is a tale of good intentions, rising cultural anxieties about teen sex and family breakdown, and the uses—and misuses—of social science.

HOW TEEN PREGNANCY BECAME AN ISSUE

Prior to the mid-1970s, few people talked about "teen pregnancy." Pregnancy was defined as a social problem primarily when a woman was unmarried; no one thought anything amiss when an 18- or 19-year-old got married and had children. And concern about pregnancies among unmarried women certainly did not stop when the woman turned twenty.

But in 1975, when Congress held the first of many hearings on the issue of adolescent fertility, expert witnesses began to speak of an "epidemic" of a "million pregnant teenagers" a year. Most of these witnesses were drawing on statistics supplied by the Alan Guttmacher Institute, which a year later published the data in an influential booklet, *Eleven Million Teenagers.* Data from that document were later cited—often down to the decimal point—in most discussions of the teenage pregnancy "epidemic."

Many people hearing these statistics must have assumed that the "million pregnant teenagers" a year were all unmarried. The Guttmacher Institute's figures, however, included married 19-year-olds along with younger, unmarried teenage girls. In fact, almost two-thirds of the "million pregnant teenagers" were 18- and 19-year-olds; about 40 percent of them were married, and about two-thirds of the married women were married prior to the pregnancy.

Moreover, despite the language of epidemic, pregnancy rates among teenagers were not dramatically increasing. From the turn of the century until the end of World War II, birth rates among teenagers were reasonably stable at approximately 50 to 60 births per thousand women. Teen birth rates, like all American birth rates, increased dramatically in the period after World War II, doubling in the baby boom years to a peak of about 97 births per thousand teenaged women in 1957. Subsequently, teen birth rates declined, and by 1975 they had gone back down to their traditional levels, where, for the most part, they have stayed (see Figure 1).

Were teen births declining in recent decades only because of higher rates of abortion? Here, too, trends are different from what many people suppose. The legalization of abortion in January of 1973 made it possible for the first time to get reliable statistics on abortions for women, teenagers and older. The rate among teenagers rose from about 27.0 to 42.9 abortions per 1,000 women between 1974 and 1980. Since 1980 teen abortion rates have stabilized, and may even have declined somewhat. Moreover, teenagers account for a declining proportion of all abortions: in the years just after *Roe v. Wade,* teenagers obtained almost a third of all abortions in the country; now they obtain about a quarter. A stable teen birth rate and a stabilizing teen abortion rate means that pregnancy rates, which rose modestly in the 1970s, have in recent years leveled off.

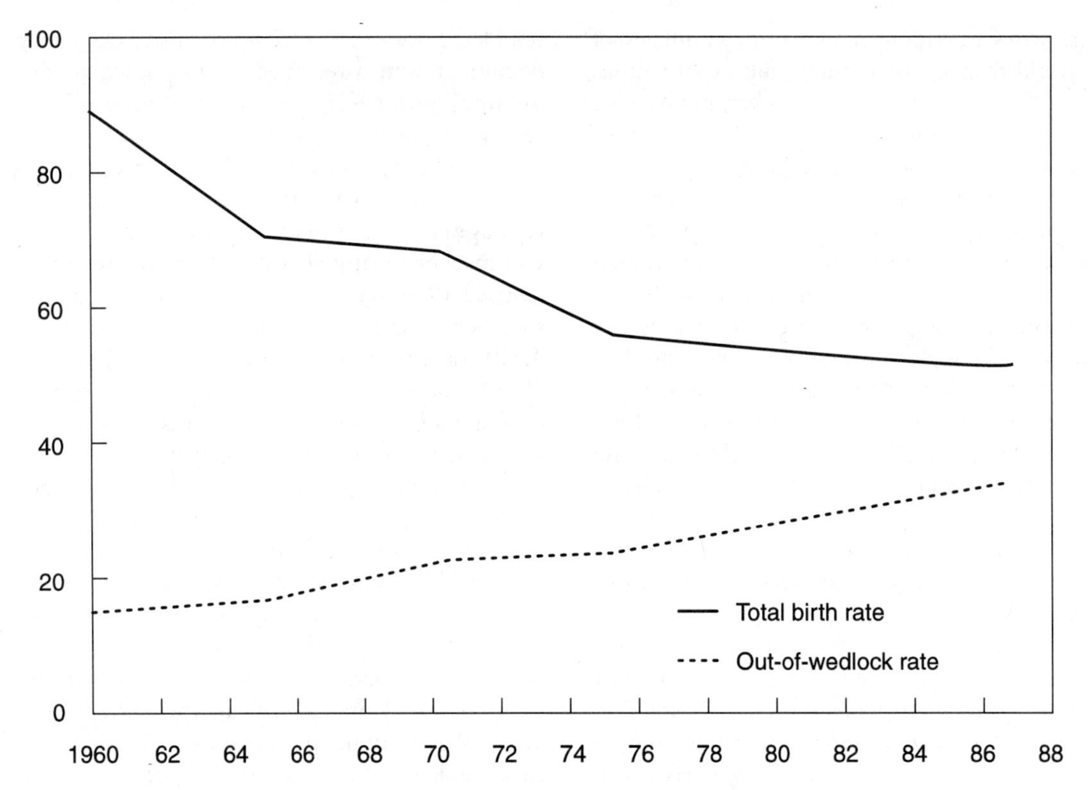

FIGURE 1 **Trends in Teen Birth Rates Total birth rate equals births per 1,000 women, ages 15–19. Out-of-wedlock rate equals births per 1,000 unmarried women, ages 15–19.**

Sources: National Center for Health Statistics, *Annual Vital Statistics,* and *Monthly Vital Statistics Reports;* U.S. DHEW, Vital and Health Statistics, "Trends in Illegitimacy, U.S. 1940–1965."

What has been increasing—and increasing dramatically—is the percentage of teen births that are out-of-wedlock (Figure 1). In 1970 babies born out of wedlock represented about a third of all babies born to teen mothers. By 1980 out-of-wedlock births were about half; and by 1986 almost two-thirds. Beneath these overall figures lie important racial variations. Between 1955 and 1988 the out-of-wedlock rate rose from 6 to 24.8 per thousand unmarried, teenage, white women, while for unmarried, nonwhite teenagers the rate rose from 77.6 to 98.3 per thousand. In other words, while the out-of-wedlock birth rate was rising 25 percent among nonwhite teens, it was actually quadrupling among white teens.

The immediate source for this rise in out-of-wedlock teen pregnancy might seem to be obvious. Since 1970 young women have increasingly postponed marriage without rediscovering the virtues of chastity. Only about 6 percent of teenagers were married in 1984, compared to 12 percent in 1970. And although estimates vary, sexual activity among single teenagers has increased sharply, probably doubling. By 1984 almost half of all American teenage women were both unmarried and sexually active, up from only one in four in 1970.

Yet the growth of out-of-wedlock births has not occurred only among teens; in fact, the increase has been more rapid among older women. In 1970 teens made up almost half of all out-of-wedlock births in America; at present they account for a little less than a third. On the other hand, out-of-wedlock births represent a much larger percentage of births to teens than of births to older women. Perhaps for that reason, teenagers have become the symbol of a problem that, to many Americans, is "out of control."

Whatever misunderstandings may have been encouraged by reports of a "million pregnant teenagers" a year, the new concept of "teen pregnancy" had a remarkable impact. By the mid-1980s, Congress had created a new federal office on adolescent pregnancy and parenting; 23 states had set up task forces; the media had published over 200 articles, including cover stories in both *Time* and *News-week*; American philanthropy had moved teen pregnancy into a high priority funding item; and a 1985 Harris poll showed that 80 percent of Americans thought teen pregnancy was a "serious problem" facing the nation, a concern shared across racial, geographic, and economic boundaries.

But while this public consensus has been taking shape, a debate has emerged about many of its premises. A growing number of social scientists have come to question whether teen pregnancy causes the social problems linked to it. Yet these criticisms have at times been interpreted as either an ivory-tower indifference to the fate of teen parents and their babies or a Panglossian optimism that teen childbearing is just one more alternate lifestyle. As a result, clarity on these issues has gotten lost in clouds of ideological mistrust. To straighten out these matters, we need to understand what is known, and not known, about the relation of teen pregnancy to poverty and other social problems.[1]

DISTINGUISHING CAUSES FROM CORRELATIONS

As the Guttmacher Institute's report made clear, numerous studies have documented an association between births to teenagers and a host of bad medical and social outcomes. Compared to women who have babies later in life, teen mothers are in poorer health, have more medically treacherous pregnancies, more still-births and newborn deaths, and more low-birthweight and medically compromised babies.

Later in life, women who have babies as teenagers are also worse off than other women. By their late 20s, women who gave birth as teenagers are less likely to have finished high school and thus not to have received any subsequent higher education. They are more likely to have routine, unsatisfying, and dead-end jobs, to be on welfare, and to be single parents either because they were never married or their marriage ended in divorce. In short, they often lead what the writer Mike Rose has called "lives on the boundary."

Yet an interesting thing has happened over the last twenty years. A description of the lives of teenage mothers and their children was transmuted into a causal sequence, and the often-blighted lives of young mothers were assumed to flow from their early childbearing. Indeed, this is what the data would show, if the women who gave birth as teenagers were the same in every way as women who give birth later. But they are not.

Although there is little published data on the social origins of teen parents, studies have documented the effects of social disadvantage at every step along the path to teenage motherhood. First, since poor and minority youth tend to become sexually active at an earlier age than more advantaged youngsters, they are "at risk" for a longer period of time, including years when they are less cognitively mature. Young teens are also less likely to use contraceptives than older teenagers. Second, the use of contraception is more common among teens who are white, come from more affluent homes, have higher educational aspirations, and who are doing well in school. And, finally, among youngsters who become pregnant, abortions and more common if they are affluent, white, urban, of higher socioeconomic status get good grades, come from two-parent families, and aspire to higher education. Thus, more advantaged youth get filtered out of the pool of young women at risk of teen parenthood.

Two kinds of background factors influence which teens are likely to become pregnant and give birth outside of marriage. First is inherited disadvantage. Young women from families that are poor, or rural, or from a disadvantaged minority, or headed by a single parent are more likely to be teen mothers than are their counterparts from more privileged backgrounds. Yet young mothers are not just disadvantaged; they are also discouraged. Studies suggest that a young woman who has other troubles—who is not doing well in school, has lower "measured ability," and lacks high aspirations for herself—is also at risk of becoming a teenaged mother.

Race plays an independent part in the route to teen motherhood. Within each racial group, according to Linda Waite and her colleagues at the Rand Corporation, teen birth rates are highest for those who have the greatest economic disadvantage and lowest academic ability. The effects of disadvantage, however, vary depending on the group. The Rand study found that among young high-ability, affluent black women from homes with two parents, only about one in a hundred become single, teenage mothers. For comparable whites, the risk was one in a thousand. By contrast, a poor, black teenager from a female-headed household who scores low on standardized tests has an astonishing one in four chance of becoming an unwed mother in her teens. Her white counterpart has one chance in twelve. Unwed motherhood thus reflects the intersecting influences of race, class and gender; race and class each has a distinct impact on the life histories of young women.

Since many, if not most, teenage unwed mothers are already both disadvantaged and discouraged before they get pregnant, the poor outcomes of their pregnancies as well as their later difficulties in life are not surprising. Consider the health issues. As the demographer Jane Menken pointed out some time ago (and as many other studies have corroborated), the medical complications associated with teen pregnancy are largely due not to age but to the poverty of young mothers. As poor people, they suffer not from some biological risk due to youth, but from restricted access to medical care, particularly to prenatal care. (To be fair, some research suggests that there may be special biological risks for the very youngest mothers, those under age 15 when they give birth, who constitute about 2 percent of all teen mothers.)

Or, to take a more complicated example, consider whether bearing a child blocks teenagers from getting an education. In the aggregate, teen mothers do get less education than women who do not have babies at an early age. But teen mothers are different from their childless peers along exactly those dimensions we would expect independently to contribute to reduced schooling. More of them are poor, come from single-parent households, and have lower aspirations for themselves, lower measured ability, and more problems with school absenteeism and discipline. Given the nature of the available data, it is difficult to sort out the effects of a teen birth apart from the personal and social factors that predispose young women to both teen motherhood and less education. Few would argue that having a baby as a teenager enhances educational opportunities, but the exact effect of teen birth is a matter of debate.

Educational differences between teen mothers and other women may also be declining, at least in terms of graduating from high school. Legislation that took effect in 1975 forbade schools to expel pregnant teens. Contrary to current skepticism about federal intervention, this regulation seems to have worked. According to a study by Dawn Upchurch and James McCarthy, only 18.6 percent of teenagers who had a baby in 1958 subsequently graduated from high school. Graduation rates among teen mothers reached 29.2 percent in 1975; by 1986 they climbed to 55 percent. Teen mothers were still not graduating at a rate equal to other women (as of 1985, about 87 percent of women ages 25 to 29 had a high school diploma or its equivalent). But over the decade prior to 1986, graduation rates had increased more quickly for teen mothers than for other women, suggesting that federal policies tailored to their special circumstances may have made a difference.

Since education is so closely tied to later status, teasing out the relationship between teen pregnancy and schooling is critical. The matter is complicated, however, because young people do many things simultaneously, and sorting out the order is no easy task. In 1984 Peter Morrison of the Rand team reported that between a half and a third of teen mothers in high school and beyond dropped out before they got pregnant. Upchurch and McCarthy, using a different and more recent sample, found that the majority of female dropouts in their study left school before they got pregnant and that teens who got pregnant while still in school were not particularly likely to drop out. On the other hand, those teens who first drop out and then get pregnant are significantly less likely to return to school than other dropouts who do not get pregnant. Thus the conventional causal view that teens get pregnant, drop out of school, and as a result end up educationally and occupationally disadvantaged simply does not match the order of events in many people's lives.

THE SEXUAL ROOTS OF PUBLIC ANXIETY

Teen pregnancy probably would not have "taken off" as a public issue quite so dramatically, were it not for the fact it intersects with other recent social changes in America, particularly the emergence of widespread, anxiety-producing shifts in teen sex. Academics debate whether there has been a genuine "sexual revolution" among adults, but there is no doubt in regard to teenagers. Today, by the time American teenagers reach age 20, an estimated 70 percent of the girls and 80 percent of the boys have had sexual experiences outside of marriage. Virtually all studies confirm that this is a dramatic historical change, particularly for young women. (As usual, much less is known about the historical experiences of young men.) For example, Sandra Hofferth and her colleagues, using nationally representative data from the 1982 National Survey of Family Growth, found that women navigating adolescence in the late 1950s had a 38.9 percent chance of being sexually active before marriage during their teenage years. Women who reached their twentieth birthday between 1979 and 1981, in contrast, had a 68.3 percent likelihood.

Yet even these statistics do not capture how profoundly different this teen sexuality is from that of earlier eras. As sources such as the Kinsey Report (1953) suggest, premarital sex for many American women before the 1960s was "engagement" sex. The women's involvement, at least, was exclusive, and she generally went on to marry her partner in a relatively short period of time. Almost half of the women in the Kinsey data who had premarital sex had it only with their fiances.

But as the age at first marriage has risen and the age at first intercourse has dropped, teen sexuality has changed. Not surprisingly, what scattered data we have about numbers of partners suggest that as the period of sexual activity before marriage has increased, so has the number of partners. In 1971, for example, almost two-thirds of sexually active teenaged women in metropolitan areas had had only one sexual partner; by 1979 fewer than half did. Data from the 1988 National Survey of Family Growth confirm this pattern for the nation as a whole, where about 60 percent of teens have had two or more partners. Similarly, for metropolitan teens, only a small fraction (about 10 percent) were engaged at the time of their first sexual experience, although about half described themselves as "going steady."

Profound changes in other aspects of American life have complicated the problem. Recent figures suggest that the average age at first marriage has increased to almost 24 years for women and over 25 years for men, the oldest since reliable data have been collected. Moreover, the age of sexual maturity over the last century has decreased by a little under six months each decade owing to nutritional and other changes. Today the average American girl has her first menstrual period at age 12½, although there are wide individual variations. (There is less research on the sexual maturity of young men.) On average, consequently, American girls and their boyfriends face over a decade of their lives when they are sexually mature and single.

As teenagers pass through this reproductive minefield; the instructions they receive on how to conduct themselves sexually are at best mixed. At least according to public opinion polls, most Americans have come, however reluctantly, to accept premarital sex. Yet one suspects that what they approve is something closer to Kinsey-era sex: sexual relations en route to a marriage. Present-day teenage sex, however, starts for many young people not when they move out of the family and into the orbit of what will be a new family or couple, but while they are still defined primarily as children.

When young people, particularly young women, are still living at home (or even at school) under the control, however nominal, of parents, sexual activity raises profound questions for adults. Many Americans feel troubled about "casual" sex, that is, sex which is not intimately tied to the process by which people form couples and settle down. Yet many teenagers are almost by definition disqualified as too young to "get serious." Thus the kinds of sexuality for which they are socially eligible—sex based in pleasure, not procreation, and in short-term relationships rather than as a prelude to marriage—challenge fundamental values about sexuality held by many adults. These ambiguities and uncertainties have given rise to broad anxieties about teen sexuality that have found expression in the recent alarm about teen pregnancy.

RAISING CHILDREN WITHOUT FATHERS

While Americans have had to confront the meaning and purpose of sexuality in the lives of teenagers, a second revolution is forcing them to think about the role—and boundaries—of marriage and family. Increasingly for Americans, childbearing and, more dramatically, childrearing have been severed from marriage. The demographer Larry Bumpass and his colleagues have estimated that under present trends; half or more of all American children will spend at least part of their childhood in a single-parent (mainly mother-only) family, due to the fact that an estimated 60 percent of recent marriages will end in divorce.

At the same time, as I indicated earlier, out-of-wedlock births are on the rise. At present, 26 percent of all births are to single women. If present trends continue, Bumpass and others estimate, almost one out of every six white women and seven out of ten black women will give birth to a child without being married. In short, single childbearing is becoming a common pattern of family formation for all American women, teenagers and older.

This reality intersects with still another fact of American life. The real value of inflation-adjusted wages, which grew 2.5 to 3.0 percent a year from the end of World War II to at least 1973, has now begun to stagnate and for certain groups decline; some recent studies point to greater polarization of economic well-being. Americans increasingly worry about their own standard of living and their taxes, and much of that worry has focused on the "underclass." Along with the elderly and the disabled, single women and their children have been the traditional recipients of public aid in America. In recent years, however, they have become especially visible among the dependent poor for at least two reasons. First, the incomes of the elderly have improved, leaving behind single mothers as a higher percentage of the poor; and second, the number of female-headed households has increased sharply. Between 1960 and 1984, households headed by women went from 9.0 percent to 12.0 percent of all white households, and from 22.0 percent to 43 percent of all black households. The incomes of about half of all families headed by women, as of 1984, fell below federal poverty levels.

Raising children as a single mother presents economic problems for women of all ages, but the problem is especially severe for teenagers with limited education and job experience. Partly for that reason, teenagers became a focus of public concern about the impact of illegitimacy and single parenthood on welfare costs. Data published in the 1970s and replicated in the 1980s suggested that about half of all families supported by Aid to Families with Dependent Children (AFDC) were started while the mother was still a teenager. One estimate calculated that in 1975 the costs for these families of public assistance alone (not including Medicaid or food stamps) amounted to $5 billion; by 1985, that figure increased to $8.3 billion.

Yet other findings—and caveats—have been ignored. For example, while about half of all AFDC cases may be families begun while the woman was still a teenager, teens represent only about 7 percent of the caseload at any one time. Moreover, the studies assessing the welfare costs of families started by teens counted any welfare family as being the result of a teen birth if the woman first had a child when under age 20. But, of course, that same woman—given her prior circumstances—might have been no less likely to draw welfare assistance if, let us say, she had a baby at age 20 instead of 19. Richard Wertheimer and Kristin Moore, the source of much of what we know about this area, have been careful to note that the relevant costs are the marginal costs—namely, how much less in welfare costs society would pay if teen mothers postponed their first births, rather than foregoing them entirely.

It turns out, not surprisingly, that calculated this way, the savings are more modest. Wertheimer and Moore have estimated that if by some miracle we could cut the teen birth rate in half, welfare costs would be reduced by 20 percent, rather than 50 percent, because many of these young women would still need welfare for children born to them when they were no longer teens.

Still other research suggests that most young women spend a transitional period on welfare, while finishing school and entering the job market. Other data also suggest that teen mothers may both enter and leave the welfare ranks earlier than poor women who postpone childbearing. Thus teen births by themselves may have more of an effect on the timing of welfare in the chain of life events than on the extent of welfare dependency. In a study of 300 teen mothers and their children originally interviewed in the mid-1960s, Frank Furstenberg and his colleagues found seventeen years later that two-thirds of those followed up had received no welfare in the previous five years, although some 70 percent of them had received public assistance at some point after the birth of their child. A quarter had achieved middle-class incomes, despite their poverty at the time of the child's birth.

None of this is to deny that teen mothers have a higher probability of being on welfare in the first place than women who begin their families at a later age, or that teen mothers may be disproportionately represented among those who find themselves chronically dependent on welfare. Given the disproportionate number of teen mothers who come from socially disadvantaged origins (and who are less motivated and perhaps less able students), it would be surprising if they were not overrepresented among those needing public assistance, whenever they had their children. Only if we are prepared to argue that these kinds of women should never have children—which is the implicit alternative at the heart of much public debate—could we be confident that they would never enter the AFDC rolls.

RETHINKING TEEN PREGNANCY

The original formulation of the teen pregnancy crisis seductively glossed over some of these hard realities. Teen motherhood is largely the province of those youngsters who are already disadvantaged by their position in our society. The major institutions of American life—families, schools, job markets, the medical system—are not working for them. But by framing the issue as teenage pregnancy, Americans could turn this reality around and ascribe the persistence of poverty and other social ills to the failure of individual teenagers to control their sexual impulses.

Framing the problem as teen pregnancy, curiously enough, also made it appear universal. Everyone is a teenager once. In fact, the rhetoric has sometimes claimed that the risk of teen pregnancy is universal, respecting no boundaries of class or race. But clearly, while teenage pregnancies do occur in virtually all walks of life, they do not occur with equal frequency. The concept of "teen pregnancy" has the advantage, therefore, of appearing neutral and universal while, in fact, being directed at people disadvantaged by class, race, and gender.

If focusing on teen pregnancy cast the problem as deceptively universal, it also cast the solution as deceptively simple. Teens just have to wait. In fact, the tacit subtext of at least some of the debate on teen pregnancy is not that young women should wait until they are past their teens, but until they are "ready." Yet in the terms that many Americans have in mind, large numbers of these youngsters will never be "ready." They have already dropped out of school and will face a marginal future in the labor market whether or not they have a baby. And as William J. Wilson has noted, many young black women in inner-city communities will not have the option of marrying because of the dearth of eligible men their age as a result of high rates of unemployment, underemployment, imprisonment, and early death.

Not long ago, Arline Geronimous, an assistant professor of public health at the University of Michigan, caused a stir when she argued that teens, especially black teens, had little to gain (and perhaps something to lose) in postponing pregnancy. The longer teenagers wait, she noted, the more they risk ill health and infertility, and the less likely their mothers are to be alive and able to help rear a child of theirs. Some observers quickly took Geronimous to mean that teen mothers are "rational," affirmatively choosing their pregnancies.

Yet, as Geronimous herself has emphasized, what sort of choices do these young women have? While teen mothers typically report knowing about contraception (which they often say they have used) and knowing about abortion, they tell researchers that their pregnancies were unplanned. In the 1988 National Survey of Family Growth, for example, a little over 70 percent of the pregnancies to teens were reported as unplanned; the teenagers described the bulk of these pregnancies as wanted, just arriving sooner than they had planned.

Researchers typically layer their own views on these data. Those who see teens as victims point to the data indicating most teen pregnancies are unplanned. Those who see teens as acting rationally look at their decisions not to use contraceptives or seek an abortion. According to Frank Furstenberg, however, the very indecisiveness of these young people is the critical finding. Youngsters often drift into pregnancy and then into parenthood, not because they affirmatively choose pregnancy as a first choice among many options, but rather because they see so few satisfying alternatives. As Laurie Zabin, a Johns Hopkins researcher on teen pregnancy, puts it, "As long as people don't have a vision of the future which having a baby at a very early age will jeopardize, they won't go to all the lengths necessary to prevent pregnancy."

Many people talk about teen pregnancy as if there were an implicit social contract in America. They seem to suggest that if poor women would just postpone having babies until they were past their teens, they could have better lives for themselves and their children. But for teenagers already at the margins of American life, this is a contract that American society may be hard put to honor. What if, in fact, they are acting reasonably? What can public policy do about teen pregnancy if many teenagers drift into childbearing as the only vaguely promising option in a life whose options are already constrained by gender, poverty, race, and failure?

The trouble is that there is little reason to think any of the "quick fixes" currently being proposed will resolve the fundamental issues involved. Liberals, for example, argue that the answer is more access to contraception, more readily available abortion, and more sex education. Some combination of these strategies probably has had some effect on teen births, particularly in keeping the teen pregnancy rate from soaring as the number of sexually active teens increased. But the inner logic of this approach is that teens and adults have the same goal: keeping teens from pregnancies they do not want. Some teens, however, do want their pregnancies, while others drift into pregnancy and parenthood without ever actively deciding what they want. Consequently, increased access to contraceptives, sex education, and abortion services are unlikely to have a big impact in reducing their pregnancies.

Conservatives, on the other hand, often long for what they imagine was the traditional nuclear family, where people had children only in marriage, married only when they could prudently afford children, and then continued to provide support for their children if the marriage ended. Although no one fully understands the complex of social, economic, and cultural factors that brought us to the present situation, it is probably safe to predict that we shall not turn the clock back to that vision, which in any event is highly colored by nostalgia.

This is not to say that there is nothing public policy can do. Increased job opportunities for both young men and young women; meaningful job training programs (which do not slot young women into traditional low-paying women's jobs); and child support programs[2] would all serve either to make marriage more feasible for those who wish to marry or to support children whose parents are not married. But older ages at first marriage, high rates of sex outside of marriage, a significant portion of all births out of wedlock, and problems with absent fathers tend to be common patterns in Western, industrialized nations.

In their attempts to undo these patterns, many conservatives propose punitive policies to sanction unmarried parents, especially unmarried mothers, by changing the "incentive structure" young people face. The new welfare reform bill of 1988, for example, made it more difficult for teens to set up their own households, at least in part because legislators were worried about the effects of welfare on the willingness to have a child out of wedlock. Other, more draconian writers have called for the children of unwed teen parents to be forcibly removed and placed into foster care, or for the reduction of welfare benefits for women who have more than one child out of wedlock.

Leave aside, for the moment, that these policies would single out only the most vulnerable in this population. The more troublesome issue is such policies often fall most heavily on the children. Americans, as the legal historian Michael Grossberg has shown, have traditionally and justifiably been leery of policies that regulate adult behavior at children's expense.

The things that public policy could do for these young people are unfortunately neither easy to implement nor inexpensive. However, if teens become parents because they lack options, public policy towards teen pregnancy and teenage childbearing will have to focus on enlarging the array of perceived options these young people face. And these must be changes in their real alternatives. Programs that seek to teach teens "future planning," while doing nothing about the futures they can expect, are probably doomed to failure.

We live in a society that continues to idealize marriage and family as expected lifetime roles for women, even as it adds on the expectation that women will also work and be self-supporting. Planning for the trade-offs entailed in a lifetime of paid employment in the labor market and raising a family taxes the skills of our most advantaged young women. We should not be surprised that women who face discrimination by race and class in addition to that of gender are often even less adept at coping with these large and contradictory demands.

Those who worry about teenagers should probably worry about three different dangers as Americans debate policies on teen pregnancy. First, we should worry that things will continue as they have and that public policy will continue to see teens as unwitting victims, albeit victims who themselves cause a whole host of social ills.

The working assumption here will be that teens genuinely do not want the children that they are having, and that the task of public policy is to meet the needs of both society and the women involved by helping them not to have babies. What is good for society, therefore, is good for the individual woman.

This vision, for all the reasons already considered, distorts current reality, and as such, is unlikely to lower the teen birth rate significantly, though it may be effective in keeping the teen birth rate from further increasing. To the extent that it is ineffective, it sets the stage for another risk.

This second risk is that the ineffectiveness of programs to lower teen pregnancy dramatically may inadvertently give legitimacy to those who want more punitive control over teenagers, particularly minority and poor teens. If incentives and persuasion do not lead teenagers to conduct their sexual and reproductive lives in ways that adults would prefer, more coercive remedies may be advocated. The youth of teen mothers may make intrusive social control seem more acceptable than it would for older women.

Finally, the most subtle danger is that the new work on teen pregnancy will be used to argue that because teen pregnancy is not the linchpin that holds together myriad other social ills, it is not a problem at all. Concern about teen pregnancy has at least directed attention and resources to young, poor, and minority women; it has awakened many Americans to their diminished life chances. If measures aimed at reducing teen pregnancy are not the quick fix for much of what ails American society, there is the powerful temptation to forget these young women altogether and allow them to slip back to their traditional invisible place in American public debate.

Teen pregnancy is less about young women and their sex lives than it is about restricted horizons and the boundaries of hope. It is about race and class and how those realities limit opportunities for young people. Most centrally, however, it is typically about being young, female, poor, and non-white and about how having a child seems to be one of the few avenues of satisfaction, fulfillment, and self-esteem. It would be a tragedy to stop worrying about these young women—and their partners—because their behavior is the measure rather than the cause of their blighted hopes.

NOTES

1. Teen pregnancy affects both young men and young women, but few data are gathered on young men. The availability of data leads me to speak of "teen mothers" throughout this article, but it is important to realize that this reflects an underlying, gendered definition of the situation.
2. Theda Skocpol, "Sustainable Social Policy: Fighting Poverty Without Poverty Programs," *TAP,* Summer 1990.

DISCUSSION QUESTIONS

1. Discuss the similarities and differences in the meaning of motherhood by sex, race, class, sexual identity, and physical ability. In other words, what similarities and differences are there in the construction of motherhood for African-Americans, Asians, Hispanics, Whites, and other ethnic groups? What similarities and differences are there in the meaning of motherhood for men and women?
2. Do you agree with Luker's argument that being young female, poor, and non-white may be related to viewing motherhood as one of the few available avenues of satisfaction, fulfillment, and self-esteem? Why or why not? Might fatherhood fill the same needs for young, poor, minority men? Why or why not?
3. If sufficient social supports were available for teen parents, would teen pregnancy cease to be a social concern? Why or why not?
4. Does our society value children? Does it value all children equally? What evidence do you have to support this view?
5. Why is there a parallel lack of concern about teen fatherhood? What does this suggest about the construction of male and female sexuality in our culture?

Negotiating Lesbian Motherhood: The Dialectics of Resistance and Accommodation

ELLEN LEWIN

As an institution, the family serves many social functions including socialization, the provision of nurturance, care, and protection, as well as the organization and regulation of sexuality and procreation. Norms and assumptions about sexuality that are embedded in the institution of the family are particularly strong because they begin almost literally from birth and are connected to some of our most fundamental psychological needs.

For many heterosexual men and women, "growing up" and becoming an adult is validated and socially affirmed by becoming a parent and starting a (nuclear) family. This is true for many homosexuals as well. Yet, to reproduce, gays and lesbians directly challenge institutional imperatives by severing the presumed links between heterosexuality, procreation, and the traditional nuclear family. Drawing on research from several in-depth interviews, Ellen Lewin examines how lesbian mothers reconcile the competing expectations of gender, motherhood, and sexual identity. Her research indicates that rather than viewing their identity negotiations as examples of resistance or accommodation of existing social strictures, it may be more useful to view these women as inventive strategists.

When I first began to assemble resources for a study of lesbian mothers in 1976, very few people were aware of the existence of such a category, and if they were, they usually saw it as an oxymoron. Lesbian mothers occasionally gained the attention of the general public when they were involved in custody cases that received publicity, but such notoriety was infrequent and typically fleeting. In fact, aside from those who had lesbian mothers in their social circles, even the wider lesbian population was aware of lesbian mothers mainly in connection with custody cases. In the early collections of articles on lesbian issues that emerged from the lesbian feminist movement, lesbian mothers were almost never mentioned except in connection with their vulnerability to custody litigation. Mothers in these cases either lost custody of their children, or won custody only under highly compromised conditions, sometimes with the stipulation that the child have no contact with the mother's partner.

Well-known custody cases in the 1970s demonstrated the likelihood that lesbian mothers would face considerable discrimination in court. . . .

When lesbian mothers found themselves in court, they necessarily had to convince the judge that they were as good at being mothers as any other women, that they were, in fact, *good* in the sense of possessing the moral attributes of altruism and nurturance that are culturally demanded of mothers in North American cultures. In these

From "Negotiating Lesbian Motherhood: The Dialectics of Resistance and Accommodation" by Ellen Lewin in *Mothering: Ideology, Experience, Agency,* ed. Evelyn Glenn et al. p. 21, 333–353. Reprinted by permission of Routledge/Taylor & Francis Books Inc..

formulations, mothers are assumed to be *naturally* equipped to place their children's interests ahead of their own, to be selfless in a way that precludes or overshadows their own sexuality,[1] such assumptions are at the heart of twentieth-century presumptions of maternal suitability for custody.[2] When mothers are lesbians, however, the courts, reflecting popular views of homosexuality as "unnatural," tend to view them as morally flawed, and thus as unfit parents. Their task in dealing with the legal system, therefore, is to demonstrate that they possess the "natural" attributes expected of mothers, and are thus worthy of receiving custody of their children. Maternal virtue, therefore, shifts from being a quality inherent to women to being a behavior one must actively demonstrate in order to pursue a claim to custody.[3]

While many lesbian mothers understood that the way to keep custody of their children was to show that they were "as good as" heterosexual mothers, they firmly believed that they would eventually be shown to be superior parents who were bringing new, nonsexist families into being. They viewed the two-parent, heterosexual, nuclear family as the arena in which the patriarchy inscribed gender expectations onto both women and men. If the power dynamics of that family form were largely responsible for the continuing devalued status of women, and for a variety of abusive practices, then a domestic arrangement based on presumably nongendered relations between two "equal" women partners would constitute a first step toward the better sort of world feminists dreamed of. Jeanne Vaughn, the coeditor of *Politics of the Heart: A Lesbian Parenting Anthology,* put it this way:

> We have an opportunity for radical social change beginning in our homes, change that requires rethinking our views of family, of kinship, of work, of social organization. We need to develop some specifically lesbian-feminist theories of family. How would/did/could we mother our children without the institution of compulsory heterosexuality?[4]

The image many lesbian mothers conjured up was utopian, resembling the broad outlines of Charlotte Perkins Gilman's *Herland,* a fictional society of women in which motherhood and caring were elevated to the center of the inhabitants' lives. Without the need to serve and please powerful males, without the degradations of daily experience in a patriarchal society, Gilman's image suggests, women might be free to express their true, nurturant natures. They would reveal abilities unlikely to emerge in male-dominated society, and would focus on creative, constructive projects rather than on frivolities such as fancy dress and (hetero) sexuality.[5]

The popular images of mothers and families that dominated the lesbian community in the 1970s, then, focused on the ways in which being a mother and having a family could constitute a form of resistance to traditional, and thereby patriarchal, family forms. In particular, success at motherhood (as measured by how well one's child turned out) would demonstrate that children did not need the structure of a heterosexual family, and, most significantly, the regular contribution of a father, to develop normally. The achievement of lesbian mothers would both counteract the notion that lesbianism and motherhood are inherently contradictory and, in fact, redefine and desexualize what it means to be a lesbian.

At the same time, however, the complexities of living as a mother required lesbian mothers to reinstate the dichotomy of natural/unnatural and mother/nonmother that their redefinition of lesbianism sought to subvert. Negotiating the daily issues of being a mother and meeting obligations to one's children brought them into conflict both with the dominant heterosexist society and with lesbians who had not chosen motherhood.

FEMINIST VIEWS OF RESISTANCE

When many of us took up a feminist agenda in our scholarship, directing our attention to documenting the experience of women from their point of view, it seemed that we had no choice but to concentrate on describing a depressing history of victimization and oppression. As we examined the social and cultural lives of women, not only in familiar terrain, but also outside Western traditions, we found over and over again that women were confined to secondary social status, relegated to devalued cultural roles, and often brutalized and demeaned in their daily lives.

. . . In many instances, the best it seemed that we could offer to help remedy this situation was to produce astute, woman-centered descriptions of the conditions under which women's lives were lived, paired with analyses geared toward change. In many instances, feminist scholars directed their energies toward the documentation of women's point of view, focusing on ways to dissolve the hegemony of male-centered assumptions about the organization of social life and women's place in it. In anthropology, such work often proposed alternative views of traditionally patriarchial institutions.[6] But in other instances, feminist interpretations came to center on resistance, looking at how even clearly oppressed women might take action on their own behalf, either by directly sabotaging the instruments of male dominance, or by constituting their consciousness in a way that undermined their subordination.

Feminist scholars have most commonly applied the concept of resistance to studies of women in the work force. Bonnie Thornton Dill's research on Black women household workers, for example, focuses on the way they manage their relationships with employers to enhance their own self-respect. She documents how these workers organized "strategies for gaining mastery over work that was socially defined as demeaning and . . . actively resisted the depersonalization of household work."[7]

. . . Notions of resistance have also informed studies of women outside the workplace. Emily Martin, for example, has contrasted women's ideas about their bodies and the ideology of mainstream medicine, describing instances in which women resist medical assumptions at variance with their own experience.[8]

. . . Louise Lamphere's study of immigrant factory workers in New England also looks carefully at resistance, but frames it as one of several strategies women can mount to cope with employers' efforts to control their lives. She views women "as active strategists, weighing possibilities and devising means to realize goals, and not as passive acceptors of their situations.[9] Lamphere cautions, however, against viewing all of women's actions on their own behalf as resistance. Rather, she emphasizes the importance of distinguishing between "strategies of resistance" and "strategies of accommodation," pointing out that some strategies may best be seen as adjustments that allow women to cope with their place in the labor market by diffusing employers' control of the workplace. Such strategies ought not to be viewed, Lamphere says, as resistance only, since they may not be based in purposeful opposition to the employer, and since they may only result in continuing exploitation of the workers, and, as such, constitute a kind of consent to existing relations of domination.[10]

Taking a different approach, Judith Butler has proposed that scholars reconsider their dependence on the concept of gender, arguing that gender, as a dualistic formulation, rests on the same asymmetry that feminists seek to overturn. She urges the adoption of strategies that would "disrupt the oppositional binary itself,"[11] and suggests that calling into question the "continuity and coherence" of gender identities, sabotaging the "intelligibility" of gender, would undermine the "regulatory aims" of gender as a cultural system.[12] Butler's claim seems to be that lesbianism, or other sexual stances at odds with normative heterosexuality, could constitute a kind of resistance to the very existence of gender.

. . . All of these approaches to resistance reveal a commitment to render women as active subjects. While these scholars are reluctant to blame women for their subordination, neither are they willing to cast them as hapless victims of actions wholly beyond their control. Women are thus seen as capable of framing strategies for enhancing their situations, whether the battleground be material—as when women's resistance improves their working conditions—or symbolic—as when refusal to conform to common conventions of gender may be interpreted as constituting sabotage of the larger system.

This concern with subjectivity and agency raises significant questions for the study of women who seem to defy gender limitations in any aspect of their lives. Just as Butler has suggested that incongruent sexuality might be viewed as resistance, one might ask whether other "disorders" of sexuality and gender could also be viewed in this light. The question becomes particularly pressing when women themselves explain their behavior as subversive. We must then ask whether apparently conscious refusals by lesbian mothers, or any other group of women, to accept the strictures of gender are best understood as instances of resistance.

LESBIAN MOTHERS AND RESISTANCE TO HETEROSEXISM

By the time I was well into my research, at the end of the 1970s, the custody problems that had concerned me at the outset were no longer the only issues facing lesbian mothers. Pregnant women were starting to appear at lesbian social gatherings, at political meetings and concerts, sometimes alone and sometimes in the company of their lovers. These women were not, for the most part, new to lesbian life; most had never been married, and child custody fears did not figure prominently for them. They certainly had not become pregnant by accident. While some of the mothers and mothers-to-be had had romantic interludes with men, more explained how they had "made themselves pregnant" by arranging a sexual situation with a man, or by using some form of "insemination."

The emphasis in these women's accounts of their experiences was on how they had to overcome their earlier fears that being lesbian would preclude motherhood. Lesbians reported that they had often thought of themselves as not being suitable mothers, having internalized images of homosexuals as self-serving, immature, or otherwise not capable of the kind of altruism basic to maternal performance.

Sarah Klein,[13] a lesbian who lives with her one-year-old daughter and her lover, explained the conflict as she perceived it:

> I've always wanted to have a child. In terms of being real tied up with being gay, it was one of the reasons that for a long time I was hesitant to call myself a lesbian. I thought that automatically assumed you had nothing to do with children. . . . I felt, well, if you don't *say* you're a lesbian, you can still work with children, you can still have a kid, you can have relationships with men. But once I put this label on myself, [it would] all [be] over.

By having a child, Sarah repudiated the boundaries she had once associated with being a lesbian; she has claimed what she sees as her right to be a mother.

But other lesbians' accounts indicate that not all perceive themselves as having had a lifelong desire for motherhood. Among those who claim not to remember wanting children when they were younger was Kathy Lindstrom. She had a child by insemination when she was in her early 30s, but says that she never considered the possibility until a few years earlier. She could only explain her behavior as arising from some sort of "hormonal change."

> It just kind of came over me. It wasn't really conscious at first. It was just a need.

Kathy's understanding of her desire to be a mother as something "hormonal," that is, natural, suggests an implicit assertion that this is something so deep and so essentially part of her that nothing, including her lesbianism, can undermine it. Her account indicates that she refuses to allow the associations others have with her status as a lesbian to interfere with her own perception of herself and her needs.

Other lesbian mothers view their urge to have a child as stemming from a desire to settle down, to achieve adulthood, and to counteract forces toward instability in their lives. Ruth Zimmerman, who had a five-year-old son from a relationship with a man she selected as a "good" father, had ended the relationship soon after she became pregnant.

> I definitely felt like I was marking time, waiting for something. I wasn't raised to be a career woman. I was raised to feel like I was grown up and finished growing up and living a regular normal life when I was married and had kids. And I knew that the married part wasn't going to happen. I feel like I've known that for a long time.

Like Kathy, Ruth defined her progress as a human being, and as a woman, in terms that are strikingly conventional and recall traditional feminine socialization. While clearly accepting motherhood as a marker of adulthood and "living a normal life," Ruth tried to overcome the equally conventional limits placed on lesbians in order to have her child.

The notion that having a child signifies adulthood, the acceptance of social responsibility, and demonstrates that one has "settled down" appears in the accounts of many lesbian mothers. Most often, lesbian mothers speak of their lives before motherhood as empty and aimless, and see the birth of their children as having centered them emotionally.

. . . Louise describes herself as living a marginal, disorganized existence until she finally decided that she would have a child. She did not consider using mainstream medicine to get pregnant, assuming that such resources would never be available to her, both for financial reasons and because she would be viewed with hostility by medical professionals. Instead, she went about asking men she met whether they would like to be sperm donors; she finally located a willing prospect and obtained a sperm sample from him. Louise never told this man her real name, and once she had conceived she left the area, concerned that he could somehow pose a threat to her relationship with her child.

Louise's account focuses on conception and birth as spiritual transitions to a higher and better existence. She became pregnant on her first attempt, which she explained as evidence that mystical forces "meant" for this to happen.

. . . Louise did not allow either her counterculture life-style or her status as a lesbian to interfere with the spiritual agenda she felt destined to complete. She says the mystical process she underwent in becoming a mother has permitted her to become more fully herself, to explore aspects of her being that would have remained hidden if she allowed lesbianism alone to define who she is.

> [After] I had [my daughter] I felt it was okay to do these things I've been wanting to do real bad. One of them is to paint my toenails red. I haven't done it yet, but I'm going to do it. I felt really okay about wearing perfume and I just got a permanent in my hair. . . . I feel like I'm robbing myself of some of the things I want to do by trying to fit this lesbian code. I feel like by my having this child, it has already thrown me out in the sidelines.

Louise has used the process of becoming a mother to construct her identity in a way that includes being a lesbian but also draws from other sources. She sees her need to do this as essential and intended, and has moved along her path with the assurance that she is realizing her destiny.

Not all lesbians become mothers as easily as Louise. On a purely practical level, of course, the obstacles to a lesbian becoming pregnant can be formidable. Even if she knows a man who is interested in such a venture, she might not contemplate a heterosexual liaison with enthusiasm and might be equally reluctant to ask him to donate sperm. Mainstream medicine may not seem like an option either, because of financial considerations, or because of fears that doctors will be unwilling to inseminate a lesbian or even a single woman—a realistic concern, of course.

Once one has defined oneself as a lesbian, the barriers to becoming a mother are so significant, in fact, that many of the formerly married lesbian mothers I interviewed explained that they had gone through with marriages (sometimes of long duration) because this seemed the only way to realize their dream of being mothers and being normal in the eyes of their families and communities.

Harriet Newman, an artist who lives with her two daughters in a rural area north of San Francisco, fell in love with another woman during her first year in college. Her parents discovered the affair and forced her to leave school and to see a psychiatrist. The experience convinced her that it would be safer "to be a regular person in the world." When she met a gay man who also wanted to live more conventionally, they married, and almost immediately had their two children.

> The main thing that made us decide to get married was that we very much wanted to part of the mainstream of life, instead of on the edges. We wanted to be substantial . . . part of the common experience.

For lesbians who become mothers through insemination or some other method then, conscious resistance to rigid formulations of "the lesbian" seems to be central to their intentions. Unwilling to deny their identity as lesbians, they also demand the right to define what that identity constitutes. The intrinsic benefits of motherhood—the opportunity to experience birth and child development—are experiences they do not want to forego.

In some instances, women explained that their age made having a child imperative. Laura Bergeron, who had two sons from a relationship prior to coming out, decided to find a donor for a third child when she entered her late 30s.

> I really did want to have a girl, and I was getting older. . . . I was feeling that I didn't really want to have children past the appropriate childbearing age. I had been doing too much reading about retardation and mongoloids and everything else . . . so I put some ads [for donors] in the paper.

. . . As Louise Green's narrative indicated, lesbians often characterize their transformation into mothers as a spiritual journey, an experience that gives them access to special knowledge and that makes them worthier than they otherwise could have been.

. . . Bonnie Peters . . . told me that being a mother connected her with sources of honesty and worthiness.

> I've become more at peace with me [since having my daughter]. She's given me added strength; she's made me—it's like looking in the mirror in many ways; she's made me see myself for who I am. She's definitely given me self-worth. I've become, I think, a more honest person.

Motherhood, then, can draw a woman closer to basic truths, sensitizing her to the feelings of others and discovering a degree of altruism they had not perceived in themselves prior to having a child. It may provide the opportunity for a woman to make clear her involvement with a kind of authenticity, a naturalness, that brings her closer to profound, but ineffable, truths.

MANAGING LESBIAN MOTHERHOOD

While the accounts given by some lesbian mothers suggest that they have resisted the cultural opposition between "mother" and "lesbian" and demanded the right to be both, the ongoing management of being a lesbian mother may depend on separating these two statuses, thus intensifying their dichotomization. Lesbian mothers frequently speak of these two dimensions of their identities as competing or interfering with each other; conflicts with lesbians who are not mothers sometimes further solidify these divisions.

Tanya Petroff, who lives with her seven-year-old daughter in an East Bay city, speaks evocatively of how being a mother overshadows her identity as a lesbian.

> The mothering thing, the thing about being a mother seems to be more important to me than my sexual orientation. . . . I've had [lesbians] tell me that I had chosen a privileged position in having a child and if it was going to be difficult for me then it was too goddam bad.

For Tanya, the conflict is most acute when she is developing a new relationship with another woman. She must then make clear that she views herself and her daughter as an indivisible social unit that takes precedence over other attachments.

> I'm definitely part of a package deal. I come with my daughter and people who can't relate to both of us are not people I want to relate to for very long.

What this means in terms of other relationships is that Tanya sees other mothers, regardless of whether they are gay or straight, as the people with whom she has the most in common. Since relocating to the Bay Area from a town in the Midwest, Tanya has tended to minimize her contact with what she calls the "lesbian community" in favor of socializing with other mothers. She feels that she is better able to resist pressures to raise her daughter to be a "little amazon," an expectation she believes common to lesbians who are not mothers. Beyond this, Tanya feels that there are simply too many practical obstacles to meaningful friendships with women who are not mothers.

> . . . There is a difference between people who have children and people who don't have children. People who don't have children, to my way of thinking, are very selfish. . . . They needn't consider anyone other than themselves. They can do exactly what they want to do at any given time. And though I admire that, it's not possible for me to do that and I guess for that reason most of my friends are single mothers, because it's hard for me to coordinate my needs and my time with someone who's in a completely different head set. "Why can't you get a sitter for the kid?"—that kind of thing. . . . I just prefer being with people who have some sense of what it's like to be me, and I understand where they are too.

Tanya's belief that she can only find truly supportive friends among those whose situations closely mirror her own with respect to single motherhood grows not only out of her very real need for material assistance, but also from the importance she places on having friends who affirm or validate her identity. The most essential aspect of her identity, by this account, is that of being a mother. It supersedes her sexual orientation, her ethnicity, her job.

For some lesbian mothers, difficult experiences with lovers parallel disappointment with the wider lesbian community. Leslie Addison, who lives alone with her twelve-year-old daughter, describes a long series of conflicts with lesbian community groups over support for mothers. While she can easily explain the failure of these women to be conscious about mothers as stemming from their being "single," she has had a harder time dealing with lovers and prospective lovers who do not understand or are unwilling to accommodate her needs as a mother.

> . . . I expected there would be more sharing between women of the child. But I found it's really not, because another woman has a role identity crisis. She can't be the mother, because you're already the mother. She can't be the father, because she's not the father, whereas the men sort of played that role. It was easier for them to fall into it. They could just play daddy, I could play momma, and everybody'd be happy.

The stark separation between "mother" and "lesbian" as elements of identity may be even more sharply drawn for women concerned with maintaining secrecy about their sexual orientation. In these instances, daily life is segregated into time when they are "mothers" and time when they are "lesbians," creating constant concern about information management and boundary maintenance. While some mothers who voice these concerns are motivated by fears about custody, others seem to be more worried by what they understand to be broad community standards. Segregation may seem the best way to protect children from being stigmatized, but in addition, lesbian mothers know that motherhood itself tends to preclude their being suspected of homosexuality. As one mother explained, "Of course, I have the mask. I have a child. I'm accepted [as heterosexual] because I have a child and that kind of protection."

. . . For some women who maintain strict separation between their identities as mothers and as lesbians, the threat of custody litigation is more than an abstract fear. Theresa Baldocchi, whose son is nine years old, survived a protracted custody trial at the time she divorced her former husband, John. Her legal expenses and liability for debts incurred by John during their marriage left her virtually bankrupt, and it has taken years for her to solidify her financial situation. Theresa was not a lesbian at the time of the divorce, but John made allegations that she was. Now that she has come out, she is convinced that she must carefully separate her life as a mother and as a lesbian, lest her former husband decide to institute another custody case against her. Despite the fact that John has an extensive history of psychiatric hospitalization, and that she is a successful professional, she is sure that her chances of winning in such a trial would be slim.

> Now that I'm gay, I'd lose. There's just no way in the world I would win, after having had my fitness questioned when I was Lady Madonna, let alone now.

Theresa has decided that living in a middle-class suburban area and arranging her home in an impeccably conventional fashion help shield her from suspicion of being anything other than a typical "mom." The Bay Bridge, which she must cross each day between her home and San Francisco, where she works and socializes with her lesbian friends, symbolizes her strategy. She feels that each trip involves a palpable transition, as she prepares herself to meet the requirements of her destination—home or San Francisco. Most crucial for her strategy is not telling her son that she is a lesbian, since she feels it would be inappropriate to expect him to maintain her secret.

If Theresa was concerned only with managing information about her homosexuality, she would probably avoid seeing her former husband, and thus be able to relax, at least, at home. But Theresa firmly believes that being a good mother demands that she take every opportunity to maximize her son's contact with John, a model father in her eyes. Because John is not regularly employed, he has offered to take care of their son each day while Theresa is at work. This arrangement has meant both that Theresa does not have to obtain paid child care during these hours, and that her son has daily contact with his father. It also means that she has virtually no privacy. She must control the kinds of friends who visit her, and must make sure that nothing that might reveal her sexual orientation can be found in her home. Most poignantly, she must limit her lover's access to her home for fear that her presence would somehow make the situation transparent. She consigns her most reliable potential source of support to the background, leaving herself isolated and anxious much of the time.

. . . Other mothers explain the separation of motherhood from other dimensions of their lives, and the centrality of being a mother, to framing their identities more practically, citing the weighty and unrelenting obligations faced by parents. Peggy Lawrence, who lives with her lover, Sue Alexander, her ten-year-old daughter, and Sue's two sons, spoke at length about the effects of being a mother on her personal freedom. Being a mother means that she must be concerned about continuity and stability in ways that constrain her spontaneity, and earning money must be a priority no matter how oppressive her work. Peggy and Sue live in a neighborhood close to their children's school, and have chosen to live in San Francisco because they think their children will encounter less discrimination here as the children of lesbians than in the Midwest, where they would prefer to live. Peggy explains what being a mother means to her:

> Being a mother, to me—being a mother is more consuming than any other way that I could possibly imagine identifying myself . . . any other way that I identify myself is an identification of some part of my being a mother. I am a lesbian mother, I am a working mother—"mother" hardly ever modifies any other thing. Mother is always the primary—it's always some kind of mother, but it's never a mother-anything. Mother is—mother, for mothers, is always the thing that is more consuming.

But others understand motherhood to mean the uniquely intense feelings that exist between mother and child. Lisa Stark, who describes the weightiness of single parenthood as almost unbearable, has come to see her children as the reason she can continue to struggle with her obligations, paradoxically the explanation for both her suffering and her very survival.

> I've . . . never had to live for myself. The only reason I get up in the morning is to get them off to school. For me to trot off to work in order to earn the money to support them. I don't know what I'd do if I didn't have them. They're everything I've got. . . . I love them so much that it really is painful.

Having a child or being a mother may be said to create and reinforce meaningful ties with the world, and to make struggle worthwhile. While being a lesbian mother can be difficult, and may make a woman's life complicated and stressful, children offer significant intrinsic rewards—most importantly, a way to experience feelings of special intimacy, and to be connected to higher-order, spiritual values. Motherhood allows lesbians to be more like other women, at least with respect to the most defining feminine role expectation, but segregating these two dimensions of the self becomes the most efficient way to manage practical obligations, and intensifies the dichotomization of "lesbian" and "mother."

LESBIAN MOTHERHOOD: RESISTANCE OR ACCOMMODATION?

The goals motherhood allows lesbians to enhance are, of course, no different from those heterosexual women describe for themselves. Being a mother, in particular, becoming a mother, is perceived as a transformative experience, an accomplishment that puts other achievements in their proper perspective. It is also construed as an individual achievement, something a woman can "do" to make herself a mother, that is, to transform herself into an altruistic, spiritually-aware human being. In a culture that elevates what has been characterized as "mythic individualism" as a central value, individuals idealize autonomy, self-reliance, and the notion that one must "find oneself" and "make something" of oneself.[14]

Women in America have particular difficulty living up to this cultural ideal. Individualistic and assertive behaviors valued in men are discouraged in women. Dependency, particularly through marriage, is represented as a specifically feminine sort of success. I have discussed elsewhere the remarkable congruences I observed in accounts both lesbian and heterosexual women offered of their divorces, and the similarities between these stories and lesbians' coming-out narratives.[15] These narratives are constructed around themes of agency, independence, and individuality, and celebrate women's ability to define their own lives, to decide how to represent their identities, and to achieve adulthood and autonomy. Despite the fact that both divorce and coming out as a lesbian are popularly understood to be problematic, and, indeed, have historically been defined as stigmatized statuses, women represent them as odysseys of self-discovery leading to more authentic formulations of the self.

Accounts of becoming a mother, in similar fashion, focus on the power of the individual to construct herself as a mother, to negotiate the formation of her self and to bring something good into her life. For lesbians, particularly for lesbians who decided to become mothers once their identification as lesbians was firm, the process of becoming a mother demands agency. At the same time, to the extent that wanting to be a mother is perceived as a *natural* desire, one unmediated by culture or politics, then becoming a mother permits a lesbian to move into a more natural or normal status than she would otherwise achieve. In this sense, becoming a mother represents a sort of conformity with conventional gender expectations. At the same time, to the extent that becoming a mother means overcoming the equation of homosexuality with *unnaturalness,* then this transformation allows the lesbian mother to resist gendered constructions of sexuality. This act of resistance is paradoxically achieved through compliance with conventional expectations for women, so it may also be construed as a gesture of accommodation.

Placing motherhood at the center of one's identity often involves, as we have seen, simultaneously placing other aspects of the self, most notably lesbianism, at the margins. Demanding the right to be a mother suggests a repudiation of gender conventions that define "mother" and "lesbian" as inherently incompatible identities, the former natural and intrinsic to women, organized around altruism, the latter unnatural, and organized around self-indulgence. But living as a mother means making other choices, and these choices reinscribe the opposition between "mother" and "lesbian." Subversion of orderly gender expectations is hypothetical, at best, in the lives of many lesbian mothers, at the same time that knowledge of their existence can only be imagined by the wider public as a rebellion of the most fundamental sort.

The model I would suggest based on the accounts presented here is that lesbian mothers are neither resisters nor accommodators—or perhaps that they are both. A more accurate way of framing their narratives is that they are strategists, using the cultural resources offered by motherhood to achieve a particular set of goals. That these are the goals framed by past experience in a heterosexist and perhaps patriarchal society, and that these resources are culturally constrained and shaped by the exigencies of gender, does not simplify the analysis. While such women are often conscious resisters, others gladly organize their experience as a reconciliation with what they view as traditional values. At the same time that some outsiders may see their behavior as transgressive (and thereby label them resisters or subversives), others perceive lesbian motherhood (along with other indications of compliance with conventional behaviors, such as gay/lesbian marriage) as evidence that lesbians (and other "deviants") can be domesticated and tamed.

The search for cultures of resistance continues to be a vital dimension of the feminist academic enterprise. At the same time that we cannot limit our analyses of women's lives to accounts of victimization, we cannot be complacent when we discover evidence of resistance and subversion. Either interpretation may fail to reveal the complex ways in which resistance and accommodation, subversion and compliance, are interwoven and interdependent, not distinct orientations, but mutually reinforcing aspects of a single strategy. Lesbian mothers are, in some sense, both lesbians and mothers, but they shape identity and renegotiate its meanings at every turn, reinventing themselves as they make their way in a difficult world.

NOTES

1. Not only lesbians, but heterosexual mothers whose sexual activity comes to the attention of the authorities, may be vulnerable in cases where their custody is challenged. See Nancy D. Polikoff, "Gender and Child Custody Determinations: Exploding the Myths," in Irene Diamond, ed., *Families, Politics, and Public Policy: A Feminist*

Dialogue on Women and the State (New York: Longman, 1983), pp. 183-202.

2. Nan Hunter and Nancy D. Polikoff, "Custody Rights of Lesbian Mothers: Legal Theory and Litigation Strategy," *Buffalo Law Review* 25, (1976), p. 691; Lewin, "Lesbianism and Motherhood."
3. Ellen Lewin, "Claims to Motherhood: Custody Disputes and Maternal Strategies," in Faye Ginsburg and Anna Lowenhaupt Tsing, eds., *Uncertain Terms: Negotiating Gender in American Culture* (Boston: Beacon Press, 1990), pp. 199–214.
4. Jeanne Vaughn, "A Question of Survival," in Sandra J. Pollack and Jeanne Vaughn, eds., *Politics of the Heart: A Lesbian Parenting Anthology* (Ithaca, N.Y.: Firebrand Books, 1987), p. 26.
5. Charlotte Perkins Gilman, *Herland* (1915, New York: Pantheon Books, 1979).
6. See, for example, Jane Goodale, *Tiwi Wives: A Study of the Women of Melville Island, North Australia* (Seattle: University of Washington Press, 1971); Annette B. Weiner, *Women of Value, Men of Renown: New Perspectives in Trobriand Exchange* (Austin: University of Texas Press, 1976); Margery Wolf, *Women and the Family in Rural Taiwan* (Stanford: Stanford University Press, 1972).
7. Bonnie Thornton Dill. "Domestic Service and the Construction of Personal Dignity," in Ann Bookman and Sandra Morgen, eds., *Women and the Politics of Empowerment* (Philadelphia: Temple University Press, 1988), p. 33.
8. Emily Martin, *The Woman in the Body: A Cultural Analysis of Reproduction* (Boston: Beacon Press, 1987), p. 110.
9. Louise Lamphere, *From Working Daughters to Working Mothers: Immigrant Women in a New England Industrial Community* (Ithaca, N.Y.: Cornell University Press, 1987), pp. 29–30.
10. Lamphere, *From Working Daughters to Working Mothers,* p. 30.
11. Judith Butler, *Gender Trouble: Feminism and the Subversion of Identity* (New York: Routledge, 1990), p. 27.
12. Butler, *Gender Trouble,* p. 17.
13. Names and some other details have been changed to preserve the anonymity of women whom I interviewed. For a detailed account of the methods used in this research, see Lewin, *Lesbian Mothers.*
14. Robert Bellah, et al., *Habits of the Heart: Individualism and Commitment in American Life* (Berkeley: University of California Press, 1985), p. 65.
15. Lewin, *Lesbian Mothers.*

DISCUSSION QUESTIONS

1. Is gay fatherhood socially equivalent to lesbian motherhood (particularly in terms of gender identity and social acceptance)? Why or why not?
2. How does the normal presumption of heterosexuality affect lesbian women's lives? Heterosexual women's? Bisexual? Transgendered? Asexual?
3. Is there evidence of internalized homophobia and/or heterosexism among the interviewees? If so, provide examples. What effects, if any, do you think this has on the political agenda of homosexual parents?
4. How is the negotiation of lesbian motherhood affected by race, class, and physical ability?
5. What normative structures of the family are challenged by lesbian motherhood? In other words, what norms or assumptions about the family does lesbian motherhood challenge? What normative structures of the family are challenged by gay fatherhood?
6. How would you assess the actions and motives of these women? Are they resisters, accommodators, strategists, or some other type of conscious actor? Provide examples to help defend your position.

In Pursuit of the Perfect Penis

LEONORE TIEFER

As previous works in this volume have demonstrated, much of "being a man" is tied up with sexuality. "Real" men have sex; driving, penetrative sex—and lots of it. Because of this, a well-functioning penis is a crucial resource for the attainment of hegemonic masculinity. In this classic reading (which predates "Viagra" and other erectile dysfunction medications), Leonore Tiefer critiques the use of penile prostheses to help men overcome sexual dysfunction. She contends that the appeal of this rather extreme surgical solution is based, at least in part, in the increased medicalization of sexuality and the rising importance of sexuality in personal life. Tiefer contends that the vision of "perfect potency" is illusory and obscures the larger underlying problem of the construction of male sexual scripts.

Sexual virility—the ability to fulfill the conjugal duty, the ability to procreate, sexual power, potency—is everywhere a requirement of the male role, and thus "impotence" is everywhere a matter of concern. Although the term has been used for centuries to refer specifically to partial or complete loss of erectile ability, the first definition in most dictionaries never mentions sex but refers to a general loss of vigor, strength, or power. Sex therapists, concerned about these demeaning connotations, have written about the stigmatizing impact of the label: "The word *impotent* is used to describe the man who does not get an erection, not just his penis. If a man is told by his doctor that he is impotent, and the man turns to his partner and says he is impotent, they are saying a lot more than that the penis cannot become erect" (Kelley 1981, p. 126). . . .

MALE SEXUALITY

Sexual competence is part—some would say the *central* part—of contemporary masculinity. This is true whether we are discussing the traditional man, the modern man, or even the "new" man:

> What so stokes male sexuality that clinicians are impressed by the force of it? Not libido, but rather the curious phenomenon by which sexuality consolidates and confirms gender. . . . An impotent man always feels that his masculinity, and not just his sexuality, is threatened. In men, gender appears to "lean" on sexuality . . . the need for sexual performance is so great. . . . In women, gender identity and self-worth can be consolidated by other means. (Person 1980, pp. 619, 626)

John Gagnon and William Simon (1973) explained how, during adolescent masturbation, genital sexuality (i.e., erection and orgasm) acquires nonsexual motives—such as the desire for power, achievement, and peer approval—that have already become important during preadolescent gender role training. "The capacity for erection," they wrote, "is an important sign element of masculinity and control" (Gagnon and Simon 1973, p. 62) without which a man is not a man. Allan Gross (1978) argued that by adulthood few men can accept other successful aspects of masculinity in lieu of adequate sexual performance.

Masculine sexuality assumes the ability for potent function, but the performances that earn acceptance and status often occur far from the bedroom. Observing behavior in a Yorkshire woolen mill, British sociologist Dennis Marsden described how working-class men engage in an endless performance of sexual stories, jokes, and routines: "As a topic on which most men could support a conversation and as a source of jokes, sexual talk and gesture were inexhaustible. In the machine noise a gesture suggestive of masturbation, intercourse, or homosexuality was enough to raise a conventional smile and re-establish a bond over distances too great for talking" (Marsden, quoted in Tolson 1977, p. 60). Andrew Tolson argued that this type of ritualized sexual exchange validates working men's bond of masculinity in a situation that otherwise emasculates them and illustrates the enduring homosocial function of heterosexuality that develops from the adolescent experience (Gagnon and Simon 1973).

Psychologically, then, male sexual performance may have as much or more to do with male gender role confirmation and homosocial status as with pleasure, intimacy, or tension release. This assessment may explain why men express so many rules concerning proper sexual performance: Their agenda relates not merely to personal or couple satisfaction but to acting "like a man" during intercourse in order to qualify for the title elsewhere, where it *really* counts.

I have drawn on the writings of several authorities to compile an outline of ten sexual beliefs to which many men subscribe (Doyle 1983; Zilbergeld 1978; LoPiccolo 1978; LoPiccolo 1985): (1) Men's sexual apparatus and needs are simple and straightforward, unlike women's; (2) most men are ready, willing, and eager for as much sex as they can get; (3) most men's sexual experiences approximate ecstatic explosiveness (the standard by which individual men compare their own experience, thus becoming disappointed over suspicions that they are not doing as well as others); (4) it is the responsibility of the man to teach and lead his partner to experience pleasure and orgasm(s); (5) sexual prowess is a serious, task-oriented business, no place for experimentation, unpredictability, or play; (6) women prefer intercourse, particularly "hard-driving" intercourse, to other sexual activities; (7) all really good and normal sex must end in intercourse; (8) any physical contact other than a light touch is meant as an invitation to foreplay and intercourse; (9) it is the responsibility of the man to satisfy both his partner and himself; (10) sexual prowess is never permanently earned; each time it must be reproven.

Many of these demands directly require—and all of them indirectly require—an erection. James Nelson (1985) pointed out that male sexuality is dominated by a genital focus in several ways: Sexuality is isolated from the rest of life as a unique experience with particular technical performance requirements; the subjective meaning for the man arises from genital sensations first practiced and familiar in adolescent masturbation and directly transferred without thought to the interpersonal situation; and the psychological meaning primarily depends on the confirmation of virility that comes from proper erection and ejaculation.

It is no surprise, then, that any difficulty in getting the penis to do what it "ought" can become a source of profound humiliation and despair, both in terms of the immediate blow to self-esteem and in terms of the eventual destruction of masculine reputation that is assumed will follow. Two contemporary observers expressed men's fears of impotence in these terms:

> Few sexual problems are as devastating to a man as his inability to achieve or sustain an erection long enough for successful sexual intercourse. For many men the idea of not being able to "get it up" is a fate worse than death. (Doyle 1983, p. 205)

BIOMEDICAL APPROACHES TO MALE SEXUAL PROBLEMS

. . . Although the physiological contributions to adequate sexual functioning can be theoretically specified in some detail, there are as yet few diagnostic tests that enable specific identification of different types of pathophysiological contributions. Moreover, there are few medical treatments available for medically caused erectile disorders aside from changing medications (particularly in the case of hypertension) or correcting an underlying disease process. The most widely used medical approach is an extreme one: surgical implantation of a device into the penis to permit intromission. This is the penile prosthesis.

The history of these devices is relatively short (Melman 1978). Following unsuccessful attempts with bone and cartilage, the earliest synthetic implant (1948) consisted of a plastic tube placed in the middle of the penis of a patient who had had his urethra removed for other reasons. Today, several different manufacturers produce slightly different versions of two general types of implant.

One type is the "inflatable" prosthesis. Inflatable silicone cylinders are placed in the *Corpora cavernosa* of the penis, the cylindrical bodies of erectile tissue that normally fill with blood during erection. The cylinders are connected to a pump placed in the scrotum that is connected to a small, saline-filled reservoir placed in the abdomen. Arnold Melman (1978) described how the prosthesis works: "When the patient desires a tumescent phallus, the bulb is squeezed five or six times and fluid is forced from the reservoir into the cylinder chambers. When a flaccid penis is wanted, a deflation valve is pressed and the fluid returns to the reservoir" (p. 278). The other type of prosthesis consists of a pair of semirigid rods, now made of silicone, with either bendable silver cores or hinges to allow concealment of the erection by bending the penis down or up against the body when the man is dressed.

. . . Past reports have encouraged the belief that the devices function mechanically, that men and their partners are able to adjust to them without difficulty, and that the prostheses result in satisfactory sexual function and sensation. But recent papers challenge these simple conclusions. One review of the postoperative follow-up literature was so critical of methodological weaknesses (brief follow-up periods, rare interviews with patients' sexual partners, few objective data or even cross-validation of subjective questions about sexual functioning, among others) that the authors could not summarize the results in any meaningful way (Collins and Kinder 1984). Another recent summary criticized the implants' effectiveness:

> First, recent reports indicate that the percentage of surgical and mechanical complications from such prosthetic implants is much higher than might be considered acceptable. Second, despite claims to the contrary by some surgeons, it appears likely that whatever degree of naturally occurring erection a man is capable of will be disrupted, and perhaps eliminated by the surgical procedures and scarring involved in prosthetic implants. Finally, it has been my experience that, although patients are typically rather eager to have a prosthesis implanted and report being very happy with it at short-term surgical follow-up, longer term behavioral assessment indicates poor sexual adjustment in some cases. (LoPiccolo 1985, p. 222)

Three recent urologist papers report high rates of postoperative infection and mechanical failure of the inflatable prosthesis, both of which necessitate removal of the device (Apte, Gregory, and Purcell 1984; Joseph, Bruskewitz, and Benson 1984; Fallon, Rosenberg, and Culp 1984). . . .

THE ALLURE OF MEDICALIZED SEXUALITY

Men are drawn to a technological solution such as the penile prosthesis for a variety of personal reasons that ultimately rest on the inflexible central place of sexual potency in the male sexual script. Those who assume that "normal" men must always be interested in sex and who believe that male sexuality is a simple system wherein interest leads easily and directly to erection (Zilbergeld 1978) are baffled by any erectile difficulties. Their belief that "their penis is an instrument immune from everyday problems, anxieties and fears" (Doyle 1983, p. 207) conditions them to deny the contribution of psychological or interpersonal factors to male sexual responsiveness. This denial, in turn, results from fundamental male gender role prescriptions for self-reliance and emotional control (Brannon, 1976).

Medicalized discourse offers an explanation of impotence that removes control over, and therefore responsibility and blame for, sexual failure from the man and places it on his physiology. . . . Medical treatments not only offer tangible evidence of nonblameworthiness but also allow men to avoid psychological treatments such as marital or sex therapy, which threaten embarrassing self-disclosure and admissions of weakness men find aversive (Peplau and Gordon 1985).

The final allure of a technological solution such as the penile prosthesis is its promise of permanent freedom from worry. One of Masters and Johnson's (1970) major insights was their description of the self-conscious self-monitoring that men with erectile difficulties develop in sexual situations. "Performance anxiety" and "spectatoring," their two immediate causes of sexual impotence, generate a self-perpetuating cycle that undermines a man's confidence about the future even as he recovers from individual episodes. Technology seems to offer a simple and permanent solution to the problem of lost or threatened confidence, as doctors cited from *Vogue* to the *Journal of Urology* have noted.

THE RISING IMPORTANCE OF SEXUALITY IN PERSONAL LIFE

Even though we live in a time when the definition of masculinity is moving away from reliance on physical validation (Pleck 1976), there seems to be no apparent reduction in the male sexual focus on physical performance. Part of the explanation for this lack of change must rest with the increasing importance of sexuality in contemporary relationships. Recent sociocultural analyses have suggested that sexual satisfaction grows in importance to the individual and couple as other sources of personal fulfillment and connection with others wither. Edmund White (1980) wrote,

> I would say that with the collapse of other social values (those of religion, patriotism, the family, and so on), sex has been forced to take up the slack, to become our sole mode of transcendence and our only touchstone of authenticity. . . . In our present isolation we have few ways besides sex to feel connected with each other. (p. 282)

. . . The increasing pressure on intimate relationships to provide psychological support and gratification comes at the same time that traditional reasons for these relationships (i.e., economics and raising families) are declining. Both trends place pressure on compatibility and companionship to maintain relationships. Given that men have been raised "not to be emotionally sensitive to others or emotionally expressive or self-revealing" (Pleck 1981, p. 140), much modern relationship success would seem to depend on sexual fulfillment. Although some contemporary research indicates that marriages and gay relationships can be rated successful despite the presence of sexual problems (Frank, Anderson, and Rubinstein 1978; Bell and Weinberg 1978), popular surveys suggest that the public believes sexual satisfaction to be essential to relationship success.

The importance of sexuality also increases because of its use by consumption-oriented capitalism (Altman 1982). The promise of increased sexual attractiveness is used to sell products to people of all ages. Commercial sexual meeting places are popular in both gay and heterosexual culture, and a whole system of therapists, books, workshops, and magazines sells advice on improving sexual performance and enjoyment. Restraint and repression are inappropriate in a consumer culture in which the emphasis is on immediate gratification.

The expectation that sexuality will provide ever-increasing rewards and personal meaning has also been a theme of the contemporary women's movement, and women's changing attitudes have affected many men, particularly widowed and divorced men returning to the sexual "market." Within the past decade, sexual advice manuals have completely changed their tone regarding the roles of men and women in sexual relations (Weinberg, Swensson, and Hammersmith 1983). Women are advised to take more responsibility for their own pleasure, to possess sexual knowledge and self-knowledge, and to expect that improved sexual functioning will pay off in other aspects of life. Removing responsibility from the man for being the sexual teacher and leader reduces the definition of sexual masculinity to having excellent technique and equipment to meet the "new woman" on her "new" level.

Finally, the new importance of sexual performance has no upper age limit. . . . Sex is a natural act, Masters and Johnson said over and again, and there is no "natural" reason for ability to decline or disappear as one ages. Erectile difficulties become "problems" that can be corrected with suitable treatment, and aging provides no escape from the male sexual role.

THE MEDICALIZATION OF IMPOTENCE: PART OF THE PROBLEM OR PART OF THE SOLUTION?

. . . Men view physical explanations for their problems as less stigmatizing and are better able to maintain their sense of masculinity and self-esteem when their problems are designated as physical. Accepting medicine as a source of authority and help reassures men who feel that they are under immense pressure from role expectations but who are unable to consult with or confide in either other men or women because of pride, competitiveness, or defensiveness. That "inhibited sexual excitement . . . in males, partial or complete failure to attain or maintain erection until completion of the [*sic*] sexual act," is a genuine disorder (APA 1980, p. 279) legitimates an important aspect of life that physicians previously dismissed or made jokes about. . . . Permanent mechanical solutions to sexual performance worries are seen as a gift from heaven; one simple operation erases a source of anxiety dating from adolescence about failing as a man.

The disadvantages to medicalizing male sexuality, however, are numerous and subtle. (My discussion here is informed by Catherine Riessman's 1983 analysis of the medicalization of many female roles and conditions.) First, dependence on medical remedies for impotence has led to the escalating use of treatments that may have unpredictable long-term effects. Iatrogenic ("doctor-caused") consequences of new technology and pharmacology are not uncommon and seem most worrisome when medical treatments are offered to men with no demonstrable organic disease. Second, the use of medical language mystifies human experience, increasing the public's dependence on professionals and experts. If sexuality becomes fundamentally a matter of vasocongestion and myotonia (as in Masters and Johnson's famous claim, 1966, p. 7), personal experience requires expert interpretation and explanation. Third, medicalization spreads the moral neutrality of medicine and science over sexuality, and people no longer ask whether men "should" have erections. If the presence of erections is healthy and their absence (in whole or part) is pathological, then healthy behavior is correct behavior and vice versa. This view again increases the public's dependence on health authorities to define norms and standards for conduct.

The primary disadvantage of medicalization is that it denies, obscures, and ignores the social causes of whatever problem is under study. Impotence becomes the problem of an individual man. This effect seems particularly pertinent in the case of male sexuality, an area in which the social demands of the male sexual role are so related to the meaning of erectile function and dysfunction. Recall the list of men's beliefs about sexuality, the evaluative criteria of conduct and performance. Being a man depends on sexual adequacy, which depends on potency. A rigid, reliable erection is necessary for full compliance with the script. The medicalization of male sexuality helps a man conform to the script rather than analyzing where the script comes from or challenging it. In addition,

"Medicine attracts public resources out of proportion to its capacity for health enhancement, because it often categorizes problems fundamentally social in origin as biological or personal deficits, and in so doing smothers the impulse for social change which could offer the only serious resolution" (Stark and Flitcraft, quoted in Riessman, 1983, p. 4). Research and technology are directed only toward better and better solutions. Yet the demands of the script are so formidable, and the pressures from the sociocultural changes I have outlined so likely to increase, that no technical solution will ever work—certainly not for everyone.

PREVENTIVE MEDICINE: CHANGING THE MALE SEXUAL SCRIPT

Men will remain vulnerable to the expansion of the clinical domain so long as masculinity rests heavily on a particular type of physical function. As more research uncovers subtle physiological correlates of genital functioning, more men will be "at risk" for impotence. Fluctuations of physical and emotional state will become cues for impending impotence in any man with, for example, diabetes, hypertension, or a history of prescription medication usage.

One of the less well understood features of sex therapy is that it "treats" erectile dysfunction by changing the individual man's sexual script. Sex therapists have described the process as follows:

> This approach is primarily educational—you are not curing an illness but learning new and more satisfactory ways of getting on with each other. (Bancroft 1989, p. 537)
>
> Our thesis is that the rules and concepts we learn [about male sexuality] are destructive and a very inadequate preparation for a satisfying and pleasurable sex life. . . . Having a better sex life is in large measure dependent upon your willingness to examine how the male sexual mythology has trapped you. (Zilbergeld 1978, p. 9)

Sexuality can be transformed from a rigid standard for masculine adequacy to a way of being, a way of communicating, a hobby, a way of being in one's body—and *being* one's body—that does not impose control but rather affirms pleasure, movement, sensation, cooperation, playfulness, relating. Masculine confidence cannot be purchased because there can never be perfect potency. Chasing its illusion may line a few pockets, but for most men it will only exchange one set of anxieties and limitations for another.

REFERENCES

Altman, D. (1982). *The Homosexualization of America: The Americanization of the Homosexual,* New York: St. Martin's Press.

American Psychiatric Association. (1980). *Diagnostic and Statistical Manual of Mental Disorders,* 3rd ed. Washington, D.C.: APA.

Apte, S. M., Gregory, J. G., and Purcell, M. H. (1984). The inflatable penile prosthesis, reoperation and patient satisfaction: A comparison of statistics obtained from patient record review with statistics obtained from intensive followup search. *Journal of Urology* 131, 894–895.

Bancroft, (1989). *Human Sexuality and Its Problems,* 2nd ed. Edinburgh: Churchill- Livingstone. (1st ed. published in 1983.)

Bell, A. P., and Weinberg, M. S. (1978). *Homosexualities: A study of diversity among men and women.* New York: Simon and Schuster.

Brannon, R. (1976). The male sex role: Our culture's blueprint of manhood, and what it's done for us lately, In D. David and R. Brannon, eds., *The Forty-Nine Percent Majority: The male sex role.* Reading, Mass.: Addison-Wesley.

Collins, G. F., and Kinder, B. N. (1984). Adjustment following surgical implantation of a penile prosthesis: A critical overview. *Journal of Sex and Marital Therapy* 10, 255–271.

Doyle, J. A. (1983). *The Male Sexual Experience.* Dubuque, Iowa: William C. Brown.

Fallon, B., Rosenberg, S., and Culp, D. A. (1984). Long-term follow-up in patients with an inflatable penile prosthesis. *Journal of Urology* 132, 270–271.

Frank, E., Anderson, C., and Rubinstein, D. (1978). Frequency of sexual dysfunction in "normal" couples. *New England Journal of Medicine* 299, 111–115.

Gagnon, J. H., and Simon, W. (1973). *Sexual Conduct: The social sources of human sexuality*. Chicago: Aldine.

Gross, A. E. (1978). The male role and heterosexual behavior. *Journal of Social Issues* 34, 87–107.

Joseph, D. B., Bruskewitz, R. C., and Benson, R. C. (1984). Long term evaluation of the inflatable penile prosthesis. *Journal of Urology* 131, 670–673.

Kelley, S. (1981). Some social and psychological aspects of organic sexual dysfunction in men. *Sexuality and Disability* 4, 123–128.

LoPiccolo, J. (1978). Direct treatment of sexual dysfunction. In J. LoPiccolo and L. LoPiccolo, eds., *Handbook of Sex Therapy*. New York: Plenum Press.

———. (1985). Diagnosis and treatment of male sexual dysfunction. *Journal of Sex and Marital Therapy* 11, 215–232.

Masters, W. H. (1970). *Human Sexual Inadequacy*, Boston: Little, Brown.

Melman, A. (1978). Development of contemporary surgical management for erectile impotence. *Sexuality and Disability* 1, 272–281.

Nelson, J. (1985). Male sexuality and masculine spirituality. *Siecus Report* 13, 1–4.

Peplau, L. A., and Gordon, S. L. (1985). Women and men in love: Gender differences in close heterosexual relationships. In V. E. O'Leary, R. K. Unger, and B. S. Wallston, eds., *Women, Gender and Social Psychology*. Hillsdale, N.J.: Lawrence Erlbaum Associates.

Person, E. S. (1980). Sexuality as the mainstay of identity: Psychoanalytic perspectives, *Signs* 5, 605–630.

Pleck, J. H. (1976). The male sex role: Definitions, problems and sources of change. *Journal of Social Issues* 32, 155–164.

———. (1981). *The Myth of Masculinity*. Cambridge: MIT Press.

Riessman, C. K. (1983). Women and medicalization: A new perspective. *Social Policy* 14, 3–18.

Tolson, A. (1977). *The Limits of Masculinity*. New York: Harper and Row.

Weinberg, M. S., Swensson, R. G., and Hammersmith, S. K. (1983). Sexual autonomy and the status of women: Models of female sexuality in U.S. sex manuals from 1950 to 1980. *Social Problems* 30, 312–324.

White, E. (1980). *States of Desire*. New York: E. P. Dutton.

Zilbergeld, B. (1978). *Male Sexuality*. Boston: Little, Brown.

———. (1992). The man behind the broken penis: Social and psychological determinants of erectile failure. In R. C. Rosen and S. R. Leiblum, eds., *Erectile Disorders: Assessment and Treatment*. New York: Guilford.

DISCUSSION QUESTIONS

1. What are the differential means that men and women have to secure gender identity and self-worth? Why do they differ? Can you link these differences to elements of the social structure?
2. This article was written almost 20 years ago. Are the 10 sexual beliefs listed by Tiefer still valid today? If yes, what contributes to their persistence? If not, what has changed (and why)? Are there any beliefs you would add to this list? Construct a similar list of sexual beliefs for women. What relationship, if any, do these lists have with one another?
3. Compare and contrast the acquisition of a penile prosthesis with cosmetic surgery. In what ways are they similar? In what ways do they differ? How should we determine medical necessity? Who should make these determinations? Does it matter? Why?
4. Is it necessarily problematic to focus on the penis as the central point of sexual pleasure? Why or why not? How does an erect penis as the focal point of sexual pleasure relate to contemporary constructions of sexuality?
5. What are the problems with medicalizing impotency and/or sexual dysfunction? What are the benefits? What does it say about the importance of the penis to men, that some are willing to undergo painful surgery in an attempt to correct perceived problems with its performance? Is it for their partner's benefit? Their own? How important is male erection to sexual pleasure? To masculinity?
6. How do you think Tiefer would analyze the development of Viagra and other medications for erectile dysfunctions? What does their popularity suggest about the role of erections in modern sexuality?

The following Bonus Reading can be accessed through the *InfoTrac College Edition Online Library* by typing in the name of the author or keywords in the title online at *http://www.infotrac-college.com*.

Cloning, Sex, and New Kinds of Families

GLENN MCGEE

Reproductive technologies have given birth to a host of challenges for social institutions. In this evaluation of human cloning, bio-ethicist Glenn McGee avers that too little attention has been paid to how cloning may affect social understandings (constructions) of sexuality, reproduction, and family. He indicates that the role of sexuality, in particular, has not been sufficiently considered. McGee calls for a thoughtful reassessment of the enterprise of "making children" so that the role of sex, as part of being human, can be adequately assessed.

DISCUSSION QUESTIONS

1. What constitutes a family? Should we call progenitors parents? Should we call clones children? Will cloning affect how we view families? Will it affect how we structure a family?
2. If "clone" emerges as new social category, how would you expect it to compare with the existing categories of sex, gender, and sexuality?
3. What kind of changes in sexuality and sexual norms do you envision if cloning becomes a socially-accepted method of human reproduction?
4. Does cloning threaten the viability of the nuclear family as a social construct?
5. How might cloning affect social stratification?
6. What effect, if any, do you think cloning will have on the relative importance of pleasure, love, and reproduction in interpersonal relationships? What do you think Seidman, Cancian, and Jackson might say (consider each separately)?

PART III

Hot Topics: Contemporary Issues, Debates, and Controversies

INTRODUCTION

Contemporary Issues, Debates, and Controversies locates social issues as instructive sites of discourse in which struggles between cultural hegemony and individual resistance are publicly articulated and contested. As evidenced in the previous section of this book, while social institutions are important in the production, transmission, and regulation of social constructions, resistance and opposition to hegemonic constructs is a regular feature of social existence. Though institutions may establish and reinforce specific modes of behavior and/or values, this by no means equates with general social compliance—indeed, it is often the case that active opposition exists within and between institutional constituencies. Resistance can emanate from many sources including, but not limited to, single individuals who simply disagree with existing normative structures as well as groups of similarly-situated individuals whose interests have been subordinated.

Opposition, resistance, and competition are not the sole sources of tension within the social structure. Another crucial dynamic is the fact that social institutions tend to react slowly to social changes and technological innovations often cause a time lag between the introduction of change and the establishment of clear institutional structures governing them. In the interim, competition between interested social groups over control of normative constructs and public discourse can be fiercely contested with multiple parties claiming to represent the "truth" or "reality" of a particular matter. Even after normative dominion over an issue has been established it is likely that resistance and opposition will endure.

Paying particular attention to highly contested public issues, controversies, and debates reveals telling fractures in the social foundation. These fractures not only help expose normative assumptions that are hidden and embedded in the framework, they can also help reveal alternate constructions of what is and what could be. In addition, they can aid in illuminating the situated interests of the groups that control, or are competing to control, how *truth* and *reality* are constructed.

Topics selected for this anthology include: sex, the body, and sexual attractiveness (Chapter 11); sex, pornography and sexual objectification (Chapter 12); sex and violence (Chapter 13); sex and the construction of sexually transmitted diseases (Chapter 14); as well as sex, abortion, contraception, and procreation (Chapter 15). These topics have been selected because they are not only highly contested areas of public debate, but also because of the outstanding contributions that research in these areas has made to a broader understanding of sexuality. Articles include Kristin Luker's well-respected work on the abortion debate titled "Motherhood and Morality in America" (Chapter 15); Jacquelyn Dowd Hall's inspired analysis of lynching and rape in "The Mind that Burns in Each Body" (Chapter 13); and Susan Bordo's cogent critique of the construction of male sexuality and impotence in "Pills and Power Tools" (Chapter 11).

11

Sex, the Body, and Sexual Attractiveness

Medicalization of Racial Features: Asian-American Women and Cosmetic Surgery

EUGENIA KAW

Eugenia Kaw's interviews with five cosmetic surgeons and eleven Asian-American female patients demonstrates the power of hegemonic constructions of attractiveness and desirability in modern Western society. Her research identifies cultural and institutional forces that combine to make cosmetic surgery the preferred solution for some Asian women who are willing to surgically alter their bodies to overcome the racial and gender stereotypes that brand specific Asian features as inferior to those of Whites.

Throughout history and across cultures, humans have decorated, manipulated, and mutilated their bodies for religious reasons, for social prestige, and for beauty (Brain 1979). In the United States, within the last decade, permanent alteration of the body for aesthetic reasons has become increasingly common. By 1988, 2 million Americans, 87% of them female, had undergone cosmetic surgery, a figure that had tripled in two years (Wolf 1991, 218). The cosmetic surgery industry, a $300 million per year industry, has been able to meet an increasingly wide variety of consumer demands. Now men, too, receive services ranging from enlargement of calves and chests to the liposuction of cheeks and necks (Rosenthal 1991). Most noticeably, the ethnic composition of consumers has changed so that in recent years there are more racial and ethnic

From "Medicalization of Racial Features: Asian-American Women and Cosmetic Surgery" by Eugenia Kaw in *Medical Anthropology Quarterly* 7(1) March 1993.

minorities. In 1994, 14% of cosmetic surgery patients were Latinos, African Americans, and Asian Americans (American Society of Plastic and Reconstructive Surgeons, http: *www.plasticsurgery.org*). Not surprisingly, within every racial group, women still constitute the overwhelming majority of cosmetic surgery patients, an indication that women are still expected to identify with their bodies in U.S. society today, just as they have across cultures throughout much of human history (Turner 1987, 85).[1]

The types of cosmetic surgery sought of women in the United States are racially specific. Like most white women, Asian women who undergo cosmetic surgery are motivated by the need to look their best as women. White women, however, usually opt for liposuction, breast augmentation, or wrinkle removal procedures, whereas Asian-American women most often request "double-eyelid" surgery, whereby folds of skin are excised from across their upper eyelids to create a crease above each eye that makes the eyes look wider. Also frequently requested is surgical sculpting of the nose tip to create a more chiseled appearance, or the implantation of silicone or cartilage bridge in the nose for a more prominent appearance. . . . While the features that white women primarily seek to alter through cosmetic surgery (i.e., the breasts, fatty areas of the body, and facial wrinkles) do not correspond to conventional markers of racial identity, those features that Asian-American women primarily seek to alter (i.e., "small, narrow" eyes and a "flat" nose) do correspond to such markers.[2]

My research focuses on the cultural and institutional forces that motivate Asian-American women to alter surgically the shape of their eyes and noses. I argue that Asian-American women's decision to undergo cosmetic surgery is an attempt to escape persisting racial prejudice that correlates their stereotyped genetic physical features ("small, slanty" eyes and a "flat" nose) with negative behavioral characteristics, such as passivity, dullness, and a lack of sociability. With the authority of scientific rationality and technological efficiency, medicine is effective in perpetuating these racist notions. The medical system bolsters and benefits from the larger consumer-oriented society not only by maintaining the idea that beauty should be every woman's goal but also by promoting a beauty standard that requires that certain racial features of Asian-American women be modified. Through the subtle and often unconscious manipulation of racial and gender ideologies, medicine, as a producer of norms, and the larger consumer society of which it is a part encourage Asian-American women to mutilate their bodies to conform to an ethnocentric norm.

Social scientific analyses of ethnic relations should include a study of the body. As evident in my research, racial minorities may internalize a body image produced by the dominant culture's racial ideology and, because of it, begin to loathe, mutilate, and revise parts of their bodies. Bodily mutilation and adornment are symbolic mediums most directly and concretely concerned with the construction of the individual as social actor or cultural subject (Turner 1980). Yet social scientists have only recently focused on the body as a central component of social self-identity (Blacking 1977; Brain 1979; Daly 1978; Lock and Scheper-Hughes 1990; O'Neill 1985; Turner 1987). Moreover, social scientists, and sociocultural anthropologist in particular, have not yet explored the ways in which the body is central to the everyday experience of racial identity.

MUTILATION OR A CELEBRATION OF THE BODY?

The decoration, ornamentation, and scarification of the body can be viewed from two perspectives. On the one hand, such practices can be seen as celebrations of the social and individual bodies, as expressions of belonging in society and an affirmation of oneness with the body (Brain 1979; Scheper-Hughes and Lock 1991; Turner 1980). On the other hand, they can be viewed as acts of mutilation, that is, as expressions of alienation in society and a negation of the body induced by unequal power relationships (Bordo 1990; Daly 1978; O'Neill 1985).

Although it is at least possible to imagine race-modification surgery as a *rite de passage* or a bid for incorporation into the body and race norms of the "dominant" culture, my research findings lead me to reject this as a tenable hypothesis. Here I argue that the surgical alteration by many Asian-American women of the shape of their eyes and nose is a potent form of self, body, and society alienation. Mutilation, according to *Webster's*, is the act of maiming, crippling, cutting up, or altering radically so as to damage seriously essential parts of the body. Although the women in my study do not view their cosmetic surgeries as acts of mutilation, an examination of the cultural and institutional forces that influence them to modify their bodies so radically reveals a rejection of their "given" bodies and feelings of marginality. On the one hand, they feel they are exercising their Americanness in their use of the freedom of individual choice. Some deny that they are conforming to any standard—feminine, Western, or otherwise—and others express the idea that they are, in fact, molding their own standards of beauty. Most agreed, however, that their decision to alter their features was primarily a result of their awareness that as women they are expected to look their best and that this meant, in a certain sense, less stereotypically Asian. Even those who stated that their decision to alter their features was personal, based on individual aesthetic preference, also expressed hope that their new appearance would help them in such matters as getting a date, securing a mate, or getting a better job.

For the women in my study, the decision to undergo cosmetic surgery was never purely or mainly for aesthetic purposes, but almost always for improving their social status as women who are racial minorities. Cosmetic surgery is a means by which they hope to acquire "symbolic capital" (Bourdieu 1984 [1979]) in the form of a look that holds more prestige. For example, "Jane," who underwent double-eyelid and nose-bridge procedures at the ages of 16 and 17, said that she thought she should get her surgeries "out of the way" at an early age since as a college student she has to think about careers ahead:

> Especially if you go into business, whatever, you kind of have to have a Western facial type and you have to have like their features and stature—you know, be tall and stuff. In a way you can see it as an investment in your future.

Such a quest for empowerment does not confront the cultural and institutional structures that are the real cause of the women's feelings of distress. Instead, this form of "body praxis" (Scheper-Hughes and Lock 1991) helps to entrench these structures by further confirming the undesirability of "stereo-typical" Asian features. Therefore, the alternation by many Asian-American women of their features is a "disciplinary" practice in the Foucauldian sense; it does not so much benignly transform them as it "normalizes" (i.e., qualifies, classifies, judges, and enforces complicity in) the subject (Foucault 1977). The normalization is a double encounter, conforming to patriarchal definitions of femininity and to Caucasian standards of beauty (Bordo 1990).

Gramsci anticipated Foucault in considering subjected peoples' complicity and participation in, as well as reproduction of, their own domination in everyday practice. In examining such phenomena as Asian-American women undergoing cosmetic surgery in the late 20th-century United States, however, one must emphasize, as Foucault does, how mechanisms of domination have become much more insidious, overlapping, and pervasive in everyday life as various forms of "expert" knowledge such as plastic surgery and surgeons have increasingly come to play the role of "traditional" intellectuals (Gramsci 1971) or direct agents of the bourgeois state (Scheper-Hughes 1992, 171) in defining commonsense reality.

Particularly in Western, late capitalist societies (where the decoration, ornamentation, and scarification of the body have lost much meaning for the individual in the existential sense of "Which people do I belong to? What is the meaning of my life?" and have instead become commoditized by the media, corporations, and even medicine in the name of fashion), the normalizing elements of such practices as cosmetic surgery can become obscured. Rather than celebrations of the body, they are mutilations of the body, resulting from a devaluation of the self and induced by historically determined relationships among social groups and between the individual and society.

INTERNALIZATION OF RACIAL AND GENDER STEREOTYPES

The Asian-American women in my study are influenced by a gender ideology that states that beauty should be a primary goal of women. They are conscious that because they are women, they must conform to certain standards of beauty. "Elena," a 20-year-old Korean-American, said, "People in society, if they are attractive, are rewarded for their efforts . . . especially girls. If they look pretty and neat, they are paid more attention to. You can't deny that." "Annie," another Korean-American who is 18 years old, remarked that as a young woman, her motivation to have cosmetic surgery was "to look better" and "not different from why [other women] put on makeup." In fact, all expressed the idea that cosmetic surgery was a means by which they could escape the task of having to put makeup on every day. As "Jo," a 28-year-old Japanese American who is thinking of enlarging the natural fold above her eyes, said, "I am self-conscious about leaving the house without any makeup on, because I feel just really ugly without it. I feel like it's the mask that enables me to go outside." Beauty, more than character and intelligence, often signifies social and economic success for them as for other women in U.S. society (Lakoff and Scherr 1984; Wolf 1991).

The need to look their best as women motivates the Asian-American women in my study to undergo cosmetic surgery, but the standard of beauty they try to achieve through surgery is motivated by a racial ideology that infers negative behavioral or intellectual characteristics from a group's genetic physical features. All of the women said that they are "proud to be Asian-American" and that they "do not want to look white." But the standard of beauty they admire and strive for is a face with larger eyes and a more prominent nose. They all stated that an eyelid without a crease and a nose that does not project indicate a certain "sleepiness," "dullness," and "passivity" in a person's character. "Nellee," a 21-year-old Chinese American, said she seriously considered surgery for double eyelids in high school so that she could "avoid the stereotype of the 'Oriental bookworm'" who is "*dull* and doesn't know how to have fun." Elena, who had double-eyelid surgery two years ago from a doctor in my study, said, "When I look at Asians who have no folds and their eyes are slanted and closed, I think of how they would look better more *awake*." . . . Annie, who had an implant placed in her nasa dorsum to build up her nose bridge at age 15, said:

> I guess I always wanted that *sharp* look—a look like you are smart. If you have a roundish kind of nose, it's like you don't know what's going on. If you have that sharp look, you know, with black eyebrows, a pointy nose, you look more *alert*. I always thought that was cool. [emphasis added]

Clearly, the Asian-American women in my study seek cosmetic surgery for double eyelids and nose bridges because they associate the features considered characteristic of their race with negative traits.

These associations that Asian-American women make between their features and personality characteristics stem directly from stereotypes created by the dominant culture in the United

States and by Western culture in general which historically has wielded the most power and hegemonic influence over the world. Asians are rarely portrayed in the U.S. popular media and then only in such roles as Charlie Chan, Suzie Wong, and "Lotus Blossom Babies" (a.k.a., China Doll, Geisha Girl, and shy Polynesian beauty). They are depicted as stereotypes with dull, passive, and nonsociable personalities (Kim 1986; Tajima 1989). Subtle depictions by the media of individuals' minutest gestures in everyday social situations can socialize viewers to confirm certain hypotheses about their own natures (Goffman 1979). At present, the stereotypes of Asians as a "model minority" serve a similar purpose. In the model minority stereotype, the concepts of dullness, passivity, and stoicism are elaborated to refer to a person who is hard-working and technically skilled but desperately lacking in creativity and sociability (Takaki 1989, 477).

Similar stereotypes of the stoic Asian also exist in East and Southeast Asia, and since many Asian-Americans are immigrants or children of recent immigrants from Asia, they are likely to be influenced by these stereotypes as well. U.S. magazines and films have been increasingly available in many parts of Asia since World War II. Also, multinational corporations in Southeast Asian countries consider their work force of Asian women to be biologically suited for the most monotonous industrial labor because the "Oriental girl" is "diligent" and has "nimble fingers" and a "slow wit" (Ong 1987, 151). Racial stereotypes of Asians as docile, passive, slow witted, and unemotional are internalized by many Asian-American women, causing them to consider the facial features associated with these negative traits as defiling.

Undergoing cosmetic surgery, then, becomes a means by which the women can attempt to permanently acquire not only a feminine look considered more attractive by society, but also a certain set of racial features considered more prestigious. For them, the daily task of beautification entails creating the illusion of features they, as members of a racial minority, do not have. Nellee, who has not yet undergone double-eyelid surgery, said that at present she has to apply makeup every day "to give my eyes an illusion of a crease. When I don't wear makeup I feel my eyes are small." . . . The enormous constraints the women in my study feel with regard to their Asian features are apparent in the meticulous detail with which they describe their discontent, as apparent in a quote from Jo who already has natural folds but wants to enlarge them: "I want to make an even bigger eyelid [fold] so that it doesn't look slanted. I think in Asian eyes this inside corner fold [she was drawing on my notebook] goes down too much."

The women expressed hope that the results of cosmetic surgery would win them better acceptance by society. Ellen said that she does not think her double-eyelid surgery "makes me look too different," but she nonetheless expressed the feeling that now her features will "make a better impression on people because I got rid of that sleepy look." She says that she will encourage her daughter, who is only 12 years old, to have double-eyelid surgery as she did, because "I think having less-sleepy-looking eyes would help her in the future with getting jobs." The aesthetic results of surgery are not an end in themselves but rather a means for these women as racial minorities to attain better socioeconomic status. Clearly, their decisions to undergo cosmetic surgery do not stem from a celebration of their bodies.

MEDICALIZATION OF RACIAL FEATURES

Having already been influenced by the larger society's negative valuation of their natural "given" features, Asian-American women go to see plastic surgeons in half-hour consultation sessions. Once inside the clinic, they do not have to have the doctor's social and medical views "thrust" on them, since to a great extent, they, like their doctors, have already entered into a more general social consensus (Scheper-Hughes 1992, 199). Nonetheless, the Western medical system is a most effective promoter of the racial stereotypes that influence Asian-American women, since medical knowledge is legitimized by scientific rationality and technical efficiency, both of which hold prestige in the West and increasingly all over the world. Access to a scientific body of knowledge has given Western medicine considerable social power in defining reality (Turner 1987, 11). According to my Asian-American informants who had undergone cosmetic surgery, their plastic surgeons used several medical terms to problematize the shape of their eyes so as to define it as a medical condition. For instance, many patients were told that they had "excess fat" on their eyelids and that it was "normal" for them to feel dissatisfied with the way they looked. "Lots of Asians have the same puffiness over their eyelids and they often feel better about themselves after the operation," the doctor would assure their Asian-American patients.

The doctors whom I interviewed shared a similar opinion of Asian fact features with many of the doctors of the patients in my study. Their descriptions of Asian features verged on ideological racism, as clearly seen in the following quote from "Dr. Smith."

> The social reasons [for Asian Americans to want double eyelids and nose bridges] are undoubtedly continued exposure to Western culture and the realization that the upper eyelid *without* a fold tends to give a *sleepy* appearance, and therefore a more *dull* look to the patient. Likewise, the *flat* nasal bridge and *lack of* nasal projection can signify *weakness* in one's personality and by *lack of* extension, *lack of force* in one's character. [emphasis added]

By using words like "without," "lack of," "flat," "dull," and "sleepy" in her description of Asian features, Dr. Smith perpetuates the notion that Asian features are inadequate. Likewise, "Dr. Khoo" said that many Asians should have surgery for double eyelids since "the eye is the window to your soul and having a more open appearance makes you look a bit brighter, more inviting." "Dr. Gee" agreed:

> I would say 90% of people look better with double eyelids. It makes the eye look more spiritually alive. . . . With a single eyelid frequently they would have a little fat pad underneath [which] can half bury the eye and so the eye looks small and unenergetic.

Such powerful associations of Asian features with negative personality traits by physicians during consultations can become a medical affirmation of Asian American women's sense of disdain toward their own features.

Medical books and journals as early as the 1950s and as recent as 1990 abound with similar metaphors of abnormality in describing Asian features. The texts that were published before 1970 contain more explicit association of Asian features with dullness and passivity. In an article published in 1954 in the *American Journal of Ophthalmology*, the author, a doctor in the Philippine armed forces, wrote the following about a man on whom he performed double-eyelid surgery:

> [He] was born with mere slits for his eyes. Everyone teased him about his eyes with the comment that as he looked constantly sleepy, so his business too was just as sleepy. For this reason, he underwent the plastic operation and, now that his eyes are wider, he has lost the sleepy look. His business, too, has picked up. [Sayoc 1954, 556]

The doctor clearly saw a casual link between the shape of his patient's eyes and his patient's intellectual and behavioral capacity to succeed in life. . . . Medical texts published after 1970 are more careful about associating Asian features with negative behavioral or intellectual characteristics, but they still describe Asian features with metaphors of inadequacy or excess. For instance, in the introductory chapter to a 1990 book devoted solely to medical techniques for cosmetic surgery of the Asian face, a white American plastic surgeon begins by cautioning his audience not to stereotype the physical traits of Asians.

> Westerners tend to have a stereotyped conception of the physical traits of

> Asians: yellow skin pigmentation . . . a flat face with high cheek bones; a broad, flat nose; and narrow slit-like eyes showing characteristic epicanthal folds. While this stereotype may loosely apply to central Asian groups (i.e., Chinese, Koreans, and Japanese), the facial plastic surgeon should appreciate that considerable variation exists in all of these physical traits (McCurdy 1990, 1).

Yet, on the same page, he writes that the medicalization of Asian features is valid because Asians usually have eyes that are too narrow and a nose that is too flat.

> However, given an appreciation of the physical diversity of the Asian population, certain facial features do form a distinct basis for surgical intervention. . . . These facial features typically include the upper eyelid, characterized by an absent or poorly defined superior palpebral fold . . . and a flattened nose with poor lobular definition (McCurdy 1990, 1).

Thus, in published texts, doctors write about Asians' eyes and noses as abnormal even when they are careful not to associate negative personality traits with these features. In the privacy of their clinics, they freely incorporate both metaphors of abnormality and the association of Asian features with negative characteristics into medical discourse, which has an enormous impact on the Asian-American patients being served.

The doctors' scientific discourse is made more convincing by the seeming objective manner in which they behave and present themselves in front of their patients in the clinical setting. . . . The sterile appearance of their clinics, with white walls and plenty of medical instruments, as well as the symbolism of the doctor's white coat with its many positive connotations (e.g., purity, life, unaroused sexuality, superhuman power, and candor) reinforce in the patient the doctor's role as technician and thus his sense of objectivity (Blumhagen 1979). One of my informants, Elena, said that, sitting in front of her doctor in his office, she felt sure that she needed eyelid surgery. "[Dr. Smith] made quite an impression on me. I thought he was more than qualified—that he knew what he was talking about."

With its authority of scientific rationality and technical efficiency, medicine effectively "normalizes" not only the negative feelings of Asian-American women about their features but also their ultimate decision to undergo cosmetic surgery. . . . By changing the patients' bodies the way they would like them, she feels she provides them with an immediate and concrete solution to their feelings of inadequacy.

Dr. Jones and the other doctors say that they only turn patients away when patients expect results that are technically impossible, given such factors as the thickness of the patient's skin and the bone structure. "I turn very few patients away," said Dr. Khoo. And "Dr. Kwan" notes:

> I saw a young girl [a while back] whose eyes were beautiful but she wanted a crease . . . She was gorgeous! Wonderful! But somehow she didn't see it that way. But you know, I'm not going to tell a patient every standard I have of what's beautiful. If they want certain things and it's doable, and if it is consistent with a reasonable look in the end, then I don't stop them. I don't really discuss it with them.

Like the other doctors in my study, Dr. Kwan sees himself primarily as the technician whose main role is to correct his patient's features in a way that he thinks would best contribute to the patient's satisfaction. It does not bother him that he must expose an individual, whom he already sees as pretty and not in need of surgery, to an operation that is at least an hour long, entails the administering of local anesthesia with sedation, and involves the following risks: "bleeding," "hematoma," "hemorrhage," formation of a "gaping wound," "discoloration," "scarring," and "asymmetry in lid fold" (Sayoc 1974, 162–166). He finds no need to try to change his patients' minds.

. . . Though most of my Asian-American woman informants who underwent cosmetic surgery recovered fully within six months to a year, with only a few minor scars from their surgery, they nonetheless affirmed that the psychologically traumatic aspect of the operation was something their doctors did not stress during consultation. Elena said of her double-eyelid surgery: "I thought it was a simple procedure. He [the doctor] should have known better. It took at least an hour and a half. . . . And no matter how minor the surgery was, I bruised! I was swollen."

. . . By focusing on technique and subordinating human emotions and motivations to technical ends, medicine is capable of normalizing Asian-American women's decision to undergo cosmetic surgery.

MUTUAL REINFORCEMENT: MEDICINE AND THE CONSUMER-ORIENTED SOCIETY

The medical system bolsters and benefits from the larger consumer-oriented society by perpetuating the idea that beauty is central to women's sense of self and also by promoting a beauty standard for Asian-American women that requires the alteration of features specific to Asian-American racial identity. All of the doctors in my study stated that a "practical" benefit for Asian-American women undergoing surgery to create or enlarge their eyelid folds is that they can put eye makeup on more appropriately. Dr. Gee said that after double-eyelid surgery it is "easier" for Asian-American women to put makeup on because "they now have two instead of just one plane on which to apply makeup." Dr. Jones agreed that after eyelid surgery Asian-American women" can do more dramatic things with eye makeup." The doctors imply that Asian-American women cannot usually put on makeup adequately, and thus, they have not been able to look as beautiful as they can be with makeup. By promoting the idea that a beautiful woman is one who can put makeup on adequately, they further the idea that a woman's identity should be closely connected with her body and, particularly, with the representational problems of the self. By reinforcing the makeup industry, they buttress the cosmetic surgery industry of which they are a part. A double-eyelid surgery costs patients $1,000 to $3,000.

The medical system also bolsters and benefits from the larger consumer society by appealing to the values of American individualism and by individualizing the social problems of racial inequality. Dr. Smith remarked that so many Asian-American women are now opting for cosmetic surgery procedures largely because of their newly gained rights as women and as racial minorities:

> Asians are more affluent than they were 15 years ago. They are more knowledgeable and Americanized, and their women are more liberated. I think in the past many Asian women were like Arab women. The men had their foot on top of them. Now Asian women do pretty much what they want to do. So if they want to do surgery, they do it.

Such comments by doctors encourage Asian-American women to believe that undergoing cosmetic surgery is merely a way of beautifying themselves and that it signifies their ability to exercise individual freedom.

Ignoring the fact that the Asian-American women's decision to undergo cosmetic surgery has anything to do with the larger society's racial prejudice, the doctors state that their Asian-American women patients come to cosmetic surgeons to mold their own standards of beauty. The doctors point out that the specific width and shape the women want their creases to be or the specific shape of nose bridges they want are a matter of personal style and individual choice.

. . . In fact, the doctors point out that both they and their Asian-American patients are increasingly getting more sophisticated about what the patients want. As evidence, they point to the fact that as early as a decade ago, doctors used to provide very wide creases to every Asian-American patient who came for double eyelids, not knowing that not every Asian wanted to look exactly Caucasian. The doctors point out that today many Asian-American cosmetic surgery patients explicitly request that their noses and eyelids not be made too look too Caucasian.

Recent plastic surgery literature echoes these doctors' observations. A 1991 press release from the American academy of Cosmetic Surgery quotes a prominent member as saying, "The procedures they [minorities, including Asian Americans] seek are not so much to look 'Western' but to refine their features to attain facial harmony." The double-eyelid surgery, he says, is to give Asian eyes "a more open appearance," not a Western look.

In saying that their Asian-American women patients are merely exercising their freedom to choose a personal style or look, the doctors promote the idea that human beings have an infinite variety of needs that technology can endlessly fulfill, an idea at the heart of today's U.S. capitalism. As Susan Bordo explains, the United States has increasingly become a "plastic" culture, characterized by a "disdain for material limits, and intoxication with freedom, change, and self-determination" (Bordo 1990, 654). She points out that many consumer products that could be considered derogatory to women and racial minorities are thought by the vast majority of Americans to be only some in an array of consumer choices to which every individual has a right. She explains:

> Any different self would do, it is implied. Closely connected to this is the construction of all cosmetic changes as the same: perms for white women, corn rows on Bo Derek, tanning, makeup, changing hair styles, blue contacts for black women. [Bordo 1990, 659]

CONCLUSION

Cosmetic surgery on Asian-American women for nose bridges and double eyelids is very much influenced by gender and racial ideologies. My research has shown that by the conscious or unconscious manipulation of gender and racial stereotypes, the American medical system, along with the larger consumer-oriented society of which it is a part, influences Asian-American women to alter their features through surgery. With the authority of scientific rationality and technological efficiency, medicine is effectively able to maintain a gender ideology that validates women's monetary and time investment in beauty even if this means making their bodies vulnerable to harmful and risky procedures such as plastic surgery. Medicine is also able to perpetuate a racial ideology that states that Asian features signify "dullness," "passivity and "lack of emotions" in the Asian person. The medicalization of racial features, which reinforces and normalizes Asian-American women's feelings of inadequacy, as well as their decision to undergo surgery, help to bolster the consumer-oriented society of which medicine is a part and from which medicine benefits.

Given the authority with which fields of "expert" knowledge such as biomedicine have come to define the commonsense reality today, racism and sexism no longer need to rely primarily on physical coercion to legal authority. Racial stereotypes influence Asian-American women to seek cosmetic surgery. Yet, through its highly specialized and validating forms of discourse and practices, medicine, along with a culture based on endless self-fashioning is able to motivate women to view their feelings of inadequacy as individually motivated, as opposed to socially induced, phenomena, thereby effectively convincing them to participate in the production and reproduction of the larger structural inequalities that continue to oppress them.

NOTES

1. In a 1989 study of 80 men and women, men reported many more positive thoughts about their bodies than did women (Goleman 1991).

   According to the American Society of Plastic Surgeons, 87% of all cosmetic surgery patients in 1990 were women. In my study, in one of the two doctors' offices from which I received statistical data on Asian-American patients, 65% of Asian-American cosmetic surgery patients in 1990 were women; in the other 62%.

2. The shapes of eyes and noses of Asians are not meant in this article to be interpreted as categories that define an objective category called Asians. Categories of racial groups are arbitrarily defined by society. Likewise, the physical traits by which people are recognized as belonging to a racial group have been determined to be arbitrary (Molnar 1983).

   Also, I use the term "Asian-American" to collectively name the women in this study who have undergone or are thinking of undergoing cosmetic surgery. Although I realize their ethnic diversity, people of Asian ancestry in the United States share similar experiences in that they are subject to many of the same racial stereotypes (Takaki 1989).

REFERENCES

Blacking, John. 1977. *The Anthropology of the Body.* London: Academic Press.

Blumhagen, Dan. 1979. The doctor's white coat: The image of the physician in modern America. *Annuals of Internal Medicine* 91:111–116.

Bordo, Susan. 1990. Material girl: The effacements of postmodern culture. *Michigan Quarterly Review* 29:635–676.

Bourdieu, Pierre. 1984. *Distinction: A Social Critique of the Judgement of Taste.* R. Nice, trans. Cambridge, MA: Harvard University Press.

Brain, Robert. 1979. *The Decorated Body.* New York: Harper and Row.

Daly, Mary. 1978. *Gyn/Ecology: The Metaethics of Radical Feminism.* Boston: Beacon Press.

Foucault, Michel. 1977. *Discipline and Punish: The Birth of the Prison.* A. Sheridan, trans. New York: Vintage Books.

Goffman, Erving. 1979. *Gender Advertisement.* Cambridge, MA: Harvard University Press.

Goleman, Daniel. 1991: When ugliness is only in the patient's eye, body image can reflect a mental disorder. *New York Times* 2 October:B9.

Gramsci, Antonio. 1971. *Selections from the Prison Notebooks of Antonio Gramsci.* New York: International.

Kim, Elaine. 1986. Asian Americans and American popular culture. In *Dictionary of Asian American History*, edited by Hyung-Chan Kim. New York: Greenwood Press.

Lakoff, Robin Tolmach, and Raquel L. Scherr. 1984. *Face Value: The Politics of Beauty.* Boston: Routledge and Kegan Paul.

Lock, Margaret, and Nancy Scheper-Hughes, 1990. A critical-interpretive approach in medical anthropology: Rituals and routines of discipline and dissent. In *Medical Anthropology: Contemporary Theory and Method*, edited by Thomas M. Johnson and Carolyn F. Sargent. New York: Praeger.

Molnar, Stephen. 1983. *Human Variation: Races, Types, and Ethnic Groups.* Englewood Cliffs, NJ: Prentice-Hall.

O'Neill, John. 1985. *Five Bodies.* Ithaca, NY: Cornell University Press.

Ong, Aihwa. 1987. *Spirits of Resistance and Capitalist Discipline: Factory Women In Malaysia.* Albany: State University New York Press.

Rosenthal, Elisabeth. 1991. Cosmetic surgeons seek new frontiers. *New York Times* 24 September: B5–B6.

Sayoc, B. T. 1954. Plastic construction of the superior palpebral fold. *American Journal of Ophthalmology* 38:556–559.

———. 1974. Surgery of the Oriental eyelid. *Clinics in Plastic Surgery* 1:157–171.

Scheper-Hughes, Nancy. 1992. *Death Without Weeping.* Berkeley: University of California Press.

Scheper-Hughes, Nancy, and Margaret M. Lock. 1991. The message in the bottle: Illness and the micropolitics of resistance. *Journal of Psychohistory* 18: 409–432.

Tajima, Renee E. 1989. Lotus blossoms don't bleed: Images of Asian women. In *Making Waves: An Anthology of Writings by and about Asian Women*, edited by Diane Yeh-Mei Wong. Boston: Beacon Press.

Takaki, Ronald. 1989. *Strangers from a Different Shore.* Boston: Little, Brown.

Turner, Bryan. 1987. *Medical Knowledge and Social Power.* London: Sage Publications.

Turner, Terence. 1980. The social skin. In *Not Work Alone*, edited by J. Cherfas and R. Lewin. London: Temple Smith.

Wolf, Naomi. 1991. *The Beauty Myth: How Images of Beauty Are Used Against Women.* New York: William Morrow.

DISCUSSION QUESTIONS

1. Is it "rational" to alter one's body surgically to attain cultural ideals? Why or why not? What are the costs? What are the benefits?
2. How far is "too far" to go in altering the body to attain the cultural ideal of attractiveness?
3. Do you expect rates of cosmetic surgeries to increase or decrease in the coming years? For which segments of the population? Why?
4. In what ways do social stratification shape and affect the occurrence of cosmetic surgery?
5. How "free" is the "choice" of women and minorities to have cosmetic surgery in a social structure that is white and patriarchal?
6. What are the sources of attractiveness for men in this culture? Are they the same as women? Are they different? If so, how? Do you expect this to change anytime in the near future? Why or why not?
7. In what ways do social institutions contribute to the Westernization of non-Western bodies?

Size 6: The Western Women's Harem

FATIMA MERNISSI

For many Westerners, the practice of veiling is a social directive that is powerfully symbolic of the subordination of women in many Middle-Eastern countries. The custom of Purdah, the seclusion of women, is also viewed as living testament to just how anachronistic and hyper-patriarchal such societies are said to be. Yet there are many Western practices that are, arguably, just as detrimental to the status of women. This work reflects Fatima Mernissi's efforts to take the Western reader beyond a typical (and ethnocentric) frame of reference to appreciate how Western norms and practices also operate to subvert and limit female autonomy and power.

It was during my unsuccessful attempt to buy a cotton skirt in an American department store that I was told my hips were too large to fit into a size 6. That distressing experience made me realize how the image of beauty in the West can hurt and humiliate a woman as much as the veil does when enforced by the state police in extremist nations such as Iran, Afghanistan, or Saudi Arabia. Yes, that day I stumbled onto one of the keys to the enigma of passive beauty in Western harem fantasies. The elegant saleslady in the American store looked at me without moving from her desk and said that she had no skirt my size. "In this whole big store, there is no skirt for me?" I said. "You are joking." I felt very suspicious and thought that she just might be too tired to help me. I could understand that. But then the saleswoman added a condescending judgment, which sounded to me like an Imam's *fatwa*. It left no room for discussion:

"You are too big!" she said.

"I am too big compared to what?" I asked, looking at her intently, because I realized that I was facing a critical cultural gap here.

"Compared to a size 6," came the saleslady's reply.

From "Size 6: The Western Women's Harem" in *Scheherazade Goes West: Different Cultures, Different Harems* by Fatima Mernissi, pp. 208–220.

Her voice had a clear-cut edge to it that is typical of those who enforce religious laws. "Size 4 and 6 are the norm," she went on, encouraged by my bewildered look. "Deviant sizes such as the one you need can be bought in special stores."

That was the first time that I had ever heard such nonsense about my size. In the Moroccan streets, men's flattering comments regarding my particularly generous hips have for decades led me to believe that the entire planet shared their convictions. It is true that with advancing age, I have been hearing fewer and fewer flattering comments when walking in the medina, and sometimes the silence around me in the bazaars is deafening. But since my face has never met with the local beauty standards, and I have often had to defend myself against remarks such as *zirafa* (giraffe), because of my long neck, I learned long ago not to rely too much on the outside world for my sense of self-worth. In fact, paradoxically, as I discovered when I went to Rabat as a student, it was the self-reliance that I had developed to protect myself against "beauty blackmail" that made me attractive to others. My male fellow students could not believe that I did not give a damn about what they thought about my body. "You know, my dear," I would say in response to one of them, "all I need to survive is bread, olives, and sardines. That you think my neck is too long is your problem, not mine."

In any case, when it comes to beauty and compliments, nothing is too serious or definite in the medina, where everything can be negotiated. But things seemed to be different in that American department store. In fact, I have to confess that I lost my usual self-confidence in that New York environment. Not that I am always sure of myself, but I don't walk around the Moroccan streets or down the university corridors wondering what people are thinking about me. Of course, when I hear a compliment, my ego expands like a cheese soufflé, but on the whole, I don't expect to hear much from others. Some mornings, I feel ugly because I am sick or tired; others, I feel wonderful because it is sunny out or I have written a good paragraph. But suddenly, in that peaceful American store that I had entered so triumphantly, as a sovereign consumer ready to spend money, I felt savagely attacked. My hips, until then the sign of a relaxed and uninhibited maturity, were suddenly being condemned as a deformity.

"And who decides the norm?" I asked the saleslady, in an attempt to regain some self-confidence by challenging the established rules. I never let others evaluate me, if only because I remember my childhood too well. In ancient Fez, which valued round-faced plump adolescents, I was repeatedly told that I was too tall, too skinny, my cheekbones were too high, my eyes were too slanted. My mother often complained that I would never find a husband and urged me to study and learn all that I could, from storytelling to embroidery, in order to survive. But I often retorted that since "Allah had created me the way I am, how could he be so wrong, Mother?" That would silence the poor woman for a while, because if she contradicted me, she would be attacking God himself. And this tactic of glorifying my strange looks as a divine gift not only helped me to survive in my stuffy city, but also caused me to start believing the story myself. I became almost self-confident. I say almost, because I realized early on that self-confidence is not a tangible and stable thing like a silver bracelet that never changes over the years. Self-confidence is like a tiny fragile light, which goes off and on. You have to replenish it constantly.

"And who says that everyone must be a size 6?" I joked to the saleslady that day, deliberately neglecting to mention size 4, which is the size of my skinny twelve-year-old niece.

At that point, the saleslady suddenly gave me an anxious look. "The norm is everywhere, my dear," she said. "It's all over, in the magazines, on television, in the ads. You can't escape it. There is Calvin Klein, Ralph Lauren, Gianni Versace, Giorgio Armani, Mario Valentino, Salvatore Ferragamo, Christian Dior, Yves Saint-Laurent, Christian Lacroix, and Jean-Paul Gaultier. Big department stores go by the norm." She paused and then concluded, "If they sold size 14 or 16, which is probably what you need, they would go bankrupt."

She stopped for a minute and then stared at me, intrigued. "Where on earth do you come from? I am sorry I can't help you. Really, I am." And she looked it too. She seemed, all of a sudden, interested, and brushed off another woman who was seeking her attention with a cutting, "Get someone else to help you, I'm busy." Only then did I notice that she was probably my age, in her late fifties. But unlike me, she had the thin body of an adolescent girl. Her knee-length, navy blue, Chanel dress had a white silk collar reminiscent of the subdued elegance of aristocratic French Catholic schoolgirls at the turn of the century. A pearl-studded belt emphasized the slimness of her waist. With her meticulously styled short hair and sophisticated makeup, she looked half my age at first glance.

"I come from a country where there is no size for women's clothes," I told her. "I buy my own material and the neighborhood seamstress or craftsman makes me the silk or leather skirt I want. They just take my measurements each time I see them. Neither the seamstress nor I know exactly what size my skirt is. We discover it together in the making. No one cares about my size in Morocco as long as I pay taxes on time. Actually, I don't know what my size is, to tell you the truth."

The saleswoman laughed merrily and said that I should advertise my country as a paradise for stressed working women. "You mean you don't watch your weight?" She inquired, with a tinge of disbelief in her voice. And then, after a brief moment of silence, she added in a lower register, as if talking to herself: "Many women working in highly paid fashion-related jobs could lose their positions if they didn't keep to a strict diet."

Her words sounded so simple, but the threat they implied was so cruel that I realized for the first time that maybe "size 6" is a more violent restriction imposed on women than is the Muslim veil. Quickly I said good-bye so as not to make any more demands on the saleslady's time or involve her in any more unwelcome, confidential exchanges about age-discriminating salary cuts. A surveillance camera was probably watching us both.

Yes, I thought as I wandered off, I have finally found the answer to my harem enigma. Unlike the Muslim man, who uses space to establish male domination by excluding women from the public arena, the Western man manipulates time and light. He declares that in order to be beautiful, a woman must look fourteen years old. If she dares to look fifty, or worse, sixty, she is beyond the pale. By putting the spotlight on the female child and framing her as the ideal of beauty, he condemns the mature woman to invisibility. In fact, the modern Western man enforces Immanuel Kant's nineteenth-century theories: To be beautiful, women have to appear childish and brainless. When a woman looks mature and self-assertive, or allows her hips to expand, she is condemned as ugly. Thus, the walls of the European harem separate youthful beauty from ugly maturity.

These Western attitudes, I thought, are even more dangerous and cunning than the Muslim ones because the weapon used against women is time. Time is less visible, more fluid than space. The Western man uses images and spotlights to freeze female beauty within an idealized childhood, and forces women to perceive aging—that normal unfolding of the years—as a shameful devaluation. "Here I am, transformed into a dinosaur," I caught myself saying aloud as I went up and down the rows of skirts in the store, hoping to prove the saleslady wrong—to no avail. This Western time-defined veil is even crazier than the space-defined one enforced by the Ayatollahs.

The violence embodied in the Western harem is less visible than in the Eastern harem because aging is not attacked directly, but rather masked as an aesthetic choice. Yes, I suddenly felt not only very ugly but also quite useless in that store, where, if you had big hips, you were simply out of the picture. You drifted into the fringes of nothingness. By putting the spotlight on the prepubescent female, the Western man veils the older, more mature woman, wrapping her in shrouds of ugliness. This idea gives me the chills because it tattoos the invisible harem directly onto a woman's skin. Chinese foot-binding worked the same way: Men declared beautiful only those women who had small, childlike feet. Chinese men did not force women to bandage their feet to keep them from developing normally—all they did was to define the beauty ideal. In feudal China, a beautiful woman was the one who voluntarily sacrificed her right to unhindered physical movement by mutilating her own feet, and thereby proving that her main goal in life was to please men. Similarly, in the Western world, I was expected to shrink my hips into a size 6 if I wanted to find a decent skirt tailored for a beautiful woman. We Muslim women have only one month of fasting, Ramadan, but the poor Western woman who diets has to fast twelve months out of the year. "*Quelle horreur*", I kept repeating to myself while looking around at the American women shopping. All those my age looked like youthful teenagers.

According to the writer Naomi Wolf, the ideal size for American models decreased sharply in the 1990s. "A generation ago, the average model weighed 8 percent less than the average American woman, whereas today she weighs 23 percent less. . . . The weight of Miss America plummeted, and the average weight of Playboy Playmates dropped from 11 percent below the national average in 1970 to 17 percent below it in eight years."[1] The shrinking of the ideal size, according to Wolf, is one of the primary reasons for anorexia and other health-related problems: "Eating disorders rose exponentially, and a mass of neurosis was promoted that used food and weight to strip women of . . . a sense of control."[2]

Now, at last, the mystery of my Western harem made sense. Framing youth as beauty and condemning maturity is the weapon used against women in the West just as limiting access to public space is the weapon used in the East. The objective remains identical in both cultures: to make women feel unwelcome, inadequate, and ugly.

The power of the Western man resides in dictating what women should wear and how they should look. He controls the whole fashion industry, from cosmetics to underwear. The West, I realized, was the only part of the world where women's fashion is a man's business. In places like Morocco, where you design your own clothes and discuss them with craftsmen and -women, fashion is your own business. Not so in the west. As Naomi Wolf explains in *The Beauty Myth*, men have engineered a prodigious amount of fetish-like, fashion-related paraphernalia: "Powerful industries—the $33-billion-a-year diet industry, the $20-billion cosmetic industry, the $300-million cosmetic surgery industry, and the $7-billion pornography industry—have arisen from the capital made out of unconscious anxieties, and are in turn able, through their influence on mass culture, to use, stimulate, and reinforce the hallucination in a rising economic spiral."[3]

But how does the system function? I wondered. Why do women accept it?

Of all the possible explanations, I like that of the French sociologist, Pierre Bourdieu, the best. In his latest book, *La Domination Masculine*, he proposes something he calls "*la violence symbolique*": "Symbolic violence is a form of power which is hammered directly on the body, and as if by magic, without any apparent physical constraint. But this magic operates only because it activates the codes pounded in the deepest layers of the body."[4] Reading Bourdieu, I had the impression that I finally understood Western man's psyche better. The cosmetic and fashion industries are only the tip of the iceberg, he states, which is why women are so ready to adhere to their dictates. Something else is going on on a far deeper level. Otherwise, why would women belittle themselves spontaneously? Why, argues Bourdieu, would women make their lives more difficult, for example, by preferring men who are taller or older than they are? "The majority of French women wish to have a husband who is older and also, which seems consistent, bigger as far as size is concerned," writes Bourdieu.[5] Caught in the enchanted submission characteristic of the symbolic violence inscribed in the mysterious layers of the flesh, women relinquish what he calls "les signes ordinaries de la hiérarchie sexuelle," the ordinary signs of sexual hierarchy, such as old age and a larger body. By so doing, explains Bourdieu, women spontaneously accept the subservient position. It is this spontaneity Bourdieu describes as magic enchantment.[6]

Once I understood how this magic submission worked, I became very happy that the conservative Ayatollahs do not know about it yet. If they did, they would readily switch to its sophisticated methods, because they are so much more effective. To deprive me of food is definitely the best way to paralyze my thinking capabilities.

Both Naomi Wolf and Pierre Bourdieu come to the conclusion that insidious "body codes" paralyze Western women's abilities to compete for power, even though access to education and professional opportunities seem wide open, because the rules of the game are so different according to gender. Women enter the power game with so much of their energy deflected to their physical appearance that one hesitates to say the playing field is level. "A cultural fixation on female thinness is not an obsession about female beauty," explains Wolf. It is "an obsession about female obedience. Dieting is the most potent political sedative in women's history; a quietly mad population is a tractable one."[7] Research, she contends, "confirmed what most women know too well—that concern with weight leads to a 'virtual collapse of self-esteem and sense of effectiveness' and that . . . prolonged and periodic caloric restriction' resulted in a distinctive personality whose traits are passivity, anxiety, and emotionality."[8] Similarly, Bourdieu, who focuses more on how this myth hammers its inscriptions onto the flesh itself, recognizes that constantly reminding women of their physical appearance destabilizes them emotionally because it reduces them to exhibited objects. "By confining women to the status of symbolical objects to be seen and perceived by the other, masculine domination . . . puts women in a state of constant physical insecurity. . . . They have to strive ceaselessly to be engaging, attractive, and available."[9] Being frozen into the passive position of an object whose very existence depends on the eye of its beholder turns the educated modern Western woman into a harem slave.

"I thank you, Allah, for sparing me the tyranny of the 'size 6 harem,' " I repeatedly said to myself while seated on the Paris-Casablanca flight, on my way back home at last. "I am so happy that the conservative male elite does not know about it. Imagine the fundamentalists switching from the veil to forcing women to fit size 6."

How can you stage a credible political demonstration and shout in the streets that your human rights have been violated when you cannot find the right skirt?

NOTES

1. Naomi Wolf, *The Beauty Myth: How Images of Beauty Are Used Against Women* (New York: Anchor Books, Doubleday, 1992), p. 185.
2. Ibid., p. 11.
3. Ibid., p. 17.
4. Pierre Bourdieu: "La force symbolique est une forme de pouvoir qui s'exerce sur les corps, directement, et comme par magie, en dehors de toute contraine physique, mais cette magie n'opère qu'en s'appuyant sur des dispositions déposées, tel des ressorts, au plus profond des corps." In *La Domination Masculine* (Paris: Editions du Seuil, 1998), op. cit. p. 44.

   Here I would like to thank my French editor, Claire Delannoy, who kept me informed of the latest debates on women's issues in Paris by sending me Bourdieu's book and many others. Delannoy has been reading this manuscript since its inception in 1996 (a first version was published in Casablanca by Edition Le Fennec in 1998 as "Êtes-Vous Vacciné Contre le Harem").
5. *La Domination Masculine*, op. cit., p.41.
6. Bourdieu, op.cit., p. 42.
7. Wolf, op. cit., p.187.
8. Wolf, quoting research carried out by S. C. Woolly and O. W. Woolly, op. cit., pp. 187–188.
9. Bourdieu, *La Domination Masculine*, p. 73.

DISCUSSION QUESTIONS

1. In what ways do appearance norms in the West and the East similarly restrict women's social status? In what way do these norms differ between the West and the East (include a consideration of Chinese foot binding here as well)? Do you agree with the author's position that the size 6 norm is more restrictive (and insidious) than the harem? Why or why not?
2. Do Western appearance norms embody violence? If so, how?
3. By aspiring to fulfill dominant appearance norms do women accept subservience? Why or why not? In recent years men have begun paying more attention to their physical appearance; does this indicate a similar kind of subservience? If so, subservience to what, or whom? If not, why not?
4. If you were to conduct a cross-cultural study of female self-esteem in Morocco and the United States, in which country would you expect rates of self-esteem to be higher? Why? If men from both countries were added to the study what would you expect the relative self-esteem ranking to be?
5. Do you feel any of the discussed practices empower women? Why or why not?
6. How voluntary is compliance to appearance norms? What are the benefits of conformity? What are the costs of deviation, defiance, and/or failure?

Pills and Power Tools

SUSAN BORDO

Susan Bordo's analysis of the popularity of Viagra reveals that men's sense of themselves as sexual beings is strongly tied to the reliability and performance of a single, all too (ph)allible, *organ.*

Viagra. When it went on sale in April of 1998, it broke all records for "fastest takeoff of a new drug" that the Rite Aid drugstore chain had ever seen. It was all over the media. Users were jubilant, claiming effects that lasted through the night, youth restored, better-"quality" erections.

Some even viewed Viagra as a potential cure for social ills. Bob Guccione, publisher of *Penthouse,* hails the drug as "freeing the American male libido" from the emasculating clutches of feminism. This diagnosis doesn't sit very comfortably with current medical wisdom, which has declared impotence to be a physiological problem. I, like Guccione, am skeptical of that declaration—but would suggest a deeper meditation on what's put the squeeze on male libido.

Think, to begin with, of the term: *Impotence.* It rings with disgrace, humiliation—and it's not feminists who invented it. Writer Philip Lopate, in an essay on this body, says that merely to say the word out loud makes him nervous.

Unlike other disorders, impotence implicates the whole man, not merely the body-part. *He is impotent.* Would we ever say about a person with a headache, "*He is a headache*"? Yet this is just

From "Pills and Power Tools" by Susan Bordo in *Men & Masculinities*, 1, pp. 87–90.

what we do with impotence, as Warren Farrell notes. "We make no attempt to separate impotence from the total personality," writes Farrell. "Then, we expect the personality to perform like a machine." "Potency" means power. So I guess it's correct to say that the machine we expect men to perform like is a power tool.

That expectation of men is embedded throughout our culture. To begin with, we encourage men to think of their sexuality as residing in their penises, and give them little encouragement to explore the rest of their bodies. The beauty of the male body has finally been brought out of the cultural closet by Calvin Klein, Versace, and other designers. But notice how many of those new underwear ads aggressively direct our attention to the (often extraordinary) endowments of the models. Many of them stare coldly, challengingly at the viewer, defying the viewer's "gaze" to define them in any way other than how they have chosen to present themselves: powerful, armored, emotionally impenetrable. "I am a rock." their bodies seem to proclaim. Commercial advertisements depict women stroking their necks, their faces, their legs, lost in sensual reverie, taking pleasure in touching themselves—all over. Similar poses with men are very rare. Touching one-self languidly, lost in the sensual pleasure of the body, is too feminine, too "soft," for a real man. Crotch grabbing, thrusting, putting it "in your face"—that's another matter.

There's a fascinating irony in the fact that although it is women whose bodies are most sexually "objectified" by this culture, women's bodies are permitted much greater sexual expression in our cultural representations than men's. In sex scenes, the moaning and writhing of the female partner have become the conventional cinematic code for heterosexual ecstasy and climax. The male's participation largely gets represented via caressing hands, humping buttocks, and—on rare occasions—a facial expression of intense concentration. She's transported to another world; he's the pilot of the ship that takes her there. When men are shown being transported themselves, it's usually been played for comedy (as in Al Pacino's shrieks in *Frankie and Johnny*, Eddie Murphy's moaning in *Boomerang*, Kevin Kline's contortions in *A Fish Called Wanda*), or it's coded to suggest that something is not quite normal with the man—he's sexually enslaved, for example (as with Jeremy Irons in *Damage*). Men's bodies in the movies are action-hero toys—wind them up and watch them perform.

Think of our slang-terms, so many which encase the penis, like a cyborg, in various sorts of metal or steel armor. Big rig. Blow torch. Bolt. Cockpit. Crank. Crowbar. Destroyer. Dipstick. Drill. Engine. Hammer. Hand tool. Hardware. Hose. Power tool. Rod. Torpedo. Rocket. Spear. Such slang—common among teen-age boys—is violent in what it suggests the machine penis can do to another, "softer" body. But the terms are also metaphorical protection against the failure of potency. A human organ of flesh and blood is subject to anxiety, ambivalence, uncertainty. A torpedo or rocket, on the other hand, would never let one down.

Contemporary urologists have taken the metaphor of man the machine even further. Erectile functioning is "all hydraulics," says Irwin Goldstein of the Boston University Medical Center, scorning a previous generation of researchers who stressed psychological issues. Goldstein was quoted in a November 1997 *Newsweek* cover story called "The New Science of IMPOTENCE," announcing the dawn of the age of Viagra. At the time, the trade name meant little to the casual reader. What caught my eye were the contradictory messages. On the one hand, that ugly shame-inducing word *IMPOTENCE* was emblazoned throughout the piece. On the other hand, we were told in equally bold letters that science was "REBUILDING THE MALE MACHINE." If it's all a matter of fluid dynamics, I thought, why keep the term *impotent*, whose definitions (according to *Webster's Unabridged*) are: "want of power," "weakness," "lack of effectiveness, helplessness" and (only lastly) "lack of ability to engage in sexual intercourse"?

In keeping the term *impotence*, I figured, the drug companies would get to have it both ways: reduce a complex human condition to a matter of chemistry, while keeping the old shame-machine working, helping to assure the flow of men to their doors.

It's remarkable, really, when you think about it, that *impotence* remained a common nomenclature among medical researchers (instead of the more forgiving, if medicalized, *erectile dysfunction*) for so long. *Frigidity*—with its suggestion that the woman is "cold," like some barren tundra—went by the board a long while ago. But *impotence*, no less loaded with ugly gender implications, remained the term of choice—not only for journalists but also for doctors—throughout all of the early reportage on Viagra. "THE POTENCY PILL," *Time* magazine called it, in its May 4th, 1998, issue three weeks after Viagra went on sale. At the same time, inside the magazine, fancy charts with colored arrows, zigzags, triangles, circles, and boxes show us "How Viagra Works," a cartoonlike hot dog the only suggestion that a penis is involved in any of this.

The drug companies eventually realized that *impotence* was as politically incorrect as *frigidity*. They also, apparently, began to worry about the reputation that Viagra was getting as a magic bullet that could produce rampant erections out of thin air. Pfizer's current ad for Viagra announces "A pill that helps men with erectile dysfunction respond again." *Respond*. The word attempts to create a counter-image not only to the early magic-bullet hype, but also to the curious absence of partners in men's descriptions of the effects of the drug. *It's* "Stronger." *It's* "Harder." "Longer-lasting." "Better quality." *It's* "Firmer." "The characters in the drama of Viagra were three: a man, his blessed power pill, and his restored power tool.

The way Viagra is supposed to work—as the Pfizer ad goes on to say—is by helping you "achieve erections the natural way—in response to sexual stimulation." *Natural. Response.* It illustrates its themes with a middle-aged man in a suit dipping his gray-haired partner, smiling ecstatically. *Partners. Happy partners.* A playful, joyous, moment. "Let the dance begin," announces Pfizer at the very bottom of the ad, as though it were orchestrating a timeless, ritual coupling. The way men *had* been talking about the effects of Viagra, that dance was entirely between them and their members.

It wasn't playful rumba though. More like a march performed to the finale of the "1812 Overture," accompanied with cannon-blasts. "This little pill is like a package of dynamite," says one user. "Turned into a monster" (says another, with pleasure). "You just keep going all night. The performance is unbelievable," said one. I'm not making fun of these responses; I find them depressing. The men's explosive pride, to me, is indicative of how small and snail-like these men had felt before, and of the extravagant relief now felt at becoming a "real man," imagined in these comments as some kind of monster Energizer Bunny (pardon me, *Rabbit.*).

Something else is revealed, too, by the absence of partners in these descriptions of the effects of Viagra. The first "sex life" of most men in our culture—and a powerful relationship that often continues throughout their lives—involves a male, his member, and a magazine (or some other set of images seemingly designed with male libido in mind). Given the fast-trigger nature of adolescent sexuality, it doesn't take much; indeed, it sometimes seems to the teenage boy as though everything female has been put on the face of the earth just to get men hot. Philip Roth's descriptions of Alex Portnoy masturbating at the sight of his sister's bra, capable of getting a hard-on even at the sound of the *word* "panties," are hilarious—and true to life. Despite myths to the contrary, it's sometimes not so different for adolescent girls, either. But when we grow up, those "hard-on" moments (if, to make a point, I may use that metaphor in a unisexual way) aren't transformed into launch-off preparations for a sexual "performance."

It's often been noted that women's sexual readiness can be subtle to read. We are not required to cross a dramatic dividing line in order to engage in intercourse. And we aren't expected—as men are expected, as men seem to expect of themselves—to retain that hair-trigger sexuality of adolescence. Quite the opposite, in fact; the mythology about women is that we're "slow-cookers" when it comes to sex. Men, in contrast, get hit with a double whammy: they feel that they have to perform and they expect themselves to do so at the mere sight of a fancy brassiere! It's one reason, I think, why so many men "trade up" for younger wives as they get older; they're looking for that quick sexual fix of adolescence. It's their paradigm of sexual response, their criterion (ironically, since it represents the behavior of a fifteen-year-old) of manliness.

It comes as no surprise then, to learn that as sexual "performers," many men seem to expect no tactile help from the audience except—hopefully—applause at the end. Gail Sheehy (who, by the way, has cleaned up her own terminology; in an article from the early nineties, her term for "male menopause" was *viropause*, now she calls it *manopause*) reports that Leonore Tiefer's interviews with hundreds of cops, firemen, sanitation workers, and blue-color workers at Montefiore Medical Center in the Bronx revealed that most of these men expect, even in their fifties, to be able to get an erection just from paging through *Playboy*. Sheehy goes on: "When the sexologist suggests that at this age a man often needs physical stimulation they balk: 'C'mon, Doc, it's not *masculine* for a woman to have to get it up for me.' Their wives often echo that rigid code: 'He should get it up.' "

Some dance, huh?

Most studies of Viagra's "effectiveness" leave partners out of the picture, too. When you put them in, you get a somewhat different picture of the "success" of the drug. In England, they used something called a "RigiScan" to measure the penis's "resistance" against a cloth-covered ring while Viagra-treated men watched porn movies. In the United States, the 69 percent "success rate" that Pfizer submitted to the F.D.A. was based on questionnaires filled out by patients. "Real soft data, no pun intended, "William Steers, Chief of Urology at the University of Virginia was quoted
New Yorker. He we
study included no s
cheers to him, *did* a
when you ask wo
"Viagra-enhanced"
the success of the
men—is about 48 p
matter what measure of "success" you use.

Steers does not provide detail as to what those different measures were. Perhaps "monsters" were not what partners were looking for in bed. Perhaps they didn't appreciate the next-day chafing that usually accompanies "going all night." (I once knew a man who could stay erect all night; it had its upside—so to speak—but I couldn't escape the feeling that a bit of sadism, at the very least a control complex, was fueling his unbending passion.) Perhaps partners didn't want *just* a proud member, but the kind of romantic attention that goes along with the proud member in romance novels. "We're a very meat-and-potatoes culture," says Karen Martin, a sex therapist in upstate New York. "In other cultures, they toss in a few mushrooms." Maybe the wives of Viagrans wanted a few mushrooms tossed in with the beef.

They are less apt to be served them, however, if the couple has had sexual problems. Such couples may have grown distant from each other, may no longer know how to communicate intimately with each other, physically or verbally. Since the initial wave of enthusiasm about Viagra, therapists have begun to worry that Viagra is providing couples with a way to sidestep dealing not only with relationship problems that may have *contributed* to their sexual difficulties ("just because there is a physiological problem doesn't mean there is no psychological cause," reminds Eileen Palace, director of the Center for Sexual Health at Tulane) but also patterns of alienation, resentment, and anger that may develop *because* of those difficulties. A number of studies have found that when men begin to have erectile difficulties, a common response is to turn away from *all* romantic and affectionate gestures—kissing, ugging—so as not to (as one said) gs up." Many don't offer manual or oral ation in place of intercourse, because that ld be to admit to themselves that they can't perform" the "way a man should." They're often uncomfortable talking about the situation with their wives (and even their doctors, who report that most of the men who are asking for prescriptions for Viagra never mentioned their dysfunction before.) "I would tend to kind of brush the problem under the rug," says one man. "It isn't an easy topic to deal with. It goes to the heart of your masculinity."

Into the middle of all this distance, confusion, anxiety, and strain walks Viagra, and with it the news that the problem is only a malfunctioning hydraulic system—which, like any broken machinery, can be fixed. And "let the dance begin!" Many couples, unsurprisingly, don't know the steps, stumble, and step all over each other's toes.

Let me make it clear that I have no desire to withhold Viagra from the many men who have been deprived of the ability to get an erection by accidents, diabetes, cancer, and other misfortunes to which the flesh—or psyche—is heir. I would, however, like CNN and *Time* to spend a fraction of the time they devote to describing "how Viagra cures" to thinking about that gentleman's astute comment—"It goes to the heart of your masculinity"—and perhaps devoting a few features to exploring the functioning of *that* body part as well.

Such an exploration is, thankfully, occurring in other cultural arenas. Paul Thomas Anderson's *Boogie Nights* told the story of the rise and fall (so to speak) of a mythically endowed young porn star, Dirk Diggler, who does fine so long as he's the most celebrated stallion in the stable but loses his grip in the face of competition. On the surface, the film is about a world far removed from the lives of most men, a commercial underground where men pray for "wood" and lose their jobs unless they can achieve erection on command. On a deeper level, however, the world of the porn actor is simply the most literalized embodiment—and a perfect metaphor—for a masculinity that demands constant performance from men.

Even before he takes up a career that depends on it, Diggler's sense of self is constellated around his penis; he pumps up his ego by looking in the mirror and—like a coach mesmerizing his team before a game—intoning mantras about his superior gifts. That works well, so long as he believes it. But unlike a real power tool, the motor of male self-worth can't simply be switched on and off. In the very final shot of the movie, we see Diggler's fabled organ itself. It's a prosthesis, actually (a fact that annoyed several men I know until I pointed out that it was no more a cheat than implanted breasts passing for the real thing). But prosthesis or not and despite its dimensions, it's no masterful tool. It points downward, weighted with expectation, with shame, looking tired and used. What's ground-breaking about *Boogie Nights* is not that it displays a nude penis, but that it so unflinchingly exposes the job that the mythology of unwavering potency does on the male body. As long as the fortress holds, the sense of power may be intoxicating; but when it cracks—as it's bound to at some point—the whole structure falls to pieces. Those of whom such constancy is expected (or who require it of themselves) are set up for defeat and humiliation.

Unless, of course, he pops his little pill whenever "failure" threatens.

Some of what we now call "impotence" may indeed be physiological in origin. Some may be grounded in deep fears and anxieties. But whatever the cause of a man's sexual problems, the "heart of masculinity" isn't a mechanical pump, and in imagining the penis as such, Viagran science actually administers more of the poison it claims to counteract. Dysfunction is no longer defined as "inability to get an erection" but inability to get an erection that is adequate for "satisfactory sexual performance." *Performance.* Not pleasure. Not feeling. *Performance.* Eighty-five-year-old men are having Viagra heart attacks trying to keep those power tools running.

Sometimes, perhaps, a man's "impotence" may simply be his penis instructing him that his *feelings* are not in synch with the job he's supposed to do—or with the very fact that it's a "job." So, I like Philip Lopate's epistemological metaphor for the penis much better than the machine images. Over the years, he has come to appreciate, he writes, that his penis has its "own specialized form of intelligence." The penis knows that it is not a torpedo, no matter what a culture expects of it or what drugs are relayed to its blood vessels. If we accepted that, the notion that a man requires understanding and "tolerance" when he doesn't "perform" would go by the wayside. ("It's O.K. It happens" still assumes that there is some thing to be excused.) So, too, would the idea that there ought to be one model for understanding nonarousal. Sometimes, the penis's "specialized intelligence" should be listened to rather than cured.

Viagra, unfortunately, seems to be encouraging rather than deconstructing the expectation that men perform like power tools with only one switch—on or off. Until this expectation is replaced by a conception of manhood that permits men *and* their penises a full range of human feeling, we won't yet have the kind of "cure" we really need.

REFERENCES

Broder, David. (1998). "Side Effect of Viagra May Be End of a Great Stupidity," *Lexington Herald Leader*, July 27.

Cameron, Deborah. (1992). "Naming of Parts Gender, Culture, and Terms for the Penis Among American College Students," *American Speech* Vol. 67, No. 4 pp. 367–382.

Cowley, Geoffrey. (1998). "Is Sex a Necessity," *Newsweek*, May 11, pp. 62–63.

Farrell, Warren. (1986). *Why Men Are the Way They Are* New York: Berkley Books.

Goldstein, Irwin. (1997). "The New Science of IMPOTENCE," *Newsweek*, November.

Handy, Bruce. (1998). "The Viagra Craze," *Time*, May 4, pp. 50–57.

Hendren, John. (1998). "Pfizer Presses Insurers on Viagra," "*Washington Post*, July 7, p. E3.

Hitchens, Christopher. (1998). "Viagra Falls," *The Nation*, May 25, p. 8.

Leland, John. (1997). "A Pill for Impotence?" *Newsweek*, November 17, pp. 62–68.

Lopate, Phillip. (1993). "Portrait of My Body," *Michigan Quarterly Review*, Volume XXXII, Number 4, Fall, pp. 656–665.

Martin, Douglas. (1998). "Thanks a Bunch, Viagra," *New York Times*, May 3.

Risher, Michael T. (1998). "Controlling Viagra-Mania," *New York Times*, July 20, p. A19.

Safire, William. (1998). "Is There a Right to Sex?" *New York Times*, July 13, p. A21.

Sharpe, Rochelle. (1998). "FDA Received Data on Adverse Effects from Using Viagra," *Wall Street Journal*, June 29, p. B5.

Steinhauer, Jennifer. (1998). "Viagra's Other Side Effect: Upsets in Many a Marriage," *New York Times*, June 23, pp. B9, B11.

Tiefer, Leonore. (1994). "The Medicalization of Impotence," *Gender and Society*, Vol. 8, No. 3, September, pp. 363–377.

Time, May 4, 1998.

DISCUSSION QUESTIONS

1. In what ways does the marketing of Viagra reflect, reify, and naturalize gendered constructions of masculinity and male sexuality?
2. What, if anything, does the marketing of Viagra suggest about female sexuality?
3. Define sexual pleasure. How is it achieved? Is it tied to specific acts? Do you think this definition differs for men and women? Does Viagra enhance sexual pleasure? If so, for whom? What is the role of pleasure vis-à-vis masculinity in the popularity of Viagra?
4. How is the call for a female version of Viagra similar to and different from the demand for Viagra?
5. Can you think of any forms of resistance that might be effective for constructing alternate conceptualizations of male sexuality?
6. What are the consequences of a phallic/performance-centered sexuality for men? What are the consequences for their sexual partners?
7. On what points concerning male sexuality do you think Bordo and Tiefer would agree? Disagree? Why?

Jane Fonda, Barbara Bush, and Other Aging Bodies: Femininity and the Limits of Resistance

MYRA DINNERSTEIN AND ROSE WEITZ

In a culture in which beauty is synonymous with youth, and women are characteristically evaluated by their appearance, Myra Dinnerstein and Rose Weitz examine the challenges faced by the aging female body. In this piece, the authors examine the divergent strategies used by two powerful female public figures to manage their physical appearances. Dinnerstein and Weitz's cogent analysis reveals the complexities and contradictions demonstrated by both women's strategies and brings into question the limits of individual resistance to oppressive beauty norms.

Although separated by only half a generation, Jane Fonda and Barabara Bush present us with almost diametrically opposed images of how women can age. Fonda, born in 1937, boasts a "relentlessly improved" body—muscular and nearly fat-free, with dyed hair and surgically enhanced face and breasts. In contrast, Bush, born in 1925, exhibits a "resolutely natural" look, with her matronly figure, white hair and wrinkled countenance.[1]

The striking contrast between the appearances of these two highly visible women led us to select them for a study of how women manage their aging bodies. Their dissimilar appearances seemed to reflect two different approaches to aging and to the prevailing cultural discourse which equates femininity with a youthful appearance (Freedman 1986, 200; Seid 1989; Sontag 1972; Woodward 1991, 161). Moreover, both describe their behavior and appearance as forms of resistance to these cultural pressures. As we will show, however, their lives testify more to the limits of individual resistance than to its possibilities.

As public figures, with sufficient financial resources to employ all that the beauty and fashion industries can offer, Fonda and Bush differ in important ways from most women. In addition, as white heterosexuals, issues of beauty and aging might have different cultural meanings for them than for many lesbians and minority women. Nevertheless, the narratives of their aging reveal a dilemma that most American women confront as they age: how to handle an aging body in a culture in which aging challenges acceptable notions of femininity. . . .

AGENCY, SOCIAL CONTROL, AND THE FEMALE BODY

Part of the problem women face in resisting cultural definitions of femininity is that these definitions influence us in ways that we do not fully recognize. While many women are aware of the barrage of messages from the media about appropriate feminine appearance, few realize

From "Jane Fonda, Barbara Bush, and Other Aging Bodies: Femininity and the Limits of Resistance" by Myra Dinnerstein and Rose Weitz, *Feminist Issues* 14(3): 3–24, 1994.
Reprinted by permission of Transaction Publishers.

how insidiously these notions have entered our individual psyches or think to question their legitimacy. Newspaper articles might report on studies showing that a high proportion of ten-year-old girls diet and that an epidemic of anorexia and bulimia has swept the country, but few women respond to this evidence of the dire consequences of preoccupation with body size by changing their own attitudes and behaviors.

To illuminate the covert and powerful ways that cultural ideas about femininity shape women's conscious and unconscious attitudes toward their bodies, some feminist scholars have drawn on the work of Michel Foucault (1979, 136). Foucault describes the body in some of his writings as an "object and target of power," a field on which the hierarchies of power are displayed and inscribed, and has shown how various institutions such as the army and the school deploy their power by controlling their members' time, space and movement. Feminists have elaborated his argument, describing how definitions of femininity act, in Foucault's terms, as a "discipline" regulating women's bodies (Bartky 1988, 63–64; Bordo 1989, 13–14; Foucault 1979, 136–139). As army regulations control a soldier's gestures, walk, and posture, cultural definitions of femininity control female bodies by setting standards that specify what is considered appropriate and "normal" (Bartky 1988, 61–63, 75; Foucault 1979, 136–138, 150–153). These feminist scholars have suggested that women have so internalized the rewards and dictates of femininity that they appear to conform willingly (Bartky 1988). . . .

In contrast, other feminist scholars contend that this argument overstates the power of cultural norms to discipline women. Critics such as McNay (1991) argue that Foucault and his followers see women only as "docile bodies," ignoring the many ways that women historically have resisted societal prescriptions, while even those who emphasize oppression such as Bartky (1988, 82) concede that there are "pockets of resistance" and "oppositional discourses" to normative femininity, such as among radical lesbians and female body builders.

Our Analysis of Fonda and Bush joins this discussion by highlighting the interplay between agency and social control in the reactions of heterosexual white women to their aging appearance, demonstrating how Fonda and Bush simultaneously have attempted to resist the dominant discourse on aging and have been constrained by it. Their experiences delineate the difficulties of resistance and underscore the struggle required to establish oppositional discourses.

For this paper, we draw on all articles written about Bush and Fonda in women's magazines (including fashion magazines) indexed in the *Readers' Guide to Periodical Literature* beginning in 1977, when Fonda turned forty. . . .

AGING FEMALE BODIES: A BRIEF HISTORY

What constitutes a culturally appropriate appearance for aging women has changed considerably over time (Banner 1983; Schwartz 1986; Seid 1989). Until the mid-twentieth century, American society expected older women to have what was termed a "mature figure" and to a wear "mature" fashions. Beginning in the 1950s, however, this demarcation between youthful and aging appearances began to break down. The emphasis on youth accelerated in the 1960s and early 1970s, as both medical and fashion experts, bolstered by the highly visible youth culture, declared the youthful, slim body the standard for all (Seid 1989).

By the late 1970s, as the baby boomers who had fostered the youth culture aged, the struggle to maintain a youthful appearance had fostered a nationwide fitness craze.

The admonition to become slim and fit, which intensified in the 1980s, ignored the biological realities of aging—the typically unavoidable weight gain, increased ratio of fat to muscle, and, for women, thickening waists and sagging breasts (Seid 1989, 265). Instead, both medical and popular experts redefined "mature figures" as symbols of self-indulgence and irresponsibility. Failure to take up the new ethic of fitness became a sign of social or even moral failure, with the unfit deemed the secular equivalent of sinners and the fit promised youth and health (Crawford 1979; Stein 1982, 168–169, 174; Tesh 1988; Turner 1984, 202; Waitzkin 1981; Zola 1972).

. . . The increased cultural focus since the 1970s on controlling women's bodies has led several critics to label it a "backlash" to the rising power and visibility of women. These commentators suggest that keeping women involved with controlling their bodies diverts their energies from striving to achieve more control in the public arena (Bordo 1989, 14; Bordo 1990, 105; Faludi 1991, 203–204; Wolf 1991, 9–12). An analysis of Jane Fonda, who both rode the wave of the new fitness craze and helped to create it, and who promotes women's control of the body as a form of "liberation," illuminates the contradictions of preoccupation with the body.

MANAGING THE AGING BODY

Jane Fonda: Relentlessly Improved

Jane Fonda has undergone many metamorphoses in her lifetime, summarized by one magazine writer "as sex kitten in the 50s, antiwar radical in the late 60s, feminist in the 70s, successful entrepreneur in the 80s," and someone who focused on "personal fulfillment" in the 90s (Ball 1992, 96). Throughout however, Fonda has worked diligently to maintain a shapely body.

Ever since she turned forty, Fonda has garnered particular attention and admiration as a woman who has aged yet retained her beauty. With each successive year, the women's magazines have marked Jane Fonda's chronological aging by announcing her age and marveling at how well and, specifically, how young she looks (e.g., Andersen 1989, 112; Davis 1990, 165). Fonda fascinates aging women, one journalist suggests, "because she has a great body, an over-40 body that offers hope and promise. Along with each book and tape comes convincing evidence that it's possible . . . to remain beautiful and sexy in midlife" (Levin 1987, 27).

Ironically, for those who recall her 1968 film role as the sex machine "Barbarella" and her pleasure-filled lifestyle as the wife of French film director Roger Vadim, by the 1970s, Fonda had begun describing herself as a feminist role model (e.g., Robbins 1977). Fonda began this new phase of her life—as not only a feminist but also as an anti-Vietnam War activist—at about the time that her marriage to Vadim was ending and shortly before meeting and marrying political activist Tom Hayden (Anderson 1990, 214–265).

[S]he described how her feminist awareness had grown when Hayden showed her pictures of Vietnamese women who had plastic surgery to make their eyes round and enlarge their breasts. Fonda was "stunned and I thought my God that same phony Playboy image that made me wear falsies for ten years, that made billions of American women dissatisfied with their own bodies, has been transported thousands of miles to another culture and made these women too hate their bodies, made them willing to mutilate themselves" (Lear 1976, 15).

. . . In the 1980s, as Fonda emerged as an exercise guru she at first extolled the virtues of intervening only minimally in the aging process. At age 47, writing about aging skin and cosmetic surgery in *Women Coming of Age* (1984, 71), her book on midlife, she advised, "The course I prefer: making peace with the growing numbers of fine (and some not so fine) lines you see on your face. . . . Wrinkles are part of who we are, of where we've been. Not to have wrinkles means never having laughed or cried or expressed passion, never having squinted into the sun or felt the bite of winter's wind—never having fully lived!"

. . . Despite this rhetoric, however, most of Fonda's pronouncements even during these years centered on changing and controlling rather than accepting the body. Yet in these statements as well, Fonda used feminist rhetoric such as "freedom" and "liberation" to frame this bodily control as a form of resistance to traditional expectations of female emotional and physical weakness. Bodily control, she has argued, "gives me a sense of freedom" (Kaplan 1987, 417), for "discipline is freedom" (*Mademoiselle* 1980, 38). Similarly, she has described herself as "liberated" from the cultural pressures that led her to engage in binge/purge eating until age 35.

Magazine writers, too, have described Fonda's life in feminist terms. One reporter, for example, comments that "There is a sense she's on the right side, making a political statement in warm-ups and running shoes. On one level, her tapes are about women taking control of their bodies, gaining physical confidence in front of their VCRs, then striding out of their homes to flex their new power" (Kaplan 1987, 417).

As she began to promote her fitness methods, Fonda also argued that working to develop bodily control empowers women by improving their health. . . . Similarly, Fonda sees exercise as a means toward mental health: "Exercise does a tremendous amount for emotional and mental stability. . . . That is what satisfies me most, . . . the people who come up after a week or two of classes and say, 'Hey, I'm not depressed anymore' " (*Harper's Bazaar* 1980, 82).

Given the prevailing discourse on femininity and on aging, it is not surprising that Fonda and her admirers have characterized her as feminist. In a culture which characterizes women as weak, Fonda's goal of strengthening the body offers an alternative to fragile womanhood. It also disputes the idea that the aging body is a rundown machine (Martin 1987, 40–46).

. . . The benefits of a reasonable exercise program as a way to enhance emotional and physical health, maintain a healthy body weight, and protect against osteorporosis and heart disease have been well established and should not be minimized. What Fonda and her admirers have overlooked, however, in their litany of the benefits of exercise, is the contrast between the obsessive and punitive self-control and self-surveillance implicit in the exercise and beauty regimens she advocates and the liberation she claims to derive from them.

. . . In one interviewer's words, Fonda is "an exercise addict, like few others" (Levin 1987, 28). Her language of war, struggle, and labor indicate the amount of effort and self-surveillance she devotes to bodily control: "I have a constant weight problem. . . . It's just a constant struggle" (*Mademoiselle* 1980, 40). Similarly: "I constantly fight against weight, and I don't necessarily have a good figure. I have to work for it, and I do work for it" (Harrison 1978, 40).

. . . It seems, then, that Fonda's desire for a culturally appropriate body has tyrannized rather than liberated her. Fonda's life demonstrates that maintaining an "acceptable" body requires constant vigilance. As feminist critics have pointed out, such constant demands for self-surveillance keep women in line (Faludi 1991; Wolf 1991).

At the same time, Fonda's insistence on the importance of bodily self-control can lead to denigrating those who appear to lack control. In Fonda's worldview, care of the body is not only a personal responsibility but also a moral one. Bodily control signifies self-control, and hence moral superiority. This belief, which permeates the health and fitness movement, has been labeled a "new asceticism" or "new Puritanism" (Kilwein 1989, 9–10; Turner 1984). If in the past individuals disciplined the body to control their passions and submit to God, nowadays individuals discipline the body to extend their lives and increase their pleasures (Turner 1984, 156, 161–163, 172). Yet the value placed on self-discipline continues to resonate with its religious origins, for a fit body still announces a good character.

In Fonda's secularized world, personal "salvation" (of the character and body, if not the soul) is possible through self-abnegation—denying oneself food and exercising even to the point of pain (cf., Bordo 1990, 83). She has waxed lyrical about the physical benefits of denying herself food: "Going to bed on an empty stomach and waking up hungry is the greatest thing you can do for yourself. If you go to bed hungry and wake up hungry, you've got unbelievable energy" (*Mademoiselle* 1980, 40).

. . . Fonda explicitly rejects the notion that the ability to control one's body might depend upon such factors as genetics, leisure time, or financial resources. When, for example, a reporter suggested in 1980 that Fonda might have inherited her father's slimness, Fonda bristled and replied, "I don't have my father's body at all. He has absolutely no ass. I like to think a lot of my body is my own doing and my own blood and guts. It's my responsibility" (*Mademoiselle* 1980, 40).

By extension, those who do not take responsibility for their bodies have only themselves to blame for their problems. . . . Despite Fonda's rhetoric of empowerment, therefore, her emphasis on self-control disheartens and disempowers more women than it empowers.

A similar disempowerment results from Fonda's stance on sexuality. Fonda appears to be rescuing aging women from their usual portrayal as asexual by arguing that is possible for aging women to remain active, sexual beings. "I think," she says, "that when you're healthy, no matter what age you are, you have more sexual stamina and desire and flexibility and the things that go into an active sex life" (Orth 1984, 416). In Fonda's worldview, however, a woman can retain sexuality only by retaining a youthful body. No alternative vision of aging sexuality is presented or considered. All that remains is the impossible goal of remaining young.

An emphasis on youth is not the only drawback to Fonda's ideal of femininity. Despite her emphasis on fitness, Fonda has not created a new and liberating model of femininity so much as she has adopted a model that unrealistically combines stereotypically masculine and feminine elements. Fonda began her fitness quest by arguing for the virtues of a muscled and virtually far-free body. When asked why she added bodybuilding exercises, Fonda remarked "I didn't have any muscles. I like to see muscle, I like to see sinews, and after taking this class, I could see definition in my arms, shoulders and back. There's no fat anywhere, not anywhere" (*Mademoiselle* 1980, 38). This near-fatless ideal corresponds to an archetypical male body. More recently, as her relationship with Hayden faded, Fonda additionally has strived to achieve the archetypical female hallmark—large breasts—by submitting to cosmetic surgery (Anderson 1990, 327–328; Ball 1992, 97). Since her marriage to cable television owner Ted Turner in 1991, she is now, one magazine writer notes, "playing the part of the glamorous wife—blond, bejeweled and dressed to kill" (Messina 1993, 65).

Fonda's body thus has come to symbolize the duality of current femininity: tight muscles but with large breasts. The desire for a more masculine body, Bordo argues, appeals to some women because it symbolizes power in the public arena and a revolt against maternity and restrictive definitions of femininity (Bordo 1990, 105). Simultaneously, the inflated breasts serve as a reminder that a woman's body is there for male desires. Mainstream American culture idealizes women who are both supermen (in their muscles) and superwomen (in their breasts). These women, their bodies suggest, won't be dependent but will continue to seek the "male gaze" to affirm themselves. Fonda's enthusiastic embrace of this dual-function body represents not resistance but submission to the demands of femininity.

Fonda's use of cosmetic surgery (which did not end with her breast implants) suggests that the value she places on having a culturally acceptable body overrides any philosophical commitment to achieving a fit body through hard work.

. . . The image of Fonda passively lying on a table while receiving breast implants appears to contradict her active and aggressive advocacy of exercise as a means of achieving self-improvement. Yet all her efforts are really of one piece, aimed at maintaining a youthful and culturally acceptable appearance.

Fonda's struggles to control her inevitably aging body find resonance in many women's lives as they face a culture that rejects the aging female. As one writer said sympathetically about Fonda's "fanatical" physical fitness, "Wasn't it really just the panic of an aging actress with a terror of growing old? Children of Hollywood probably know more than the rest of us what extra wrinkles and pounds mean—and the rest of us know plenty" (Davis 1990, 165).

. . . Over the years, Fonda has expressed the desire to construct a new image for aging women, saying "You can be big and still be in proportion and well-toned" (Bachrach 1989, 82) and that the point is to be the best you can be (Messina 1993, 65). Ultimately, however, she has not done so. Her fit and muscled body and her unwrinkled face offer a standard that few women can attain, suggesting that only by remaining young and fit can women be sexual, strong, and good. As one writer observed, Fonda "at every age . . . has managed to sell her youth. And, most important, we have bought it" (Bachrach 1989). . . .

Barbara Bush: Resolutely "Natural"

In contrast to Jane Fonda, Barbara Bush at least appears to resist and reject the cultural ideal of femininity that valorizes youthfulness. Bush's white hair and wrinkles seem to challenge the prevailing feminine ideal and signal her rejection of technological "fixes" for aging (Morgan 1991, 28).

Even more than Fonda, Bush frames her appearance as resistance to cultural strictures about women's bodies. In Bush's worldview, her unwillingness to change her appearance beyond dressing well, using light makeup, and exercising regularly and moderately underscores her interest in focusing on what she considers important rather than frivolous and narcissistic concerns. She thus draws on a spiritual tradition that disregards the body and looks to the state of the soul for moral value (Turner 1984, 164).

To emphasize this philosophy, Bush often and not very subtly contrasts herself with Nancy Reagan, her ulta-stylish predecessor. Bush told one interviewer that while she "admires" Reagan's "perfectionism," she herself is "more interested in people than in perfection" (Cook 1990, 229). . . . Bush appears not only to present aging positively but also to scorn and mock those who seem more concerned about their looks.

Bush appears to counterpose the value of "naturalness" against the contrivances of anti-aging technology. "What you see is what you get" (*Vogue* 1988, 442) is one of her favorite self-descriptions, implying that her relative lack of artifice in appearance means that she stands revealed for who she is without the "disguises" of cosmetics, plastic surgery, or even a diet-improved shape. . . . Bush's discourse reflects an older view of aging that made allowances for the weight gain and other changes that can accompany aging and that excused women above a certain age from the burdens of maintaining sexual attractiveness (Banner 1983; Schwartz 1986; Seid 1989). Despite Bush's rhetoric, however, a close reading of her interviews reveals the great efforts she makes to meet standards of feminine attractiveness and to show herself as reasonably concerned about her appearance. In doing so she reveals her exquisite awareness of the demands of femininity and of her need at least partially to comply. . . .

Bush also pays attention to her wardrobe, wearing fashionable designer clothes. As Judith Viorst astutely noted, "though Barbara Bush seems to define herself publicly as a slightly schlumpy Everywoman, she is considerably more than that. . . . Her grooming, from well-coiffed white hair to well-shod toe, is what anyone would call 'impeccable' " (Viorst 1991, 40), an observation repeated by other magazine writers (*Vogue* 1988, 444; Cook 1990, 230). As one points out, "Her hairdresser attends her regularly, sometimes as often as three times a day" (McClellan 1992, 191).

Bush's use of self-deprecating humor further indicates her felt need to explain her deviation from cultural expectations. Self-deprecating humor, in which individuals mock themselves for not meeting social expectations, allows individuals both to demonstrate their commitment to those expectations and to frame their deviations as humorous rather than serious flaws (Coser 1960; Goffman 1963, 100–101; Haig 1988; Koller 1988; Ungar 1984; Walker 1988). As a result, such humor is most common among relatively powerless groups. . . .

It seems, then, that Bush, like many women, uses self-deprecating humor to try to turn her departures from appearance norms into an unimportant, humorous matter. At least one reporter has recognized this, describing her humor as "a preemptive strike" (Cook 1990, 230), in which Bush makes comments about her appearance before interviewers can. Thus, Bush is not, despite her protests, rejecting femininity standards, but is acknowledging their power and her failure to meet them.

Bush's statements about her sexuality also reveal the limits of her resistance. In her self-descriptions, she often desexualizes herself, referring to herself not merely as a grandmother but as "everybody's grandmother." She uses this exceptional status to place herself outside the bounds of femininity. "I mean," she says, "kissing me is like kissing your grandmother" (Reed 1989, 314).

Bush's depiction of herself as outside the definition of normative femininity does not help to construct a positive, alternative model of female aging. Rather, she has escaped the bonds of cultural standards only by forfeiting claims to sexuality—a high price to pay. Thus her model is a bleak one, which denies the sensual possibilities of the aging female body.

. . . To "normalize" Bush's appearance, the magazines, like Bush, have stressed her grandmotherly qualities, labeling her "every American's favorite grandmother" (Mower 1922) and an honorific grandmother of us all. They have focused on her work with children and her role as the matriarch of the large Bush clan (Reed 1989, 314). In this role as super-grandmother, the magazines allow Bush to remain outside the normal discipline of femininity, a disembodied maternal archetype.

The magazines also deal with Bush's appearance, as she does herself, by highlighting and praising her apparent "naturalness." For example, in an article entitled "The Natural," the headline notes that "Barbara Bush Remains Doggedly Herself" (Reed 1989, 312), while another article proclaims "Barbara Bush is real" (*Vogue* 1988, 218). Like Bush, the magazines assume that such an entity as a "natural body" exists and disregard the considerable role that culture plays in the construction of this "naturalness." Focusing on her "naturalness" as indicative of a praiseworthy inner self allows the magazines to downplay her body, while underscoring for readers how very noteworthy and thus unnatural it is.

Despite these efforts to normalize Bush's appearance, the magazines continue to demonstrate considerable ambivalence toward her looks. For example, in an article published in *Ladies Home Journal* (Avery 1989), the author praises Bush for her apparent acceptance of aging but suggests that other women should hesitate before adopting Bush as a role model, citing the numerous studies that show that attractive women do better in the job market. The writer concludes with the admonition that even "exceptional women" like Bush need to recognize the benefits of attractiveness, and predicts that "perhaps after she's had a year or two squarely in the public eye, we'll begin to notice subtle changes—a wrinkle smoothed here, a pound or two dieted off their." Bush, in other words, will see the light and conform to cultural standards of femininity.

The author and magazine reveal their real attitudes in two full-length pictures of Bush which dominate the article's first two pages (Avery 1989, 120–121). On the first page is the actual Barbara Bush. Attached to various parts of her in balloon-like fashion are boxes advising her how to deal with her wrinkes, her clothes, and her shape. On the opposite page stands a retouched Bush with all the suggestions put into practice: slimmer, with fluffed out hair, wrinkles surgically removed, wearing cosmetics, and fashionable clothes. It is the ultimate make-over! This desire to change Bush so that she fits in with prevailing notions of femininity is a frequent theme in women's magazines.

Women's Power and Women's Bodies

The treatment of Jane Fonda and Barbara Bush by women's magazines explicates the discourse on femininity and illustrates the difficulties women face in resisting cultural dictates regarding their aging bodies. Women's magazines praise Fonda only because she conforms to cultural ideals. The magazines' more ambivalent attitude toward Bush reflects their consternation and ultimate disapproval of those who do not appear to comply with appearance norms, even when such individuals in various ways acknowledge their deviance. Women do not have to be filmstars or political wives to feel the pressures imposed by cultural definitions of femininity. Those pressures are everywhere, and it is the rare woman who can evade them . . .

The pressures to conform to cultural standards of femininity disempower women in insidious ways—insidious because we have internalized them in ways we are hardly aware of and at costs we have not calculated. Cultural standards encourage women's sense of inadequacy and promote frantic use of expensive, time-consuming, and sometimes dangerous technologies in a futile effort to check aging and increase one's "femininity." Furthermore, such a focus on the body as a privatistic concern turns people away from focusing on social issues (Stein 1982, 176–177). This is particularly relevant during this current period of backlash to the feminist movement in which many women now find it easier to focus on their individual selves than to continue struggling for the social betterment of women. As Fonda herself admits about exercising "in a world that is increasingly out of control, it's something you can control" (Ball 1992, 143). Thus an emphasis on the self diverts women from resisting current power relations.

Moreover, as women get older and attempt to move into senior ranks in business, government, and the professions, their employers, clients, and colleagues expect them to maintain norms of femininity. These norms create barriers to advancement for aging women. While on men, gray hair, wrinkles, even a widening waist signify experience, wisdom, maturity, and sometimes sexiness (as in the cases of Clint Eastwood and Sean Connery, for example), on women they denote decline and asexuality. Women's power is diminished, therefore, at the very moment when they might otherwise begin to move into more powerful positions.

Given the difficulties of contesting the discourse on women's bodies, many of which are explicated in the experiences of Fonda and Bush, can we still conclude that effective resistance is possible? We would argue, perhaps optimistically, that it is, but only if women become more aware of the insidious, internalized ways that the discipline of femininity disempowers women and join together to fight it. . . .

Changing cultural standards of femininity will not be easy, and we concur with Bordo that it would be a mistake to minimize the power that the discourse on femininity has to regulate women's lives. To do so is to underestimate the amount of struggle that is required for change and to minimize the difficulty women experience when they try to accomplish change (Bordo 1989, 13). Yet, as the history of women has shown, it would also be a mistake to dismiss the possibility of resistance. While the experiences of Barbara Bush and Jane Fonda demonstrate how difficult it is for any woman, struggling alone, no matter how visible, respected, and in some ways powerful, to fight against entrenched cultural notions, the history of women has taught us that, even in the face of seemingly implacable hegemonic discourses, women can make change when they join together. Perhaps now, with the graying of the feminist movement, feminists will begin to devote more energy to challenging the cultural construction of women's aging bodies. At the same time, the increasing if slow movement of women in their forties and above into positions of power may help a new cultural construction to evolve.

Making changes in our attitudes toward our bodies will not be easy, but it will surely be an important step in empowering women. In such a world, Jane Fonda might be able to moderate her painful and obsessive regimens and Barbara Bush would not have to be so defensive about her appearance.

NOTE

1. We have borrowed the terms "resolutely natural" and "relentlessly improved" from an article by Avery (1989) on Barbara Bush. Bush is referred to as "resolutely natural" while, in this article, it is Cher and not Jane Fonda who is referred to as "relentlessly improved."

REFERENCES

Andersen, Christopher 1990. *Citizen Jane.* New York: Henry Holt.

Andersen, Christopher P. October 1989. Jane Fonda: I'm stronger than ever. *Ladies Home Journal.*

Avery, Caryl S. June 1989. How good should you look? *Ladies Home Journal.*

Bachrach, Judy. October 1989. Feel the burn. *Savvy Woman.*

Ball, Aimee Lee. March 1992. How does Jane do it? *McCall's.*

Banner, Lois W. 1983. *American Beauty.* New York: Alfred A. Knopf.

Bartky, Sandra Lee. 1988. Foucault, femininity, and the modernization of patriarchal power. In *Feminism and Foucault: Reflections on Resistance* edited by Irene Diamond and Lee Quinby. Boston: Northeastern University Press.

Bordo, Susan. 1989. The body and the reproduction of femininity: A feminist appropriation of Focault. In *Gender/Body/Knowledge*, edited by Alison M. Jaggar and Susan R. Bordo. New Brunswick, NJ: Rutgers University Press.

———, 1990. Reading the slender body. In *Body/Politics: Women and the Discourses of Science*, edited by Mary Jacobus, Evelyn Fox Keller, and Sally Shuttleworth. New York: Routledge.

Chernin, Kim. 1981. *The Obsession: Reflections on the Tyranny of Slenderness.* New York: Harper & Row.

Cook, Alison. March 1990. At home with Barbara Bush. *Ladies Home Journal.*

Coser, Rose Loeb. 1960. Laughter among colleagues. *Psychiatry* 23:81–95.

Crawford, Robert. 1979. Individual responsibility and health politics. In *Health Care in America: Essays in Social History*, edited by Susan Reverby and David Rosner. Philadelphia: Temple University Press.

Davis, Sally Ogle. January 1990. Jane Fonda bounces back. *Cosmopolitan.*

———. March 1993. Hollywood marriages: The good, the bad, and the disasters. *Ladies Home Journal.*

Faludi, Susan. 1991. *Backlash: The Undeclared War Against American Women.* New York: Crown.

Fonda, Jane. 1984. *Women Coming of Age.* New York: Simon and Schuster.

———. 1986. *Jane Fonda's Workout Book.* New York: Simon and Schuster.

Foucault, Michel. 1979. *Discipline and Punish.* New York: Vintage Books.

Freedman, Rita. 1986. *Beauty Bound.* New York: D. C. Heath and Company.

Goffman, Erving. 1963. *Stigma: Notes on the Management of Spoiled Identity.* Englewood Cliffs, NJ: Prentice-Hall.

Haig, Robin A. 1988. *Anatomy of Humor.* Springfield, IL: Charles C. Thomas.

Harper's Bazaar. January 1980, The California workout.

Harrison, Barbara Grizzuti. April 1978. Jane Fonda: Trying to be everywoman. *Ladies Home Journal.*

———. November 1987. The fitness queen. *Vogue.*

Kilwein, John H. 1989. No pain, no gain: A Puritan legacy. *Health Education Quarterly* 16:9–12.

Kolter, Marvin R. 1988. *Humor and Society: Explorations in the Sociology of Humor.* Houston: Cap and Gown Press, Inc.

Lear, Martha Weinman. June 1976. Jane Fonda: A long way from yesterday. *Redbook.*

Levin, Susanna. December 1987. Jane Fonda: From Barbarella to barbells. *Women's Sports and Fitness.*

Mademoiselle. March 1980. Fitness.

Martin, Emily. 1987. *The Woman in the Body.* Boston: Beacon Press.

McClellan, Diana. October 1992. Barbara Bush: The final battle. *Ladies Home Journal.*

McNay, Lois. 1991. The Foucauldian body and the exclusion of experience. *Hypatia* 6:125–139.

Messina, Andrea. January 12, 1993. Fonda's workouts that work. *Family Circle.*

Mower, Joan. September 1992. What kind of First Lady do we really want? *McCall's.*

Orth, Maureen. February 1984. Fonda: Driving passions. *Vogue.*

Reed, Julia. August 1989. The Natural. *Vogue.*

Robbins, Fred. November 1977. Jane Fonda, the woman. *Vogue.*

Schwartz, Hillel. 1986. *Never Satisfied: A Cultural History of Diets, Fantasies, and Fat.* New York: Free Press.

Seid, Roberta Pollack. 1989. *Never Too Thin.* New York: Prentice-Hall.

Sontag, Susan. October 1972. The double standard of aging. *Saturday Review.*

Stein, Howard F. 1982. Neo-Darwinism and survival through fitness in Reagan's America. *The Journal of Psychohistory* 10:163–187.

Tech, Sylvia. 1988. *Hidden Arguments: Political Ideology and Disease Prevention Policy.* New Brunswick, NJ: Rutgers University Press.

Turner, Bryan S. 1984. *The Body and Society: Explorations in Social Theory.* New York: Basil Blackwell.

Unger, Sheldon. 1984. Self-mockery: An alternative form of self-presentation. *Symbolic Interaction* 7:121–133.

Viorst, Judith. May 1991. It's time to bring back the family. *Redbook.*

Vogue. October 1988. First Ladies, first impressions.

———. November 1988. Winning style: Kitty Dukakis and Barbara Bush on first lady dressing.

Waitzkin, Howard. 1981. The social origins of illness: A neglected history. *International Journal of Health Services* 11:77–103.

Walker, Nancy A. 1988. *A Very Serious Thing: Women's Humor and American Culture.* Minneapolis: University of Minnesota.

Wolf, Naomi. 1991. *The Beauty Myth.* New York: William Morrow.

Woodward, Kathleen. 1991. *Aging and Its Discontents.* Bloomington, IN: Indiana University Press.

Zola, Irving K. 1972. Medicine as an institution of social control. *Sociological Review* 20:487–504.

DISCUSSION QUESTIONS

1. How difficult is it to maintain a positive sense of self while resisting dominant norms of attractiveness?
2. How do issues of attractiveness differ for males? How does aging affect issues of attractiveness for men? Are male standards of attractiveness changing? If so, how? If not, why not? What kinds of differential resources can men and women draw on to attain these standards? Do you envision a time when men and women are held to the same standards? Why or why not?
3. Are norms of attractiveness ableist? That is, do they privilege people who are "able-bodied"? If so how? If not, why not? What kinds of differential resources do people who are able-bodied and people who have disabilities draw on to attain these standards? Do you envision this changing in the future? Why or why not?
4. Can a woman be a feminist and participate in activities to enhance her physical appearance? If so, to what extent? If not, why not? Can a man be masculine/a "real man" and participate in activities to enhance his physical appearance? If so, to what extent? If not, why not?
5. How do cultural constructions of aging and attractiveness vary by race?
6. Is resistance futile?

The following Bonus Reading can be accessed through the InfoTrac *College Edition Online Library* by typing in the name of the author or keywords in the title online at *http://www.infotrac-college.com*

Visibly Queer: Body Technologies and Sexual Politics

VICTORIA PITTS

Body modifications such as tattooing, piercing, and branding have been strategically adopted by some members of the Queer community for purposes that satisfy personal as well as political motives. For example, many body modifiers argue that politically, modifications can be invaluable transgressive resources that challenge and potentially subvert dominant social and sexual norms. Victoria Pitts' interviews with six body modifiers examine both the potential and limits of corporeal revision given existing systems of power, privilege, and social control.

DISCUSSION QUESTIONS

1. Do you agree with the assertion made by David that "I ought to be able to do this. It may be a struggle but it's my choice and my body"? What, if any, role should the state have in relation to practices such as branding, scarring, tattooing, piercing, and corseting? Should it always allow such practices? Allow some but not others? Regulate them? Make them illegal? Why? Should the state have the right to limit the pursuit of pleasure or self-expression? Why or why not?
2. Is it necessarily deviant or pathological to seek pleasure through pain? Why or why not?
3. Do you agree with Mark that the relative meaning or transgressive power of body modification differs when engaged in by members of subordinated social groups as compared to dominant social groups? Why or why not? What, if any differences are there between a black heterosexual fraternity member branding himself and a white heterosexual fraternity member? A lesbian who engages in genital piercing and a heterosexual female? A gay man who has had both legs amputated who engages in corseting and a heterosexual man who engages in the same practice? Is it coincidental that in contemporary American society most marginalized forms of body modification are imported by the larger culture from socially subordinated groups? Why or why not?
4. Are body modifications effective ways of recapturing and controlling one's sexual agency? Why or why not? How important is *public display* of body modifications to your position on this issue? Does concealment nullify the subversive power of body modification? Why or why not?
5. Do you agree with the author's critique that many body modifiers may be "romancing the 'primitive' as a redress to Western cultural problems"? Does it matter? Why or why not?
6. Does lesbian engagement in sadomasochistic practices perpetuate patriarchy? What do you think Stoltenberg would have to say about this argument?

1 2

Sex, Pornography, and Sexual Objectification

Pornography, Civil Rights, and Speech

CATHARINE MACKINNON

Catharine MacKinnon is a renowned feminist legal scholar who is perhaps best known for her staunch opposition to pornography. In this classic essay, MacKinnon presents the argument that pornography sexualizes inequality and is a key contributor to the victimization and social inequality of women. She argues that pornography operates as an institution of gender inequality that sexualizes acts of abuse including rape, battery, sexual harassment, prostitution, and the sexual abuse of children. Toward this end, she presents her platform for a human rights ordinance co-authored with Andrea Dworkin for the city of Minneapolis, which includes provisions for civil redress of harms caused by pornography.

There is a belief that this is a society in which women and men are basically equals. Room for marginal corrections is conceded, flaws are known to exist; attempts are made to correct what are conceived as occasional lapses from the basic condition of sex equality. Sex discrimination law has concentrated most of its focus on these occasional lapses. It is difficult to overestimate the extent to which this belief in equality is an article of faith for most people, including most women, who wish to live in self-respect in an internal universe, even (perhaps especially) if not in the world. It is also partly an expression of natural law thinking: if we are inalienably equal, we can't "really" be degraded.

This is a world in which it is worth trying. In this world of presumptive equality, people make money based on their training or abilities or diligence or qualifications. They are employed and advanced on the basis of merit. In this world of just deserts, if someone is abused, it is thought to violate the basic rules of the community. If it doesn't, victims are seen to have done something they could have chosen to do differently, by

From "Pornography, Civil Rights, and Speech" by Catharine MacKinnon in *The Harvard Civil Rights Civil Liberties Law Review,* vol. 20, no. 1, pp. 1–17, 1985.

exercise of will or better judgment. Maybe such people have placed themselves in a situation of vulnerability to physical abuse. Maybe they have done something provocative. Or maybe they were just unusually unlucky. In such a world, if such a person has an experience, there are words for it. When they speak and say it, they are listened to. . . . If certain experiences are never spoken about, if certain people or issues are seldom heard from, it is supposed that silence has been chosen. The law, including much of the law of sex discrimination and the First Amendment, operates largely within the realm of these beliefs. . . . Defining feminism in a way that connects epistemology with power as the politics of women's point of view, . . . can be summed up by saying that women live in another world: specifically a world of *not* equality, a world of inequality.

Feminism is the discovery that women do not live in this world, that the person occupying this realm is a man, so much more a man if he is white and wealthy. This world of potential credibility, authority, security, and just rewards, recognition of one's identity and capacity, is a world that some people do inhabit as a condition of birth, with variations among them. It is not a basic condition accorded humanity in this society, but a prerogative of status, a privilege, among other things, of gender.

. . . Looking at the world from this point of view, a whole shadow world of previously invisible silent abuse has been discerned. Rape, battery, sexual harassment, forced prostitution, and the sexual abuse of children emerge as common and systematic. We find that rape happens to women in all contexts, from the family, including rape of girls and babies, to students and women in the workplace, on the streets, at home, in their own bedrooms by men they do not know and by men they do know, by men they are married to, men they have had a social conversation with, and, least often, men they have never seen before. Sexual harassment of women by men is common in workplaces and educational institutions. Based on reports in one study of the federal workforce, up to 85 percent of women will experience it, many in physical forms. Between a quarter and a third of women are battered in their homes by men. Thirty-eight percent of little girls are sexually molested inside or outside the family. Until women listened to women, this world of sexual abuse was *not spoken* of. It was the unspeakable. What I am saying is, if you *are* the tree falling in the epistemological forest, your demise doesn't make a sound if no one is listening. Women did not "report" these events, and overwhelmingly do not today, because no one is listening, because no one believes us. This silence does not mean nothing happened, and it does not mean consent.

. . . In pornography, there it is, in one place, all of the abuses that women had to struggle so long even to begin to articulate, all the *unspeakable* abuse: the rape, the battery, the sexual harassment, the prostitution, and the sexual abuse of children. Only in the pornography it is called something else: sex, sex, sex, sex, and sex, respectively. Pornography sexualizes rape, battery, sexual harassment, prostitution, and child sexual abuse; it thereby celebrates, promotes, authorizes, and legitimizes them. More generally, it eroticizes the dominance and submission that is the dynamic common to them all. It makes hierarchy sexy and calls that "the truth about sex" or just a mirror of reality. Through this process pornography constructs what a woman is as what men want from sex. This is what the pornography means.

Pornography constructs what a woman is in terms of its view of what men want sexually, such that acts of rape, battery, sexual harassment, prostitution, and sexual abuse of children become acts of sexual equality. Pornography's world of equality is a harmonious and balanced place. Men and women are perfectly complementary and perfectly bipolar. Women's desire to be fucked by men is equal to men's desire to fuck women. All the ways men love to take and violate women, women love to be taken and violated. The women who most love this are most men's equals, the most liberated; the most participatory child is the most grown-up, the most equal to an adult. Their consent merely expresses or ratifies these preexisting facts.

The content of pornography is one thing. There, women substantively desire dispossession and cruelty. We desperately want to be bound, battered, tortured, humiliated, and killed. Or, to be fair to the soft core, merely taken and used. This is erotic to the male point of view. Subjection itself, with self-determination ecstatically relinquished, is the content of women's sexual desire and desirability. Women are there to be violated and possessed, men to violate and possess us, either on screen or by camera or pen on behalf of the consumer. On a simple descriptive level, the inequality of hierarchy, of which gender is the primary one, seems necessary for sexual arousal to work. Other added inequalities identify various pornographic genres or subthemes, although they are always added through gender: age, disability, homosexuality, animals, objects, race (including anti-Semitism), and so on. Gender is never irrelevant.

What pornography *does* goes beyond its content: it eroticizes hierarchy, it sexualizes inequality. It makes dominance and submission into sex. Inequality is its central dynamic; the illusion of freedom coming together with the reality of force is central to its working. Perhaps because this is a bourgeois culture, the victim must look free, appear to be freely acting. Choice is how she got there. Willing is what she is when she is being equal. It seems equally important that then and there she actually be forced and that forcing be communicated on some level, even if only through still photos of her in postures of receptivity and access, available for penetration. Pornography in this view is a form of forced sex, a practice of sexual politics, an institution of gender inequality.

From this perspective, pornography is neither harmless fantasy nor a corrupt and confused misrepresentation of an otherwise natural and healthy sexual situation. It institutionalizes the sexuality of male supremacy, fusing the erotization of dominance and submission with the social construction of male and female. To the extent that gender is sexual, pornography is part of constituting the meaning of that sexuality. Men treat women as who they see women as being. Pornography constructs who that is. Men's power over women means that the way men see women defines who women can be. Pornography is that way. Pornography is not imagery in some relation to a reality elsewhere constructed. It is not a distortion, reflection, projection, expression, fantasy, representation, or symbol either. It is a sexual reality.

. . . Pornography defines women by how we look according to how we can be sexually used. Pornography codes how to look at women, so you know what you can do with one when you see one. Gender is an assignment made visually, both originally and in everyday life. A sex object is defined on the basis of its looks, in terms of its usability for sexual pleasure, such that both the looking—the quality of the gaze, including its point of view—and the definition according to use become eroticized as part of the sex itself.

. . . To give a set of rough epistemological translations, to defend pornography as consistent with the equality of the sexes is to defend the subordination of women to men as sexual equality. What in the pornographic view is love and romance looks a great deal like hatred and torture to the feminist. Pleasure and eroticism become violation. Desire appears as lust for dominance and submission. The vulnerability of women's projected sexual availability, that acting we are allowed (that is, asking to be acted upon), is victimization. Play conforms to scripted roles. Fantasy expresses ideology, is not exempt from it. Admiration of natural physical beauty becomes objectification. Harmlessness becomes harm. Pornography is a harm of male supremacy made difficult to see because of its pervasiveness, potency, and principally, because of its success in making the world a pornographic place. Specifically, its harm cannot be discerned, and will not be addressed, if viewed and approached neutrally, because it *is* so much of "what is." In other words, to the extent pornography succeeds in constructing social reality, it becomes invisible as harm. If we live in a world that pornography creates through the power of men in a male-dominated situation, the issue is not what the harm of pornography is, but how that harm is to become visible.

. . . At the request of the city of Minneapolis, Andrea Dworkin and I conceived and designed a local human rights ordinance in accordance with our approach to the pornography issue. We define pornography as a practice of sex discrimination, a violation of women's civil rights, the opposite of sexual equality. Its point is to hold those who profit from and benefit from that injury accountable to those who are injured. It means that women's injury—our damage, our pain, our enforced inferiority—should outweigh their pleasure and their profits, or sex equality is meaningless.

We define pornography as the graphic sexually explicit subordination of women through pictures or words that also includes women dehumanized as sexual objects, things, or commodities; enjoying pain or humiliation or rape; being tied up, cut up, mutilated, bruised, or physically hurt; in postures of sexual submission or servility or display; reduced to body parts, penetrated by objects or animals, or presented in scenarios of degradation, injury, torture; shown as filthy or inferior; bleeding, bruised, or hurt in a context that makes these conditions sexual. Erotica, defined by distinction as not this, might be sexually explicit materials premised on equality. We also provide that the use of men, children, or transsexuals in the place of women is pornography. The definition is substantive in that it is sex-specific, but it covers everyone in a sex-specific way, so is gender neutral in overall design. . . .

This law aspires to guarantee women's rights consistent with the First Amendment by making visible a conflict of rights between the equality guaranteed to all women and what, in some legal sense, is now the freedom of the pornographers to make and sell, and their consumers to have access to, the materials this ordinance defines. Judicial resolution of this conflict, if the judges do for women what they have done for others, is likely to entail a balancing of the rights of women arguing that our lives and opportunities, including our freedom of speech and action, are constrained by—and in many cases flatly precluded by, in and through—pornography, against those who argue that the pornography is harmless, or harmful only in part but not in the whole of the definition; or that it is more important to preserve the pornography than it is to prevent or remedy whatever harm it does.

. . . The harm of pornography, broadly speaking, is the harm of the civil inequality of the sexes made invisible as harm because it has become accepted as the sex difference. Consider this analogy with race: if you see Black people as different, there is no harm to segregation; it is merely a recognition of that difference. To neutral principles, separate but equal was equal. The injury of racial separation to Blacks arises "solely because [they] choose to put that construction upon it." Epistemologically translated: how you see it is not the way it is. Similarly, if you see women as just different, even or especially if you don't know that you do, subordination will not look like subordination at all, much less like harm. It will merely look like an appropriate recognition of the sex difference.

Pornography does treat the sexes differently, so the case for sex differentiation can be made here. But men as a group do not tend to be (although some individuals may be) treated the way women are treated in pornography. As a social group, men are not hurt by pornography the way women as a social group are. Their social status is not defined as *less* by it. So the major argument does not turn on mistaken differentiation, particularly since the treatment of women according to pornography's dictates makes it all too often accurate. The salient quality of a distinction between the top and the bottom in a hierarchy is not difference, although top is certainly different from bottom; it is power. So the major argument is: subordinate but equal is not equal.

. . . [O]ur law makes trafficking in pornography—production, sale, exhibition, or distribution—actionable. Under the obscenity rubric, much legal and psychological scholarship has centered on a search for the elusive link between harm and pornography defined as obscenity. Although they were not very clear on what obscenity was, it was its harm they truly could not find. They looked high and low—in the mind of the male consumer, in society or in its "moral fabric," in correlations between variations in levels of antisocial acts and liberalization of obscenity laws. The only harm they have found has been harm to "the social interest in order and morality." Until recently, no one looked very persistently for harm to women, particularly harm to women through men. The rather obvious fact that the sexes *relate* has been overlooked in the inquiry into the male consumer and his mind. The pornography doesn't just drop out of the sky, go into his head, and stop there. Specifically, men rape, batter, prostitute, molest, and sexually harass women. Under conditions of inequality, they also hire, fire, promote, and grade women, decide how much or whether we are worth paying and for what, define and approve and disapprove of women in ways that count, that determine our lives.

If women are not just born to be sexually used, the fact that we are seen and treated as though that is what we are born for becomes something in need of explanation. If we see that men relate to women in a pattern of who they see women as being, and that forms a pattern of inequality, it becomes important to ask where that view came from or, minimally, how it is perpetuated or escalated. Asking this requires asking different questions about pornography than the ones obscenity law made salient.

Now I'm going to talk about causality in its narrowest sense. Recent experimental research on pornography shows that the materials covered by our definition cause measurable harm to women through increasing men's attitudes and behaviors of discrimination in both violent and nonviolent forms. Exposure to some of the pornography in our definition increases the immediately subsequent willingness of normal men to aggress against women under laboratory conditions. It makes normal men more closely resemble convicted rapists attitudinally, although as a group they don't look all that different from them to start with. Exposure to pornography also significantly increases attitudinal measures known to correlate with rape and self-reports of aggressive acts, measures such as hostility toward women, propensity to rape, condoning rape, and predicting that one would rape or force sex on a woman if one knew one would not get caught. On this latter measure, by the way, about a third of all men predict that they would rape, and half would force sex on a woman.

As to that pornography covered by our definition in which normal research subjects seldom perceive violence, long-term exposure still makes them see women as more worthless, trivial, nonhuman, and objectlike, that is, the way those who are discriminated against are seen by those who discriminate against them. Crucially, all pornography by our definition acts dynamically over time to diminish the consumer's ability to distinguish sex from violence. The materials work behaviorally to diminish the capacity of men (but not women) to perceive that an account of a rape is an account of a rape. The so-called sex-only materials, those in which subjects perceive no force, also increase perceptions that a rape victim is worthless and decrease the perception that she was harmed. The overall direction of current research suggests that the more expressly violent materials accomplish with less exposure what the less overtly violent—that is, the so-called sex-only materials—accomplish over the longer term. Women are rendered fit for use and targeted for abuse. The only thing that the research cannot document is which individual women will be next on the list.

. . . Show me an abuse of women in society, I'll show it to you made sex in the pornography. If you want to know who is being hurt in this society, go see what is being done and to whom in pornography and then go look for them other places in the world. You will find them being hurt in just that way. We did in our hearings.

In our hearings women spoke, to my knowledge for the first time in history in public, about the damage pornography does to them. We learned that pornography is used to break women, to train women to sexual submission, to season women, to terrorize women, and to silence their dissent. It is this that has previously been termed "having no effect." The way men inflict on women the sex they experience through the pornography gives women no choice about seeing the pornography or doing the sex. Asked if anyone ever tried to inflict unwanted sex acts on them that they knew came from pornography, 10 percent of women in a recent random study said yes. Among married women, 24 percent said yes. That is a lot of women. A lot more don't know. Some of those who do testified in Minneapolis. One wife said of her ex-husband, "He would read from the pornography like a textbook, like a journal. In fact when he asked me to be bound, when he finally convinced me to do it, he read in the magazine how to tie the knots." Another woman said of her boyfriend, "[H]e went to this party, saw pornography, got an erection, got me . . . to inflict his erection on. . . . There is a direct causal relationship there." One woman, who said her husband had rape and bondage magazines all over the house, discovered two suitcases full of Barbie dolls with rope tied on their arms and legs and with tape across their mouths. Now think about the silence of women. She said, "He used to tie me up and he tried those things on me." A therapist in private practice reported:

> Presently or recently I have worked with clients who have been sodomized by broom handles, forced to have sex with over 20 dogs in the back seat of their car, tied up and then electrocuted on their genitals. These are children, [all] in the ages of 14 to 18, all of whom [have been directly affected by pornography,] [e]ither where the perpetrator has read the manuals and manuscripts at night and used these as recipe books by day or had the pornography present at the time of the sexual violence.[1]

One woman, testifying that all the women in a group of exprostitutes were brought into prostitution as children through pornography, characterized their collective experience: "[I]n my experience there was not one situation where a client was not using pornography while he was using me or that he had not just watched pornography or that it was verbally referred to and directed me to pornography." "Men," she continued, "witness the abuse of women in pornography constantly and if they can't engage in that behavior with their wives, girl friends or children, they force a whore to do it."

Men also testified about how pornography hurts them. One young gay man who had seen *Playboy* and *Penthouse* as a child said of such heterosexual pornography: "It was one of the places I learned about sex and it showed me that sex was violence. What I saw there was a specific relationship between men and women. . . . [T]he woman was to be used, objectified, humiliated and hurt; the man was in a superior position, a position to be violent. In pornography I learned that what it meant to be sexual with a man or to be loved by a man was to accept his violence." For this reason, when he was battered by his first lover, which he described as "one of the most profoundly destructive experiences of my life," he accepted it.

. . . Pornography stimulates and reinforces, it does not cathect or mirror, the connection between one-sided freely available sexual access to women and masculine sexual excitement and sexual satisfaction. The catharsis hypothesis is fantasy. The fantasy theory is fantasy. Reality is: pornography conditions male orgasm to female subordination. It tells men what sex means, what a real woman is, and codes them together in a way that is behaviorally reinforcing.

It is worth considering what evidence has been enough when other harms involving other purported speech interests have been allowed to be legislated against. By comparison to our trafficking provision, analytically similar restrictions have been allowed under the First Amendment, with a legislative basis far less massive, detailed, concrete, and conclusive. Our statutory language is more ordinary, objective, and precise and covers a harm far narrower than the legislative record substantiates. Under *Miller*, obscenity was allowed to be made criminal in the name of the "danger of offending the sensibilities of unwilling recipients or exposure to juveniles." Under our law, we have direct evidence of harm, not just a conjectural danger, that unwilling women in considerable numbers are not simply offended in their sensibilities, but are violated in their persons and restricted in their options. Obscenity law also suggests that the applicable standard for legal adequacy in measuring such connections may not be statistical certainty. The Supreme Court has said that it is not their job to resolve empirical uncertainties that underlie state obscenity legislation. Rather, it is for them to determine whether a legislature could reasonably have determined that a connection might exist between the prohibited material and harm of a kind in which the state has legitimate interest. Equality should be such an area. The Supreme Court recently recognized that prevention of sexual exploitation and abuse of children is, in their words, "a governmental objective of surpassing importance." This might also be the case for sexual exploitation and abuse of women, although I think a civil remedy is initially more appropriate to the goal of empowering adult women than a criminal prohibition would be.

Other rubrics provide further support for the argument that this law is narrowly tailored to further a legitimate governmental interest consistent with the goals underlying the First Amendment. Exceptions to the First Amendment—you may have gathered from this—exist. The reason they exist is that the harm done by some speech outweighs its expressive value, if any. In our law a legislature recognizes that pornography, as defined and made actionable, undermines sex equality. One can say—and I have—that pornography is a causal factor in violations of women; one can also say that women will be violated so long as pornography exists; but one can also say simply that pornography violates women. . . . *Chaplinsky v. New Hampshire* recognized the ability to restrict as "fighting words" speech which, "by [its] very utterance inflicts injury." Perhaps the only reason that pornography has not been "fighting words"—in the sense of words that by their utterance tend to incite immediate breach of the peace—is that women have seldom fought back, yet.

. . . For those of you who still think pornography is only an idea, consider the possibility that obscenity law got one thing right. Pornography is more actlike than thoughtlike. The fact that pornography, in a feminist view, furthers the idea of the sexual inferiority of women, which is a political idea, doesn't make the pornography itself into a political idea. One can express the idea a practice embodies. That does not make the practice into an idea. Segregation expresses the idea of the inferiority of one group to another on the basis of race. That does not make segregation an idea. A sign that says "Whites Only" is only words. Is it therefore protected by the First Amendment? Is it not an act, a practice, of segregation because what it means is inseparable from what it does? *Law* is only words.

I take one of two penultimate points from Andrea Dworkin, who has often said that pornography is not speech for women, it is the silence of women. Remember the mouth taped, the woman gagged, "Smile, I can get a lot of money for that." The smile is not her expression, it is her silence. It is not her expression not because it didn't happen, but because it *did* happen. The screams of the women in pornography are silence, like the screams of Kitty Genovese, whose plight was misinterpreted by some onlookers as a lovers' quarrel. The flat expressionless voice of the woman in the New Bedford gang rape, testifying, is silence. She was raped as men cheered and watched, as they do in and with the pornography. When women resist and men say, "Like this, you stupid bitch, here is how to do it" and shove their faces into the pornography, this "truth of sex" is the silence of women. When they say, "If you love me, you'll try," the enjoyment we fake, the enjoyment we learn is silence. Women who submit because there is more dignity in it than in losing the fight over and over live in silence. Having to sleep with your publisher or director to get access to what men call speech is silence. Being humiliated on the basis of your appearance, whether by approval or disapproval, because you have to look a certain way for a certain job, whether you get the job or not, is silence. The absence of a woman's voice, everywhere that it cannot be heard, is silence. And anyone who thinks that what women say in pornography is women's speech—the "Fuck me, do it to me, harder," all of that—has never heard the sound of a woman's voice.

The most basic assumption underlying First Amendment adjudication is that, socially, speech is free. The First Amendment says Congress shall not abridge the freedom of speech. Free speech, get it, *exists*. Those who wrote the First Amendment *had* speech—they wrote the Constitution. *Their* problem was to keep it free from the only power that realistically threatened it: the federal government. They designed the First Amendment to prevent government from constraining that which, if unconstrained by government, was free, meaning *accessible to them*. At the same time, we can't tell much about the intent of the framers with regard to the question of women's speech, because I don't think we crossed their minds. It is consistent with this analysis that their posture toward freedom of speech tends to presuppose that whole segments of the population are not systematically silenced socially, prior to government action. If everyone's power were equal to theirs, if this were a nonhierarchical society, that might make sense. But the place of pornography in the inequality of the sexes makes the assumption of equal power untrue.

This is a hard question. It involves risks. Classically, opposition to censorship has involved keeping government off the backs of people. Our law is about getting some people off the backs of other people. The risks that it will be misused have to be measured against the risks of the status quo. Women will never have that dignity, security, compensation that is the promise of equality so long as the pornography exists as it does now. The situation of women suggests that the urgent issue of our freedom of speech is not primarily the avoidance of state intervention as such, but getting affirmative access to speech for those to whom it has been denied.

NOTE

1. *Public Hearings on Ordinances to Add Pornography as Discrimination Against Women,* Committee on Governmental Operations, City Council, Minneapolis, MN, December 12–13, 1983.

DISCUSSION QUESTIONS

1. Can you think of examples to support MacKinnon's claim that dominance is eroticized in our culture (do not restrict yourself to a consideration of only gender)? Can you think of any evidence that would contradict this statement? On the whole, do you agree or disagree? Why or why not?
2. If MacKinnon is right, what are the implications for female agency and sexuality in this culture? Can women truly know and act upon their sexuality freely under patriarchy?
3. How is pornography constructed in modern American culture? What impression would outsiders have about sexual relations in this country if pornography were their only source of information? What other sources of information about sexuality are available? Do these sources' messages generally contradict, support, mitigate, or exacerbate the images and messages about sexuality depicted in pornography?
4. Extend MacKinnon's arguments to gay and lesbian pornography. How would you evaluate the social harm it generates? How does that harm compare with heterosexual pornography? Is gay male pornography different than lesbian pornography with respect to harm? Why or why not?
5. If pornography were outlawed would the rates or nature of sexual violence against women change? Why (and how) or why not? Would violence against social minorities change? Why (and how) or why not?

Pornography and the Alienation of Male Sexuality

HARRY BROD

Unlike most contemporary critiques of pornography, Harry Brod's treatment is concerned with the harmful effects it has on male *sexuality. Brod uses a Marxist framework to argue that while the system of patriarchy operates to privilege men, contradictions inherent in this system simultaneously serve to disadvantage men and to alienate them from their own sexuality. For Brod, pornography represents capitalism's colonization of the body. It constructs a commercialized performance-based sexuality for men that tends to prevent the fulfillment of emotional needs and creates an over-preoccupation with large, overused, genitally based organs.*

This chapter is intended as a contribution to an ongoing discussion. It aims to augment, not refute or replace, what numerous commentators have said about pornography's role in the social construction of sexuality. I have several principal aims in this chapter. My primary focus is to examine pornography's model of male sexuality. Furthermore, in the discussion of pornography's role in the social construction of sexuality, I wish to place more emphasis than is common on the social construction of pornography. As I hope to show, these are related questions. One reason I focus on the image of male sexuality in pornography is that I believe this aspect of the topic has been relatively neglected. In making this my topic here, I do not mean to suggest that this is the most essential part of the picture. Indeed, I am clear it is not. It seems clear enough to me that the main focus of discussion about the effects of pornography is and should be the harmful effects of pornography on women, its principal victims. Yet, there is much of significance which needs to be said about pornography's representation, or perhaps I should more accurately say misrepresentation, of male sexuality. My focus shall be on what is usually conceived of as "normal" male sexuality, which for my purposes I take to be consensual, non-violent heterosexuality, as these terms are conventionally understood. I am aware of analyses which argue that this statement assumes distinctions which are at least highly problematic, if not outright false, which argue that this "normal" sexuality is itself coercive, both as compulsory heterosexuality and as containing implicit or explicit coercion and violence. My purpose is not to take issue with these analyses, but simply to present an analysis of neglected aspects of the links between mainstream male sexuality and pornography. I would argue that the aspect of the relation between male sexuality and pornography usually discussed, pornography's incitement to greater extremes of violence against women, presupposes such a connection with the more accepted mainstream. Without such a link, pornography's messages would be rejected by, rather than assimilated into, male culture. My intention is to supply this usually missing link.

My analysis proceeds from both feminist and Marxist theory. These are often taken to be theories which speak from the point of view

Adapted from "Pornography and the Alienation of Male Sexuality" by Harry Brod, *Social Theory and Practice* 14, no. 3 (1988): 256–84. Reprinted by permission.

of the oppressed, in advocacy for their interests. That they indeed are, but they are also more than that. For each claims not simply to speak for the oppressed in a partisan way, but also to speak a truth about the social whole, a truth perhaps spoken in the name of the oppressed, but a truth objectively valid for the whole. That is to say, Marxism is a theory which analyzes the ruling class as well as the proletariat, and feminism is a theory which analyzes men as well as women. It is not simply that Marxism is concerned with class, and feminism with gender, both being united by common concerns having to do with power. Just as Marxism understands class as power, rather than simply understanding class differences as differences of income, lifestyle, or opportunities, so the distinctive contribution of feminism is its understanding of gender as power, rather than simply as sex role differentiation. Neither class nor gender should be reified into being understood as fixed entities, which then differentially distribute power and its rewards. Rather, they are categories continually constituted in ongoing contestations over power. The violence endemic to both systems cannot be understood as externalized manifestations of some natural inner biological or psychological drives existing prior to the social order, but must be seen as emerging in and from the relations of power which constitute social structures. Just as capitalist exploitation is caused not by capitalists' excess greed but rather by the structural imperatives under which capitalism functions, so men's violence is not the manifestation of some inner male essence, but rather evidence of the bitterness and depth of the struggles through which genders are forged.[1]

For my purposes here, to identify this as a socialist feminist analysis is not, in the first instance, to proclaim allegiance to any particular set of doctrinal propositions, though I am confident that those I subscribe to would be included in any roundup of the usual suspects, but rather to articulate a methodological commitment to make questions of power central to questions of gender, and to understand gendered power in relation to economic power, and as historically, materially structured.[2] If one can understand the most intimate aspects of the lives of the dominant group in these terms, areas which would usually be taken to be the farthest afield from where one might expect these categories to be applicable, then I believe one has gone a long way toward validating claims of the power of socialist feminist theory to comprehend the totality of our social world. This is my intention here. I consider the analysis of male sexuality I shall be presenting part of a wider socialist feminist analysis of patriarchal capitalist masculinity, an analysis I have begun to develop elsewhere.[3]

As shall be abundantly clear, I do not take a "sexual liberationist" perspective on pornography. I am aware that many individuals, particularly various sexual minorities, make this claim on pornography's behalf. I do not minimize or negate their personal experiences. In the context of our society's severe sexual repressiveness, pornography may indeed have a liberating function for certain individuals. But I do not believe an attitude of approval for pornography follows from this. Numerous drugs and devices which have greatly helped individual women have also been medical and social catastrophes—the one does not negate the other.

I shall be claiming that pornography has a negative impact on men's own sexuality. This is a claim that an aspect of an oppressive system, patriarchy, operates, at least in part, to the disadvantage of the group it privileges, men. This claim does not deny that the overall effect of the system is to operate in men's advantage, nor does it deny that the same aspect of the system under consideration, that is, male sexuality and pornography under patriarchy, might not also contribute to the expansion and maintenance of male power even as it also works to men's disadvantage. Indeed, I shall be arguing precisely for such complementarity. I am simply highlighting one of the "contradictions" in the system. My reasons for doing so are in the first instance simply analytic: to, as I said, bring to the fore relatively neglected aspects of the issue. Further, I also have political motivations for emphasizing this perspective. I view raising consciousness of the prices of male power as part of a strategy through which one could at least potentially mobilize men against pornography's destructive effects on both women and men.

It will aid the following discussion if I ask readers to call to mind a classic text in which it is argued that, among many other things, a system of domination also damages the dominant group, and prevents them from realizing their full humanity. The argument is that the dominant group is "alienated" in specific and identifiable ways. The text I have in mind is Marx's "Economic and Philosophic Manuscripts of 1844." Just as capitalists as well as workers are alienated under capitalism according to Marxist theory (in a certain restricted sense, even more so), so men, I shall argue, and in particular male modes of sexuality, are also alienated under patriarchy. In the interests of keeping this paper a manageable length, I shall here assume rather than articulate a working familiarity with Marx's concept of alienation, the process whereby one becomes a stranger to oneself and one's own powers come to be powers over and against one. . . . While much of this paper presents an analysis of men's consciousness, I should make clear that while alienation may register in one's consciousness (as I argue it does), I follow Marx in viewing alienation not primarily as a psychological state dependent on the individual's sensibilities or consciousness but as a condition inevitably caused by living within a system of alienation. I should also note that I consider what follows an appropriation, not a systematic interpretation, of some of Marx's concepts.

Alienated pornographic male sexuality can be understood as having two dimensions, what I call the objectification of the body and the loss of subjectivity. I shall consider each in greater detail, describing various aspects of pornographic male sexuality under each heading in a way which I hope brings out how they may be conceptualized in Marx's terms. Rather than then redoing the analysis in Marx's terms, I shall then simply cite Marx briefly to indicate the contours of such a translation.

OBJECTIFICATION OF THE BODY

In terms of both its manifest image of and its effects on male sexuality, that is, in both intrinsic and consequentialist terms, pornography restricts male sensuality in favor of a genital, performance oriented male sexuality. Men become sexual acrobats endowed with oversized and overused organs which are, as the chapter title of a fine book on male sexuality describes, "The Fantasy Model of Sex: Two Feet Long, Hard as Steel, and Can Go All Night."[4] To speak non-euphemistically, using penile performance as an index of male strength and potency directly contradicts biological facts. There is no muscle tissue in the penis. Its erection when aroused results simply from increased blood flow to the area. All social mythology aside, the male erection is physiologically nothing more than localized high blood pressure. Yet this particular form of hypertension has attained mythic significance. Not only does this focusing of sexual attention on one organ increase male performance anxieties, but it also desensitizes other areas of the body from becoming what might otherwise be sources of pleasure. A colleague once told me that her favorite line in a lecture on male sexuality I used to give in a course I regularly taught was my declaration that the basic male sex organ is not the penis, but the skin.

The predominant image of women in pornography presents women as always sexually ready, willing, able, and eager. The necessary corollary to pornography's myth of female perpetual availability is its myth of male perpetual readiness. Just as the former fuels male misogyny when real-life women fail to perform to pornographic standards, so do men's failures to similarly perform fuel male insecurities. Furthermore, I would argue that this diminishes pleasure. Relating to one's body as a performance machine produces a split consciousness wherein part of one's attention is watching the machine, looking for flaws in its performance, even while one is supposedly immersed in the midst of sensual pleasure. This produces a self-distancing self-consciousness which mechanizes sex and reduces pleasure. (This is a problem perpetuated by numerous sexual self-help manuals, which treat sex as a matter of individual technique for fine-tuning the machine rather than as human interaction. I would add that men's sexual partners are also affected by this, as they can often intuit when they are being subjected to rote manipulation.)

LOSS OF SUBJECTIVITY

In the terms of discourse of what it understands to be "free" sex, pornographic sex comes "free" of the demands of emotional intimacy or commitment. It is commonly said as a generalization that women tend to connect sex with emotional intimacy more than men do. Without romantically blurring female sexuality into soft focus, if what is meant is how each gender consciously thinks or speaks of sex, I think this view is fair enough. But I find it takes what men say about sex, that it doesn't mean as much or the same thing to them, too much at face value. I would argue that men do feel similar needs for intimacy, but are trained to deny them, and are encouraged further to see physical affection and intimacy primarily if not exclusively in sexual terms. This leads to the familiar syndrome wherein, as one man put it:

> Although what most men want is physical affection, what they end up thinking they want is to be laid by a Playboy bunny.[5]

This puts a strain on male sexuality. Looking to sex to fulfill what are really non-sexual needs, men end up disappointed and frustrated. Sometimes they feel an unfilled void, and blame it on their or their partner's inadequate sexual performance. At other times they feel a discomfitting urgency or neediness to their sexuality, leading in some cases to what are increasingly recognized as sexual addiction disorders (therapists are here not talking about the traditional "perversions," but behaviors such as what is coming to be called a "Don Juan Syndrome," an obsessive pursuit of sexual "conquests"). A confession that sex is vastly overrated often lies beneath male sexual bravado. I would argue that sex seems overrated because men look to sex for the fulfillment of nonsexual emotional needs, a quest doomed to failure. Part of the reason for this failure is the priority of quantity over quality of sex which comes with sexuality's commodification. As human needs become subservient to market desires, the ground is laid for an increasing multiplication of desires to be exploited and filled by marketable commodities.[6]

For the most part the female in pornography is not one the man has yet to "conquer," but one already presented to him for the "taking." The female is primarily there as sex object, not sexual subject. Or, if she is not completely objectified, since men do want to be desired themselves, hers is at least a subjugated subjectivity. But one needs another independent subject, not an object or a captured subjectivity, if one either wants one's own prowess validated, or if one simply desires human interaction. Men functioning in the pornographic mode of male sexuality, in which men dominate women, are denied satisfaction of these human desires.[7] Denied recognition in the sexual interaction itself, they look to gain this recognition in wider social recognition of their "conquest."

To the pornographic mind, then, women become trophies awarded to the victor. For women to serve this purpose of achieving male social validation, a woman "conquered" by one must be a woman deemed desirable by others. Hence pornography both produces and reproduces uniform standards of female beauty. Male desires and tastes must be channeled into a single mode, with allowance for minor variations which obscure the fundamentally monolithic nature of the mold. Men's own subjectivity becomes masked to them, as historically and culturally specific and varying standards of beauty are made to appear natural and given. The ease with which men reach quick agreement on what makes a woman "attractive," evidenced in such things as the "1–10" rating scale of male banter and the reports of a computer program's success in predicting which of the contestants would be crowned "Miss America," demonstrates how deeply such standards have been internalized, and consequently the extent to which men are dominated by desires not authentically their own.

Lest anyone think that the analysis above is simply a philosopher's ruminations, too far removed from the actual experiences of most men, let me just offer one recent instantiation, from among many known to me, and even more, I am sure, I do not know. The following is from the *New York Times Magazine*'s "About Men" weekly column. In an article titled "Couch Dancing," the author describes his reactions to being taken to a place, a sort of cocktail bar, where women "clad only in the skimpiest of bikini underpants" would "dance" for a small group of men for a few minutes for about twenty-five or thirty dollars, men who "sat immobile, drinks in hand, glassy-eyed, tapping their feet to the disco music that throbbed through the room."

> Men are supposed to like this kind of thing, and there is a quite natural part of each of us that does. But there is another part of us—of me, at least—that is not grateful for the traditional male sexual programming, not proud of the results. By a certain age, most modern men have been so surfeited with images of unattainably beautiful women in preposterous contexts that we risk losing the capacity to respond to the ordinarily beautiful women we love in our bedrooms. There have been too many times when I have guiltily resorted to impersonal fantasy because the genuine love I felt for a woman wasn't enough to convert feeling into performance. And in those sorry, secret moments, I have resented deeply my lifelong indoctrination into the esthetic of the centerfold.[8]

ALIENATION AND CRISIS

I believe that all of the above can be translated without great difficulty into a conceptual framework paralleling Marx's analysis of the alienation experienced by capitalists. The essential points are captured in two sentences from Marx's manuscripts:

1. *All* the physical and intellectual senses have been replaced by the simple alienation of *all* these senses; the sense of *having.*[9]
2. The wealthy man is at the same time one who *needs* a complex of human manifestations of life, and whose own self-realization exists as an inner necessity, a need.[10]

Both sentences speak to a loss of human interaction and self-realization. The first articulates how desires for conquest and control prevent input from the world. The second presents an alternative conception wherein wealth is measured by abilities for self-expression, rather than possession. Here Marx expresses his conceptualization of the state of alienation as a loss of sensuous fulfillment, poorly replaced by a pride of possession, and a lack of self-consciousness and hence actualization of one's own real desires and abilities. One could recast the preceding analysis of pornographic male sexuality through these categories. In Marx's own analysis, these are more properly conceived of as the results of alienation, rather than the process of alienation itself. This process is at its basis a process of inversion, a reversal of the subject-object relationship, in which one's active powers become estranged from one, and return to dominate one as an external force. It is this aspect which I believe is most useful in understanding the alienation of male sexuality of which pornography is part and parcel. How is it that men's power turns against them, so that pornography, in and by which men dominate women, comes to dominate men themselves?

To answer this question I shall find it useful to have recourse to two other concepts central to Marxism, the concept of "crisis" in the system and the concept of "imperialism."[11] Marx's conception of the economic crisis of capitalism is often misunderstood as a prophecy of a cataclysmic doomsday scenario for the death of capitalism. Under this interpretation, some look for a single event, perhaps like a stock market crash, to precipitate capitalism's demise. But such events are for Marx at most triggering events, particular crises, which can shake the system, if at all, only because of the far more important underlying structural general crisis of capitalism. This general crisis is increasingly capitalism's ordinary state, not an extraordinary occurrence. It is manifest in the ongoing fiscal crisis of the state as well as recurring crises of legitimacy, and results from basic contradictory tensions within capitalism. One way of expressing these tensions is to see them as a conflict between the classic laissez-faire capitalist market mode, wherein capitalists are free to run their own affairs as individuals, and the increasing inability of the capitalist class to run an increasingly complex system without centralized management. The result of this tension is that the state increasingly becomes a managerial committee for the capitalist class, and is increasingly called upon to perform functions previously left to individuals. As entrepreneurial and laissez-faire capitalism give way to corporate capitalism and the welfare state, the power of capitalism becomes increasingly depersonalized, increasingly reft from the hands of individual capitalists and collectivized, so that capitalists themselves come more and more under the domination of impersonal market forces no longer under their direct control.

To move now to the relevance of the above, there is currently a good deal of talk about a perceived crisis of masculinity, in which men are said to be confused by contradictory imperatives given them in the wake of the women's movement. Though the male ego feels uniquely beleaguered today, in fact such talk regularly surfaces in our culture—the 1890s in the United States, for example, was another period in which the air was full of a "crisis of masculinity" caused by the rise of the "New Woman" and other factors.[12] Now, I wish to put forward the hypothesis that these particular "crises" of masculinity are but surface manifestations of a much deeper and broader phenomenon which I call the "general crisis of patriarchy," paralleling Marx's general crisis of capitalism. Taking a very broad view, this crisis results from the increasing depersonalization of patriarchal power which occurs with the development of patriarchy from its precapitalist phase, where power really was often directly exercised by individual patriarchs, to its late capitalist phase where men collectively exercise power over women, but are themselves as individuals increasingly under the domination of those same patriarchal powers.[13] I would stress that the sense of there being a "crisis" of masculinity arises not from the decrease or increase in patriarchal power as such. Patriarchal imperatives from men to retain power over women remain in force throughout. But there is a shift in the mode of that power's exercise, and the sense of crisis results from the simultaneous promulgation throughout society of two conflicting modes of patriarchal power, the earlier more personal form and the later more institutional form. The crisis results from the incompatibility of the two conflicting ideals of masculinity embraced by the different forms of patriarchy, the increasing conflicts between behavioral and attitudinal norms in the political/economic and the personal/familial spheres.

FROM PATRIARCHY TO FRATRIARCHY

To engage for a moment in even broader speculation than that which I have so far permitted myself, I believe that much of the culture, law, and philosophy of the nineteenth century in particular can be reinterpreted as marking a decisive turn in this transition. I believe the passing of personal patriarchal power and its transformation into institutional patriarchal power in this period of the interrelated consolidation of corporate capitalism is evidenced in such phenomena as the rise of what one scholar has termed "judicial patriarchy," the new social regulation of masculinity through the courts and social welfare agencies, which through new support laws, poor laws, desertion laws and other changes transformed what were previously personal obligations into legal duties.[14] . . .

I would like to tentatively and preliminarily propose a new concept to reflect this shift in the nature of patriarchy caused by the deindividualization and collectivization of male power. Rather than speak simply of advanced capitalist patriarchy, the rule of the *fathers,* I suggest we speak of fratriarchy, the rule of the *brothers.* For the moment, I propose this concept more as a metaphor than as a sharply defined analytical tool, much as the concept of patriarchy was used when first popularized. I believe this concept better captures what I would argue is one of the key issues in conceptualizing contemporary masculinities, the disjunction between the facts of public male power and the feelings of individual male powerlessness. As opposed to the patriarch, who embodied many levels and kinds of authority in his single person, the brothers stand in uneasy relationships with each other, engaged in sibling rivalry while trying to keep the power of the family of man as a whole intact. I note that one of the consequences of the shift from patriarchy to fratriarchy is that some people become nostalgic for the authority of the benevolent patriarch, who if he was doing his job right at least prevented one of the great dangers of fratriarchy, fratricide, the brothers' killing each other. Furthermore, fratriarchy is an intragenerational concept, whereas patriarchy is intergenerational. Patriarchy, as a father-to-son transmission of authority, more directly inculcates traditional historically grounded authority, whereas the dimension of temporal continuity is rendered more problematic in fratriarchy's brother-to-brother relationships. I believe this helps capture the problematic nature of modern historical consciousness as it emerged from the nineteenth century, what I would argue is the most significant single philosophical theme of that century. If taken in Freudian directions, the concept of fratriarchy also speaks to the brothers' collusion to repress awareness of the violence which lies at the foundations of society.

To return to the present discussion, the debate over whether pornography reflects men's power or powerlessness, as taken up recently by Alan Soble in his book *Pornography: Marxism, Feminism, and the Future of Sexuality,* can be resolved if one makes a distinction such as I have proposed between personal and institutional male power. Soble cites men's use of pornographic fantasy as compensation for their powerlessness in the real world to argue that "pornography is therefore not so much an expression of male power as it is an expression of their lack of power."[15] In contrast, I would argue that by differentiating levels of power one should more accurately say that pornography is both an expression of men's public power and an expression of their lack of personal power. The argument of this paper is that pornography's image of male sexuality works to the detriment of men personally even as its image of female sexuality enhances the powers of patriarchy. It expresses the power of alienated sexuality or, as one could equally well say, the alienated power of sexuality.

With this understanding, one can reconcile the two dominant but otherwise irreconcilable images of the straight male consumer of pornography: on the one hand the powerful rapist, using pornography to consummate his sexual violence, and on the other hand the shy recluse, using it to consummate his masturbatory fantasies. Both images have their degree of validity, and I believe it is a distinctive virtue of the analysis presented here that one can understand not only the merits of each depiction, but their interconnection.

EMBODIMENT AND EROTICA

In the more reductionist and determinist strains of Marxism, pornography as ideology would be relegated to the superstructure of capitalism. I would like to suggest another conceptualization: that pornography is not part of patriarchal capitalism's superstructure, but part of its infrastructure. Its commodification of the body and interpersonal relationships paves the way for the ever more penetrating ingression of capitalist market relations into the deepest reaches of the individual's psychological makeup. The feminist slogan that "The Personal is Political" emerges at a particular historical moment, and should be understood not simply as an imperative declaration that what has previously been seen solely as personal should now be viewed politically, but also as a response to the real increasing politicization of personal life.

This aspect can be illuminated through the Marxist concept of imperialism. The classical Marxist analysis of imperialism argues that it is primarily motivated by two factors: exploitation of natural resources and extension of the market. In this vein, pornography should be understood as imperialism of the body. The greater public proliferation of pornography, from the "soft-core" pornography of much commercial advertising to the greater availability of "hard-core" pornography, proclaims the greater colonization of the body by the market.[16] The increasing use of the male body as a sex symbol in contemporary culture is evidence of advanced capitalism's increasing use of new styles of masculinity to promote images of men as consumers as well as producers.[17] Today's debates over the "real" meaning of masculinity can be understood in large part as a struggle between those espousing the "new man" style of masculinity more suited to advanced corporate, consumerist patriarchal capitalism and those who wish to return to an idealized version of "traditional" masculinity suited to a more production-oriented, entrepreneurial patriarchal capitalism.[18]

In a more theoretical context, one can see that part of the reason the pornography debate has been so divisive, placing on different sides of the question people who usually find themselves allies, is that discussions between civil libertarians and feminists have often been at cross purposes. Here one can begin to relate political theory not to political practice, but to metaphysical theory. The classical civil liberties perspective on the issue remains deeply embedded in a male theoretical discourse on the meaning of sexuality. The connection between the domination of nature and the domination of women has been argued from many Marxist and feminist points of view.[19] The pivot of this connection is the masculine overlay of the mind-body dualism onto the male-female dichotomy. Within this framework, morality par excellence consists in the masculinized mind restraining the feminized body, with sexual desires seen as the crucial test for these powers of restraint. From this point of view, the question of the morality of pornography is primarily the quantitative question of how much sexual display is allowed, with full civil libertarians opting to uphold the extreme end of this continuum, arguing that no sexual expression should be repressed. But the crucial question, for at least the very important strain of feminist theory which rejects these dualisms which frame the debate for the malestream mainstream, is not *how much* sexuality is displayed but rather *how* sexuality is displayed. These theories speak not of mind-body dualism, but of mind/body wholism, where the body is seen not as the limitation or barrier for the expression of the free moral self, but rather as the most immediate and intimate vehicle for the expression of that self. The question of sexual morality here is not that of restraining or releasing sexual desires as they are forced on the spiritual self by the temptations of the body, but that of constructing spirited and liberating sexual relationships with and through one's own and others' bodies. Here sexual freedom is not the classical liberal freedom *from* external restraint, but the more radical freedom *to* construct authentically expressive sexualities.

I have argued throughout this paper that pornography is a vehicle for the imposition of socially constructed sexuality, not a means for the expression of autonomously self-determined sexuality. (I would add that in contrasting imposed and authentic sexualities I am not endorsing a sexual essentialism, but simply carving out a space for more personal freedom.) Pornography is inherently about commercialized sex, about the eroticization of power and the power of eroticization. One can look to the term's etymology for confirmation of this point. It comes from the classical Greek *"pornographos,* meaning 'writing (sketching) of harlots,'" sometimes women captured in war.[20] Any distinction between pornography and erotica remains problematic, and cannot be drawn with absolute precision. Yet I believe some such distinction can and must be made. I would place the two terms not in absolute opposition, but at two ends of a continuum, with gray areas of necessity remaining between them. The gradations along the continuum are marked not by the explicitness of the portrayal for sexuality or the body, nor by the assertiveness vs. passivity of persons, nor by any categorization of sexual acts or activities, but the extent to which autonomous personhood is attributed to the person or persons portrayed. Erotica portrays sexual subjects, manifesting their personhood in and through their bodies. Pornography depicts sex objects, persons reduced to their bodies. While the erotic nude presents the more pristine sexual body before the social persona is adopted through donning one's clothing, the pornographic nude portrays a body whose clothing has been more or less forcibly removed, where the absence of that clothing remains the most forceful presence in the image. Society's objectification remains present, indeed emphasized, in pornography, in a way in which it does not in erotica. Erotica, as sexual art, expresses a self, whereas pornography, as sexual commodity, markets one. The latter "works" because the operation it performs on women's bodies resonates with the "pornographizing" the male gaze does to women in other areas of society.[21] These distinctions remain problematic, to say the least, in their application, and disagreement in particular cases will no doubt remain. Much more work needs to be done before one would with any reasonable confidence distinguish authentic from imposed, personal from commercial, sexuality. Yet I believe this is the crucial question, and I believe these concepts correctly indicate the proper categories of analysis. Assuming a full definition of freedom as including autonomy and self-determination, pornography is therefore incompatible with real freedom.

CONCLUSIONS

It has often been noted that while socialist feminism is currently a major component of the array of feminisms one finds in academic feminism and women's studies, it is far less influential on the playing fields of practical politics.[22] While an analysis of male sexuality may seem an unlikely source to further socialist feminism's practical political agenda, I hope this paper's demonstration of the interconnections between intimate personal experiences and large-scale historical and social structures, especially in what may have initially seemed unlikely places, may serve as a useful methodological model for other investigations.

. . . I would like to conclude with some remarks on the practical import of this analysis. First of all, if the analysis of the relationship between pornography and consumerism and the argument about pornography leading to violence are correct, then a different conceptualization of the debate over the ethics of the feminist anti-pornography movement emerges. If one accepts, as I do, the idea that this movement is not against sex, but against sexual abuse, then the campaign against pornography is essentially not a call for censorship but a consumer campaign for product safety. The proper context for the debate over its practices is then not issues of free speech or civil liberties, but issues of business ethics. Or rather, this is the conclusion I reach remaining focused on pornography and male sexuality. But we should remember the broader context I alluded to at the beginning of this chapter, the question of pornography's effects on women. In that context, women are not the consumers of pornography, but the consumed. Rather than invoking the consumer movement, perhaps we should then look to environmental protection as a model.[23] Following this line of reasoning, one could in principle then perhaps develop under the tort law of product liability an argument to accomplish much of the regulation of sexually explicit material some are now trying to achieve through legislative means, perhaps developing a new definition of "safe" sexual material.

Finally, for most of us, most of our daily practice as academics consists of teaching rather than writing or reading in our fields. If one accepts the analysis I have presented, a central if not primary concern for us should therefore be how to integrate this analysis into our classrooms. I close by suggesting that we use this analysis and others like it from the emerging field of men's studies to demonstrate to the men in our classes the direct relevance of feminist analysis to their own lives, at the most intimate and personal levels, and that we look for ways to demonstrate to men that feminism can be personally empowering and liberating for them without glossing over, and in fact emphasizing, the corresponding truth that this will also require the surrender of male privilege.

NOTES

1. I am indebted for this formulation to Tim Carrigan, Bob Connell, and John Lee, "Toward a New Sociology of Masculinity," in Harry Brod, ed., *The Making of Masculinities: The New Men's Studies* (Boston: Allen & Unwin, 1987).
2. For the *locus classicus* of the redefinition of Marxism as method rather than doctrine, see Georg Lukács, *History and Class Consciousness: Studies in Marxist Dialectics,* trans. Rodney Livingstone (Cambridge, MA: MIT Press, 1972).
3. See my Introduction to Brod, *The Making of Masculinities.* For other recent books by men I consider to be engaged in essentially the same or a kindred project, see Jeff Hearn, *The Gender of Oppression: Men, Masculinity, and the Critique of Marxism* (New York: St. Martin's Press, 1987) and R. W. Connell, *Gender and Power* (Stanford, CA: Stanford University Press, 1987), particularly the concept of "hegemonic masculinity," also used in Carrigan, Connell, and Lee, "Toward a New Sociology of Masculinity." Needless to say, none of this work would be conceivable without the pioneering work of many women in women's studies.
4. Bernie Zilbergeld, *Male Sexuality: A Guide to Sexual Fulfillment* (Boston: Little, Brown and Company, 1978).
5. Michael Betzold, "How Pornography Shackles Men and Oppresses Women," in *For Men against Sexism: A Book of Readings,* ed. Jon Snodgrass (Albion, CA: Times Change Press, 1977), p. 46.
6. I am grateful to Lenore Langsdorf and Paula Rothenberg for independently suggesting to me how this point would fit into my analysis.
7. See Jessica Benjamin, "The Bonds of Love: Rational Violence and Erotic Domination," *Feminist Studies* 6 (1980): 144–174.
8. Keith McWalter, "Couch Dancing," *New York Times Magazine,* December 6, 1987, p. 138.
9. Karl Marx, "Economic and Philosophic Manuscripts: Third Manuscript," in *Early Writings,* ed. and trans. T. B. Bottomore (New York: McGraw-Hill, 1964), pp. 159–160.
10. Marx, pp. 164–165.
11. An earlier version of portions of the following argument appears in my article "Eros Thanatized: Pornography and Male Sexuality" with a "1988 Postscript," forthcoming in Michael Kimmel, ed., *Men Confronting Pornography* (New York: Crown, 1989). The article originally appeared (without the postscript) in *Humanities in Society* 7 (1984) pp. 47–63.
12. See the essays by myself and Michael Kimmel in Brod, *The Making of Masculinities.*
13. Compare Carol Brown on the shift from private to public patriarchy: "Mothers, Fathers, and Children: From Private to Public Patriarchy" in Lydia Sargent, ed., *Women and Revolution* (Boston: South End Press, 1981).
14. According to Martha May in her paper "'An Obligation on Every Man': Masculine Breadwinning and the Law in Nineteenth Century New York," presented at the American Historical Association, Chicago, Illinois, 1987, from which

I learned of these changes, the term "judicial patriarchy" is taken from historian Michael Grossberg *Governing the Hearth: Law and the Family in Nineteenth Century America* (Chapel Hill: University of North Carolina Press, 1985) and "Crossing Boundaries: Nineteenth Century Domestic Relations Law and the Merger of Family and Legal History," *American Bar Foundation Research Journal* (1985): 799–847.

15. Alan Soble, *Pornography: Marxism, Feminism, and the Future of Sexuality* (New Haven: Yale University Press, 1986), p. 82. I agree with much of Soble's analysis of male sexuality in capitalism, and note the similarities between much of what he says about "dismemberment" and consumerism and my analysis here.
16. See John D'Emilio and Estelle B. Freedman, *Intimate Matters: A History of Sexuality in America* (New York: Harper & Row, 1988).
17. See Barbara Ehrenreich, *The Hearts of Men: American Dreams and the Flight from Commitment* (New York: Anchor-Doubleday, 1983); and Wolfgang Fritz Haug, *Critique of Commodity Aesthetics: Appearance, Sexuality, and Advertising in Capitalist Society*, trans. Robert Bock (Minneapolis: University of Minnesota Press, 1986).
18. See my "Work Clothes and Leisure Suits: The Class Basis and Bias of the Men's Movement," originally in *Changing Men* 11 (1983) 10–12 and 38–40, reprint forthcoming in *Men's Lives: Readings in the Sociology of Men and Masculinity*, ed. Michael Kimmel and Michael Messner (New York: Macmillan, 1989).
19. This features prominently in the work of the Frankfurt school as well as contemporary ecofeminist theorists.
20. Rosemarie Tong, "Feminism, Pornography and Censorship," *Social Theory and Practice* 8 (1982): 1–17.
21. I learned to use "pornographize" as a verb in this way from Timothy Beneke's "Introduction" to his *Men on Rape* (New York: St. Martin's Press, 1982).
22. See the series of ten articles on "Socialist-Feminism Today" in *Socialist Review* 73–79 (1984–1985).
23. I am indebted to John Stoltenberg for this point.

DISCUSSION QUESTIONS

1. How does one determine when an image connotes sexual subjectivity and when it connotes sexual objectivity (or is this a false dichotomy)?
2. In what ways is Brod's analysis consistent with that of MacKinnon's? In what ways does it differ? Imagine that MacKinnon and Brod were in charge of drafting federal pornography regulation statutes. What would be the key points of their legislation? On what issues do you think they would have the most difficulty coming to agreement? What would your own pornography regulation statutes look like?
3. Do you think the alienation of male sexuality has increased or decreased in the time since this article was written (1988)? Why? What do you predict will happen in this regard in the next 10 years? 20? 50? Why?
4. Will an increase in women's social and economic power lessen the alienation of male sexuality? Why or why not? How will it affect women's relationship with sexuality?
5. What are the costs of male sexual alienation? What is its relationship, if any, to female sexual victimization?
6. Will the marketing of products like Viagra help or hinder male sexual alienation? Why?
7. Are gay, bisexual, and transgendered men alienated in the same way as heterosexual men? Why or why not?

Odyssey of a Feminist Pornographer

MARCY SHEINER

In this compelling autobiographical account, Marcy Sheiner recounts the evolution of her career as a phone-sex operator and writer of pornography. Her essay focuses on the turmoil and shame that typically accompany this vocation and the difficulties she has faced in trying to reconcile her sexual tastes and love of pornography with her identity as a self-proclaimed feminist.

Since the mid-eighties, I've been writing and selling pornographic stories; and since the early nineties, I've worked sporadically as a phone-sex operator. Like many people who work in the sex industry, these are not things I planned or expected to be doing in my life—I've always seen myself as a "serious" writer, with aspirations along the lines of Doris Lessing or Virginia Woolf. Sex work is something I drifted into, and when I learned that I could not only earn money at it but also enjoy it immensely, I continued.

My first intimate contact with pornography occurred when I "caught" my husband masturbating into a *Playboy* centerfold. I was all of eighteen at the time, and my budding sexual ego was seriously bruised by the discovery that he had chosen an air-brushed bunny rather than me on which to spill his precious seed. With not a little guilt, he tried to convince me that, even were he "married to Raquel Welch," he would still feel compelled to participate in this all-American male sport.

Thereafter I tried to ignore what I saw as his perversion, but one day I stumbled upon a collection of girlie magazines in the back of the file cabinet in his study. (Men in early '60s' suburbia, no matter what their profession, always had "studies.") With a sinking heart, I sat down to peruse the images of naked women, expecting to feel degraded and repulsed. But as I turned the pages to view one exquisite female creature after another, a funny thing began to happen: I got excited. Tentatively I touched the perfectly shaped breast on the page before me, letting my fingers wander down to a feathery vulva, imagining what it would feel like in real life. I found myself reading the silly captions with interest: "Lola loves to ride horses and play tennis, but her main ambition is to fulfill her man's every fantasy." "Monique works out every day to keep in shape for her lover." The not-so-subtle theme of these biographical sketches was that the women existed solely to bring men pleasure. As a budding feminist, I was confused by the arousal I felt from such a notion—but that didn't prevent me from lying right down to masturbate.

. . . [While] I believed that women were more than the sum of their physical parts, the fact that they so wantonly displayed themselves for money made me positively feverish. At the time I had no idea why this excited me, but now I do: power is an aphrodisiac, and in our culture money is the ultimate expression of power. This element was and remains, for me, one of pornography's key turn-ons: its sheer existence. The fact that some people earn money by sexually entertaining those who can afford to pay for it has always been in and of itself exciting to me. That it was usually women entertaining men went unexamined (and was indeed part of the turn-on) until much later, when feminist theory

From "Odyssey of a Feminist Pornographer" by Marcy Sheiner in *Whores and Other Feminists*, ed. by Jill Nagle, p. 36–43. Reprinted by permission of Routledge/Taylor & Francis Books, Inc.

compelled me to view the selling of sex as the foundation of women's oppression. Later still, I did another about face, becoming newly inspired by of actually being the object for sale, of knowing that men pay for the privilege of hearing my voice or reading my stories.

And that's where I am today, without apology. All across the country men masturbate to the memory of my voice on the telephone wire, whispering nasty things in their ear, and their hot cocks throb in their hands. They close their eyes and visualize the woman I've described myself to be (entirely different from who I really am) and they moan with desire. I cannot overemphasize the positive effect this has had upon my psyche: It makes me feel terrifically lush and powerful.

Similarly, somewhere in this world a couple is lying in bed perusing a sex rag. They read from a story I have written. . . .

. . . The man and woman reading my words turn to each other and embrace. She flings her legs around his hips and his cock slides into her. Spurred on by my story, they fuck for hours.

Is this power? It sure feels like it to me. But it did not always. My journey to self-acceptance as a pornographer has been fraught with anxiety. Those masturbatory moments in my husband's study turned out to be only the beginning of a long, confusing journey.

After leaving my six-year marriage, for reasons having nothing to do with pornography—or, for that matter, sex— . . . I became involved in consciousness-raising groups and the women's liberation movement. Porn was not a hot topic among feminists then; we were too preoccupied with more basic concerns, ranging from legalizing abortion to who did the dishes, and pornography did not command much of my conscious attention. The issue resurfaced when I fell in love with Marco Vassi, one of the most articulate erotic writers of our time.

I met Marco in Woodstock, New York, during the postfestival halcyon days of hippiedom. . . .

I was still traveling in feminist circles, organizing "Take Back the Night" marches, writing tirades for equality in the weekly newspaper. I took pains to keep my relationship with Marco hidden from my feminist friends. When Women Against Pornography (WAP) brought pornography to the forefront of public debate, I became positively schizophrenic, arguing with Marco on the one hand and with feminists on the other.

I saw some validity in the position taken by WAP: it was perfectly obvious that the porn industry was one in which women serviced men. But to Marco, as to most male pornographers, there was no debate: he regarded the antiporn women as pathetic morons who possessed no aesthetic sensibility whatsoever. With feminists I argued the other side: that the portrayal of sexuality per se wasn't necessarily bad, and that perhaps we as women might even learn something about our sexuality from pornography. To my feminist friends, the case was just as cut and dried as it was to Marco, but their position was that pornography exploited women, period.

I felt that neither side understood the complicated nuances of the issue—the gestalt of the sexual/political dichotomy. I was in a difficult place, a place many women are still in today: I bought the feminist rhetoric that pornography demeaned women, yet I was undeniably aroused by it.

I felt completely isolated in my confusion. . . . The appearance of *On Our Backs,* the first pornography I'd ever seen created specifically by and for women, only served to deepen my confusion.

It wasn't that I'd had no lesbian experience: one of the hottest relationships of my life had been with a woman. . . . During the year we were together I assumed I was a lesbian, but when we broke up I slept with men again. And then another woman. Then a man: Woman. Man. Woman. Who/what was I? Lesbians called me a traitor for "sleeping with the enemy." Male lovers wanted me to take them to dyke bars and choreograph threesomes. Nobody back then thought that being bisexual was natural or even hip. But bisexual was apparently what I was.

And I was more attracted to men than to women—another demerit on my feminist report card. So when *On Our Backs* didn't grab my crotch with the same intensity as *Playboy,* it only increased my sense of shame. My response to pornography seemed to be entwined with the very idea I opposed: women servicing men. I decided that my sexuality was hopelessly "male identified." For it wasn't just my attraction to male pornography that troubled me: it was also my politically incorrect fantasies.

Since childhood I have had an extremely vivid sexual imagination; I can't recall a single orgasm that hasn't been induced in part by movielike images and long involved plots spinning through my brain, no matter who my partner. . . .

For a while I tried to alter my fantasies, to "reprogram" my responses as advised by some feminists. I would masturbate to visions of egalitarian sex in meadows, or try not fantasizing at all. Like the good Christian who fights with the impulse to masturbate, I fought what I saw as weakness, but my weakness won out. Time and again, hovering on the brink of orgasm, I would allow some image of myself as groveling sex slave to carry me blissfully across the abyss. After several months of these reprogramming attempts, I decided that, like my long-gone husband who would masturbate to *Playboy* centerfolds even were Raquel Welch his wife, I would be fantasizing scenarios at odds with my political beliefs for the rest of my natural days.

. . . Today I am no longer ashamed of my fantasies; in fact, I exult in having been blessed with such a rich imagination. It has brought me enormous joy, a joy that increased considerably when I discovered my fantasies could be financially lucrative. The way I began writing porn was wholly accidental: I was lying in bed one night fantasizing about my neighbor, who was driving me nuts by warming up his truck at five o'clock every morning. I imagined confronting him, and the confrontation becoming sexual. As I began to masturbate, it occurred to me that I might put my fingers to better use with a pen, so I got out of bed and simply wrote out the fantasy.

. . . Aiming high, I sent "A Neighborly Compromise" off to *Penthouse.* In a few weeks I received an offer of $350 and publication in *Hot Talk,* one of that empire's many magazines. Let me tell you, after years of papering my walls with rejections from *The New Yorker* and *Ms.,* I nearly fell off my chair.

"Writing," said Moliére, "is like prostitution. First we do it for love. Then we do it for a few friends. Then we do it for money." During the course of the next year I sold four stories to *Penthouse Forum* at $1,000 a shot. When the editor I'd been working with left, the gig was up, but by then I'd started to connect with the sex network, and eventually I found other markets for my work.

. . . As I shyly confessed my extracurricular activities to more and more friends, I found, to my surprise, that they were amused rather than disapproving. I even told my mother, who was actually delighted that I'd finally found a means of earning money through writing. But while friends and family accepted what I was doing, they certainly did not consider it valuable work, nor did they have any notion that I put as much thought and effort into writing porn as I did into poetry, or that I was as proud of one as of the other. In other words, while I didn't encounter disapproval, neither did I gain the kind of recognition every artist, or person, wants and needs for their work.

This, of course, remains a persistent problem. My mother saves all my boring news stories, but never even sees my witty porn. My friends praise my analytical book reviews but laugh dismissively at the unread erotica. In my pornography I've traveled to Paris, Greece, and Hawaii, but the writing my peers are likely to see is confined to a few square miles. To them, the fact that I write porn is some kind of a lark or a way to support the more "creative" work they imagine I do. To me, porn *is* the creative work.

Since I moved to the San Francisco Bay Area, though, things have improved by leaps and bounds, in fact, it is here that I've been able to move from shame to self-acceptance. (I'm still working on pride.)

. . . I said that I drifted into sex work, and that's true; but actually this drifting has been a very organic process. I do sex work because it interests me, because I'm good at it, and because I can make money at it; but unlike most sex radicals, I'm not in this as a political cause. I don't really give a shit if Molly in Idaho reads one of my stories and has her first orgasm, though it's fine with me if she does. Sure, if a phone sex customer expresses guilt and shame about masturbating, I give him the sex-positive pep talk. But I don't like debating with the prudish about the ethics of what I do, and I'm not out to convert anyone; I just don't want others telling me I can't do the work I love.

Don't get me wrong; I fully support the folks who are out there on the front lines defending sexual freedom; I've even done a bit of it myself, in the natural course of my work. The sex-negative culture surrounding this little oasis of sanity known as the Bay Area does tend to push those of us who dare seek and promote pleasure into an aggressive stance.

But for me, the discourse among sex radicals—the exchange of information and ideas—all feels utterly natural, the way things ought to be. My life is filled with a variety of activities, many of them having to do with sex and many of them absolutely mundane. I don't feel all that unusual, or think of my friends and our talks as extraordinary. I'm apt to forget that mainstream America regards us as perverts, and that I'd do well to clean up my act any time I decide to leave the Bay Area.

Just how extraordinary we really are, and how perverted we appear to the mainstream, was recently brought home when I was interviewed for an article in *Elle* magazine on—what else?—"the new women's erotica."

"So you all sleep with each other?" the reporter asked, convinced that my life was one long orgy.

"It's just amazing," he continued, ignoring my negative reply. "Here you are, a nice Jewish girl, a mother, and you're living this life full of sex!"

That old sense of being an oddball returned full force. What did not accompany it, however, was the feeling of being "inappropriate." In fact, I began to view my inquisitor as the true oddball. Can he really, I wondered, be so ignorant? Can he really still hold these archaic attitudes?"

"None of my friends talk about sex," he told me. "We would never say the things you say."

I actually felt sorry for him.

Meanwhile, the feminist debates continue, aided and abetted by writers for *Elle*. As far as I can see, the level of discussion hasn't evolved much since the early days of WAP. As I said, I try to stay away from the debate. I admire women like Susie Bright who can articulate reasoned arguments in the face of picket lines and red-faced hecklers, but this is something I am unable to do, even one-on-one, even with questioning friends. I grow weary of the oh-so-rational anti-censorship tack. Sexual expression is about more than anticensorship. It's about, well, sexual expression!

Contrary to the opinions of the conservative antisex faction of the women's movement, I no longer doubt that I am a feminist. Ironically, this "F" word has fallen into almost as much disrepute as the original "F" word, but I consider feminism a core part of my identity as well as an honorable philosophy to espouse. To me feminism means a belief in a very basic goal: full equality of women. I don't see pornography or sexual pleasure as undermining that belief or that goal. Like everything else, it can certainly be used destructively—but it isn't an inherently negative force.

Au contraire. Writing and reading pornography has been, for me, fun, exciting, creative, illuminating, empowering, and lucrative. And I feel no qualms or contradiction when I say that I am both a feminist and a pornographer.

DISCUSSION QUESTIONS

1. How central are themes of domination and submission to the social construction of sexuality in our culture? How central are they to gender? To gendered sexuality?

2. How do individuals resolve contemporary tensions between constructions of sex as romance and love on the one hand, and domination and control on the other?
3. Does pornography undermine the full equality of women? Why or why not?
4. Can a person legitimately be considered a feminist if they are sexually aroused by thoughts of sexual submission? Does your answer to this question differ for men and women? Should it? Why or why not?
5. Is there value is pornography? Is there value in *studying* pornography? Why or why not?
6. People are often surprised by, and even ashamed of, their erotic tastes and fantasies. For many, these tastes include pornography. Yet our fantasies and desires are shaped within cultural contexts. What is society's role in shaping our erotic desires? How much control do *we* have over what "turns us on"?
7. How do you think D'Emilio, Brod, and MacKinnon would respond to Sheiner's arguments? (Address each separately.)

13

Sexuality, Aggression, and Violence

Men on Rape

TIM BENEKE

Women are socialized from a very early age to be vigilantly mindful of the possibility of sexual victimization. Tim Beneke's analysis begins by assessing how this constant vigilance may affect the lives of women throughout society. His analysis also reveals many of the insidious ways that some men may construct blame so as to hold women responsible for their own victimization.

Rape may be America's fastest growing violent crime; no one can be certain because it is not clear whether more rapes are being committed or reported. It *is* clear that violence against women is widespread and fundamentally alters the meaning of life for women; that sexual violence is encouraged in a variety of ways in American culture; and that women are often blamed for rape.

Consider some statistics:

- In a random sample of 930 women, sociologist Diana Russell found that 44 percent had survived either rape or attempted rape. Rape was defined as sexual intercourse physically forced upon the women, or coerced by threat of bodily harm, or force upon the woman when she was helpless (asleep, for example). The survey included rape and attempted rape in marriage in its calculations. (Personal communication)
- In a September 1980 survey conducted by *Cosmopolitan* magazine to which over 106,000 women anonymously responded, 24 percent had been raped at least once. Of these, 51 percent had been raped by friends, 37 percent by strangers, 18 percent by relatives, and 3 percent by husbands. 10 percent of the women in the survey had been victims of incest. Seventy-five percent of the women had been "bullied into making love." Writer Linda Wolfe, who reported on the survey, wrote in reference

to such bullying: "Though such harassment stops short of rape, readers reported that it was nearly as distressing."

- An estimated 2–3 percent of all men who rape outside of marriage go to prison for their crimes.[1]
- The F.B.I. estimates that if current trends continue, one woman in four will be sexually assaulted in her lifetime.[2]
- An estimated 1.8 million women are battered by their spouses each year.[3] In extensive interviews with 430 battered women, clinical psychologist Lenore Walker, author of *The Battered Women,* found that 59.9 percent had also been raped (defined as above) by their spouses. Given the difficulties many women had in admitting they had been raped, Walker estimates the figure may well be as high as 80 or 85 percent (Personal communication). If 59.9 percent of the 1.8 million women battered each year are also raped, then a million women may be raped in marriage each year. And a significant number are raped in marriage without being battered.
- Between one in two and one in ten of all rapes are reported to the police.[4]
- Between 300,000 and 500,000 women are raped each year outside of marriage.[5]

What is often missed when people contemplate statistics on rape is the effect of the *threat* of sexual violence on women. I have asked women repeatedly, "How would your life be different if rape were suddenly to end?" (Men may learn a lot by asking this question of women to whom they are close.) The threat of rape is an assault upon the meaning of the world; it alters the feel of the human condition. Surely any attempt to comprehend the lives of women that fails to take issues of violence against women into account is misguided.

Through talking to women, I learned: *The threat of rape alters the meaning and feel of the night.* Observe how your body feels, how the night feels, when you're in fear. The constriction in your chest, the vigilance in your eyes, the rubber in your legs. What do the stars look like? How does the moon present itself? What is the difference between walking late at night in the dangerous part of a city and walking late at night in the country, or safe suburbs? When I try to imagine what the threat of rape must do to the night, I think of the stalked, adrenalated feeling I get walking late at night in parts of certain American cities. Only, I remind myself, it is a fear different from any I have known, a fear of being raped.

It is night half the time. If the threat of rape alters the meaning of the night, it must alter the meaning and pace of the day, one's relation to the passing and organization of time itself. For some women, the threat of rape at night turns their cars into armored tanks, their solitude into isolation. And what must the space inside a car or an apartment feel like if the space outside is menacing?

I was running late one night with a close woman friend through a path in the woods on the outskirts of a small university town. We had run several miles and were feeling a warm, energized serenity.

"How would you feel if you were alone?" I asked.

"Terrified!" she said instantly.

"Terrified that there might be a man out there?" I asked, pointing to the surrounding moonlit forest, which had suddenly been transformed into a source of terror.

"Yes."

Another women said, "I know what I can't do and I've completely internalized what I can't do. I've built a viable life that basically involves never leaving my apartment at night unless I'm directly going some place to meet somebody. It's unconsciously built into what it occurs to women to do." When one is raised without freedom, one may not recognize its absence.

The threat of rape alters the meaning and feel of nature. Everyone has felt the psychic nurturance of nature. Many women are being deprived of that nurturance, especially in wooded areas near cities. They are deprived either because they cannot experience nature in solitude because of threat, or because, when they do choose solitude in nature, they must cope with a certain subtle but nettlesome fear.

Women need more money because of rape and the threat of rape makes it harder for women to earn money. It's simple: if you don't feel safe walking at night, or riding public transportation, you need a car. And it is less practicable to live in cheaper, less secure, and thus more dangerous neighborhoods if the ordinary threat of violence that men experience, being mugged, say, is compounded by the threat of rape. By limiting mobility at night, the threat of rape limits where and when one is able to work, thus making it more difficult to earn money. An obvious bind: women need more money because of rape, and have fewer job opportunities because of it.

The threat of rape makes women more dependent on men (or other women). One woman said: "If there were no rape I wouldn't have to play games with men for their protection." The threat of rape falsifies, mystifies, and confuses relations between men and women. If there were no rape, women would simply not need men as much, wouldn't need them to go places with at night, to feel safe in their homes, for protection in nature.

The threat of rape makes solitude less possible for women. Solitude, drawing strength from being alone, is difficult if being alone means being afraid. To be afraid is to be in need, to experience a lack; the threat of rape creates a lack. Solitude requires relaxation; if you're afraid, you can't relax.

The threat of rape inhibits a woman's expressiveness. "If there were no rape," said one woman, "I could dress the way I wanted and walk the way I wanted and not feel self-conscious about the responses of men. I could be friendly to people. I wouldn't have to wish I was ugly. I wouldn't have to make myself small when I got on the bus. I wouldn't have to respond to verbal abuse from men by remaining silent. I could respond in kind."

If a woman's basic expressiveness is inhibited, her sexuality, creativity, and delight in life must surely be diminished.

The threat of rape inhibits the freedom of the eye. I know a married couple who live in Manhattan. They are both artists, both acutely sensitive and responsive to the visual world. When they walk separately in the city, he has more freedom to look than she does. She must control her eye movements lest they inadvertently meet the glare of some importunate man. What, who, and how she sees are restricted by the threat of rape.

The following exercise is recommended for men.

> Walk down a city street. Pay a lot of attention to your clothing; make sure your pants are zipped, shirt tucked in, buttons done. Look straight ahead. Every time a man walks past you, avert your eyes and make your face expressionless. Most women learn to go through this act each time we leave our houses. It's a way to avoid at least some of the encounters we've all had with strange men who decided we looked available.[6]

To relate aesthetically to the visual world involves a certain playfulness, spirit of spontaneous exploration. The tense vigilance that accompanies fear inhibits that spontaneity. The world is no longer yours to look at when you're afraid.

I am aware that all culture is, in part, restriction, that there are places in America where hardly anyone is safe (though men are safer than women virtually everywhere), that there are many ways to enjoy life, that some women may not be so restricted, that there exist havens, whether psychic, geographical, economic, or class. But they are *havens,* and as such, defined by threat.

Above all, I trust my experience: no woman could have lived the life I've lived the last few years. If suddenly I were restricted by the threat of rape, I would feel a deep, inexorable depression. And it's not just rape; it's harassment, battery, Peeping Toms, anonymous phone calls, exhibitionism, intrusive stares, fondlings—all contributing to an atmosphere of intimidation in women's lives. And I have only scratched the surface; it would take many carefully crafted short stories to begin to express what I have only hinted at in the last few pages. I have not even touched upon what it might mean for a woman to be sexually assaulted. Only women can speak to that. Nor have I suggested how the threat of rape affects marriage.

Rape and the threat of rape pervade the lives of women, as reflected in some popular images of our culture.

SHE ASKED FOR IT—BLAMING THE VICTIM[7]

Many things may be happening when a man blames a woman for rape.

First, in all cases where a women is said to have asked for it, her appearance and behavior are taken as a form of speech. "Actions speak louder than words" is a widely held belief; the woman's actions—her appearance may be taken as action—are given greater emphasis than her words; an interpretation alien to the women's intentions is given to her actions. A logical extension of "she asked for it" is the idea that she wanted what happened to happen; if she wanted it to happen, she *deserved* for it to happen. Therefore, the man is not to be blamed. "She asked for it" can mean either that she was consenting to have sex and was not really raped, or that she was in fact raped but somehow she really deserved it. "If you ask for it, you deserve it," is a widely held notion. If I ask you to beat me up and you beat me up, I still don't deserve to be beaten up. So even if the notion that women asked to be raped had some basis in reality, which it doesn't, on its own terms it makes no sense.

Second, a mentality exists that says: a woman who assumes freedoms normally restricted to a man (like going out alone at night) and is raped is doing the same thing as a woman who goes out in the rain without an umbrella and catches a cold. Both are considered responsible for what happens to them. That men will rape is taken to be a legitimized given, part of nature, like rain or snow. The view reflects a massive abdication of responsibility for rape on the part of men. It is so much easier to think of rape as natural than to acknowledge one's part in it. So long as rape is regarded as natural, women will be blamed for rape.

A third point. The view that it is natural for men to rape is closely connected to the view of women as commodities. If a woman's body is regarded as a valued commodity by men, then of course, if you leave a valued commodity where it can be taken, it's just human nature for men to take it. If you left your stereo out on the sidewalk, you'd be asking for it to get stolen. Someone will just take it. (And how often men speak of rape as "going out and *taking* it.") If a woman walks the streets at night, she's leaving a valued commodity, her body, where it can be taken. So long as women are regarded as commodities, they will be blamed for rape.

Which brings us to a fourth point. "She asked for it" is inseparable from a more general "psychology of the dupe." If I use bad judgment and fail to read the small print in a contract and later get taken advantage of, "screwed" (or "fucked over") then I deserve what I get; bad judgment makes me liable. Analogously, if a woman trusts a man and goes to his apartment, or accepts a ride hitchhiking, or goes out on a date and is raped, she's a dupe and deserves what she gets. "He didn't *really* rape her" goes the mentality—"he merely took advantage of her." And in America it's okay for people to take advantage of each other, even expected and praised. In fact, you're considered dumb and foolish if you don't take advantage of other people's bad judgement. And so, again, by treating them as dupes, rape will be blamed on women.

Fifth, if a woman who is raped is judged attractive by men, and particularly if she dresses to look attractive, then the mentality exists that she attacked him with her weapon so, of course, he counter-attacked with his. The preview to a popular movie states: "She was the victim of her own *provocative beauty*." Provocation: "There is a line which, if crossed, will *set me off* and I will lose control and no longer be responsible for my behavior. If you punch me in the nose then, of course, I will not be responsible for what happens: you will have provoked a fight. If you dress, talk, move, or act a certain way, you will have provoked me to rape. If your appearance *stuns* me, *strikes* me, *ravishes* me, *knocks me out,* etc., then I will not be held responsible for what happens; you will have asked for it." The notion that sexual feeling makes one helpless is part of a cultural abdication of responsibility for sexuality. So long as a woman's appearance is viewed as a weapon and sexual feeling is believed to make one helpless, women will be blamed for rape.

Sixth, I have suggested that men sometimes become obsessed with images of women, that images become a substitute for sexual feeling, that sexual feeling becomes externalized and out of control and is given an undifferentiated identity in the appearance of women's bodies. It is a process of projection in which one blurs one's own desire with her imagined, projected desire. If a woman's attractiveness is taken to signify one's own lust and a woman's lust, then when an "attractive" woman is raped, some men may think she wanted sex. Since they perceive their own lust in part projected onto the woman, they disbelieve women who've been raped. So long as men project their own sexual desires onto women, they will blame women for rape.

And seventh, what are we to make of the contention that women in dating situations say "no" initially to sexual overtures from men as a kind of pose, only to give in later, thus revealing their true intentions? And that men are thus confused and incredulous when women are raped because in their sexual experience women can't be believed? I doubt that this has much to do with men's perceptions of rape. I don't know to what extent women actually "say no and mean yes"; certainly it is a common theme in male folklore. I have spoken to a couple of women who went through periods when they wanted to be sexual but were afraid to be, and often rebuffed initial sexual advances only to give in later. One point is clear: the ambivalence women may feel about having sex is closely tied to the inability of men to fully accept them as sexual beings. Women have been traditionally punished for being openly and freely sexual; men are praised for it. And if many men think of sex as achievement of possession of a valued commodity, or aggressive degradation, then women have every reason to feel and act ambivalent.

These themes are illustrated in an interview I conducted with a 23-year-old man who grew up in Pittsburgh and works as a file clerk in the financial district of San Francisco. Here's what he said:

"Where I work it's probably no different from any other major city in the U.S. The women dress up, in high heels, and they wear a lot of makeup, and they just look really *hot* and really sexy, and how can somebody who has a healthy sex drive not feel lust for them when you see them? I feel lust for them, but I don't think I could find it in me to overpower someone and rape them. But I definitely get the feeling that I'd like to rape a girl. I don't know if the actual act of rape would be satisfying, but the *feeling* is satisfying.

"These women look so good, and they kiss ass of the men in the three-piece suits who are *big* in the corporation, and most of them relate to me like 'Who are *you*? Who are *you* to even *look* at?' They're snobby and they condescend to me, and I resent it. It would take me a lot longer to get to first base than it would somebody with a three-piece suit who had money. And to me a lot of the men they go out with are superficial assholes who have no real feelings or substance, and are just trying to get ahead and make a lot of money. Another thing that makes me resent these women is thinking, 'How could she want to hang out with somebody like that? What does that make her?'

"I'm a file clerk, which makes me feel like a nebbish, a nerd, like I'm not making it, I'm a failure. But I don't really believe I'm a failure because I know it's just a phase, and I'm just doing it for the money, just to make it through this phase. I catch myself feeling like a failure, but I realize that's ridiculous."

What Exactly Do You Go Through When You See These Sexy, Unavailable Women?

"Let's say I *see* a woman and she looks really pretty and really clean and sexy, and she's giving off very feminine, sexy vibes. I think, 'Wow, I would love to make love to her,' but I know she's not really interested. It's a tease. A lot of times a woman knows that she's looking really good and she'll use that and flaunt it, and it makes me feel like she's laughing at me and I feel *degraded*.

"I also feel dehumanized, because when I'm being teased I just turn off, I cease to be human.

Because if I go with my human emotions I'm going to want to put my arms around her and kiss her, and to do that would be unacceptable. I don't like the feeling that I'm supposed to stand there and take it, and not be able to hug her or kiss her; so I just turn off my emotions. It's a feeling of humiliation, because the woman has forced me to turn off my feelings and react in a way that I really don't want to.

"If I were actually desperate enough to rape somebody, it would be from wanting the person, but it would be a very spiteful thing, just being able to say, 'I have power over you and I can do anything I want with you,' because really I feel that *they* have power over *me* just by their presence. Just the fact that they can come up to me and just melt me and make me feel like a dummy makes me want revenge. They have power over me so I want power them. . . .

"Society says that you have to have a lot of sex with a lot of different women to be a real man. Well, what happens if you don't? Then what are you? Are you half a man? Are you still a boy? It's ridiculous. You see a whiskey ad with a guy and two women on his arm. The implication is that real men don't have any trouble getting women."

How Does It Make You Feel Toward Women to See All These Sexy Women in Media and Advertising Using Their Looks to Try to Get You to Buy Something?

"It makes me hate them. As a man you're taught that men are more powerful than women, and that men always have the upper hand, and that it's a man's society; but then you see all these women and it makes you think, 'Jesus Christ, if we have all the power how come all the beautiful women are telling us what to buy?' And to be honest, it just makes me hate beautiful women because they're using their power over me. I realize they're being used themselves, and they're doing it for money. In *Playboy* you see all these beautiful women who look so sexy and they'll be giving you all these looks like they want to have sex so bad; but then in reality you know that except for a few nymphomaniacs, they're doing it for the money; so I hate them for being used and for using their bodies in that way.

"In this society, if you ever sit down and realize how manipulated you really are it makes you pissed off—it makes you want to take control. And you've been manipulated by women, and they're a very easy target because they're out walking along the streets, so you can just grab one and say, 'Listen, you're going to do what I want you to do,' and it's an act of revenge against the way you've been manipulated.

"I know a girl who was walking down the street by her house, when this guy jumped her and beat her up and raped her, and she was black and blue and had to go to the hospital. That's beyond me. I can't understand how somebody could do that. If I were going to rape a girl, I wouldn't hurt her. I might *restrain* her, but I wouldn't *hurt* her. . . .

"The whole dating game between men and women also makes me feel degraded. I hate being put in the position of having to initiate a relationship. I've been taught that if you're not aggressive with a woman, then you've blown it. She's not going to jump on *you*, so *you've* got to jump on *her*. I've heard all kinds of stories where the woman says, 'No! No! No!' and they end up making great love. I get confused as hell if a woman pushes me away. Does it mean she's trying to be a nice girl and wants to put up a good appearance, or does it mean she doesn't want anything to do with you? You don't know. Probably a lot of men think that women don't feel like real women unless a man tries to force himself on her, unless she brings out the 'real man,' so to speak, and probably too much of it goes on. It goes on in my head that you're complimenting a woman by actually staring at her or by trying to get into her pants. Lately, I'm realizing that when I stare at women lustfully, they often feel more threatened than flattered."

NOTES

1. Such estimates recur in the rape literature. See *Sexual Assault* by Nancy Gager and Cathleen Schurr, Grosset & Dunlap, 1976, or *The Price of Coercive Sexuality* by Clark and Lewis, The Women's Press, 1977.

2. *Uniform Crime Reports*, 1980.
3. See *Behind Closed Doors* by Murray J. Strauss and Richard Gelles, Doubleday, 1979.
4. See Gager and Schurr (above) or virtually any book on the subject.
5. Again see Gager and Schurr, or Carol V. Horos, *Rape*, Banbury Books, 1981.
6. From "Willamette Bridge" in *Body Politics* by Nancy Henley, Prentice-Hall, 1977, p. 144.
7. I would like to thank George Lakoff for this insight.

DISCUSSION QUESTIONS

1. As Beneke discusses, women are often blamed for their own victimization in the aftermath of a sexual assault. How are cultural constructions of gender and sexuality complicit in this?
2. Construct discourses of blame that would target men rather than women. In what ways, if any, do these differ from the types of attributions made toward women? Do any other themes emerge?
3. Beneke states "So long as rape is regarded as natural, women will be blamed for rape." Do you agree or disagree? Why? What kind of changes to the social structure would be necessary to dismantle the naturalization of rape?
4. In what ways do victim/survivor characteristics shape the patterns of blame attribution for the crime of sexual assault? For example, do black women experience different dynamics of blame than white? Does class matter? Age? Sex? Sexual Identity? If so, how?
5. How are the dynamics of blame attribution different for rape and sexual assault compared with other interpersonal crimes of violence (e.g., robbery, murder, etc.)?
6. In what ways does sexual violence alienate women from their own sexuality? From the "meaning of the world"?

'The Mind That Burns in Each Body': Women, Rape, and Racial Violence

JACQUELYN DOWD HALL

In this powerful analysis, Jacquelyn Dowd Hall examines the historical ties between rape and lynching in the United States and reveals them to be interlinked forms of oppression and hostility "focused on human flesh," which serve as weapons of terror to ensure the maintenance of white male supremacy. Hall draws particular attention to the ways in which cultural myths of black women's promiscuity, white women's purity, and black men's rapacious lust for white women were (and continue to be) used to legitimate and structure the subordination of women and black men.

HOSTILITY FOCUSED ON HUMAN FLESH

> Florida to burn Negro at stake: sex criminal seized from jail, will be mutilated, set afire in extra-legal vengeance for deed
>
> Dothan(Alabama) Eagle, October 26, 1934

> After taking the nigger to the woods . . . they cut off his penis. He was made to eat it. Then they cut off his testicles and made him eat them and say he liked it.
>
> Member of a lynch mob, 1934[1]

Lynching, like rape, has not yet been given its history. Perhaps it has been too easily relegated to the shadows where "poor white" stereotypes dwell. Perhaps the image of absolute victimization it evokes has been too difficult to reconcile with what we know about black resilience and resistance. Yet the impact of lynching, both as practice and as symbol, can hardly be underestimated. Between 1882 and 1946 almost 5,000 people died by lynching. The lynching of Emmett Till in 1955 for whistling at a white woman, the killing of three civil rights workers in Mississippi in the 1960s, and the hanging of a black youth in Alabama in 1981 all illustrate the persistence of this tradition of ritual violence in the service of racial control, a tradition intimately bound up with the politics of sexuality.

Vigilantism originated on the eighteenth-century frontier where it filled a vacuum in law enforcement. Rather than passing with the frontier, however, lynching was incorporated into the distinctive legal system of southern slave society.[2] In the nineteenth century, the industrializing North moved toward a modern criminal justice system in which police, courts, and prisons administered an impersonal, bureaucratic rule of law designed to uphold property rights and discipline unruly workers. The South, in contrast, maintained order through a system of deference and customary authority in which all whites had informal police power over all blacks, slave owners meted out plantation justice undisturbed by any generalized rule of law, and the state encouraged vigilantism as part of its overall reluctance to maintain a strong system of formal authority that would have undermined the planter's prerogatives. The purpose of one system was class control, of the other, control over a slave population. And each tradition continued into the period after the Civil War. In the North, factory-like penitentiaries warehoused displaced members of the industrial proletariat. The South maintained higher rates of personal violence than any other region in the country and lynching crossed over the line from informal law enforcement into outright political terrorism.

White supremacy, of course, did not rest on force alone. Routine institutional arrangements denied to the freedmen and women the opportunity to own land, the right to vote, access to education, and participation in the administration of the law. Lynching reached its height during the battles of Reconstruction and the Populist revolt; once a new system of disfranchisement, debt peonage, and segregation was firmly in place, mob violence gradually declined. Yet until World War I, the average number of lynchings never fell below two or three a week. Through the twenties and thirties, mob violence reinforced white dominance by providing planters with a quasi-official way of enforcing labor contracts and crop lien laws and local officials with a means of extracting deference, regardless of the letter of the law. Individuals may have lynched for their own twisted reasons, but the practice continued only with tacit official consent.[3]

Most importantly, lynching served as a tool of psychological intimidation aimed at blacks as a group. Unlike official authority, the lynch mob was unlimited in its capriciousness. With care and vigilance, an individual might avoid situations that landed him in the hands of the law. But a lynch mob could strike anywhere, any time. Once the brush fire of rumor began, a manhunt was organized, and the local paper began putting out special editions announcing a lynching in progress, there could be few effective reprieves. If the intended victim could not be found, an innocent bystander might serve as well.

It was not simply the threat of death that gave lynching its repressive power. Even as outbreaks of mob violence declined in frequency, they were increasingly accompanied by torture and sexual mutilation. Descriptions of the first phase of Hitler's death sweep are chillingly applicable to lynching: "Killing was ad hoc, inventive, and in its dependence on imagination, peculiarly expressive . . . this was murder uncanny in its anonymous intimacy, a hostility so personally focused on human flesh that the abstract fact of death was not enough.[4]

At the same time, the expansion of communications and the development of photography in the late nineteenth and early twentieth centuries gave reporting a vividness it had never had before. The lurid evocation of human suffering implicated white readers in each act of aggression and drove home to blacks the consequences of powerlessness. Like whipping under slavery, lynching was an instrument of coercion intended to impress not only the immediate victim but all who saw or heard about the event. And the mass media spread the imagery of rope and faggot far beyond the community in which each lynching took place. Writing about his youth in the rural South in the 1920s, Richard Wright describes the terrible climate of fear:

> The things that influenced my conduct as a Negro did not have to happen to me directly; I needed but to hear of them to feel their full effects in the deepest layers of my consciousness. Indeed, the white brutality that I had not seen was a more effective control of my behavior than that which I knew. The actual experience would have let me see the realistic outlines of what was really happening, but as long as it remained something terrible and yet remote, something whose horror and blood might descend upon me at any moment, I was compelled to give my entire imagination over to it.[5]

A penis cut off and stuffed in a victim's mouth. A crowd of thousands watching a black man scream in pain. Such incidents did not have to occur very often, or be witnessed directly, to be burned indelibly into the mind.

NEVER AGAINST HER WILL

> White men have said over and over—and we have believed it because it was repeated so often—that not only was there no such thing as a chaste Negro woman—but that a Negro woman could not be assaulted, that it was never against her will.
>
> Jessie Daniel Ames (1936)

Schooled in the struggle against sexual rather than racial violence, contemporary feminists may nevertheless find familiar this account of lynching's political function, for analogies between rape and lynching have often surfaced in the literature of the anti-rape movement . . . It is the suggestion of this essay, however, that there is significant resonance between these two forms of violence. We are only beginning to understand the web of connections among racism, attitudes toward women, and sexual ideologies. The purpose of looking more closely at the dynamics of repressive violence is not to reduce sexual assault and mob murder to static equivalents but to illuminate some of the strands of that tangled web.

The association between lynching and rape emerges most clearly in their parallel use in racial subordination. As Diane K. Lewis has pointed out, in a patriarchal society, black men, as men, constituted a potential challenge to the established order.[6] Laws were formulated primarily to exclude black men from adult male prerogatives in the public sphere, and lynching meshed with these legal mechanisms of exclusion. Black women represented a more ambiguous threat. They too were denied access to the politico-jural domain, but since they shared this exclusion with women in general, its maintenance engendered less anxiety and required less force. Lynching served primarily to dramatize hierarchies among men. In contrast, the violence directed at black women illustrates the double jeopardy of race and sex. The records of the Freedmen's Bureau and the oral histories collected by the Federal Writers' Project testify to the sexual atrocities endured by black women as whites sought to reassert their command over the newly freed slaves. Black women were sometimes executed by lynch mobs, but more routinely they served as targets of sexual assault.

Like vigilantism, the sexual exploitation of black women had been institutionalized under slavery. Whether seized through outright force or voluntarily granted within the master-slave relation, the sexual access of white men to black women was a cornerstone of patriarchal power in the South. It was used as a punishment or demanded in exchange for leniency. Like other forms of deference and conspicuous consumption, it buttressed planter hegemony. And it served the practical economic purpose of replenishing the slave labor supply.

After the Civil War, the informal sexual arrangements of slavery shaded into the use of rape as a political weapon, and the special vulnerability of black women helped shape the ex-slaves' struggle for the prerequisites of freedom. . . . [T]he sharecropping system that replaced slavery . . . grew in part from the desire of blacks to withdraw from gang labor and gain control over their own work, family lives, and bodily integrity. The sharecropping family enabled women to escape white male supervision, devote their productive and reproductive powers to their own families, and protect themselves from sexual assault.[7]

Most studies of racial violence have paid little attention to the particular suffering of women.[8] Even rape has been seen less as an aspect of sexual oppression than as a transaction between white and black men. Certainly Claude Lèvi-Strauss's insight that men use women as verbs with which to communicate with one another (rape being a means of communicating defeat to the men of a conquered tribe) helps explain the extreme viciousness of sexual violence in the post-emancipation era.[9] Rape *was* in part a reaction to the effort of the freedmen to assume the role of patriarch, able to provide for and protect his family. Nevertheless, as writers like Susan Griffin and Susan Brownmiller and others have made clear, rape is first and foremost a crime against women.[10] Rape sent a message to black men, but more centrally, it expressed male sexual attitudes in a culture both racist and patriarchal.

. . . In the United States, the fear and fascination of female sexuality was projected onto black women; the passionless lady arose in symbiosis with the primitively sexual slave. House slaves often served as substitute mothers; at a black woman's breast white men experienced absolute dependence on a being who was both a source of wish-fulfilling joy and of grief-producing disappointment. In adulthood, such men could find in this black woman a ready object for the mixture of rage and desire that so often underlies male heterosexuality. The black woman, already in chains, was sexually available, unable to make claims for support or concern; by dominating her, men could replay the infant's dream of unlimited access to the mother.[11] The economic and political challenge posed by the black patriarch might be met with death by lynching, but when the black woman seized the opportunity to turn her maternal and sexual resources to the benefit of her own family, sexual violence met her assertion of will. Thus rape reasserted white dominance and control in the private arena as lynching reasserted hierarchical arrangements in the public transactions of men.

LYNCHING'S DOUBLE MESSAGE

> The crowds from here that went over to see [Lola Cannidy, the alleged rape victim in the Claude Neal lynching of 1934] said he was so large he could not assault her until he took his knife and cut her, and also had either cut or bit one of her breast [sic] off.
>
> Letter to Mrs. W. P. Cornell, October 29, 1934, Association of Southern Women for the Prevention of Lynching Papers

> . . . more than rape itself, the fear of rape permeates our lives . . . and the best defense against this is not to be, to deny being in the body, as a self, to . . . avert your gaze, make yourself, as a presence in the world, less felt.
>
> Susan Griffin
> *Rape: The Power of Consciousness* (1979)

In the 1920s and 1930s, the industrial revolution spread through the South, bringing a demand for more orderly forms of law enforcement. Men in authority, anxious to create a favorable business climate, began to withdraw their tacit approval of extralegal violence. Yet lynching continued, particularly in rural areas, and even as white moderates criticized lynching in the abstract, they continued to justify outbreaks of mob violence for the one special crime of sexual assault. For most white Americans, the association between lynching and rape called to mind not twin forms of white violence against black men and women, but a very different image: the black rapist, "a monstrous beast, crazed with lust";[12] the white victim—young, blond, virginal; her manly Anglo-Saxon avengers. Despite the pull of modernity, the emotional logic of lynching remained: only swift, sure violence, unhampered by legalities, could protect white women from sexual assault.

The "protection of white womanhood" was a pervasive fixture of racist ideology. In 1839, for example, a well-known historian offered this commonly accepted rationale for lynching: black men find "something strangely alluring and seductive . . . in the appearance of the white woman; they are aroused and stimulated by its foreignness to their experience of sexual pleasures, and it moves them to gratify their lust at any cost and in spite of every obstacle." . . . Despite its tenacity, however, the myth of the black rapist was never founded on objective reality. Less than a quarter of lynch victims were even accused of rape or attempted rape. Down to the present, almost every study has underlined the fact that rape is overwhelmingly an intraracial crime, and the victims are more often black than white.[13]

A major strategy of anti-lynching reformers, beginning with Ida B. Wells in the 1880s and continuing with Walter White of the NAACP and Jessie Daniel Ames of the Association of Southern Women for the Prevention of Lynching, was to use such facts to undermine the rationalizations for mob violence. But the emotional circuit between interracial rape and lynching lay beyond the reach of factual refutation. A black man did not literally have to attempt sexual assault for whites to perceive some transgression of caste mores as a sexual threat. White women were the forbidden fruit, the untouchable property, the ultimate symbol of white male power. To break the racial rules was to conjure up an image of black over white, of a world turned upside down.

Again, women were a means of communication and, on one level, the rhetoric of protection, like the rape of black women, reflected a power struggle among men. But impulses toward women as well as toward blacks were played out in the drama of racial violence. The fear of rape was more than a hypocritical excuse for lynching; rather, the two phenomena were intimately intertwined. The "southern rape complex" functioned as a means of both sexual and racial suppression.[14]

For whites, the archetypal lynching for rape can be seen as a dramatization of cultural themes, a story they told themselves about the social arrangements and psychological strivings that lay beneath the surface of everyday life. The story such rituals told about the place of white women in southern society was subtle, contradictory, and demeaning. The frail victim, leaning on the arms of her male relatives, might be brought to the scene of the crime, there to identify her assailant and witness his execution. This was a moment of humiliation. A woman who had just been raped, or who had been apprehended in a clandestine interracial affair, or whose male relatives were pretending that she had been raped, stood on display before the whole community. Here was the quintessential Woman as Victim: polluted, "ruined for life," the object of fantasy and secret contempt. Humiliation, however mingled with heightened worth as she played for a moment the role of the Fair Maiden violated and avenged. For this privilege—if the alleged assault had in fact taken place—she might pay with suffering in the extreme. In any case, she would pay with a lifetime of subjugation to the men gathered in her behalf.

Only a small percentage of lynchings, then revolved around charges of sexual assault; but those that did received by far the most attention and publicity—indeed, they gripped the white imagination far out of proportion to their statistical significance. Rape and rumors of rape became the folk pornography of the Bible Belt. As stories spread the rapist became not just a black man but a ravenous brute, the victim a beautiful young virgin.

The lynch mob in pursuit of the black rapist represented the trade-off implicit in the code of chivalry: for the right of the southern lady to protection presupposed her obligation to obey. The connotations of wealth and family background attached to the position of the lady in the antebellum South faded in the twentieth century, but the power of "ladyhood" as a value construct remained. . . . If a woman passed the tests of ladyhood, she could tap into the reservoir of protectiveness and shelter known as southern chivalry. Women who abandoned secure, if circumscribed, social roles forfeited the claim to personal security. Together the practice of ladyhood and the etiquette of chivalry controlled white women's behavior even as they guarded caste lines.

. . . In general, the law of rape expressed profound distrust of women, demanding evidence of "utmost resistance," corroboration by other witnesses in addition to the victim's word, and proof of the victim's chastity—all contrary to the rules of evidence in other forms of violent crime. In sharp contrast, however, when a black man and a white woman were concerned intercourse was prima facie evidence of rape. The presiding judge in the 1931 Scottsboro trial, in which nine black youths were accused of rape, had this to say:

> Where the woman charged to have been raped, as in this case is a white woman, there is a very strong presumption under the law that she would not and did not yield voluntarily to intercourse with the defendant, a Negro; and this is true, whatever the station in life the prosecutrix may occupy, whether she be the most despised, ignorant and abandoned woman of the community, or the spotless virgin and daughter of a prominent home of luxury and learning.[15]

Lynching, then, like laws against intermarriage, masked uneasiness over the nature of white women's desires. It aimed not only to engender fear of sexual assault but also to prevent voluntary unions. It upheld the comforting fiction that at least in relation to black men, white women were always objects and never agents of sexual desire.

. . . When women in the 1920s and 1930s did begin to assert their right to sexual expression, . . . inheritors of the plantation legend responded with explicit attacks that revealed the sanctions at the heart of the chivalric ideal. William Faulkner's *The Sanctuary,* published in 1931, typified a common literary reaction to the fall of the lady. The corncob rape of Temple Drake—a "new woman" of the 1920s—was the ultimate revenge against the abdicating white virgin. Her fate represented the "desecration of cult object," the implicit counterpoint to the idealization of women in a patriarchal society.[16]

LADY INSURRECTIONISTS

> The lady insurrectionists gathered together in one of our southern cities. . . . They said calmly that they were not afraid of being raped; as for their sacredness, they would take care of it themselves; they did not need the chivalry of lynching to protect them and did not want it.
>
> Lillian Smith
>
> Killers of the Dream (1949)

On November 1, 1930, twenty-six white women from six southern states met in Atlanta to form the Association of Southern Women for the Prevention of Lynching. Organized by Texas suffragist Jessie Daniel Ames, the association had a central, ideological goal: to break the circuit between the tradition of chivalry and the practice of mob murder. The association was part of a broader interracial movement; its contribution to the decline of lynching must be put in the perspective of the leadership role played by blacks in the national anti-lynching campaign. But it would be a mistake to view the association simply as a white women's auxiliary to black-led struggles. Rather, it represented an acceptance of accountability of racist mythology that white women had not created but that they nevertheless served, a point hammered home by black women's admonitions that "when Southern white women get ready to stop lynching, it will be stopped and not before."[17]

Jessie Ames, the association's leader, stood on the brink between two worlds. Born in 1883 in a small town in East Texas, a regional hotbed of mob violence, she directed the anti-lynching campaign from Atlanta, capital of the New South. . . .

Ames had come to maturity in a transitional phase of the women's movement, when female reformers used the group consciousness and Victorian sense of themselves as especially moral beings to justify a great wave of female institution building. . . . Ames's strategy for change called for enfranchised women to exercise moral influence over the would-be lynchers in their own homes, political influence over the public officials who collaborated with them, and cultural influence over the editors and politicians who created an atmosphere where mob violence flourished. Like Frances Willard and the temperance campaign, she sought to extend women's moral guardianship into the most quintessentially masculine affairs.

Ames's tenacity and the emotional energy of her campaign derived from her perception that lynching was a women's issue: not only an obstacle to regional development and an injustice to blacks, but also an insult to white women. Along with black women leaders before her, who had perceived that the same sexual stereotyping that allowed black women to be exploited caused black men to be feared, she challenged both racist and patriarchal ideas.[18] Disputing the notion that blacks provoked mob action by raping white women, association members traced lynching to its roots in white supremacy.[19] More central to their campaign was an effort to dissociate the image of the lady from its connotations of sexual vulnerability and retaliatory violence. If lynching held a covert message for white women as well as an overt one for blacks, then the anti-lynching association represented a woman-centered reply. Lynching, it proclaimed, far from offering a shield against sexual assault, served as a weapon of both racial and sexual terror, planting fear in women's minds and dependency in their hearts. It thrust them in the role of personal property or sexual objects, ever threatened by black men's lust, ever in need of white men's protection. Asserting their identity as autonomous citizens, requiring not the paternalism of chivalry but the equal protection of the law, association members resisted the part assigned to them.

If, as Susan Brownmiller claims, the larger anti-lynching movement paid little attention to lynching's counterpart, the rape of black women, the women's association could not ignore the issue. For one thing, black women in the interracial movement continually brought it to their attention, prodding them to take responsibility for stopping both lynching and sexual exploitation. For another, from slavery on, interracial sex had been a chronic source of white women's discontent.[20] In 1920, for example, a white interracialist and women's rights leader, who had come to her understanding of racial issues through pioneering meetings with black women, warned a white male audience:

> The race problem can never be solved as long as the white man goes unpunished [for interracial sex], while the Negro is burned at the stake. I shall say no more, for I am sure you need not have anything more said. When the white men of the South have come to that position, a single standard for both men and women, then you will accomplish something in this great problem.[21]

In the winter of 1931, Jessie Daniel Ames called a meeting of black and white women for an explicit discussion of the split female image and the sexual double standard. . . . White male attitudes, the group concluded, originated in a slave system where black women "did not belong to themselves but were in effect the property of white men." They went on to explore the myths of black women's promiscuity and white women's purity, and noted how this split image created a society that "considers an assault by a white man as a moral lapse upon his part, better ignored and forgotten, while an assault by a Negro against a white woman is a hideous crime punishable with death by law or lynching." Relationships among women interracialists were far from egalitarian, nor could they always overcome the impediments to what Ames called "free and frank" discussion.[22] Yet on occasions like this one the shared experience of gender opened the way for consciousness-raising communication across the color line.

. . . Even more treacherous was the question of sex between black men and white women. In 1892, Memphis anti-lynching reformer and black women's club leader Ida B. Wells was threatened with death and run out of town for proclaiming that behind many lynchings lay consensual interracial affairs. Over sixty years later, in the wake of the famous Scottsboro case, Jessie Daniel Ames began delving beneath the surface of lynchings in which white women were involved. Like Barnett, she found that black men were sometimes executed not for rape but for interracial sex. And she used that information to disabuse association members of one of the white South's central fictions: that, as a Mississippi editor put it, there had never been a southern white woman so depraved as to "bestow her favors on a black man."[23]

But what of lynching cases in which rape actually had occurred? Here association leaders could only fall back on a call for law and order, for they knew from their own experience that the fear engendered in their constituency by what some could bring themselves to call only "the unspeakable crime" was all too real. . . . It would be left to a future generation to point out that the chief danger to white women came from white men and to see rape in general as a feminist concern. Association leaders could only exorcise their own fears of male aggression by transferring the means of violence from mobs to the state and debunking the myth of the black rapist.

In the civil rights movement of the 1960s, white women would confront the sexual dimension of racism and racial violence by asserting their right to sleep with black men. Anti-lynching reformers of the 1930s obviously took a very different approach. They abhorred male violence and lynching's eroticism of death, and asserted against them a feminine standard of personal and public morality. They portrayed themselves as moral beings and independent citizens rather than vulnerable sexual objects. And the core of their message lay more in what they were than in what they said: southern ladies who needed only their own rectitude to protect them from interracial sex and the law to guard them from sexual assault. When Jessie Ames referred to "the crown of chivalry that has been pressed like a crown of thorns on our heads," she issued a cry of protest that belongs to the struggle for both racial and sexual emancipation.[24]

THE DECLINE OF CHIVALRY

> As male supremacy becomes ideologically untenable, incapable of justifying itself as protection, men assert their domination more directly, in fantasies and occasionally in acts of raw violence.
>
> Christopher Lasch
> *Marxist Perspectives* (1978)

In the 1970s, for the second time in the nation's history, rape again attracted widespread public attention. The obsession with interracial rape, which peaked at the turn of the nineteenth century but lingered from the close of the Civil War into the 1930s, became a magnet for racial and sexual oppression. Today the issue of rape has crystallized important feminist concerns.

Rape emerged as a feminist issue as women developed an independent politics that made sexuality and personal life a central arena of struggle. First in consciousness-raising groups, where autobiography became a politicizing technique, then in public "speakouts," women broke what in retrospect seems a remarkable silence about a pervasive aspect of female experience. From that beginning flowed both an analysis that held rape to be a political act by which men affirm their power over women and strategies for change that ranged from the feminist self-help methods of rape crisis centers to institutional reform of the criminal justice and medical care systems. After 1976, the movement broadened to include wife-battering, sexual harassment, and, . . . media images of women.

By the time Susan Brownmiller's *Against Our Will: Men, Women and Rape* gained national attention in 1975, she could speak to and for a feminist constituency already sensitized to the issue by years of practical, action-oriented work. . . . [It] was an important milestone, pointing the way for research into a subject that has consistently been trivialized and ignored. Many grass-roots activists would demur from Brownmiller's assertion that all men are potential rapists, but they share her understanding of the continuum between sexism and sexual assault.[25]

The demand for control over one's own body—control over whether, when, and with whom one has children, control over how one's sexuality is expressed—is central to the feminist project. . . . It is this right to bodily integrity and self-determination that rape, and the fear of rape, so thoroughly undermines. Rape's devastating effect on individuals derives not so much from the sexual nature of the crime . . . as from the experience of helplessness and loss of control, the sense of one's self as an object of rage. And women who may never be raped share, by chronic attrition, in the same helplessness, "otherness," lack of control. The struggle against rape, like the anti-lynching movement, addresses not only external danger but also internal consequences: the bodily muting, the self-censorship that limits one's capacity to "walk freely in the world."[26]

The focus on rape, then, emerged from the internal dynamics of feminist thought and practice. But it was also a response to an objective increase in the crime. From 1969 to 1974, the number of rapes rose 49 percent, a greater increase than for any other violent crime. Undoubtedly rape statistics reflect general demographic and criminal trends, as well as a greater willingness of victims to report sexual attacks (although observers agree that rape is still the most underreported of crimes).[27] But there can be no doubt that rape is a serious threat and that it plays a prominent role in women's subordination. Using recent high-quality survey data, Allan Griswold Johnson has estimated that, at a minimum, 20 to 30 percent of girls now twelve years old will suffer a violent attack sometime in their lives. A woman is as likely to be raped as she is to experience a divorce or to be diagnosed as having cancer.[28]

In a recent anthology on women and pornography, Tracey A. Gardner has drawn a parallel between the wave of lynching that followed Reconstruction and the increase in rapes in an era of anti-feminist backlash.[29] Certainly, as women enter the workforce, postpone marriage, live alone or as single heads of households, they become easier targets for sexual assault. But observations like Gardner's go further, linking the intensification of sexual violence directly to the feminist challenge. Such arguments come dangerously close to blaming the victim for the crime. . . . Yet it seems clear that just as lynching ebbed and flowed with new modes of racial control, rape—both as act and idea—cannot be divorced from changes in the sexual terrain.

In 1940, Jessie Ames released to the press a statement that, for the first time in her career, the South could claim a "lynchless year," and in 1948 . . . she allowed the Association . . . to pass quietly from the scene. The women's efforts, the larger, black-led anti-lynching campaign, black migration from the rural South, the spread of industry—these and other developments contributed to the decline of vigilante justice. Blacks continued to be victimized by covert violence and routinized court procedures that amounted to "legal lynchings." But after World War II, public lynchings, announced in the papers, openly accomplished, and tacitly condoned, no longer haunted the land, and the black rapist ceased to be a fixture of political campaigns and newspaper prose.

This change in the rhetoric and form of racial violence reflected new attitudes toward women as well as toward blacks. By the 1940s few southern leaders were willing, as Jessie Ames put it, to "lay themselves open to ridicule" by defending lynching on the grounds of gallantry, in part because gallantry itself had lost conviction.[30] . . . Industrial capitalism on the one hand and women's assertion of independence on the other weakened paternalism and with it the conventions of protective deference.[31] This is not to say that the link between racism and sexism was broken; relations between white women and black men continued to be severely sanctioned, and black men, to the present, have drawn disproportionate punishment for sexual assault. The figures speak for themselves: of the 455 men executed for rape since 1930, 405 were black, and almost all the complainants were white.[32] . . .

The social feminist mainstream, of which Jessie Ames and the anti-lynching association were a part, thus chipped away at a politics of gallantry that locked white ladies in the home under the guise of protecting them from the world. But because such reformers held to the genteel trappings of their role even as they asserted their autonomous citizenship, they offered reassurance that women's influence could be expanded without mortal danger to male prerogatives and power. Contemporary feminists have eschewed some of the comforting assumptions of their nineteenth-century predecessors: women's passionlessness, their limitation to social housekeeping, their exclusive responsibility for childrearing and housekeeping. They have couched their revolt in explicit ideology and unladylike behavior. Meanwhile, as Barbara Ehrenreich has argued, Madison Avenue has perverted the feminist message into the threatening image of the sexually and economically liberated women. The result is a shift toward the rapaciousness that has always mixed unstably with sentimental exaltation and concern. Rape has emerged more clearly into the sexual domain, a crime against women most often committed by men of their own race rather than a right of the powerful over women of a subordinate group or a blow by black men against white women's possessors.[33]

It should be emphasized, however, that the connection between feminism and the upsurge of rape lies not so much in women's gains but in their assertion of rights within a context of economic vulnerability and relative powerlessness. In a perceptive article published in 1901, Jane Addams traced lynching in part to "the feeling of the former slave owner to his former slave, whom he is now bidden to regard as this fellow citizen."[34] Blacks in the post-Reconstruction era were able to express will and individuality, to wrest from their former masters certain concessions and build for themselves supporting institutions. Yet they lacked the resources to protect themselves from economic exploitation and mob violence. Similarly, contemporary feminist efforts have not yet succeeded in overcoming women's isolation, their economic and emotional dependence on men, their cultural training toward submission. There are few restraints against sexual aggression, since up to 90 percent of rapes go unreported, 50 percent of assailants who are reported are never caught, and seven out of ten prosecutions end in acquittal.[35] Provoked by the commercialization of sex, cut loose from traditional community restraints, and "bidden to regard as his fellow citizen" a female being whose subordination has deep roots in the psyches of both sexes, men turn with impunity to the use of sexuality as a means of asserting dominance and control. Such fear and rage are condoned when channeled into right-wing attacks on women's claim to a share in public power and control over their bodies. Inevitably they also find expression in less acceptable behavior. Rape, like lynching, flourishes in an atmosphere in which official policies toward members of a subordinate group give individuals tacit permission to hurt and maim.

In 1972 Anne Braden, a southern white woman and long-time activist in civil rights struggles, expressed her fear that the new anti-rape movement might find itself "objectively on the side of the most reactionary social forces" unless it heeded a lesson from history. In a pamphlet titled *Open Letter to Southern White Women*—much circulated in regional women's liberation circles at the time—she urged anti-rape activists to remember the long pattern of racist manipulation of rape fears. . . . She went on to discuss her own politicization through left-led protests against the prosecution of black men on false rape charges. Four years later, she joined the chorus of black feminist criticism of *Against Our Will,* seeing Brownmiller's book as a realization of her worst fears.[36]

Since this confrontation between Old Left and the New, between a white woman who placed herself in a southern tradition of feminist anti-racism and a radical feminist from the North, a black women's movement has emerged, bringing its own perspectives to bear. White activists at the earliest "speakouts" had acknowledged "the racist image of black men as rapists," pointed out the large number of black women among assault victims, and debated the contradictions involved in looking for solutions to a race and class-biased court system. But not until black women had developed their own autonomous organizations and strategies were true alliances possible across racial lines.

. . . As the anti-rape movement broadens to include Third World women, analogies between lynching and rape and the models of women like Ida B. Wells and Jessie Daniel Ames may become increasingly useful. Neither lynching nor rape is the "aberrant behavior of a lunatic fringe."[37] Rather, both grow out of everyday modes of interaction. The view of women as objects to be possessed, conquered, or defiled fueled racial hostility; conversely, racism has continued to distort and confuse the struggle against sexual violence. Black men receive harsher punishment for raping white women, black rape victims are especially demeaned and ignored, and, until recently, the different historical experience of black and white women has hindered them from making common cause. Taking a cue from the women's anti-lynching campaign of the 1930s as well as from the innovative tactics of black feminists, the anti-rape movement must not limit itself to training women to avoid rape or depending on imprisonment as a deterrent, but must aim its attention at changing the behavior and attitudes of men. Mindful of the historical connection between rape and lynching, it must make clear its stand against *all* uses of violence in oppression.

NOTES

1. Quoted in Howard Kester, *The Lynching of Claude Neal* (New York: National Association for the Advancement of Colored People, 1934).
2. Michael Stephen Hindus, *Prison and Plantation: Crime, Justice, and Authority in Massachusetts and South Carolina, 1767–1878* (Chapel Hill: University of North Carolina Press, 1980), pp. xix, 31, 124, 253.
3. For recent overviews of lynching, see Robert L. Zangrando, *The NAACP Crusade Against Lynching, 1909–1950* (Philadelphia: Temple University Press, 1980), McCovern, *Anatomy of a Lynching;* and Hall, *Revolt Against Chivalry.*
4. Terrence Des Pres, "The Struggle of Memory," *The Nation,* 10 April 1982, p. 433.
5. Quoted in William H. Chafe, *Women and Equality: Changing Patterns in American Culture* (New York: Oxford University Press, 1977), p. 60.
6. Diane K. Lewis, "A Response to Inequality: Black Women, Racism, and Sexism, *Signs* 3 (Winter 1977): 341–342.
7. Jacqueline Jones, *Freed Women?: Black Women, Work, and the Family During the Civil War and Reconstruction,* Working Paper No. 61, Wellesley College, 1980, Roger L. Ransom and Richard Sutch, *One Kind of Freedom: The Economic Consequences of Emancipation* (New York: Cambridge University Press, 1977), pp. 87–103.

8. Gerda Lerner, *Black Women in White America: A Documentary History* (New York: Random House, 1972), is an early and important exception.
9. Robin Morgan, "Theory and Practice: Pornography and Rape," *Take Back the Night: Women on Pornography*, ed. Laura Lederer (New York: William Morrow, 1980), p. 140.
10. Susan Griffin, "Rape: The All-American Crime," *Ramparts* (September 1971): 26–35; Susan Brownmiller, *Against Our Will: Men, Women, and Rape* (New York: Simon and Schuster, 1975). See also Kate Millet, *Sexual Politics* (Garden City, N.Y.: Doubleday & Co., 1970).
11. Dorothy Dinnerstein, *The Mermaid and the Minotaur: Sexual Arrangements and Human Malaise* (New York: Harper and Row, 1977). See also Phyllis Marynick Palmer, "White Women/Black Women: The Dualism of Female Identity and Experience," unpublished paper presented at the American Studies Association, September 1979, pp. 15–17. Similarly, British Victorian eroticism was structured by class relations in which upper-class men were nursed by lower-class country women.
12. A statement made in 1901 by George T. Winston, president of the University of North Carolina, typifies these persistent images: "The southern woman with her helpless little children in a solitary farm house no longer sleeps secure. . . . The black brute is lurking in the dark, a monstrous beast, crazed with lust. His ferocity is almost demoniacal. A mad bull or a tiger could scarcely be more brutal" (quoted in Charles Herbert Stember, *Sexual Racism: the Emotional Barrier to an Integrated Society* [New York: Elsevier, 1976], p. 23).
13. For a contradictory view, see, for example, S. Nelson and M. Amir, "The Hitchhike Victim of Rape: A Research Report," in *Victimology: A New Focus. Vol. 5. Exploiters and Exploited,* ed. M. Agopian, D. Chappell, and G. Geis, and I. Drapkin and E. Viano (1975), p. 47; and "Black Offender and White Victim: A Study of Forcible Rape in Oakland, California," in *Forcible Rape: The Crime, The Victim, and the Offender* (New York: Columbia University Press, 1977).
14. Winthrop Jordan, *White over Black: American Attitudes Toward the Negro, 1550–1812* (Baltimore: Penguin Books, 1969); W. J. Cash, *The Mind of the South* (New York: Knopf, 1941), p. 117.
15. Dan T. Carter, *Scottsboro: An American Tragedy* (Baton Rouge: Louisiana State University Press, 1969), p. 36.
16. Leslie Fiedler, *Love and Death in the American Novel* (New York: Delta, 1966), pp. 320–24.
17. Rich, "Disloyal to Civilization"; Jessie Daniel Ames to Mary McLeod Bethune, 9 March 1938, Association of Southern Women for the Prevention of Lynching (ASWPL) Papers, Atlanta University, Atlanta, Georgia. (Henceforth cited as ASWPL Papers.)
18. Deb Friedman, "Rape, Racism—and Reality," *Aegis* (July/August, 1978): 17–26.
19. Jessie Daniel Ames to Miss Doris Loraine, 5 March 1935, ASWPL Papers.
20. Anne Firor Scott, "Women's Perspective on the Patriarchy in the 1850's," *Journal of American History* 6 (June 1974): 52–64.
21. Carrie Parks Johnson Address, Commission on Interracial Cooperation (CIC), CIC Papers, Atlanta University, Atlanta, Georgia.
22. Jessie Daniel Ames to Nannie Burroughs, 24 October 1931; Burroughs to Ames, 30 October 1931, ASWPL Papers; "Appendix F, Digest Discussion," n.d. [November 20, 1931], Jessie Daniel Ames Papers, University of North Carolina at Chapel Hill.
23. Jackson (Mississippi) *Daily News,* February 1931, ASWPL Papers.
24. Quoted in Wilma Dykeman and James Stokely, *Seeds of Southern Change: The Life of Will Alexander* (Chicago: University of Chicago Press, 1962), p. 143.
25. Interview with Janet Colm, director of the Chapel Hill-Carrboro (North Carolina) Rape Crisis Center, April 1981. Two of the best recent analyses of rape are Ann Wolbert Burgess and Lynda Lytle Holmstrom, *Rape: Crisis and Recovery* (Bowie, Md.: Robert J. Brady Co., 1979) and Lorenne M. G. Clark and Debra J. Lewis, *Rape: The Price of Coercive Sexuality* (Toronto: Canadian Women's Educational Press, 1977).
26. Adrienne Rich, "Taking Women Students Seriously," in *Lies, Secrets and Silences,* p. 242.
27. Vivian Berger, "Man's Trial, Women's Tribulation: Rape Cases in the Courtroom," *Columbia Law Review* 1 (1977): 3–12. Thanks to Walter Dellinger for this reference.
28. Allan Griswold Johnson, "On the Prevalence of Rape in the United States," *Signs* 6 (Fall 1980): 136–146.
29. Tracey A. Gardner, "Racism in Pornography and the Women's Movement," in *Take Back the Night,* p. 111.
30. Jessie Daniel Ames, "Editorial Treatment of Lynching," *Public Opinion Quaterly* 2 (January 1938): 77–84.

31. For a statement of this theme, see Christopher Lasch, "The Flight from Feeling Sociopsychology of Sexual Conflict," *Marxist Perspectives* 1 (Spring 1978): 74–95.
32. Berger, "Man's Trial, Woman's Tribulation," p. 4. For a recent study indicating that the harsher treatment accorded black men convicted of raping white women is not limited to the South and has persisted to the present, see Gary D. LaFree, "The Effect of Sexual Stratification by Race on Official Reactions to Rape," *American Sociological Review* 45 (October 1980): 842–854. Thanks to Darnell Hawkins for this reference.
33. Barbara Ehrenrich, "The Women's Movement: Feminist and Antifeminist," *Radical America* 15 (Spring 1981): 93–101; Lasch, "Flight from Feeling." Because violence against women is so inadequately documented, it is impossible to make accurate racial comparisons in the incidence of the crime. Studies conducted by Menachen Amir in the late 1950s indicated that rape was primarily intraracial, with 77 percent of rapes involving black victims and black defendants and 18 percent involving whites. More recent investigations claim a somewhat higher percentage of interracial assaults. Statistics on reported rapes show that black women are more vulnerable to assault than white women. However, since black women are more likely than white women to report assaults, and since acquaintance rape, most likely to involve higher status white men, is the most underreported of crimes, the vulnerability of white women is undoubtedly much greater than statistics indicate (Berger, "Man's Trial, Woman's Tribulation," p. 3, n. 16; LaFree, "Effect of Sexual Stratification," p. 845, n. 3; Johnson, "On the Prevalence of Rape," p. 145).
34. Quoted in Aptheker. *Lynching and Rape*, pp. 10–11.
35. Berger, "Woman's Trial, Man's Tribulation," p. 6; Johnson, "On the Prevalence of Rape," p. 138.
36. Anne Braden, "A Second Open Letter to Southern White Women," *Generations: Women in the South,* a special issue of *Southern Exposure* 4 (Winter 1977), edited by Susan Angell, Jacquelyn Dowd Hall, and Candace Waid.
37. Johnson, "On the Prevalence of Rape," p. 137.

DISCUSSION QUESTIONS

1. How have gendered and racial constructions of sexuality changed since the era of Southern lynchings? How have they remained the same? What, if anything, does this suggest about who holds power in contemporary American society?
2. What evidence might MacKinnon draw from this research to support her contention that hierarchy and dominance are sexualized and eroticized in this culture?
3. Given the arguments presented in this article, what predictions, if any, could you make about the social characteristics of individuals who engage in rape and lynching as forms of social control (e.g., class, social status, etc.)? Why?
4. Does Hall's work support a "backlash" explanation for violence against women and minorities? Why or why not?
5. What is gained by including sympathetic oppressors in political struggles (e.g., men in anti-rape campaigns and whites in anti-lynching campaigns)? What is lost? Is it necessary for social change? Politically beneficial? Naïve? Counterproductive?

Trains

NATHAN MCCALL

Nathan McCall's autobiographical account of participation in a gang rape raises disturbing questions about relationships between the male-peer group, masculinity, and sexual violence.

It was the first day of summer vacation. I was fourteen years old and had just completed the eighth grade, marking the end of my junior high school days. I was sitting at home, watching TV, when the telephone rang. "Hello," I said.

"Yo, Nate, this is Lep!"

"Yo, Lep, what's up?"

"We got one. She phat as a motherfucka! Got nice titties, too! We at Turkey Buzzard's crib. You better come on over and get in on it!"

"See you in a heartbeat."

When I got to Turkey Buzzard's place a few blocks away, Bimbo, Frog Dickie, Shane, Lep, Cooder, almost the whole crew—about twelve guys in all—were already there, grinning and joking like they had stolen something. Actually, they *had* stolen something: They were holding a girl captive in one of the back bedrooms. Turkey Buzzard's parents were away at work. I learned that the girl was Vanessa, a black beauty whose family had recently moved into our neighborhood, less than two blocks from where I lived. She seemed like a nice girl. When I first noticed her walking to and from school, I had wanted to check her out. Now it was too late. She was about to have a train run on her. No way she could be somebody's straight-up girl after going through a train.

Vanessa was thirteen years old and very naive. She thought she had gone to Turkey Buzzard's crib just to talk with somebody she had a crush on. A bunch of the fellas hid in closets and under beds. When she stepped inside and sat down, they sprang from their hiding places and blocked the door so that she couldn't leave. When I got there, two or three dudes were in the back room, trying to persuade her to give it up. The others were pacing about the living room, joking and arguing about the lineup, about who would go first.

Half of them were frontin', pretending they were more experienced than they really were. Some had never even had sex before, yet they were trying to act like they knew what to do. I fronted, too. I acted like I was eager to get on Vanessa, because that's how everybody else was acting.

I went back to the room and joined the dudes trying to persuade Vanessa to let us jam her. She wouldn't cooperate. She said she was a virgin. That forced us to get somebody to play the crazyman role, act like he was gonna go off on her if she didn't give it up. That way, she'd get scared and give in. That's how the older boys did it. We figured it would work for us.

Lep played the heavy. He started talking loud enough for her to hear. His eyes got wide, like Muhammad Ali's used to do when he was talking trash. As guys pretended to hold Lep back, he struggled wildly, like he was fighting to get into the room. "If that bitch don't give me some, she ain't never leavin' this house!"

Frog played the good guy, acting like he was fighting to hold Lep off. "Come on, man, let the girl go. She said she don't wanna do nothing, so you can't make her."

In the ruckus, Turkey Buzzard stood over a deathly silent Vanessa as she sat on the bed, horrified, looking at the doorway where the staged struggle was going on. Buzzard talked soothingly to Vanessa, trying to convince her that he, too, was a nice guy who was on her side. "Look, baby, if you let one of us do it, then the rest of them will be satisfied and they'll let you go. But if you don't let at least one of us do it, then them other dudes gon' get mad and they ain't gonna wanna let you leave. . . . Lep is crazy. We can't keep holdin' him off like that. If we let him come in this room, it's gonna be all over for you."

Vanessa seemed in a daze, like she couldn't believe what was happening to her. She looked up at Buzzard, glanced at the doorway, then looked back at Buzzard. I stood off to the side, studying her. I could see the wheels turning in her head. She knew she was cornered. She had never been in a situation like this before and she didn't have a clue how to handle it. I could tell by the way she kept looking at Buzzard, searching his face for something she could trust, that she was on the verge of cracking. She *had* to trust Buzzard. There was really no choice. It was either that or risk having that crazy Lep burst into the room and pounce on her like he was threatening to do. I could see her thinking about it, adding things up in her mind, trying to figure out if there was something she could do or say to get herself out of that jam. Then a look of resignation washed over her face. It was a sad, fearful look.

She looked so sad that I started to feel sorry for her. Something in me wanted to reach out and do what I knew was right—do what we all instinctively knew was right: Lean down, grab Vanessa's hand, and lead her from that room and out of that house; walk her home and apologize for our temporary lapse of sanity; tell her, "Try, as best as possible, to forget any of this ever happened."

But I couldn't do that. It was too late. This was our first train together as a group. All the fellas were there and everybody was anxious to show everybody else how cool and worldly he was. If I jumped in on Vanessa' behalf, they would accuse me of falling in love. They would send word out on the block that when it came to girls, I was a wimp. Everybody would be talking at the basketball court about how I'd caved in and got soft for a bitch. There was no way I was gonna put that pressure on myself. I thought, *Vanessa got her stupid self into this. She gonna have to get herself out.*

Turkey Buzzard put his hand on her shoulder and said, "What you gonna do, girl? You gonna let one of us do it?"

Vanessa's eyes filled with water. Her lips parted momentarily, as if she intended to speak, but no words came out. Instead, she swallowed hard and nodded her head up and down, indicating that she'd give in. Slowly, she lay back on the bed, like Buzzard told her, and closed her eyes tight. Moving quickly, Buzzard slipped her pants off. When he grabbed the waistband of her panties, she rose up suddenly and grabbed his hands, as if she'd changed her mind. She gripped his hands tight for a second, looking pleadingly into his eyes. Then she turned and looked at Frog and me as if she wanted one of us to come and rescue her. I looked away. Buzzard said, almost in a whisper, "C'mon, girl, it's gonna be all right. It'll be over in a minute."

Sitting upright, Vanessa searched his face again, then released his hand, lay back on the bed, and cupped her hands over her eyes.

By then, the fellas in the living room had grown quiet. They knew she was about to give in, and the sense of anticipation—and fear—rose. Some guys, curious, crept to the doorway and kneeled low so that she couldn't see them peeping into the room. Others stayed in the living room, smiling and whispering among themselves. "Buzzard got her pants off. He got the pants."

I don't remember who went first. I think it was Buzzard. When Vanessa tried to get up after the first guy finished, another was there to climb on top. Guys crowded into the room and hovered, wide-eyed, around the bed, like gawkers at a zoo. Then another went, and another, until the line of guys climbing on and off Vanessa became blurred.

After about the fifth guy had gone, I still hadn't taken my turn. I could have gone before then, but I was having a hard time mustering the heart to make a move. I was in no great hurry to have sex in front of a bunch of other dudes.

The whole scene brought to mind a day a few years earlier when I was sitting in the gym bleachers at my junior high school, watching an intramural basketball game. Scobie-D and some older hoods came and sat near me. At one point, Scobe got on another dude's case for failing to make a move on a girl who'd given him some play. I'll never forget it: Scobe stood up and announced loudly to the rest of the guys, "That motherfucka is scared a' pussy!" The other dude hung his head in shame.

While hovering near Vanessa, I remembered how Scobe had disgraced that guy. I wasn't about to let that happen to me. I wasn't about to let it be said that I was scared of pussy. I look a deep breath and tried to relax and free my mind. I knew I couldn't get an erection if I wasn't relaxed. That would be even more embarrassing. I didn't want to pull my meat out unless I had an erection. It would look small. Somebody might see it, shriveled up and tiny, and start laughing. Imitating some of the others in the room, I took one hand and cuffed it between my legs, caressing my meat, trying to coax it to harden so I could pull it out proudly and take my turn. Then Lep said, "Nate, have you gone yet?"

I said, "Naw, man. I'm gettin' ready to go now."

I moved forward. My heart pounded like crazy, so crazy that I feared that if somebody looked closely enough they could see it beating against my frail chest. I thought, *It's my turn. I gotta go now.*

As several other guys hung around nearby, I went and stood over Vanessa. She was stretched out forlornly on the bed with a pool of semen running between her legs. She stayed silent and kept her hands cupped over her eyes, like she was hiding from a bad scene in a horror movie. With my pants still up, I pulled down my zipper, slid on top of her and felt the sticky stuff flowing from between her legs. Half-erect and fumbling nervously, I placed myself into her wetness and moved my body, pretending to grind hard. After a few miserable minutes, I got up and signaled for the next man to take his turn.

While straightening my pants, I walked over to a corner, where two or three dudes stood, grinning proudly. Somebody whispered, "That shit is *good,* ain't it?"

I said, "Yeah, man. That shit is good." Actually, I felt sick and unclean.

After the last man had taken his turn, Vanessa got up, put on her pants, and went into the bathroom, holding her panties balled up in her hands. She came out and stood in the living room, waiting for somebody to open the front door and let her out. By then, the fellas and I were lounging around, stretched out on Buzzard's mother's couch, sitting on the floor or wherever else we could find a spot. Vanessa looked solemnly around the room at each of us. Nobody said a word. Then Lep led her out and walked with her down the street.

I felt sorry for Vanessa, knowing she would never be able to live that down. I think some of the other guys felt sorry, too. But the guilt was short-lived. It was eclipsed in no time by the victory celebration we held after she left. We burst into cheers and slapped five with each other like we'd played on the winning baseball team. We joked about who was scared, whose dick was small, and who didn't know how to put it in. Everybody had a story to tell: "Beamish couldn't even get his dick hard! He had to go in a corner and beat his meat!"

"Did you see the bed sink when Bimbo climbed his big, elephant ass up there? I thought he was gonna crush the poor girl!"

"She acted like she didn't even wanna grind when I got on it! I had to teach the bitch how to grind. . . ."

That train on Vanessa was definitely a turning point for most of us. We weren't aware of what it symbolized at the time, but that train marked our real coming together as a gang. It certified us as a group of hanging partners who would do anything and everything together. It sealed our bond in the same way some other guys consummated their alliances by rumbling together in gang wars against downtown boys. In so doing, we served notice—mostly to ourselves—that we were a group of up-and-coming young cats with a distinct identity in a specific portion of Cavalier Manor that we intended to stake out as our own.

After that first train, we perfected the art of luring babes into those kinds of traps. We ran a train at my house when my parents were away. We ran many at Bimbo's crib because both his parents worked. And we set up one at Lep's place and even let his little brother get in on it. He couldn't have been more than eight or nine. He probably didn't even have a sex drive yet. He was just imitating what he saw us do, in the same way we copied older hoods we admired.

One night, when I was sitting at home by myself, thinking about it, it occurred to me that the whole notion of running trains was weird. Even though it involved sex, it didn't seem to be about sex at all. Like almost everything else we did, it was a macho thing. Using a member of one of the most vulnerable groups of human beings on the face of the earth—black females—it was another way for a guy to show the other fellas how cold and hard he was.

It wasn't until I became an adult that I figured out how utterly confused we were. I realized that we thought we loved sisters but that we actually hated them. We hated them because they were black and we were black and, on some level much deeper than we realized, we hated the hell out of ourselves.

I didn't understand all that at the time. I don't think any of us understood. But by then, we had started doing a whole lot of crazy things we didn't understand. . . .

DISCUSSION QUESTIONS

1. How can young men most effectively resist the influence of the male-peer group? What are the costs of resistance? What are the costs of cooperation and participation?
2. What specific tenets of the social construction of masculinity and male sexuality are complicit in the occurrences described in this account? What specific tenets of femininity and female-sexual agency helped to structure the plight of the victim in this account? Note: Be sure to focus on features of social structure and organization *rather* than placing blame on individuals.
3. What analytical contributions do you think Stoltenberg, Brod, Messner, MacKinnon, and Dowd would add after reading this account?
4. What do you think was the primary motivation for this crime? Why? Is there a way to prevent crimes like this in the future? Are your proposed solutions individual or structural? Why?
5. Why is the conquest of women's bodies so often the proving ground for masculinity?

The Sexual Politics of Murder

JANE CAPUTI

Apart from the loss of life itself, a particularly disturbing aspect of many serial killings is the horrific degree of sexual abuse and mutilation inflicted upon the victims. In this essay, Jane Caputi uses the archetypical examples of Ted Bundy and Jack the Ripper to demonstrate how the phenomenon of serial sex murders is inextricably bound to the social marginalization of women. Caputi presents "sex murder" as a form of killing rooted in a system of male supremacy. She argues that it is an extreme example of the use of sexual force that is endemic to patriarchal social systems.

In her book, *The Demon Lover: On the Sexuality of Terrorism,* Robin Morgan (1989) relates an incident that occurred during a civil rights movement meeting in the early 1960s. A group composed of members of both the Congress of Racial Equality (CORE) and the Student Non-violent Coordinating Committee . . . had gathered in the wake of the disappearance of three civil rights workers in Mississippi. The FBI, local police, and the national guard had been dredging local lakes and rivers in search of the bodies. During this search, the mutilated parts of an estimated 17 unidentified bodies were found, all but one of whom were women. . . . A male CORE leader, upon hearing that news, agonized: "There's been a whole goddamned lynching we never even *knew* about. There's been some brother disappeared who never even got *reported*." when Morgan asked, Why only one lynching and what about the other 16 bodies, she was told: Those were obviously *sex* murders. Those weren't political"(pp. 223–224).

Twenty years later, that perception still holds sway. For example, in the spring of 1984, Christopher Wilder raped and murdered a still unknown number of women. About to be apprehended by the police, he shot himself. The *Albuquerque Tribune* (April 14, 1984) commented:

> Wilder's death leaves behind a mystery as to the motives behind the rampage of death and terror. With plenty of money, soft-spoken charm, a background in photography, and a part-time career on the glamorous sports car racing circuit, Wilder, 39, would have had no trouble attracting beautiful women. (p.2)

This man not only murdered women but first extensively tortured them. Although the FBI refuses to release all the details of that abuse, it was revealed that Wilder had bound, raped, repeatedly stabbed his victims, and tortured them with electric shocks. One woman (who survived the attack) had even had her eyelids glued shut. Obviously, Wilder did not wish to date, charm, or attract women; his desire was to torment and destroy. From a feminist perspective there is no mystery behind Wilder's actions. His were sexually political murders, a form of murder rooted in a system of male supremacy in the same way that lynching is based in white supremacy. Such murder is, in short, a form of patriarchal terrorism.

From "The Sexual Politics of Murder" by Jane Caputi in *Violence Against Women: The Bloody Footprints,* ed. Pauline B. Barts and Eileen Gail Moran, pp. 5–25. Reprinted by permission of Sage Publications.

That recognition, however, is impeded by longstanding tradition, for, as Kate Millett (1970) noted in her classic work, *Sexual Politics:*

> We are not accustomed to associate patriarchy with force. So perfect is its system of socialization, so complete the general assent to its values, so long and so universally has it prevailed in human society, that it scarcely seems to require violent implementation . . . And yet . . . control in patriarchal societies would be imperfect, even inoperable, unless it had the rule of force to rely upon, both in emergencies and as an ever-present instrument of intimidation. (pp. 44–45)

Early feminist analysts of rape (Brownmiller, 1975; Griffin, 1983; Russell,1975) asserted that rape is not, as the common mythology insists, a crime of frustrated attraction, victim provocation, or uncontrollable biological urges. Nor is it one perpetrated only by an aberrant fringe. Rather, rape is a direct expression of sexual politics, a ritual enactment of male domination, a form of terror that functions to maintain the status quo. MacKinnon (1982a) further maintains that rape is not primarily an act of violence but is a *sexual* act in a culture where sexuality itself is a form of power, where oppression takes sexual forms, and where sexuality is the very "linchpin of gender inequality" (p. 533).

The murders of women and children—including torture and murder by husbands, lovers, and fathers, as well as that committed by strangers—are not some inexplicable evil or the domain of "monsters" only. On the contrary, sexual murder is the ultimate expression of sexuality as a form of power. . . .

THE BOYS NEXT DOOR

> Most men just hate women. Ted Bundy killed them.
>
> Jimmy McDonough (1984, p. 3)

At some point when I was writing my book *The Age of Sex Crime* (1987), an analysis of the contemporary phenomenon of serial sex murder, I had a dream that I was back living in the white, middle-class, suburban neighborhood I grew up in and that Ted Bundy had moved in a few houses down. This was but one of several such dreams I had while engaged in the writing. Still it made a deep impression on me. Certainly, it meant that my subject was getting closer and closer to my psyche. But it also was significant that the nightmare figure was Ted Bundy, for Bundy is almost universally hailed as the killer who represented the all-American boy, the boy next door who did not marry, but rather killed, the girl next door.

Ted Bundy committed serial murder, and FBI statistics show that this new type of murder has increased drastically in the United States in the last 20 years. In addition, in 1984, the Justice Department estimated that there were at the very least 35 and possibly as many as 100 such killers roaming the country. A Justice Department official, Robert O. Heck, summed up the general situation:

> We all talk about Jack the Ripper; he killed five people [sic]. We all talk about the "Boston Strangler" who killed 13, and maybe "Son of Sam," who killed six. But we've got people [sic] out there now killing 20 and 30 people and more, and some of them don't just kill. They torture their victims in terrible ways and mutilate them before they kill them. Something's going on out there. It's an epidemic. (Lindsey, 1984, p. 7)

Although Heck's statement is superficially correct, his language obscures what actually is going on out there, for the "people" who torture, kill, and mutilate in this way are men, whereas their victims are characteristically females—women and girls—and to a lesser extent younger males. As this hierarchy indicates, these are crimes of sexually political, essentially *patriarchal*, domination. . . .

Although sexual force against women is endemic to patriarchy, the 20th century is marked by a new form of mass gynocide: the mutilation serial sex murder. This "age of sex crime" begins with the crimes of "Jack the Ripper," the still unidentified killer who in 1888 murdered and mutilated five London prostitutes. Patriarchal culture has enshrined "Jack the Ripper" as a mythic hero; he commonly appears as an immortal figure in literature, film, television, jokes, and other cultural products. The function of such mythicization is twofold: to terrorize women and to empower and inspire men.

The unprecedented pattern laid down during the Ripper's original siege is now enacted with some regularity: the single, territorial, and sensationally nicknamed killer; socially powerless and scapegoated victims; some stereotyped feature ascribed in common to the victims (e.g., all coeds, redheads, prostitutes, and so on); a "signature" style of murder or mutilation; intense media involvement; and an accompanying incidence of imitation or "copycat" killings.

The Ripper myth received renewed attention in 1988, the centennial year of the original crimes. That occasion was celebrated by multiple retellings of the Ripper legend. In England, Ripper paraphernalia, such as a computer game, T-shirts, buttons, and cocktails appeared (Cameron, 1988).

Within months of this anniversary celebration for the mythic father of sexual murder, the focus effortlessly and eerily shifted to figurative son of that very father—a man who himself was portrayed as a paradigmatic American son, the "handsome," "intelligent," and "charming" Ted Bundy. In 1979, he was convicted of 3 women's deaths and is suspected of being responsible for perhaps 47 more. Bundy, like "Jack the Ripper," was a sex criminal who spawned a distinctive legend and was attended by a distinctive revelry. In the days preceding his death, his story dominated the mass media, memorializing and further mythicizing a killer who had already been the subject of scores of articles, five books, and a made-for-TV movie (where he was played by Mark Harmon, an actor *People Weekly* once gushed over as the "world's sexiest man"). The atmosphere surrounding his execution was repeatedly described as a "carnival" or "circus." On the morning Bundy went to the electric chair, hundreds (from photographs of the event, the crowd seemed to be composed largely of men) gathered across the street from the prison. Many wore specially designed costumes, waved banners proclaiming a "Bundy BBQ" or "I like my Ted well done," and chanted songs such as 'He bludgeoned the poor girls, all over the head. Now we're all ecstatic, Ted Bundy is dead." A common journalistic metaphor for the overall scene was that of a tailgate party before a big game. Indications of a spreading Bundy cult continue to appear: A student group at the University of New Mexico in April 1989 offered a program showing a tape of Bundy's final interview. The poster advertising that event displayed a likeness of the killer under the headline: "A Man With Vision. A Man With Direction. A Prophet of Our Times . . . . Bundy: The Man, the Myth; the Legend."

. . . Just as Bundy's white, young, generally middle-class victims were stereotypically (and with marked racist and classist bias) universalized as "anyone's daughter," Bundy himself was depicted as the fatherland's (almost) ideal son—handsome, intelligent, a former law student, a rising star in Seattle's Republican party. And although that idealization falls apart upon close examination (Bundy's photographs show an ordinary face, and he had to drop out of law school due to bad grades), it provided an attractive persona for purposes of identification. As several feminist analysts (Lacy, 1982–1983; Millett, 1970; Walkowitz, 1982) have noted, a recurrent and vivid pattern accompanying episodes of sensationalized sex murder is ordinary male identification with the sex killer, as revealed in "jokes, innuendoes, veiled threats (*I might be the Strangler, you know*)" (Lacy, 1982–1983, p. 61).

. . . In the final days before his execution, Bundy spoke directly about his cultural construction as a sex killer, telling James Dobson, a psychologist and religious broadcaster, that since his youth he had been obsessed with pornography. Bundy claimed that pornography inspired him to act out his torture and murder fantasies. Five years earlier, another interviewer (Michaud & Aynesworth, 1983) had reported a similar conversation with Bundy:

> He told me that long before there was a need to kill there were juvenile fantasies fed by photos of women in skin magazines, suntan oil advertisements, or jiggly starlets on talk shows. He was transfixed by the sight of women's bodies on provocative display . . . Crime stories fascinated him. He read pulp detective magazines and gradually developed a store of knowledge about criminal techniques—what worked and what didn't. That learning remained incidental to the central thrill of reading about the abuse of female images, but nonetheless he was schooling himself. (p. 117)

Bundy also spoke for himself (although in the third person, as he had not yet decided to openly admit his guilt):

> Maybe he focused on pornography as a vicarious way of experiencing what his peers were experiencing in reality . . . .Then he got sucked into the more sinister doctrines that are implicit in pornography—the use, the abuse, the possession of women as objects. . . A certain percentage of it [pornography] is devoted toward literature that explores situations where a man, in the context of a sexual encounter, in one way or another engages in some sort of violence toward a woman, or the victim. There are, of course, a whole host of substitutions that could come under that particular heading. Your girlfriend, your wife, a stranger, children—whatever—a whole host of victims are found in this literature. And in this literature, they are treated as victims. (p. 117)

Bundy's self-confessed movement from pornography (reportedly introduced to him at an early age by a grandfather who beat his wife, regularly assaulted other people, and tormented animals) to actual sexual assault is consistent with testimony from other sex offenders, including sex murderers, who claim that viewing pornography affected their criminal behavior (Caputi, 1987; Einsiedel. 1986).

. . . Bundy's assertions released a wave of scorn, ridicule, and fury in the mainstream press, with some commentators seemingly more angry at his aspersions on pornography than at his crimes. As one columnist (Leo, 1989) fulminated:

> As Bundy told it, he was a good, normal fellow, an 'all-American boy' properly raised by diligent parents, though one would have liked to hear more about his 'diligent' mother. While nothing of this mother-son relationship is known, a hatred of women virulent enough to claim 50 lives does not usually spring full-blown from the reading of obscene magazines. (p. 53)

Once again, normalcy as well as "maleness and its socialization" are vehemently discarded as an etiological consideration for sexual murder; misogynist myth prevails and the finger of blame is pointed unswervingly at a woman. Since Bundy's execution, an extensive article has appeared in *Vanity Fair*; predictably, it absolves pornography and instead condemns Louise Bundy as responsible for the evolution of her son into a "depraved monster" (MacPherson, 1989).

A companion chorus of voices suggests that we cannot take Bundy seriously because Dobson, the fundamentalist crusader, led Bundy to his assertions to further his own agenda. Thus, once again, the feminist connection between violence against women and pornography is potentially discredited by its association with fundamentalism. Yet few feminists would agree with the religious Right's claim that pornography is the sole or root cause of violence against women. Rather, pornography (as well as its diffusions through mainstream culture) is a modern mode for communicating and constructing patriarchy's necessary fusion of sex and violence, for sexualizing torture. Clearly, that imperative has assumed other forms historically: the political operations of military dictatorships, the enslavement of Africans in the "new world," witch hunting and inquisitions by the Christian church and state, and so on. The basic elements for a gynocidal campaign—an ideology of male supremacy, a vivid imagination of (particularly female) sexual filth, loathing of eroticism, belief in the sanctity of marriage and the family, and the containment of women in male-controlled institutions—structure fundamentalism's very self-serving opposition to pornography.

Finally, it was claimed that Bundy, a characteristic manipulator, was simply manipulating and lying one last time, trying to absolve himself in his 11th hour by blaming society. Yet a feminist analysis would not accept the equation that to recognize the responsibility of society for sexually political murder is to absolve the murderer. Rather, it would point to the connection between Bundy and his society, naming Bundy as that society's henchman (albeit, like other sex criminals, a free-lancer) in the maintenance of patriarchal order through force. Indeed, we might further recognize Bundy as a martyr for the patriarchal state, one who after getting caught had to pay for his fervor, the purity of his misogyny, and his attendant celebrity with his life.

EVERYONE'S SISTERS

> There was wide public attention in the Ted [Bundy] case . . . because the victims resembled everyone's [sic] daughter . . . But not everybody relates to prostitution on the Pacific Highway.
>
> Robert Keppel
> member of the Green River Task Force
> (Starr, 1984, p. 106)

> Some of the victims were prostitutes, but perhaps the saddest part of this case is that some were not.
>
> Sir Michael Havers
> prosecuting counsel at the trial of Peter Sutcliffe, the "Yorkshire Ripper"
> (Holloway, 1981, p. 39)

> There'd be more response from the police if these were San Marino housewives . . . . If you're Black and living on the fringe, your life isn't worth much.
>
> Margaret Prescod
> founder of the Black Coalition Fighting Back Serial Murders (Uehling, 1986, p. 28)

Ted Bundy's victims were young white women and were consistently described in the press as "beautiful" with "long, brown hair." We can recognize some of this description as a fetishization meant to further eroticize the killings for the public. But although some highly celebrated killers such as Bundy or David Berkowitz, the "Son of Sam," chose victims on the basis of their correspondence to a pornographic, objectifying, and racist ideal, the majority of victims of serial killers are women who, as Steven Egger (1984) noted, "share common characteristics of what are perceived to be prestigeless, powerless, and/or lower socioeconomic groups" (p. 348), that is, prostitute women, runaways, "street women," women of color, impoverished women, single and elderly women, and so on. . . .

In the "Jack the Ripper" crimes, all of the victims were prostitute women. The killer (or far more likely, someone pretending to be the killer) wrote to the press a letter that not only originated the famous nickname, but also boasted: "I am down on whores and I shan't quit ripping them until I do get buckled." In the late 1970s, a gynocidal killer was active in northern England; the first victims were all prostitute women. Prepetuating the myth of the immortal and recurring sex criminal, the men of the press nicknamed him the "Yorkshire Ripper." As in many cases involving the serial murder of prostitute women, including those of the "Green River Killer," the "South Side Slayer," and a current series of murders of "prestigeless" women in San Diego County, a great deal of controversy has attended police handling, or rather, mishandling, of the case (Serrano, 1989). In the wake of that controversy in Yorkshire, the British press has claimed that the major problem that the police faced in the early years of that investigation was "apathy over the killing of prostitutes. . . ."

Four years after the first mutilation and murder, the killer had begun to target nonprostitute women, and West Yorkshire's Constable Jim Hobson issued an extraordinary statement as an "anniversary plea" to the killer: "He has made it clear that he hates prostitutes. Many people do. We, as a police force, will continue to arrest prostitutes." Here, Hobson matter-of-factly aligns Ripper motives and actions to larger social interests as well as police goals. He goes on, shifting voice to a direct appeal to the killer: "But the Ripper is now killing innocent girls. That indicates your mental state and that you are in urgent need of medical attention. You have made your point. Give yourself up before another innocent woman dies"(J. Smith, 1982, p. 11). From such official statements we learn that it is normal to hate prostitute women; the killer is even assured of social solidarity in this emotion. His deeds, it seems, only become socially problematic when he turns to so called innocent girls. . . .

In the mid-1980s, at least 17 women, characterized by the police as prostitutes, were murdered within a 40-mile radius in South Central Los Angeles, a primarily African-American neighborhood; all but 3 of the victims were African-Americans. The police waited until 10 women were killed before notifying the public that a serial murderer was operating and then waited until there were 4 more deaths before forming a task force. In response to police and media neglect, Margaret Prescod, a long-time public spokeswoman for US PROS (a national network of women who work in the sex industry and their supporters) founded the Black Coalition Fighting Back Serial Murders. Rachel West (1987) notes:

> The Black Coalition has stated again and again that they are not convinced that all the women murdered were prostitutes and that the police have offered little evidence to support that claim. When the police could not dig up a prostitution arrest record on victim 17, they immediately said, "but she was a street woman." This statement reflects the attitude of the police toward poor women generally, especially it they are black. We all know only too well that any of us at any time can be labeled a prostitute woman, if we dare step out of line in the way we speak and dress, in the hours we keep, the number of friends we have, or if we are "sexual outlaws" of any kind. (p. 285)

West further observes that in many other instances of serial murder, the killer might begin with prostitute women, but then moves on to women of all types (as in the "Hillside Strangler" killings). When the police or press describe the murdered women as prostitutes, it lulls nonprostitute women into a false feeling of safety. It plays upon sexist and frequently racist prejudices to mute the seriousness of the murders, and—most effectively—it diverts the blame to the victim.

In October 1888, Charles Warren, police chief in charge of the "Jack the Ripper" case, pontificated to the press: "The police can do nothing as long as the victims unwittingly connive at their own destruction. They take the murderer to some retired spot, and place themselves in such a position that they can be slaughtered without a sound being heard" (Cameron & Frazer, 1987, p. 20) . . .

Although, as far as I know, there are no national statistics kept on the number of prostitutes murdered annually, the Los Angeles police claim that there have been 69 murders of prostitute women and 30 women killed in what they call "street murders" in the past 4 years. Assuredly, those numbers register appalling danger.

. . . As I worked on the conclusion to this piece, I listened to a National Public Radio news program ("Morning Edition," June 7, 1989) reporting that nine women (all of whom were described as prostitutes or drug addicts) had been murdered in the past year in New Bedford, Massachusetts, the site of a notorious gang rape (Chancer, 1987). Two other women have been missing for months. A serial killer is suspected; "apathy" is said to be the primary response of the mainstream New Bedford community. Obviously, we have heard this story before. Yet the ascription of "apathy," so common in such cases, is really quite misleading. The reigning, though denied, mood is *hatred*, sexually political hatred. A "hate crime" is conventionally defined as "any assault, intimidation or harassment that is due to the victim's race, religion or ethnic background" (Malcolm, 1989, p. A12). That definition obviously must be expanded to include gender (as well as sexual preference). Vast numbers of women are now suffering and dying from various forms of hate crime worldwide, including neglect, infanticide, genital mutilation, battering, rape, and murder. What men might call "peacetime," researcher Lori Heise (1989) truthfully names a "global war on women."

REFERENCES

Albuquerque Tribume. (1984, April 14). Blood trail ends: Brother says he's glad Wilder's Killing is over. P.2.

Brownmiller, S. (1975). *Against our will: Men, women and rape.* New York: Simon & Schuster.

Brownmiller, S. (1989a, February 2). Hedda Nussbaum: Hardly a heroine. *New York Times,* p. A19.

Brownmiller, S. (1989b, April). Madly in love. *Ms.*, pp.56–61.

Cameron, D. (1988). That's entertainment? Jack the Ripper and the celebration of sexual violence. *Trouble and Strife,* 13, 17–19.

Cameron, D., & Frazer, E. (1987). *The lust to kill: A feminist investigation of sexual murder.* New York: New York University Press.

Caputi, J. (1987). *The age of sex crime.* Bowling Green, OH: Bowling Green State University Press.

Chancer, L.S. (1987). New Bedford, Massachusetts, March 6, 1983–March 22, 1984: The 'before and after' of a group rape. *Gender & Society,* 1, 239–260.

Egger, S. A. (1984). A working definition of serial murder and the reduction of linkage blindness. *Journal of Police Science and Administration.* 12, 348–357.

Einsiedel, E. F. (1986). *Social Science Report* (Prepared for the Attorney General's Commission on Pornography). Washington, DC: U.S. Department of Justice.

Griffin, S. (1983). Rape: The all-American crime. In *Made from the earth: An anthology of writings* (pp. 39–58). New York: Harper & Row.

Heise, L. (1989, April 9). The global war against a women. *Washington Post,* Sec. B, p. 1.

Holloway, W. (1981). "I Just Wanted to Kill a Woman" Why: The ripper and male sexuality, *Feminist Review,* 9, 33–40.

Lacy, S. (1982–1983, Fall/Winter). In mourning and in rage (with analysis aforethought). Ikon, pp. 60–67.

Leo, J. (1989, February 6). Crime: that's entertainment. *U.S. News & World Report,* p. 53.

Lindsey, R. (1984, January 21). Officials cite rise in killers who roam U.S. for victims. *New York Times,* p. 1.

MacKinnon, C. A. (1982a). Feminism, Marxism, method, and the state: An agenda for theory. *Signs: Journal of Women in Culture and Society,* 7, 515–544.

MacPherson, M. (1989, May). The roots of evil. *Vanity Fair,* pp. 140–448, 188–198.

Malcolm, A. (1989, May 12). New efforts developing against the hate crime. *New York Times,* p. A12.

Michaud, S. G., & Aynesworth, H. (1983). *The only living witness.* New York: Linden.

Millett, K. (1970). *Sexual Politics.* Garden City, NY: Doubleday Morgan, R. (1989). *The demon lover: On the sexuality of terrorism.* New York: Norton.

Russell, D. E. H. (1975). *The politics of rape: The victim's perspective.* New York: Stein & Day

Serrano, R. A. (1989, February 26). S. D. serial-killer probe mimics "error" pattern of Green River slayings. *Los Angeles Times.* (San Diego County), Part II, p. 1.

Smith, J. (1982, May/June). Getting away with murder. *New Socialist,* pp, 10–12.

Starr, M. (1984, November 26). The random killers. *Newsweek,* p. 106.

Uehling, M. D. (1986, June 9). The LA slayer. *Newseek,* p. 28.

Walkowitz, J. R. (1982). Jack the Ripper and the myth of male violence. *Feminist Studies,* 8, 543–574.

West, R. (1987). U.S. PROStitutes Collective. In F. Delacoste & P. Alexander (Eds.), *Sex work: Writings by women in the sex industry* (pp. 277–289). Pittsburgh, PA: Cleis.

DISCUSSION QUESTIONS

1. Should sex be included as a hate crime category? Why or why not? In what ways is the deliberate killing of women similar to and different from other categories of hate crime (e.g. race & religion)?

2. How do you account for society's apparent apathy regarding the fact that most serial sex killers are male?
3. From this article (as well as the others in this chapter), what evidence could you draw upon to defend the position that the Madonna/whore or good girl/bad girl dichotomy is a social construct used to maintain and justify the victimization of women and perpetuate social inequality? What evidence could you draw upon to refute this thesis?
4. Can Caputi's analysis be extended to include violence against gay, lesbian, bisexual, and transgendered individuals? Why or why not?
5. What evidence is presented in this article that indicates that violent victimization is affected by social hierarchies such as sex and class?

The following Bonus Reading can be accessed through the *InfoTrac College Edition Online Library* by typing in the name of the author or keywords in the title online at *http://www.infotrac-college.com*.

A Boy's Life: For Matthew Shepard's Killers, What Does It Take to Pass as a Man?

JOANN WYPIJEWSKI

In this account, journalist JoAnn Wypijewski examines the horrific 1998 murder of Matthew Shepard, a gay man who was savagely bludgeoned, tied to a fence post and left for dead on a roadside on the outskirts of Laramie, Wyoming. The article centers on the lives of Shepard's two killers and examines a myriad of factors—small-town life, class conflict, drugs, alcohol, masculinity, and homophobia—in a somewhat futile attempt to provide an accounting of how anyone might come to commit an act of such malevolent brutality.

DISCUSSION QUESTIONS

1. Was this a hate crime? Does it matter whether this could be classified as hate crime? Why or why not? Should hate crimes be punished more severely than crimes where the primary motive is not antipathy towards a particular social group? Why or why not?
2. In what ways is this crime similar to crimes or offenses discussed in the other articles of this anthology?
3. In what ways, if any, does the social construction of sexuality in contemporary society affect the occurrence of hate crimes?
4. What other kinds of influences, if any, do you think contributed to the occurrence of this crime? What, if anything can be done to erase or minimize the effect of such factors?
5. Using an analysis similar to Beneke's, discuss how the perpetration of hate crimes may shape the behaviors and fears of members of targeted groups.

14

Sex and the Social Construction of Sexually Transmitted Diseases

Social Control, Civil Liberties, and Women's Sexuality

BETH E. SCHNEIDER AND VALERIE JENNESS

Beth E. Schneider and Valerie Jenness take a critical look at how mechanisms of social control are enacted in efforts to manage and contain the AIDS crisis. Their analysis posits that many state policies repress and constrain particular forms of sexual expression. Their research uses the examples of prostitution and AIDS education to demonstrate that the state constructs and controls female sexuality by tacitly employing a "good girl" versus "bad girl" dichotomy in its enactment of AIDS-related laws and social policies.

"Crises and disasters have always held a special fascination for social scientists, at least in part because they expose the fundamental assumptions, institutional arrangements, social linkages, and cleavages that are normally implicit in the social order."[1] The AIDS epidemic is no exception. The biological and medical imperatives associated with HIV have been effectively translated into a moral panic.[2] This panic has in turn uncovered significant social

From "Social Control, Civil Liberties, and Women's Sexuality," by Beth E. Schneider and Valerie Jenness as it appears in *Women Resisting AIDS: Feminist Strategies of Empowerment*, edited by Beth E. Schneider and Nancy Stoller. Reprinted by permission of Temple University Press processes and arrangements related to sexuality, gender, and social control.

Not surprisingly, the AIDS epidemic has brought with it repetitive calls that "somebody do something." Historically, epidemics typically evoke demands for some form of managerial response and some mobilized effort to control identifiable, projected, and even unknown hazards.[3] In particular, epidemics inspire new public policy, as well as the reform of extant public policy. Again, the AIDS epidemic has proven to be no exception.

In this paper we focus on several public responses that have emerged purportedly to assist in the control of "the AIDS problem." Specifically, we focus on those responses that have consequences for the expansion of social control mechanisms and the potential denial of civil liberties. Although the AIDS epidemic in the United States has touched every segment of society, it is increasingly becoming an illness of women, as well as of racial, ethnic, and sexual minorities. There is no reason to presume that this trend will reverse as the epidemic continues through the nineties and into the next century.[4] Thus, our overarching concern is with how public policy responses to the multitude of threats born of AIDS are structured by gender and sexuality. . . .

THE AIDS EPIDEMIC AS A DISEASE AND AS A MORAL PANIC

The evolution of AIDS resembles the social construction of such diseases as leprosy, syphilis, tuberculosis, and cholera.[5] In each of these epidemics, the evolution and consequences of the disease were tied not only to its biological characteristics, but also to the socially constructed meanings attached to the disease. From the beginning, many interested parties, including some units of the state, have sought to make their interpretations of HIV and AIDS dominant. . . .

The melange of meanings surrounding the AIDS epidemic has merged to produce a "moral panic." In simplest terms, a moral panic can be thought of as a widespread feeling on the part of the public—or some relevant public—that something is terribly wrong in society because of the moral failure of a specific group of individuals. The result is that a subpopulation is defined as the enemy. Cohen describes the evolution and consequences of a moral panic:

> A condition, episode, person or group of persons emerges to become defined as a threat to societal values and interests: Its nature is presented in a stylized and stereotypic fashion by the mass media; moral barricades are manned by editors, bishops, politicians, and other right-thinking people; socially accredited experts pronounce their diagnoses and solutions; ways of coping evolve, or (more often) are resorted to. . . . Sometimes the panic passes over and is forgotten, except in folklore or collective memory; at other times it has more serious and long-lasting repercussions and might produce such changes as those in legal and social policy.[6]

Such "changes as those in legal and social policy" are necessarily intertwined with the negotiation of power and morality and, by extension, are consequential for major societal processes of social change.

Moral panics are inevitably linked to, and thus consequential for, formal systems of social control. They explicitly or implicitly challenge existing systems of control by defining them as failing or defunct. The consequence is that mandates for reform in legal and social policy are rendered timely and legitimate. Wars, epidemics, and other such moral panic-generating events have, at different points in history, served to justify the expansion of old or the introduction of new mechanisms of social control. These emergent forms of social control often constitute significant incursions on the rights of individuals or groups.

Calls for reform in legal and social policy that are consequential for individual civil liberties are especially pronounced when moral panics are tied, in some real or imagined way, to issues of sexuality.[7] Gayle Rubin, for example, has argued that "it is precisely at times such as these [the era of AIDS], when we live with the possibility of unthinkable destruction, that people are likely to become dangerously crazed about sexuality."[8] As a consequence, regulations emerge to control public and private spaces associated with sexuality and eroticism (such as attempts to close bathhouses frequented by gay men or refusal to perform abortions for HIV-positive women).

The AIDS crisis has generated contemporary discourses in which the social conditions attached to the epidemic serve to rationalize formal and informal social control mechanisms on sexuality and gender, ostensibly in the name of safeguarding the public's health. But this has not been done without historical precedent and without overcoming material and symbolic obstacles.

In an effort to make sense of the variety of ways in which history, culture, and politics frame responses to AIDS, Moerkerk and Aggleton[9] identify four overall approaches that nations in Europe have taken to deal with AIDS. Three of these approaches—the pragmatic, the political, and the biomedical—have particular relevance for our understanding of the mechanisms put in place in the United States to manage female prostitutes and to educate female adolescents. The *pragmatic response* emphasizes provision of crucial education and information, whatever that might be for a group of people, and the need to protect the afflicted. It relies on a cultural consensus and avoids coercive forms of social control. In contrast, the *political response* is based in judgment of what is politically possible and consistent with the beliefs of the nation's leadership. It relies on the law to regulate behavior, and consequently it interprets AIDS prevention as a mechanism for producing behavior it considers desirable. The *biomedical response* is limited; it relies exclusively on medical personnel to determine policy and shows little interest in the involvement of affected groups. As our analyses indicate, each response, often in combination with others, is evident in the United States, especially in the control of young women by educational and legal institutions.

PUBLIC HEALTH AND CIVIL LIBERTIES: A DIFFICULT DILEMMA

The extension of formal social control mechanisms by the state is not done automatically in times of epidemics because, as Brandt has documented, epidemics marshal two sets of values that are "highly prized by our culture": the fundamental civil liberties of the individual and the role of the state in assuring public welfare.[10] From a public policy point of view, individual civil liberties and public health concerns are generally conceived as values in competition with each other. There is a tension between the extension of social control in the name of "protecting public health" and the prohibition of such extension in the name of "preserving individual civil liberties," especially those related to notions of "privacy" as an aspect of personal liberty protected by the Fourth and Fourteenth Amendments of the U.S. Constitution.[11]

This tension is especially pronounced in situations or contexts where sexuality is salient. For example, the state still interferes in the practice of homosexual sodomy and other sexual acts. However, after providing an extensive review of relevant legislation at state and federal levels, Stoddard and Rieman conclude that, under recent Supreme Court decisions, the right to privacy has effectively precluded or sharply limited governmental interference with some personal decisions surrounding sexuality (such as the use of contraception or access to abortion). They warn that this trend is open to reversal in light of the many hazards posed by AIDS: "the government undoubtedly could treat persons who carry the HIV differently from others for some reasons."[12] The same holds for people who are "at risk" for HIV or presumed to be carriers of the virus. Indeed, everything from tattooing on the buttocks to more drastic segregation measures, such as forced quarantining, have been proposed, entertained, and occasionally adopted.[13] As a result, over the course of the epidemic, the rights of the individual have not invariably prevailed and privacy has, at least from a policy point of view, been redefined.[14]

Although it is clear that prostitutes, frequently considered women in need of control, have been particularly susceptible to constraints on their civil liberties, children are rarely understood in these terms. Children's civil liberties are effectively unacknowledged. Indeed, the treatment of children, especially their education and protection, rests far less securely on any right to privacy. Children and adolescents often need, by law, their parents' permission for most of what adults take for granted as sexual—to receive contraceptive devices at school, to attend sex education classes, to seek an abortion. Familial and educational institutions exercise physical and legal control over how young people learn about sexuality and gender. Schools are social control mechanisms that reinforce patriarchal relations of male domination and female submission. In the face of the AIDS epidemic, the ideological apparatus of schools extends social control over its charges by framing, constraining, and ultimately censoring what is thought about AIDS and how it comes to be understood.

PRIVATE SPACES, PUBLIC INTRUSIONS: THE CASE OF PROSTITUTION

There has been a virtual explosion in the formulation of policy designed to control the spread of HIV and the people who are infected. Laws have been passed by the U.S. Congress and state legislatures; the courts have issued various pronouncements; and businesses, government agencies, prisons, schools, hospitals, and other such public settings have developed workplace policy. Policy proposals with implications for civil liberties in general and privacy in particular include, but certainly are not limited to; requiring blood screening of prisoners or military recruits; banning people with AIDS from being restaurant workers; prohibiting seropositive persons from donating blood; closing gay bathhouses; banning homosexual sodomy; quarantining "suspect" groups, especially prostitutes; dismissing from federal jobs employees suspected of being seropositive; and refusing to care for or provide shelter for PWAs.[15] This list, of course, is not exhaustive of the measures that have been proposed or implemented to control AIDS by restricting liberty and, in some cases, redefining privacy. . . .

Throughout the AIDS epidemic, calls for mass and mandatory testing have been put forth. . . . However, such calls for testing are not equally applicable to all citizens. They have selectively targeted specific groups—usually gay men, intravenous drug users, prisoners, immigrants, pregnant women, and sex workers.

Implicating Female Prostitutes in AIDS

Stereotypes about women—especially African-American women, pregnant women, and female prostitutes—have been infused with policy proposals. Perhaps the most obvious case of social and legal policy embedded in gender and sexuality is that surrounding sex work. From the beginning, legislation supporting forced quarantining, reporting, screening, and prosecution of sex workers has been proposed and adopted. The biological characteristics of AIDS, combined with the way in which the disease has been socially constructed, almost guaranteed that prostitutes would be implicated in the social problem of AIDS. . . . AIDS has been conceived of primarily as a sexually transmitted disease and was, at least originally, connected with "promiscuous" sex and "deviant" lifestyles.[16] The historical association of prostitution with venereal disease, unfettered sex, and moral unworthiness remains strong. . . .

As early as 1984, medical authorities were investigating the possibility that prostitutes could spread AIDS into the heterosexual population. Meanwhile, the media continued to spread suspicion about prostitution as an avenue of transmission for the disease. For example, on an episode of the nationally televised "Geraldo Show" titled "Have Prostitutes Become the New Typhoid Marys?" the host offered the following introduction to millions of viewers:

> The world's oldest profession may very well have become among its deadliest. A recent study backed by the federal Centers for Disease Control found that one third of New York's prostitutes now carry the AIDS virus. If this study mirrors the national trend, then the implications are as grim as they are clear. Sleeping with a prostitute may have become a fatal attraction. . . . A quick trick may cost you $20, but you may be paying for it with the rest of your life.[17]

Supporting Rivera's introduction, a New York-based AIDS counselor appearing on the show argued:

> A high percentage of prostitutes infected with HIV pass it on to their sexual partners who are johns or the tricks, a lot of whom are married or have sex with a straight woman. I think this is how the AIDS epidemic is passed into the heterosexual population.[18]

He argued further that working prostitutes testing positive for HIV are guilty of manslaughter and/or attempted murder. In a relatively short period of time, claims like this became commonplace. Moreover, claims focusing on prostitution were and still are focused on female prostitutes but not their customers.

Legislation as Social Control

Like the media, legislators have turned their attention to female prostitution as an avenue of transmission. In the name of preventing HIV transmission, legislation that intrudes into private, consensual sexual relations has sprung up around the country. A number of proposals have been introduced and adopted that, in one way or another, make it a crime for someone who is antibody positive to engage in sex with anyone else, regardless of the degree to which the behavior is mutually voluntary, the use of condoms, and the failure of the uninfected participant to test seropositive.[19] . . .

The introduction of AIDS-related legislation has posed a significant legal threat to female prostitutes. Many governmental and medical establishments have reacted to AIDS with calls for increased regulation of prostitution in the form of registration, mandatory AIDS testing, and prison sentences for those carrying antibodies to the virus.[20] In the mid-eighties and into the nineties, many states considered legislation requiring arrested prostitutes to be tested for HIV infection. By 1988, some states had introduced and passed legislation requiring mandatory testing of arrested prostitutes. Georgia, Florida, Utah, and Nevada were among the first states to legislate the forcible testing of arrested prostitutes; those who test positive can then be subject to arrest on felony charges. These mandatory testing laws in effect create a state registry of infected prostitutes, while the felony charges could create a quarantine situation if prostitutes are kept in isolation while awaiting trial.

Coinciding with the introduction of this legislation, many judges and district attorneys began contemplating and occasionally charging arrested prostitutes who tested positive for HIV with attempted manslaughter and murder. In July 1990, for example, an Oakland, California, prostitute was arrested after *Newsweek* ran a photo of her and quoted her saying that she contracted the deadly virus from contaminated needles but continued to engage in prostitution. According to newspaper reports, Oakland police asked a judge to force the woman to be tested for HIV and pressured the district attorney's office to pursue an attempted manslaughter charge if she tested positive. The arresting officer stated, "I think her actions, with the knowledge that if you're going to get AIDS you're going to die, is a malicious act akin to firing into a crowd or at a passing bus."[21] Although the judge denied the charge, the woman was held for a number of days while the possibility was contemplated. As another example, in Orlando, Florida, an HIV-infected prostitute was charged with manslaughter even though she used a condom with all of her clients and despite the finding that all of her clients who had been tested were negative.[22]

In essence, the AIDS epidemic has led to increased social control of prostitutes, especially in the form of repressive legislation and increasingly punitive legal sanctions. Such changes reflect the commonly held belief that prostitutes constitute a "vector of transmission" for AIDS into the heterosexual population; thus, legislation and increased legal sanctions have been pursued in the name of controlling the spread of AIDS. Female, not male, prostitutes are arrested, even though male prostitutes are much more likely to be infected.[23] Social policy continues to be used to enforce select moral positions[24]—in this case, the control of female rather than male sexuality. Such laws effectively constitute a social x-ray, one that classifies individuals as mainstream or peripheral, normal or deviant.

Resistance

Some have suggested that the introduction and implementation of AIDS-related statutes and "enhanced penalties" is merely an attempt to mollify public fear of AIDS spreading into the "population at large."[25] Whatever the state's intention, this testing of certain special groups without consent, at both the state and the federal level, has not gone uncontested.[26] The existence of resistance underscores the control of women's sexuality; it is a sign that women are directly experiencing either the reality or the threat of constraints on their sexual practice.

In order to resist the scapegoating of prostitutes, COYOTE and other U.S. prostitute advocacy groups, using scientific studies and research to lend legitimacy to their assessment, went public with two main arguments: (1) that the rate of HIV infection among prostitutes, compared to that among other identifiable groups, is relatively low; and (2) that, regardless of infection rates, it is a violation of prostitutes' civil rights to selectively impose mandatory testing on prostitutes if they are arrested. Additionally, they publicly and persistently explained that sex workers are not at risk for AIDS because of sex work per se. As the codirector of the International Committee for Prostitute Rights explained in the late 1980s:

> They [prostitutes] are demanding the same medical confidentiality and choice as other citizens. . . . They are contesting policies which separate them from other sexually active people, emphasizing that charging money for sex does not transmit disease.[27]

An editorial on this issue by COYOTE's media liaison argues a kind of prostitute exceptionalism:

> Many readers are well aware that prostitutes practice safe sex techniques, using condoms for oral services as well as intercourse, and quite often restricting their activities to manual gratification. Many prostitutes emphasize massage, still others combine fantasy stimulation (S&M, etc.) with minimal physical contact. There is much a "working girl" can do to assure her health and the health of her clients, and we have done it. Most of us followed safe sex practices long before the onset of this epidemic.[28]

These assertions suggest that what separates prostitutes from women in general is higher rates of condom use. In essence, these sex work organizations hold that sex work per se is not responsible for the spread of HIV; viruses do not discriminate between those who exchange money for sex and those who do not. When prostitutes are infected, intravenous drug use is the primary cause. . . .

The AIDS epidemic poses a multitude of threats for prostitutes, their organizations, and their movement to decriminalize and legitimate prostitution. It has siphoned personnel and resources from sex work organizations; organizational agendas and activities have shifted in response to the way in which AIDS has been constructed as a social problem implicating sex workers.

But, at the same time, the AIDS epidemic has served to legitimate prostitutes' rights organizations. It has provided prostitutes and their advocates with financial, rhetorical, and institutional resources. The AIDS epidemic has also brought public officials and prostitutes' rights organizations together in direct and indirect ways. As a result of concern over the spread of AIDS into the heterosexual population, government agencies such as state legislatures, the Centers for Disease Control, and local departments of health have turned to prostitutes' rights organizations for assistance.

Those concerned with halting the spread of AIDS have enlisted the help of prostitutes in investigating the role of prostitution in the spread of the disease. For example, COYOTE applied for and received funds to begin an AIDS prevention project for prostitutes.

. . . This is obviously an outcome replete with contradictions. The state is not dealing with the problem of AIDS and prostitutes in a singular and consistent fashion. It has utilized what Moerkerk and Aggleton call the pragmatic and political responses. Prostitutes are scapegoats and criminals, but also allies with a unique constituency to educate. . . .

SEX EDUCATION: TO BE OR NOT TO BE, AND IN WHAT FORM?

Unlike prostitutes, other young women face a different set of social control mechanisms. Adolescent girls, even from the most privileged backgrounds, do not have the freedom and resources to organize on their own behalf, and rarely do they have strong advocates. Their families and schools, the primary socializing institutions of children, exercise physical and legal control over what and how they learn about sexuality and accomplish gender. In the face of the AIDS epidemic, the ideological apparatus of schools, which contributes to the perpetuation of race and gender inequality, extends social control over its charges by framing, constraining, and in fact censoring what is thought about AIDS and how it comes to be understood. In that process, it specifies some of the parameters of female sexuality.

Sex education is the vehicle through which children and adolescents learn some portion of what they know about HIV/AIDS, and that learning has not been accomplished (when it has occurred at all) unproblematically. Prevention and control of the spread of HIV requires discussions of sexual and drug-using activities, and sex/health education is a primary institution through which youth are advised whether, as well as how and when, to be sexual beings. In the process, it cues them to what is "safer," "safe," and "risky sex."

Schools respond to AIDS with a hybrid of the political, pragmatic, and biomedical approaches. An examination of how these programs come about to schools reveals a systematic preoccupation with heterosexuality and with the social control of young women's sexuality. These are evident in the multiple discourses of community political struggles over the control of the schools' curricula and over the specific content of prevention materials. Some of these extensions of social control are patently obvious, while others are more implicit. Their limitations are revealed whenever they are resisted, then teenagers provide education for each other in the form of theater, make their own AIDS videos, or join with groups like ACT-UP in the distribution of condoms.

Community Struggles, Social Control, and Censorship

AIDS education for young people was initially proposed in a climate of fear that reflected deep social and cultural anxieties about the disease and its transmissibility. Most importantly, homosexuality and its central symbolic attachment to AIDS rendered particularly problematic the matter of the structure and the content of AIDS education in the public schools.[29] School officials anticipated controversy as parents considered the prospects of such programs for their children.

. . . There has not been a consensus about the need for AIDS education programs (nor about their specific content) in the public schools. By the mid-1980s, the Christian Right's campaign to gain control of the nation's school boards was underway. Among the issues around which its efforts to gain control revolve are bilingual education, the teaching of evolution, affirmative action, and the existence or form of AIDS education and condom distribution. For the last decade, conservative parents have been particularly active in banning books and other materials they found offensive.[30] In the 1992–1993 New York City struggle over school board membership, traditional values groups, reaching out to the Latino community—usually one of their targets—argued that the city, through its "elite" school board (and its approval of condom distribution and its Rainbow curriculum), intended to turn their children into homosexuals.[31]

This sort of censorship has a chilling effect on the search by women, including young women, to understand their own sexuality. While early twentieth-century censorship shielded women from knowledge of birth control, the current round is preoccupied with gay and lesbian depictions and expressions of women's rebellion.

In spite of the real and anticipated trouble, most school systems have offered some form of AIDS education to their students. They have engaged in complicated debates and a variety of institutional maneuvers to structure a nonproblematic AIDS education curriculum.[32]

. . . Our study of a school system in California reveals a variety of these maneuvers, "deflection strategies" taken by accident or design to reduce the probability of criticism and interference.[33] These strategies include special parental permission forms, preview nights for parents, integration of the AIDS materials into already-existing curricula, cooptation of potential student and parent troublemakers, utilizing the "objective" approach in presentations, and avoiding all discussion of homosexuality. Each strategy enhances the influence of parents and diminishes accordingly the power and participation of students. It is the last two strategies that deserve more detailed attention here since they serve very directly to shape the contours and content of the material presented to young women.

Gender and the Biomedical Model in Sex Education

> The discourses of expertise position recipients of educational messages in a way that disables their ability to actually apply information to their lives, and leaves them liable for failing to have understood that they were to have appropriately responded to the "danger" of AIDS.[34]

The dominant view of adolescents' sex education and sexuality, heavily influenced by the fields of medicine and psychology, has shaped HIV prevention practice. Adolescents are often understood as "other," as strange beings, with those from racial/ethnic groups viewed as particularly so. The fields of medicine and psychology put forward essentially deterministic models of social behavior that are simply not flexible enough to capture the variation and dynamism of sexuality and social interaction. The supposedly value-free behaviorism that provides information about anatomy, reproductive physiology, contraception, and a limited variety of sexual practices, assumes that scientific knowledge about sexuality is nonjudgmental and can be used easily by students in making their own choices about sexual behavior. It conveys the message that heterosexual intercourse is normal behavior that can be engaged in responsibly and calmly with the use of contraception and abortions when the birth control fails.

This approach is, or should be, unsatisfactory to feminist parents, educators and young women themselves. We believe that the limits of the debate about what students should and should not have is premised on the view that they are not capable of critical thinking, emotional self-discipline, and intellectual self-direction. It isolates sexuality from other social relations. It focuses on organs or viruses with little link to humans' relationship to one another or the historical context. It tends to ignore the continued, persistent discrepancies between young women and men regarding attitudes towards birth control, sex, and relationships. As research on classroom interaction continues to indicate, male domination is uncritically accepted as natural, as an important topic but never as one requiring critical discussion or presented as in any way problematic.[35] Hence, most of these efforts ignore the unequal power relations between men and women that structure heterosexuality.

Moreover, the models utilized in public health campaigns teach biomedical safe sex guidelines as the basis of individuals' everyday behavior. But neither of the two prevailing models deals directly with common-sense knowledge about AIDS and its relationship to AIDS prevention. And it is how these youth understand AIDS that is crucial to what they do. For example, Maticka-Tyndale's study shows gender differences in the ways in which young women and men assess the risk of intercourse based on their views of its consequences. As she describes it:

> For women, coitus, even before AIDS, carried a variety of risks; they commonly cited both risks of pregnancy and emotional hurt. Relative to other risks, HIV was the least likely to occur. . . . For men, the experience of coitus generally lacked any prior sense of risk. Lacking a concept of risky coitus, men spoke of risk by ranking coitus against other sexual activities [such as anal intercourse].[36]

Finally, in high schools AIDS is seen as an aspect of health and family services, not as a political or historical matter. Public policy issues and historical debates are typically not discussed. Students consequently are shielded from an understanding of what is controversial in what they may be learning or doing and from a more conscious understanding of the political significance of condoms, contraception, and sex. The passive, "objective" stance of most education prevents the expression of opposing or alternative perspectives. The family life-planning classes, though the site of whatever sex education is offered in California high schools, emphasize pregnancy and disease. They are not about sex; indeed, they de-eroticize sexuality. When schools avoid presenting alternative perspectives, including those that incorporate a discussion of eroticism and pleasure, they continue to perpetuate existing class, race, gender, and sexuality hierarchies.

Nevertheless, adolescents have learned the public health information presented to them. Considerable research over the last five years has confidently concluded that adolescents have high levels of knowledge about AIDS and can voice the biomedical position about the role of condoms in AIDS prevention.[37] However, numerous studies show that the major rule they follow to prevent HIV infection is to try not to have sexual intercourse (unprotected and protected) with an infected partner, a determination requiring trust in a specific partner and/or strong faith in one's own ability to judge people based on reputation or appearance.

Gender differences in sexual scripts are also evident. Trust means something different for women and men. Young women expect young men to disclose prior risky sexual activities; men expect that women with whom they are sexual have had no prior sexual activities.[38] That is, they expect them to be "good girls." Since condoms are still seen as primarily for contraception, there are serious gender differences in the ways women and men approach the introduction of condoms. Women, presumed to be using contraception, may be queried about their lack of other birth control and/or about their own or their partners' infection status. Men may introduce condoms in the guise of protecting their partner and in fact be protecting themselves. Young heterosexual women, similar to older women who have been studied in efforts to improve AIDS interventions, fear the regular use of condoms as an insult to their male partner. Most couples—prostitutes with boyfriends included—determine a point in their relationship when they stop using condoms, a point representing deepening trust and commitment. Indeed, the decision to use safe sex is based on perceived HIV status of the partner or on quality of feelings. . . .

AIDS Videos and Social Control

Videos are a major educational tool through which adults bring children AIDS education. In many school systems, a video with an hour or two of discussion may be all the AIDS education students receive.[39] In the best situations, the program around AIDS takes several days, speakers from Planned Parenthood or the local AIDS project appear, and students get to role-play some of what they have learned. But these elaborations on a basic program are rare. The package as a whole usually fosters heterosexism, and every point in the process has its gendered content or outcome.

An examination of four of the commonly used videos available to high schools in the district we studied indicates the variety of ways in which gender inequality and gender difference is reproduced. When given a choice from among these videos, parents tend to select the ones produced by the Red Cross, "A Letter from Brian, and Don't Forget Sherrie," targeted primarily to white students, and "Don't Forget Sherrie," targeted to their African-American counterparts.[40] In each, a young woman or man learns that a former sex partner is HIV-infected and near death. The friends of the just-notified teenager try to figure out what all of them should think and do about their future sexual or drug use behavior in light of the information that one of them might be infected. The two videos carry a long disclaimer on a black screen:

> [This video deals] with teens and others discussing sexual activity and AIDS prevention in frank terms. The film deals with the threat of the disease, AIDS, for teenagers, and how they can avoid getting the disease.

It is followed by the American Red Cross position: Abstinence is highly recommended for young single teenagers, and education regarding sex should be provided within the family with supplementary materials from the schools and community organizations. "Sex education should be based on religious, ethical, legal, and moral foundations."

"Sex, Drugs, and HIV" takes a less apologetic and fearful approach. This video uses popular music, an actress familiar to young people, and an interracial cast. It is divided into three parts: "Relax, AIDS is Hard to Get," a section intended to overcome myths about mosquitoes, touching, and other forms of casual contact; "You Can Get AIDS By Sharing Needles," a brief section whose sole point is that shooting drugs is bad, "so don't shoot up"; and a last section, "AIDS Can Be a Sexually Transmitted Disease."

Each of the videos talks about the "facts" or the "truth" about AIDS. "Sex, Drugs, and HIV," a 19-minute video, goes so far as to conclude with "That's it. That's all you need to know." This is not only simplistic, it is misleading. Most videos take approximately one minute to explain that "AIDS is caused by a virus," one that infects a person and causes the failure to the immune system: And in most videos one confident speaker or the narrator refers to AIDS as a fatal disease.

Even though videos try to use the language of the teenagers to whom they are geared, the language is not believable in its description of risk behaviors. Not only do young girls and boys not use the same language to talk about sex, virtually none call their own practices vaginal, oral, or anal intercourse. Typically only penile-vaginal intercourse "counts" as "going all the way." This is particularly problematic for AIDS education. As Melese-d'Hospital found in her study with adolescents, many are staying "virgins" as a means to avoid HIV. However, the meaning of virginity, what it includes and what it does not, is consequential. The concept is a marker for what is acceptable sexual behavior. For young women, virginity was "located in the vagina" and related to pregnancy, a greater and more visible concern to them than HIV.[41]

Even in those videos that move away from simply telling the facts to more emotion-laden interactions, it is highly unusual to find serious interaction between males and females: males talk to males, females talk to females. "Sex, Drugs, and HIV" has two such scenes. Three girls are stretching in a gymnasium. They are talking about true love, about whether to have sex, about which birth control to use. One suggests that her friend, who has never before had intercourse, should use the pill; the third counters that pills do not protect against disease or AIDS. The young woman without the experience of vaginal intercourse is convinced to use a condom but worries about being rejected. The friend with the handy condom responds: "Sit and talk to him. He cares about you. . . . If you can't talk to him about birth control, you shouldn't have sex with him. . . . If you're not sure you want to have sex, you should wait." This enactment, one of the best of its kind in this genre, ignores any direct acknowledgement of the normative context of gender inequality and gendered differences in knowledge about sex.

Conversation among the males in both "Sex, Drugs, and HIV" and "A Letter from Brian" show young men trying to persuade one another to use condoms, with one attempting to best the others by claiming always to have used them. This is a positive and important effort to change the norms of the group. Yet it does not deal directly with the variety of myths surrounding condom use or the interactional and emotional matters at stake in sexual encounters.

The videos are not sex-positive. No other sexual practices are discussed though adolescents engage in a great many other activities. Consistent with the political approach to the control of AIDS, as described in the Moerkerk and Appleton study, no acknowledgment is made of the possibility of same-sex experience or of the existence of gay and lesbian students. And, in these and other videos, vague use is made of such terms as *love, respect, commitment* and *monogamy* without any attempt to operationalize them or recognize their multiple meanings. Adolescents are treated as if they are one group, in spite of the strong evidence of diverse and overlapping communities of youth even within similar schools and neighborhoods.[42]

As many women involved in AIDS prevention work have already noted, women are expected to take responsibility for what happens sexually. Many . . . videos are addressed to young women, even when they aren't explicitly targeted to them. For example, in the "Brian" sex video, the U.S. surgeon general offers this confusing statement:

> There is another way where you don't need to worry about condoms and that's to have mutually faithful relationship. In other words, find someone worthy of your love and respect. Give that person both and expect the same from him and remain as faithful to him as he is to you. [If you do this] you will never have to read about AIDS again in your life because it doesn't apply to you.

Aside from the head-in-the-sand attitude, the use of the male pronoun implies that women are the only ones likely to be "good," that they have the burden to make men safe in any relational or sexual situation, and that monogamy protects.

Finally, there are the absences. These videos ignore entrenched fears about homosexuals or unconscious fears about death, though these can surely be said to frame the reception of the facts of AIDS education "With the exception of "Sex, Drugs, and HIV," neither compassion nor "complicity with discrimination"[43] are taught in these and most other HIV-prevention videos. Hence, the strategies of deflection used to shape the content, particularly the avoidance of discussion of homosexuality, highlight the hidden curriculum of schools in their transmission of dominant values and beliefs about heterosexuality. Students are sheltered from the controversial nature of the material presented them and subsequently rendered politically ignorant.

Working from a feminist framework, a good sex education program would not be simply biological; it would be a political program to change gender relations, free women from emotionally and physically debilitating inequality, and foster positive values about sex. If social relations are bad, so are sexual ones. This kind of sex education would provide what young women need: lessons that women are not just victims of sexuality, that they can construct their own sexual identities and pleasures. This would require a significant restructuring not only of the substance of AIDS education, but also of the organizations responsible for delivering it—the schools. This is no small feat, in that it requires greater recognition of institutionalized gender inequality, including that which is routinely affirmed in the classroom.

Although this sort of sex/AIDS education is not available in any complete form anywhere, some student initiatives are moving in this direction. Students often perceive that part of adults' interest in AIDS education is its[44] potential for containing adolescent sexuality. Young women have been integrally involved in, if not actually leaders of, a number of innovative, usually nonschool-based AIDS prevention programs, such as theater groups supported by local Planned Parenthood organizations and off-campus condom distribution efforts initiated by ACT-UP. At the college level, young women are the major players in most AIDS education efforts.

DISCUSSION

While the cultural and economic implications of the AIDS epidemic are certainly far reaching, so are the consequences for the social organization of sex and gender. Through a logic born of the current epidemic, forms of regulatory intervention that might in other circumstances appear excessive can now be justified in the name of prevention. Such justifications are embedded in, and seemingly cannot be divorced from, larger social systems of gender and sexuality.

The social organization of gender and sexuality is policed by laws and public policy that oversee and regulate our sexual desires, exchanges, images, and identities.[45] The AIDS epidemic strengthens, in legal and educational discourses, the rationale for the extension of social control. Though the special treatment of prostitutes and HIV education of young women in schools may seem wildly divergent, each helps shape the contours of female sexuality, in efforts to contain the "bad girl" and to construct the "good" one. Whether through limits on mutually contracted sex by prostitutes or through limits on balanced information on safe sex for young women, the state is interrupting and forbidding certain sexual practices.

Yet, these processes are not consistent or unidirectional. The combination of political and pragmatic approaches to prostitutes by public health and legal institutions expands social control while legitimizing and in some ways normalizing the existence of prostitutes' organizations. To the extent that prostitutes are organized and in charge of their own HIV education, they are positioned to shape, if not to transform, their own sexuality in their own terms. Still, regulation of this always-suspect group of women continues.

Young women face a different, seemingly more constrained, situation: Schools are forced to confront their own failure to educate, as the rate of new infection in women and adolescents increases in the United States. Because of their use of three approaches to HIV prevention (political, pragmatic, and biomedical), the schools are faced with a continuing series of contradictions. They are the locus of community struggles over the nature and meaning of sexuality. For them, virtually every effort to educate, even in the simple case of supplying just the biological facts, results in challenges to their program, from both interested parties who want more and from those who want less.

Since young women in school are not in charge of their education and are politically disenfranchised, they have severe limits on their privacy and virtually no cultural permission to construct their own sexuality in their own terms. The protection of their "innocence" through the narrowness of the programs presented to them offers some, but rather limited, access to a fuller knowledge of sexuality and the range of sexual options a young woman might have. Moreover, while the few most expansive programs recognize young women as sexual actors and no longer force them to be chaste and modest, almost all still enforce a femininity centered in presumed heterosexuality in appearance and practice.

It is certainly possible that this crisis has the potential for the adoption of a more positive approach to the sexual in talk and in practice. Nevertheless, it will require the leadership of women of all ages, of all cultural and economic backgrounds, with strong feminist motivations, to counter the emergent forms of social control. Such controls do not stop or slow the epidemic and constitute incursions in the rights of individual and specific groups of women.

NOTES

1. Susan Shapiro, "Policing Trust," in *Private Policing*, ed. Clifford D. Shearing and Philip C. Stenning (Newbury Park, Calif.: Sage, 1987), 194–220.
2. Watney (p. 43) argues that "we are not, in fact, living through a distinct, coherent and progressing 'moral panic' about AIDS. Rather, we are witnessing the latest variation in the spectacle of the defensive ideological rearguard action which has been mounted on behalf of the 'family' for more than a century." Simon Watney, *Policing Desire: Pornography, AIDS, and the Media* (Minneapolis: University of Minnesota Press, 1987).
3. Allan Brandt, *No Magic Bullet: A Social History of Venereal Disease in the United States Since 1880* (New York: Oxford University Press, 1985); Linda Singer, *Erotic Welfare: Sexual Theory and Politics in the Age of Epidemic* (New York: Routledge, 1993). See also Susan Sontag, *Illness as Metaphor* (New York: Vintage, 1977) and *AIDS and Its Metaphors* (New York: Farrar, Straus and Giroux, 1988).
4. William Darrow, "AIDS: Socioepidemiologic Responses to an Epidemic," in *AIDS and The Social Sciences: Common Threads*, ed. Richard Ulack and William F. Skinner (Lexington: University of Kentucky Press, 1991), 83–99; Samuel V. Duh, *Blacks and AIDS: Causes and Origins* (Newbury Park, Calif.: Sage, 1991); Nan D. Hunter, "Complications of Gender: Women and HIV Disease," in *AIDS Agenda: Emerging Issues in Civil Rights,* ed. Nan D. Hunter and William B. Rubenstein (New York: The New Press, 1992), 5–39; Beth E. Schneider, "Women, Children, and AIDS: Research Suggestions," in *AIDS and the Social Sciences: Common Threads,* ed. Richard Ulack and William F. Skinner (Lexington: University of Kentucky Press, 1991), 134–148; Beth E. Schneider, "AIDS and Class, Gender, and Race Relations," in *The Social Context of AIDS,* ed. Joan Huber and Beth E. Schneider (Newbury Park, Calif.: Sage, 1992), 19–43.
5. The history of epidemics is vast. Some of the books and articles most frequently used in discussions of AIDS include, in addition to Brandt 1975, Sontag 1977, and Sontag 1998. Charles E. Rosenberg, *The Cholera Years* (Chicago: University of Chicago Press, 1962) and William H. McNeil, *Plagues and Peoples* (Garden City: Anchor Books, 1976). See also, Elizabeth Fee and Daniel M. Fox, eds., *AIDS: The Burdens of History* (Berkeley: University of California Press, 1988) and Ilse J. Volinn, "Health Professionals as Stigmatizers and Destigmatizers of Diseases: Alcoholism and Leprosy as Examples," *Social Science and Medicine* 17(1983): 385–393.
6. Nachman Ben-Yehuda, *The Politics and Morality of Deviance: Moral Panics, Drug Abuse, Deviant Science, and Reversed Stigmatization* (New York: State University of New York Press, 1990); Stanley Cohen, *Folk Devils and Moral Panics* (London: MacGibbon and Kee, 1972).
7. Brandt; Singer.
8. Gayle Rubin, "Thinking Sex: Notes for a Radical Theory of the Politics of Sexuality," in *Pleasure and Danger: Exploring Female Sexuality,* ed. Carole S. Vance (Boston: Routledge and Kegan Paul, 1984).
9. Peter Aggleton, Peter Davies and Graham Hart (eds.), *AIDS: Individual, Cultural and Policy Dimensions* (London: The Falmer Press, 1990), 181–190.
10. Brandt, 195.
11. Larry Gostin, ed., *Civil Liberties in Conflict* (New York: Routledge, 1988); Joel Feinberg, "Harmless Immoralities and Offersive Nuisances," in *AIDS: Ethics and Public Policy*, ed. Christine Pierce and Donald VandeVeer (Belmont, Calif.: Wadsworth, 1988), 92–102; Thomas B. Stoddard and Walter Rieman, "AIDS and the Rights of the Individual: Toward a More Sophisticated Understanding of Discrimination," in *A Disease of Society: Cultural and Institutional Responses to AIDS*, ed. Dorothy Nelkin, David P. Willis, and Scott V. Parris (Cambridge: Cambridge University Press, 1991), 241–271.

12. Ibid.
13. William Buckley, "Identify All the Carriers," *The New York Times*, 18 March 1986, p. 26; Ronald Elsberry, "AIDS Quarantining in England and the United States," *Hastings International and Comparative Law Journal* 10 (1986): 113–126; Mark H. Jackson, "The Criminalization of HIV" in *AIDS Agenda: Emerging Issues in Civil Rights*, ed. Nan D. Hunter and William B. Rubenstein (New York: The New Press, 1992), 239–270.
14. Privacy encompasses "those places, spaces and matters upon or into which others may not intrude without the consent of the person or organization to whom they are designated as belonging" (p. 20); Albert J. Reiss, Jr., "The Legitimacy of Intrusion Into Private Space," in *Private Policing*, ed. Clifford D. Shearing and Philip C. Stenning (Newbury Park, Calif.: Sage, 1988), 19–44.
15. Jackson; Pierce and VandeVeer.
16. Edward Albert, "Illness and/or Deviance: The Response of the Press to Acquired Immunodeficiency Syndrome," in *The Social Dimensions of AIDS; Method and Theory*, ed. Douglas A. Feldman and Tom Johnson (New York: Praeger, 1986), 163–178; Edward Albert, "AIDS and the Press: The Creation and Transformation of a Social Problem," in *Images of Issues; Typifying Contemporary Social Problems*, ed. Joel Best (New York: Aldine De Gruyter Press,1989), 39–54. See also Ann Giudici Fettner and William Check, *The Truth About AIDS: The Evolution of an Epidemic* (New York: Holt, 1985); Randy Shilts, *And the Band Played On* (New York: St, Martin's Press, 1988); Harry Schwartz, "AIDS and the Media," in *Science in the Streets* (New York: Priority Press, 1984).
17. Geraldo Rivera, "Are Prostitutes the New Typhoid Marys?" The Geraldo Show, 1989, Fox Headquarters, 10201 Pico Boulevard, Los Angeles, California.
18. John Cristallo, "Are Prostitutes the New Typhoid Marys?" The Geraldo Show, 1989, Fox Headquarters, 10201 Pico Boulevard, Los Angeles, California.
19. Jackson; Luxenburg and Guild, 1992; Joan Luxenburg and Thomas Guild, "Coercion, Criminal Sanctions and AIDS" (Paper presented at the annual meetings of the Society for the Study of Social Problems, Washington, D.C., 1990).
20. This parallels what happened to prostitutes in the first half of the twentieth century when "physicians and social reformers associated venereal disease, almost exclusively, with the vast population of prostitutes in American cities" (Brandt, p. 31). Perceived threats like these led to the increased social control of prostitution, primarily in the form of state regulation.
21. "No Murder-Try Case for Addicted Hooker," *Sacramento Bee*, 18 July 1990, p. B7.
22. Priscilla Alexander, "A Chronology of Sorts," personal files, 1988.
23. Darrow, p. 94. He observes, "To date, no HIV infections in female prostitutes or their clients can be directly linked to sexual exposure."
24. Lord Patrick Devlin, "Morals and the Criminal Law," in *Pierce and VandeVeer*, 77–86.
25. Carol Leigh, "AIDS: No Reason For A Witchhunt," *Oakland Tribune*, (17 August 1987), p. 1; Carol Leigh, "Further Violations of Our Rights," in *AIDS Cultural Analysis, Cultural Activisim*, ed. Douglas Crimp (Cambridge, MA: The MIT Press, 1998), an October book, 177–181.
26. Stoddard and Reiman, 264.
27. Gail Pheterson, *A Vindication of the Rights of Whores* (Seattle: Seal Press, 1989), 28.
28. Leigh 1987, 1.
29. Dennis Altman, *AIDS in the Mind of America* (New York: Anchor, 1987); Paula A. Treichler, "AIDS, Homophobia, and Biomedical Discourse: An Epidemic of Signification," in *AIDS: Cultural Analysis, Cultural Activism*, ed. D. Crimp (Cambridge, Mass.: MIT Press, 1988), 31–70.
30. Michael Granberry, "Besieged by Book Banners," *Los Angeles Times*, (10 May 1993), p. 1ff.
31. Donna Minkowitz, "Wrong Side of the Rainbow," *The Nation*, (28 June, 1993), pp. 901–904.
32. Douglas Kirby, "School-Based Prevention Programs: Design, Evaluation, and Effectiveness," in *Adolescents and AIDS: A Generation in Jeopardy*, ed. Ralph DiClemente, (Newbury Park, Calif.: Sage, 1992); David C. Sloane and Beverlie Conant Sloane, "AIDS in Schools: A Comprehensive Initiative," *McGill Journal of Education* 25 (1990): 205–228.
33. Schneider, Jenness, and Fenstermaker.
34. Patton, 99.
35. Susan Russell, "The Hidden Curriculum of School: Reproducing Gender and Class Hierarchies," in *Feminism and Political Economy: Women's Work, Women's Struggles*, ed. H. J. Marmey and M. Luxton (Toronto: Methuen, 1987).
36. Eleanor Maticka-Tyndale, "Social Construction of HIV Transmission and Prevention Among Heterosexual Young Adults," *Social Problems* 39 (1992): 238–252.
37. DiClemente, 1992.

38. Maticka-Tyndale.
39. Kirby.
40. These videos were produced in 1988.
41. Isabelle Melese-d' Hospital, "Still a Virgin: Adolescent Social Constructions o: Sexuality and HIV Prevention Education" (Paper presented at the annual meetings of the American Sociological Association, Miami, 1993).
42. Benjamin P. Bowser and Gina M. Wingood, "Community Based HIV-Prevention Programs for Adolescents," in DiClemente.
43. Patton 1990, 108.
44. Sloane and Conant Sloane; Schneider, Jenness, and Fenstermaker.
45. Singer; Watney. See also Jeffrey Weeks, *Sexuality and Its Discontents: Meanings Myths & Modern Sexualities* (New York: Routledge, Kegan Paul, 1985).

DISCUSSION QUESTIONS

1. In the context of the AIDS crisis, what are the consequences of defining some groups of individuals as "innocents" who are in need of protection? What groups are most likely to be labeled as "guilty" or "responsible" for social problems? What are the implications of this for social control? Social equality?
2. Should anyone be required to sacrifice their civil liberties for the sake of social welfare? If so, under what circumstances? If not, how do we protect the public? What criteria and standard(s) of evidence should we use before restricting liberty? Must we demonstrate social harm? A potential for harm? Statistical correlation between a specific group and harm? How do we avoid panic and over-restriction of liberty?
3. Who should decide the content of sex education curriculum?
4. If you were in charge of the government's AIDS policies, what social control/protection strategies would you use?
5. How do contemporary constructions of sexuality contribute to the AIDS crisis? How do contemporary constructions of sex, race, class, age, and disability interact with construction of sexuality to differentially affect AIDS-related issues such as exposure, risk, diagnosis, and treatment?

Identity and Stigma of Women with STDs

ADINA NACK

This excerpt from Adina Nack's research on women with sexually transmitted diseases explores how a small sample of women diagnosed with genital herpes and human papillomavirus attempt to manage the stigma associated with their STDs in an attempt to construct and maintain a positive and sexual sense of self. Nack found that her respondents progressed through a three-stage model of identity management in which the respondents first attempted to pass as uninfected, transferred blame to others, and finally, disclosed their status to family or sexual partners.

From "Damaged Goods: Women Managing the Stigma of STDs," by Adina Nack, *Deviant Behavior* Vol. 21, No. 2. Reproduced by permission of Taylor & Francis, Inc, http://www. routledge-ny.com.

The HIV/AIDS epidemic has garnered the attention of researchers from a variety of academic disciplines. In contrast, the study of other sexually transmitted diseases (STDs) has attracted limited interest outside of epidemiology and public health. In the United States, an estimated three out of four sexually-active adults have HPV infections (human papillomavirus—the virus that can cause genital warts); one out of five have genital herpes infections (Ackerman 1998; CDC Server 1998). . . . This article focuses on how the sexual self-concept is transformed when the experience of living with a chronic STD casts a shadow of disease on the health and desirability of a woman's body, as well as on her perceived possibilities for future sexual experiences. . . .

STIGMA AND THE SEXUAL SELF

For all but one of the 28 women, their STD diagnoses radically altered the way that they saw themselves as sexual beings. Facing both a daunting medical and social reality, the women employed different strategies to manage their new stigma. Each stigma management strategy has ramifications for the transformation of their sexual selves. . . .

Stigma Nonacceptance

Goffman (1963) proposed that individuals at risk for a deviant stigma are either "the discredited" or the "discreditable." The discrediteds' stigma was known to others either because the individuals revealed the deviance or because the deviance was not concealable. In contrast, the discreditable were able to hide their deviant stigma. Goffman found that the majority of discreditables were "passing" as non-deviants by avoiding "stigma symbols," anything that would link them to their deviance, and by utilizing "disidentifiers," props or actions that would lead others to believe they had a non-deviant status. Goffman (1963) also noted that individuals bearing deviant stigma might eventually resort to "covering," one form of which he defined as telling deceptive stories. To remain discreditable in their everyday lives, nineteen of the women employed the individual stigma management strategies of passing and/or covering. In contrast, nine women revealed their health status to select friends and family members soon after receiving their diagnoses.

Passing The deviant stigma of women with STDs was essentially concealable, though revealed to the necessary inner circle of health care and health insurance providers. For the majority, passing was an effective means of hiding stigma from others, sometimes, even from themselves.

Hillary, a 22-year-old White senior in college, recalled the justifications she had used to distance herself from the reality of her HPV infection and facilitate passing strategies.

> At the time, I was in denial about it. I told myself that that wasn't what it was because my sister had had a similar thing happen, the dysplasia. So, I just kind of told myself that it was hereditary. That was kinda' funny because I asked the nurse that called if it could be hereditary, and she said "No, this is completely sexually transmitted." . . . I really didn't accept it until a few months after my cryosurgery.

Similarly, Gloria, a Chicana graduate student and mother of four, was not concerned about a previous case of gonorrhea she had cured with antibiotics or her chronic HPV "because the warts went away." Out of sight, out of her sex life: "I never told anybody about them because I figured they had gone away, and they weren't coming back. Even after I had another outbreak, I was still very promiscuous. It still hadn't registered that I needed to always have the guy use a condom."

When the women had temporarily convinced themselves that they did not have a contagious infection, it was common to conceal the health risk with partners because the women, themselves, did not perceive the risk as real. Kayla, a lower-middle class, White college senior, felt justified in passing as healthy with partners who used condoms, even though she knew that condoms could break. . . . Tasha, a White graduate student, found out that she might have inadvertently passed as healthy when her partner was diagnosed with chlamydia. "I freaked out—I was like, 'Oh my God! I gave you chlamydia. I am so sorry!' I felt really horrible, and I felt really awful." Sara, a Jewish, upper-middle class 24-year-old, expressed a similar fear of having passed as healthy and exposed a partner to HPV. "Evan called me after we'd been broken up and told me had genital warts. And, I was with another guy at the time, doing the kinda-sorta-condom-use thing. It was like, 'Oh, my gosh, am I giving this person something?' " Even if the passing is done unintentionally, it still brings guilt to the passer.

The women also tried to disidentify themselves from sexual disease in their attempts to pass as being sexually healthy. Rather than actively using a verbal or symbolic prop or action that would distance them from the stigma, the women took a passive approach. Some gave nonverbal agreement to put downs of other women who were known to have STDs. For example, Hillary recalled such an interaction. "It's funny being around people that don't know that I have an STD and how they make a comment like, 'That girl, she's such a slut. She's a walking STD'. And how that makes me feel when I'm confronted with that, and having them have no idea that they could be talking about me." Others kept silent about their status and tried to maintain the social status of being sexually healthy and morally pure. Kayla admitted to her charade: "I guess I wanted to come across as like really innocent and everything just so people wouldn't think that I was promiscuous, just because inside I felt like they could see it even though they didn't know about the STD." Putting up the facade of sexual purity, these women distanced themselves from any suspicion of sexual disease.

Covering When passing became too difficult, some women resorted to covering to deflect family and friends from the truth. Cleo summed up the rationale by comparing her behavior to what she had learned growing up with an alcoholic father. "They would lie, and it was obvious that it was a lie. But, I learned that's what you do. Like you don't tell people those things that you consider shameful, and then, if confronted, you know, you lie."

Hillary talked to her parents about the HPV surgery, but never as treatment for an STD. She portrayed her moderate cervical dysplasia as a precancerous scare, unrelated to sex. "We never actually talked about it being a STD, and she kind of thought that it was the same thing that my sister had which wasn't sexually transmitted." When Tasha's sister helped her get a prescription for pubic lice, she actually provided the cover story for her embarrassed younger sister. "She totally took control, and made a personal inquiry: 'So, how did you get this? From a toilet seat.' And I was like, 'a toilet seat,' and she believed me." When I asked Tasha why she confirmed her sister's misconception, she replied "because I didn't want her to know that I had had sex." For Anne, a 28-year-old lower-middle class graduate student, a painful herpes outbreak almost outed her on a walk with a friend. She was so physically uncomfortable that she was actually "waddling." Noticing the strange behavior, her friend asked what was wrong. Anne told her that it was a hemorrhoid: that was only a partial truth because herpes was the primary cause of her pain. As Anne put it, telling her about the hemorrhoid "was embarrassing enough!"

Deception and Guilt The women who chose to deny, pass as normal, and use disidentifiers or cover stories shared more than the shame of having an STD—they had also told lies. With lying came guilt. Anne, who had used the hemorrhoid cover story, eventually felt extremely guilty. Her desire to conceal the truth was in conflict with her commitment to being an honest person. "I generally don't lie to my friends. And I'm generally very truthful with people and I felt like a sham lying to her." Deborah, a 32-year-old White professional from the Midwest, only disclosed to her first sexual partner after she had been diagnosed with HPV: she passed as healthy with all other partners. Deborah reflected, "I think my choices not to disclose have hurt my sense of integrity." However, her guilt was resolved during her last gynecological exam when the nurse practitioner confirmed that after years of "clean" pap smear results Deborah was not being "medically unethical" by not disclosing to her partners. In other words, her immune system had probably dealt with the HPV in such a way that she might never have another outbreak or transmit the infection to sexual partners.

When Cleo passed as healthy with a sexual partner, she started, "feeling a little guilty about not having told." However, the consequences of passing as healthy were very severe for Cleo:

> No. I never disclosed it to any future partner. Then, one day, I was having sex with Josh, my current husband, before we were married, and we had been together for a few months, maybe, and I'm like looking at his penis, and I said, "Oh, my goodness! You have a wart on your penis! Ahhh!" All of a sudden, it comes back to me.

Cleo's decision to pass left her with both the guilt of deceiving and infecting her husband.

Surprisingly, those women who had *unintentionally* passed as being sexually healthy (i.e., they had no knowledge of their STD-status at the time) expressed a similar level of guilt as those who had been purposefully deceitful. Violet, a middle class White 36-year-old, had inadvertently passed as healthy with her current partner. Even after she had preventively disclosed to him, she still had to deal with the guilt over possibly infecting him.

> It hurt so bad that morning when he was basically furious at me thinking I was the one he had gotten those red bumps from. It was the hour from hell! I felt really majorily dirty and stigmatized. I felt like "God, I've done the best I can: if this is really caused by the HPV I have, then I feel terrible."

When employing passing and covering techniques, the women strove to keep their stigma from tainting social interactions. They feared reactions that Lemert (1951) has labeled the *dynamics of exclusion:* rejection from their social circles of friends, family and—most importantly—sexual partners. For most of the women, guilt surpassed fear and became the trigger to disclose. Those who had been deceitful in passing or covering had to assuage their guilt: their options were either to remain in nonacceptance, disclose, or transfer their guilt to somebody else.

Stigma Deflection

As the women struggled to manage their individual stigma of being sexually diseased, real and imaginary social interactions became the conduit for the contagious label of "damaged goods." Now that the unthinkable had happened to them, the women began to think of their past and present partners as infected, contagious, and potentially dangerous to themselves or other women. The combination of transferring stigma and assigning blame to others allowed the women to deflect the STD stigma away from themselves.

Stigma Transference I propose the concept of *stigma transference* to capture this element of stigma management that has not been addressed by other deviance theorists. Stigma transference is a specialized case of projection deviance which . . . manifests as a clear expression of anger and fear. The women did not connect this strategy to a reduction in their levels of anxiety; in fact, several discussed it in relation to increased anxiety. . . .

Transference of stigma to a partner became more powerful when the woman felt betrayed by her partner. When Hillary spoke of the "whole trust issue" with her ex-partner, she firmly believed he had lied to her about his sexual health status and he would lie to others. Even though she had neither told him about her diagnosis nor had proof of him being infected, she fully transferred her stigma to him.

> He's the type of person who has no remorse for anything. Even if I did tell him, he wouldn't tell the people that he was dating. So it really seemed pretty pointless to me to let him know because he's not responsible enough to deal with it, and it's too bad knowing that he's out there spreading this to God knows how many other people.

Kayla also transferred the stigma of sexual disease to an ex-partner, never confronting him about whether or not he had tested positive for STDs. The auxiliary trait of promiscuity colored her view of him: "I don't know how sexually promiscuous he was, but I'm sure he had had a lot of partners." Robin, a 21-year-old White undergraduate, went so far so to tell her ex-partner that he needed to see a doctor and "do something about it." He doubted her ability to pinpoint contracting genital warts from him, and called her a "slut." Robin believed that *he* was the one with the reputation for promiscuity and decided to "trash" him by telling her two friends who hung out with him. Robin hoped to spoil his sexual reputation and scare off his future partners. In the transference of stigma, the women ascribed the same auxiliary traits onto others that others had previously ascribed to them. . . .

In all cases, it was logical to assume that past and current sexual partners may have also been infected. However, the stigma of being sexually diseased had far-reaching consequences into the imaginations of the women. The traumatic impact on their sexual selves led most to infer that future, as yet unknown, partners were also sexually diseased. Kayla summed up this feeling. "After I was diagnosed, I was a lot more cautious and worried about giving it to other people or getting something else because somebody hadn't told me." They had already been damaged by at least one partner. Therefore, they expected that future partners, ones who had not yet come into their lives, held the threat of also being *damaged goods*.

For Hillary, romantic relationships held no appeal anymore. She had heard of others who also had STDs but stayed in nonacceptance and never changed their lifestyle of having casual, unprotected sex:

> I just didn't want to have anything to do with it. A lot of it was not trusting people. When we broke up, I decided that I was not having sex. Initially, it was because I wanted to get an HIV test. Then, I came to kind of a turning point in my life and realized that I didn't want to do the one-night-stand thing anymore. It just wasn't worth it. It wasn't fun. . . .

Blame The women's uses of stigma transference techniques were attempts to alleviate their emotional burdens. First, the finger of shame and guilt pointed inward, toward the women's core sexual selves. Their sexual selves became tainted, dirty, damaged. In turn, they directed the stigma outward to both real and fictional others. Blaming others was a way for all of the women to alleviate some of the internal pressure and turn the anger outward. This emotional component of the *damaged goods* stage externalized the pain of their stigma.

Francine recalled how she and her first husband dealt with the issue of genital warts. "We kind of both ended up blaming it on the whole fraternity situation. I just remember thinking that it was not so much that we weren't clean, but that he hadn't been at some point, but now he was." Francine's husband had likely contracted genital warts from his wild fraternity parties. "We really thought of it as, that woman who did the trains [serial sexual intercourse]. It was still a girl's fault kind-of-thing." By externalizing the blame to the promiscuous women at fraternity parties, Francine exonerated not only herself, but also her husband. . . .

For Violet, it was impossible to neatly deflect the blame away from both herself and her partner. "I remember at the time just thinking, 'Oh man! He gave it to me!' While, he was thinking, 'God, [Violet]! You gave this to me!' So, we kind of just did a truce in our minds. Like, OK, we don't know who gave it—just as likely both ways. So, let's just get treated. We just kind of dropped it." Clearly the impulse to place blame was strong even when there was no easy target.

Often, the easiest targets were men who exhibited the auxiliary traits of promiscuity and deception. Tasha wasn't sure which ex-partner had transmitted the STD. However, she rationalized blaming a particular guy. "He turned out to be kind of a huge liar, lied to me a lot about different stuff. And, so I blamed him. All the other guys were, like, really nice people, really trustworthy." Likewise, when I asked Violet from whom she believed she had contracted chlamydia, she replied, "Dunno,' it could've been from one guy, because that guy had slept with some unsavory women, so therefore he was unsavory." Later, Violet contracted HPV, and the issue of blame contained more anger: "I don't remember that discussion much other than, being mad over who I got it from: 'oh it must have been Jess because he had been with all those women.' I was mad that he probably never got tested. I was OK before him." The actual guilt or innocence of these blame targets was secondary. What mattered to the women was that they could hold someone else responsible.

Stigma Acceptance

Eventually, every woman in the study stopped denying and deflecting the truth of her sexual health status by disclosing to loved ones. The women disclosed for either preventive or therapeutic reasons. That is, they were either motivated to reveal their STD status to prevent harm to themselves or others, or to gain the emotional support of confidants.

Preventive and Therapeutic Disclosures

The decision to make a preventive disclosure was linked to whether or not the STD could be cured. Kayla explained, "Chlamydia went away, and I mean it was really bad to have that, but I mean it's not something that you have to tell people later 'cause you know, in case it comes back. Genital warts, you never know." Kayla knew that her parents would find out about the HPV infection because of insurance connections. Prior to her cryosurgery, Kayla decided to tell her mom about her condition. "I just told her what [the doctor] had diagnosed me with, and she knew my boyfriend and everything, so—it was kind of hard at first. But, she wasn't upset with me. Main thing, she was disappointed, but I think she blamed my boyfriend more than she blamed me." Sara's parents also reacted to her preventive disclosure by blaming her boyfriend: they were disappointed in their daughter, but angry with her boyfriend.

Preventive disclosures to sexual partners, past and present, were a more problematic situation. The women were choosing to put themselves in a position where they could face blame, disgust, and rejection. For those reasons, the women put off preventive disclosures or partners as long as possible. For example, Anne made it clear that she would not have disclosed her herpes to a female sexual partner had they not been, "about to have sex." After "agonizing weeks and weeks and weeks before trying to figure out how to tell," Diana, a 45-year-old African-American professional, finally shared her HPV and herpes status before her current relationship became sexual. Unfortunately, her boyfriend "had a negative reaction": "he certainly didn't want to touch me anywhere near my genitals." In Cleo's case, she told her partner about her HPV diagnosis because she wasn't going to be able to have sexual intercourse for a while after her cryosurgery. Violet described the thought process that led up to her decision to disclose her HPV status to her current partner:

> That was really scary because once you have [HPV], you can't get rid of the virus. And then having to tell my new partner all this stuff. I just wanted to be totally up front with him; we could use condoms. Chances are he's probably

> totally clean. I'm like, "Oh my god, here I am tainted because I've been with, at this point, 50 guys, without condoms. Who knows what else I could have gotten (long pause, nervous laugh)?" So, that was tough. . . .

Many of the therapeutic disclosures were done to family members. The women wanted the support of those who had known them the longest. Finally willing to risk criticism and shame, they hoped for positive outcomes: acceptance, empathy, sympathy—any form of nonjudgmental support. Tasha disclosed to her mother right after she was diagnosed with chlamydia. "My family died—'Guess what, mom, I got chlamydia.' She's like, 'Chlamydia? How did you find out you got chlamydia?' I'm like. 'Well, my boyfriend got an eye infection.' (laughter) 'How'd he get it in his eye?' (laughter) So, it was the biggest joke in the family for the longest time." In contrast, Rebecca, a White professional in her mid-fifties, shared her thought process behind *not* disclosing to her adult children. "I wanted to tell my younger one . . . I wanted very much for him to know that people could be asymptomatic carriers because I didn't want him to unjustly suspect somebody of cheating on him . . . and I don't believe I ever managed to do it . . . it's hard to bring something like that up." . . .

Consequences of Disclosure With both therapeutic and preventive disclosure, the women experienced some feeling of relief in being honest with loved ones. However, they still carried the intense shame of being sexually diseased women. The resulting emotion was anxiety over how their confidants would react: rejection, disgust, or betrayal. Francine was extremely anxious about disclosing to her husband. "That was really tough on us because I had to go home and tell Damon that I had this outbreak of herpes." When asked what sorts of feelings that brought up, she immediately answered. "Fear. You know I was really fearful—I didn't think that he would think I had recently had sex with somebody else . . . but, I was still really afraid of what it would do to our relationship." . . .

Overall, disclosing intensified the anxiety of having their secret leaked to others in whom they would have never chosen to confide. In addition, each disclosure brought with it the possibility of rejection and ridicule from the people whose opinions they valued most. For Gloria, disclosing was the right thing to do but had painful consequences when her partner's condom slipped off in the middle of sexual intercourse.

> I told him it doesn't feel right. "You'd better check." And, so he checked, and he just jumped off me and screamed, "Oh fuck!" And, I just thought, oh no, here we go. He just freaked and went to the bathroom and washed his penis with soap. I just felt so dirty. . . .

Disclosures were the interactional component of self-acceptance. The women became fully grounded in their new reality when they realized that the significant people in their lives were now viewing them through the discolored lenses of sexual disease.

CONCLUSION

The women with STDs went through an emotionally difficult process, testing out stigma management strategies, trying to control the impact of STDs on both their self-concepts and on their relationships with others. In keeping with Cooley's "looking glass self" (1902/1964), the women derived their sexual selves from the imagined and real reactions of others. Unable to immunize themselves from the physical wrath of disease, they focused on mediating the potentially harmful impacts of STDs on their sexual self-concepts and on their intimate relationships.

REFERENCES

Ackerman, S. J. 1998. "HPV: Who's Got It and Why They Don't Know." *HPV News* 8(2), Summer: 1, 5–6.

Centers for Disease Control and Prevention. 1998. "Genital Herpes." *National Center for HIV, STD, & TB Prevention*. Online. Netscape Communicator. 4 February.

Cooley, Charles H. [1902] 1964. *Human Nature and the Social Order*. New York: Schocken.

Goffman, Erving, 1963. *Stigma*. Englewood Cliffs, NJ: Prentice Hall.

Lemert, Edwin. 1951. *Social Pathology*. New York: McGraw-Hill.

DISCUSSION QUESTIONS

1. How would you expect the stigma management strategies of men to compare with those of the women in this study? How would they be similar? How would they differ?
2. How do contemporary constructions of sexuality affect the way women react to sexually transmitted diseases? Men?
3. What factors do you think affect whether someone discloses their STD?
4. In what ways, if any, do STDs relate to issues of attractiveness? Should they?
5. What kind of patterns would you expect to see in the attribution of blame for STDs? Who is most likely to be blamed?
6. What kind of sex-education programs, if any, do you think would be most effective in eliminating or reducing problems such as denial, passing, and covering in relation to STDs?
7. Why, if they are medical problems, are people ashamed of STDs?
8. What kind of consequences, if any, should people face when they knowingly pass on STDs? Should the state be involved? What are the dangers of state involvement? Of state non-involvement?

Gay Villain, Gay Hero: Homosexuality and the Social Construction of AIDS

ROBERT A. PADGUG

In this classic 1989 essay, Robert A. Padgug analyzes how sexuality (and more specifically, gay male sexuality) has emerged as a central feature of the social construction of the AIDS epidemic. For Padgug, the association of the disease with gay males is largely due to the way that male homosexuality has itself been constructed through modern history (e.g., hypersexual, promiscuous). He contends that because male homosexuals have historically been categorically defined as sexually deviant moral outsiders, blame for the epidemic has typically been placed squarely on their shoulders.

I

Patterns of disease are as much the product of social, political, and historical processes as of "natural history." From a historian's perspective, the current AIDS epidemic in the United States is "socially constructed"—the product of multiple historical determinations involving the complex social interaction of human beings over time. It is not, as the National Academy of Sciences would have it, "the story of a virus."[1] The emphasis here is on *historical* determination because AIDS, an event of the present, is imagined and dealt with on the basis of ideologies and institutions developed over time. As Marx wrote, "Human beings do make their own history, but they do not make it just as they please; they do not make it under circumstances chosen by themselves, but under circumstances directly encountered, given and transmitted from the past."[2]

The circumstances of AIDS certainly involve an epidemiologic pattern as well as associated illness and their effects on individual human lives; viewed more broadly these circumstances also involve beliefs, struggles, and institutions that have developed over time. That is, social, political, ideological, economic, religious, and public health realities define the meaning and treatment of AIDS both for its victims and the entire society.

AIDS, like other life-threatening diseases whose causes, means of spreading, and ultimate trajectory are not adequately known, has become, in Susan Sontag's now-classic formulation, a "metaphor," or a set of sometimes conflicting metaphors.[3] That is, most of us can comprehend and confront AIDS only if its social meaning is extended well beyond the relatively narrow spheres of medicine and epidemiology. The metaphors that have been constructed around the AIDS epidemic appear to have an unusual power for good or for evil. This power derives from the unanticipated emergence of a deadly new contagion of uncertain course and extent in an era when fatal infectious diseases were believed to have been eliminated. It is also linked to the nature of the groups first affected by AIDS, groups already included in powerful social metaphors.

As metaphor, social struggle, or disease management, AIDS is a contemporary crisis that impels us to draw upon a wide variety of historical material in the struggle to comprehend and deal with it. At the same time, however, we should recognize that such material can be interpreted differently and can be combined to form contradictory ideological and institutional responses to AIDS.

This essay develops these themes by exploring a central feature of the social construction of AIDS—sexuality, particularly, male homosexuality in the United States. Equally important aspects of the crisis, involving medical, epidemiological, economic, class, gender, and racial issues, are slighted here, although they clearly are essential elements in a full comprehension of the epidemic.

II

Sexuality is central in the social construction of AIDS in the United States because sexuality in general and male homosexuality in particular appear to play a paramount role in its etiology and spread. This distinguishes AIDS from most other diseases. As Michel Foucault and others have argued, sexuality is immensely important in the construction of personality as well as of ethics and morality in the modern world. A vast array of competing sexual ideologies, forming the basis for complicated ethical and political positions and struggles, demonstrates this in everyday American practice. In a word, sexuality is itself a set of metaphors and the product of a complicated social history.

Closely connected to sexuality, AIDS has become "moralized" to a much greater degree than, for example, tobacco- or alcohol-related diseases despite the fact they can also be seen as "self-inflicted" and devastatingly costly when measured in lives, health, and social resources.

This heightened moralism is possible because AIDS in the United States has been constructed largely in the image of male homosexuality, as that image itself has been constructed in the scientific and popular mind from the mid-nineteenth century to the present.

During the first stages of the epidemic, essentially 1981 and 1982, few aside from gay men—certainly not the popular press and only to a limited degree the government and medical community—paid much attention to AIDS. The disease was happening to "them"—outsiders.[4] This silence—this absence from public discourse—is essentially the way homosexuality has been treated in our society except in periods of moral panic. The moral panic that brought AIDS to the attention of the so-called general public in periodic waves of hysteria, was the possibility that the disease might be spreading to "us," might be crossing that invisible, but ever-present ideological line that divides the normal from the abnormal, the moral from the immoral, the deserving from the undeserving. The definition of AIDS as a disease of homosexual men became entrenched in popular and medical attitudes in spite of growing evidence to the contrary and was used to describe both the disease and the majority of its sufferers. A complex history has shaped the beliefs about homosexuality that were marshalled for the AIDS crisis. The English historian Jeffrey Weeks summarizes those most often chosen:

> Certain forms of sexuality, socially deviant forms—homosexuality especially—have long been promiscuously classified as "sins" *and* "diseases," so that you can be born with them, seduced into them and catch them, all at the same time. . . . In the fear and loathing that AIDS evokes there is a resulting conflation between two plausible, if unproven theories—that there is an elective affinity between disease and certain sexual practices, and that certain practices cause disease—and a third, that certain types of sex *are* diseases.[5]

Implicit in Weeks' description is the nineteenth-century "medicalization" of homosexuality at the hands of physicians and psychiatrists, a conception that has become widely accepted in the twentieth century. Men and women who were categorized as sodomites, practitioners of a sinful sexuality, became "inverts" and, later, homosexuals, that is, they were seen as individuals with physically or mentally diseased personalities who had, in effect, *become* their sexuality.[6] Their sexuality, and therefore their personality, bore the features of its own corruption: it was confused as to gender, it was uncontrollable, it was irresponsible, it sought ever-new pleasures[7] and it was, above all, "promiscuous" or, in more contemporary terms, "addictive."[8]

These features were used to draw the character of the person with AIDS and the person who was thought to infect others. Diseased in mind and body, the man was also "contagious" with respect to both his sexuality and his disease. AIDS was the very mark of his inner disorder, revelatory of his homosexuality as well as the "self-inflicted" result of it. As the conservative journalist Patrick Buchanan put it, "The poor homosexuals—they have declared war upon Nature, and now Nature is exacting an awful retribution."[9] Right-wing polemicists like Buchanan, followed by many ordinary citizens, used these medicalized and moralistic views of homosexuality in their most extreme and hostile forms, proclaiming AIDS a natural or divine judgment upon all homosexuals, regardless of whether they actually had AIDS. And this punishment could spread more widely if the rest of "us" were not morally careful.

The American right wing has shown an "elective affinity" for attacking homosexuality, most notably during the McCarthy period and, in response to the 1970s gay liberation movements, even before the appearance of AIDS. Aversion to homosexuality, which can be connected with a fear of social change and modernism, has its own long and complicated history.[10] But even more temperate and sympathetic observers, including some gays, have adopted two key elements of the historical indictment of homosexuality: "willful irresponsibility" and "promiscuity." Irresponsibility was linked to disregard for the effects of immoral acts, a lack of interest in dealing with the crisis that resulted from them, and, indeed, a certain *desire* to see the crisis unresolved. Thus envisioned, those who wished to "spread" their sexuality in the manner of a contagion desired to spread their contagion in the manner of their sexuality. Elizabeth Fee captures this attitude well:

> It has proved an easy cultural move from the idea of populations at risk to that of populations guilty of harboring disease: from a gay plague to a plague of gays. A member of a population at risk is thus not only a potential victim but a potential villain of the epidemic: to be a member of a risk group is to be a dangerous person.[11]

Just as significantly, promiscuity—shorthand for that set of irresponsible sexual practices gay men stand accused of—is central to the entire construction of AIDS metaphors around homosexuality. In both popular and scientific fantasy over the last century, the characteristic feature of homosexuality was precisely its narrowing to pure sexuality—a sexuality lacking in order, in discrimination, in rules—a sexuality in some sense outside social institutions, and, therefore, dangerous. It is precisely this reduction that underlies the definition of homosexuality as quintessentially the "other," the utterly different, that which lies outside society. Well-entrenched in European and American culture during the nineteenth century, this definition had begun to dissipate by the 1970s, but it has been partly revived by the AIDS crisis. The "swishy queen" has been replaced in the popular and medical imagination by the dying AIDS victim as the exemplar of male homosexuality. What has not changed is the reduction of all homosexuals to the image of the supposedly most "visible" minority among them.

Many measures commonly suggested to deal with AIDS victims are patterned on those traditionally used to deal with homosexuals. Various forms of expulsion—real and symbolic—from society as a whole and, especially from the realm of politics and public discourse have been proposed: quarantine and other forms of isolation for AIDS sufferers, public surveillance of HIV-positive individuals, HIV antibody testing without sufficient provision for confidentiality or anonymity, the general refusal to discuss homosexual sex acts publicly, as well as the strong desire of much of the population to remove AIDS patients from schools, jobs, and housing. These measures are not only entirely congruent with the definition of homosexuals as outsiders, but threaten to reinforce it significantly. As two commentators unusually sensitive to the practical effects of the metaphors of AIDS have noted: "With AIDS now regarded as manifest 'proof' of the profoundly 'diseased' and 'decadent' nature of 'queers' and 'junkies,' reactionary calls to remove gays and JVDUS [intravenous drug users] from the midst of so-called 'civil society' have truly reached new depths."[12]

The effects of such thinking have gone beyond rhetoric to produce massive discrimination against those who have contracted AIDS, inadequate funding for AIDS research and for social welfare programs for AIDS sufferers, great resistance to public health education—including an overwhelming Congressional vote on October 14, 1987, to ban federal funding for education efforts that "promote homosexuality"—and a politics of not-so-benign neglect of the epidemic by the federal government.

Ultimately, the construction of a disease from this complex, historically elaborated imagery leaves us with a view of homosexual persons with or likely to contract AIDS as either individual victims or immoral agents; they become the bearers of a disease just as they are the bearers of a psychological, social, or biologically determined (homo)sexuality. Even as an epidemiological "risk group," they tend to be considered a "group" only insofar as they share individually determined and very narrowly defined behavioral patterns (sexual acts) that bring them into contact with a specific viral agent. Such a view is insufficient on both epidemiological and historical grounds. Disease patterns are epidemiologically meaningful only when applied to groups, not to individuals. In any case, the point is not that homosexuals form a "risk group" but that certain sexual acts are "risky."[13] And whatever the biological or psychological roots of homosexuality and heterosexuality, they are historically meaningful only when viewed as socially constructed within specific societies; as such, they cannot simply be posed as opposites nor reduced to purely physical sexuality. The obsession with homosexuality, especially as a descriptor of individual personality, is seriously misleading and leaves both homosexuals and AIDS outside history.

In reality, no group has ever been outside history altogether, without rules, without internal order of some sort, except in the minds of its enemies. (This is surely as true of drug users and other groups at "higher risk" for AIDS as it is of homosexuals.) The attitudes and fantasies discussed above stem from defining homosexuals negatively by their failure to conform rather than by their actual history. This other and more complicated history involves changing social definitions of homosexuality, the emergence of homosexuals as a special class of person, and their exclusion from the rest of society. Largely in response, homosexuals have created their own communities or subcultures with specific self-definitions, institutions, and ideologies.[14]

In *this* history the role of gay men in the AIDS crisis can be explored only insofar as we view them as a historically defined and developed group and not as predefined and ahistorical individuals. In the intersection between this history and the history of attitudes toward homosexuality it makes sense to study the AIDS epidemic as social history. Even the concept of gay men as a "risk group" ultimately is meaningful only on this wider, nonepidemiological level.

III

For gay men, the AIDS crisis has exacerbated serious problems of discrimination and homophobia. In a society that does not provide adequate funds, care, and sympathy for dealing with AIDS, the crisis means the continuation of gay men's medical and social isolation. Above all, the crisis means living with the threat and the reality of disease, death, and bereavement.

As a whole, the gay community has shown a rational fear of the disease, but it also has demonstrated a remarkable capacity to avoid panic. This relative calm undoubtedly derives from a familiarity with AIDS that is significantly greater than that of the general public, as well as the self-knowledge as a community that rejects ordinary metaphors of AIDS built on fear and loathing of homosexuality. Unlike the heterosexual world, the gay world cannot construct AIDS as a disease of "the other," but is forced to "normalize" it and to construct its own series of metaphors for so doing.

The AIDS crisis is remarkable due to the degree to which the group that appeared most affected by the disease became extensively involved in its management. Gays have been in the forefront of groups providing social aid and health care to persons with AIDS (whether homosexual or not), conducting research, lobbying for funds and other governmental intervention, creating education programs, negotiating with legislators and health insurers, and the like.[15] . . .

The gay community has demonstrated a remarkable willingness and ability to work within and outside the established governmental and private institutions that normally handle health emergencies. When gays have not succeeded in persuading or forcing established institutions to provide money, care, and compassion to deal with AIDS, they have provided the needed resources through their own institutions. This insistence on taking an active role in disease management serves as notice that gays will not be forced to remain outsiders, as the victims or villains of popular metaphor.

Ironically, despite the obvious danger of doing so, the gay community has in a sense embraced the identification of AIDS with homosexuality that is central to popular and medical metaphors of the disease, but has redefined this concept significantly to make effective self-management of the disease central to it. Comprehension of this effort requires a close look at the actual history of the community.

IV

The gay community—as a community of interacting and self-defining persons rather than as a pool of victims—has been noticeably and continually absent from the media and public discourse (except for the gay press). Silence about gay people and the realities of their lives is at the very center of homosexual oppression. Almost from its inception in the nineteenth century, the gay community has struggled for the right to control its own fate, to free itself from the interference of the state, the church, and the medical and psychiatric professions. The struggle has, therefore, always been about *power:* the power to define, to victimize, to deny entry into the public realm as a legitimate group.

This struggle is evident in the long history of the gay communities[16] that grew mainly within large urban centers, particularly in New York, San Francisco, and Los Angeles, precisely the communities where AIDS made its first and most deadly appearance. And the power struggle is equally apparent in the closely connected struggle for political identity and power, beginning before World War II and growing to national significance in the period of "gay liberation" in the 1970s and 1980s.[17]

This continued struggle for political and social power molded the struggle over AIDS; that is, gay men have fought, most often against the same groups that denied their rights, to retain some degree of control over the definition of the disease and the way it was combated. Refusing to see themselves as victims or to be expelled from society again, while political, medical, and moral "professionals" determine their fate, gay men are building on many decades of political and social organizing and using that experience to create new forms of resistance. The speed with which gay self-help and political organizations sprang up to meet the AIDS crisis, and the efficiency with which they achieved their aims, was a measure of the community's organizational and institutional sophistication. The gay community, and the so-called gay ghettos, had long since developed a wide variety of social, cultural, political and legal institutions—including a large number of newspapers and magazines—that could be enlisted in the fight against AIDS. As Michael Bronski puts it, the huge effort of AIDS organizing "is in the tradition of the gay movement—a direct response to an oppressive situation."[18] It is also a very American response to crisis as well; this society characteristically creates a wide range of voluntary organizations to meet social, welfare, and health needs of portions of the population.

The gay community, of course, has never been monolithic. Like all communities that have emerged in a context of struggle against oppression, it is highly complex and continually developing. Not all who are homosexual in "orientation" or practice belong to it; one must embrace one's homosexuality more or less publicly for that. People move in and out of the community as need and circumstance require. The character and strength of gay communities differ considerably depending upon location, with the greatest strength and development evident mainly in large urban centers. And each has its idiosyncratic features, differences largely correlated with the cities where they formed. Moreover, the gay community is subject to the same differentiation across class, political, income, gender, age, ethnic, and racial lines as the rest of American society. Insofar as it has spawned a number of political and social movements, however, the community has tended to be white and middle class, a characteristic that has led to significant tension.

Most notable, perhaps, has been the tendency for contradictions to develop between gay men and lesbians. Many lesbians find an alternative political and social practice in the women's movement and do not share gay men's way of life and attitudes toward sexuality. What has, in fact, largely brought the varied strands of the gay community together is an abstract "sexual orientation," originally defined by hostile outsiders, and, above all, a common history of struggle against oppression.

Despite these internal tensions and continued hostility from outsiders, the AIDS crisis neither destroyed nor weakened the community. In fact, its maturity enabled the community to meet the challenge of AIDS with a noticeably strengthened sense of identity, inner cohesion, and ability to work for common aims. Gay men and lesbians, for example, have worked together more closely than ever in the political struggle surrounding AIDS as well as in the care of the ill and dying, mainly gay men. Lesbians have indeed performed heroically in a struggle that on many grounds need not have been theirs at all. . . .

On the simplest level, lesbian and gay political organizations, fundamentally oriented towards a familiar kind of pressure-group activity, have been joined or replaced by many new groups that combine health care and organizing with gay politics and legal action. Society's failure to address the health and social needs of the gay community, as well as the limitations of existing institutions, forced gays and lesbians to create their own institutions to confront the crisis. Once these community-based organizations were in place (approximate 250–300 now exist), their members realized that they had to broaden their scope by adding political goals and activities, that is, they would have to combat both homophobia and discrimination as well as AIDS. As Eric Rofes, an activist in AIDS health care issues, has noted:

> If we have learned anything from AIDS, it's that you cannot separate politics from health care. They are one and the same. Women have understood that for a long time. People of color have understood that for a long, time. A lot of gay white men have. But too many haven't.[19]

The new politics of AIDS has succeeded in procuring additional funding and health resources for the struggle against AIDS, as well as in combating homophobia in the government, the health sector, and the insurance industry. It is beginning to build bridges to other groups who see health care as a central issue. However, stretching itself to the limit in the struggle against AIDS, the gay community has reduced the degree to which the remainder of society is responsible for providing the money, care, and volunteers it supplies for most other epidemics.[20] Peter Arno and Karyn Feiden observe:

> The fact that the SSA [Social Security Administration] has ruled that AIDS patients are eligible for presumptive disability illustrates the gay community's ability to influence public health policy. In cities with large gay populations, such as New York and San Francisco, the supportive care received by many AIDS patients is actually superior to that received by the victims of other severe chronic illnesses. The Federal Government, however, cannot take credit for this. The development and growth of community-based AIDS service organizations, largely through massive gay-organized volunteer efforts, is helping to create a high-quality integrated care delivery system. Whether the current level of voluntarism can continue to match the pace of the epidemic or serve the growing segment of IV [intravenous] drug users; whether volunteer care is viable outside major metropolitan areas and can serve victims of other diseases; and whether voluntarism allows the government to abdicate its obligations to its people, remains to be seen.[21]

Thus, the success of the gay community's crisis intervention might have the paradoxical effect of leaving most of the responsibility for care with the community and perpetuating the segregation of gay men and lesbians. One task for gay organizations is to devise ways to share the responsibility for combatting AIDS with the rest of society. . . .

V

Like politics, gay male sexuality has substantially changed as a result of AIDS. Those gay institutions largely devoted to sexual activity—bars, baths, backrooms, public spaces—were of great importance, although they hardly exhausted the content of gay community life. The fundamental link among gay persons was, after all, sexual, but these institutions have represented far more than sites of sexual activity. Such places have developed a far greater symbolic and social significance to the gay community than is readily understood by non-gays.[22] For decades they represented the only public spaces that could in any sense be termed homosexual and where homosexuals could discover each other as well as a wider homosexual world, in spite of frequent police raids and moral crusades against them. Outsiders probably cannot imagine the significance of these spaces in the complicated double process of "coming out"—that is, entering the homosexual world as well as publicly committing oneself to one's homosexuality.

Not surprisingly, when gay people asserted their right to exist in the gay liberation period that began in the 1960s, sexual institutions expanded astronomically. The room for sexual experimentation and creativity also expanded immensely, as an expression of gay identity, as a protest against the earlier suppression of homosexuality, and as a genuine, although sometimes utopian, attempt to fashion a society under new conditions of freedom. And the public nature of much of this sexuality became another expression of the fact that gay sex was a product of a community, not merely of a group of preexisting homosexual individuals.

Gay men, like feminists, have been quite aware for some time that sexuality has a deeply political aspect. Because of its role in their identity, this sexuality was central to their political struggles against the oppressive institutions of society. The rightists who chose the gay community as one of their primary targets appear to have comprehended this as well.

Gay identity itself may be said to have been built largely on sexual identity and sexual institutions. But in the context of the gay liberation movements of the late 1960s and 1970s, the existence of sexual institutions and identity encouraged the expansion of nonsexual institutions, including political and protest groups, self-help groups, and cultural institutions. In the 1970s both sexual and nonsexual institutions grew in importance as a real gay community emerged in the wake of gay liberation. It was, however, precisely the sexual institutions and the role of sexuality within the gay community that were most definitively shaken by AIDS, a disease spread, at least in part, through some forms of sexual intercourse. The gay community is still struggling to cope with the challenge of AIDS to its sexual beliefs and practices. But the strong emphasis on sexuality in its multiple forms is too deeply rooted in gay history simply to be abandoned. The forces—church, state, or medical—that seek to use the AIDS crisis to restore their authority will not easily banish all homosexual acts. . . . But the struggle for control of sexuality is only in part a legislative or judicial struggle; it also encompasses the community's struggle to control its own sexuality and the ability to control, through sexuality, such disparate aspects of human life as health care, the body, and the family. The gay community well recognizes the nature of this struggle and its implications for the community's ability to survive the challenge of AIDS.

Under the impetus of AIDS, sexual institutions themselves have declined in importance within the gay community. They are increasingly being replaced by nonsexual social, cultural, and political institutions, including the community-based health, social welfare, and related organizations mentioned earlier. The nature of the gay community, the institutions that provide its cohesion, and the way gay people deal with one another have all changed considerably. At the same time, gay sexuality has increasingly been reconstructed under the impetus of AIDS along the lines of sexuality found in the majority heterosexual world. This has, for example, meant a new emphasis on "dating" and on longer-term, more monogamous relationships among gay men. . . . But change has also meant a decline in the various forms of sexual experimentation, spontaneity, unorthodox relationships, community "flamboyance," willingness to cross social boundaries, and sense of "celebration"—the "Dionysian" aspects of life—that made the gay community so interesting and creative in the 1970s.

Gay men will need many years to come to terms with the current realties of sexuality and its place within their community and self-identity. In the meantime they need to undertake the serious theoretical work—lesbians and the women's movement in general have done far better in this sphere—of understanding sexuality in an age of crisis and reconstructing its role within gay identity. Broadly theoretical concerns have attached themselves to narrowly defined issues such as the debates, both within and outside the gay community in San Francisco and New York in 1984 and 1985, over whether gay baths should be allowed to operate in a time of epidemic. But the largely symbolic nature of these concerns ultimately makes them poor arenas for elaborating a new gay sexuality. The debates fed on the rhetoric of "promiscuity" among non-gays as well as old and not particularly well-conceived arguments over the role of monogamy and multiple sexual partners among gay men. In fact, only a minority of gay men regularly used the baths, and there was never substantial evidence to suggest that closing the baths would decrease the spread of HIV infection. Closing baths and similar institutions in many cities represents less a restructuring of gay sexuality than a partial defeat for the gay community's control over its own sexuality.

To date, the major changes in gay male sexual practice are mainly the product of a practical need to meet the AIDS crisis directly. A significant "safe-sex" movement, largely staffed, operated, and funded by the gay community, has been created to carry out the necessary work of educating gay men (and others) regarding sexual practices in a crisis, using pamphlets, discussion groups, safe-sex pornography, and the like. Substantial evidence suggests that this movement has helped change the behavior of the majority of gay men.[23] While some gay men are evidently abandoning sex altogether—a few even appear to have accepted the assertion that all gay sex equals death—most are adapting to the crisis by building new sexual identities around safe-sex activities.

This restructuring of gay institutions and deeply rooted sexual practices may reflect a certain malleability that enables the gay community to respond rapidly to external changes. This malleability appears to be conditioned by a long history of oppression and by the fact that gay traditions are not passed on through the family. A history that at first sight seems to demonstrate significant weaknesses may turn out in retrospect to offer important advantages as well.

And, in fact, gay men were never as "obsessed" with sex as their enemies and even many in their own community believed. A significant proportion of gay men never made sex central to their lives, and the main gay tradition, which did privilege sexual relations especially in the 1970s and early 1980s, did not sexualize the entire world or foster an obsession with sex. Ironically, sex became tame, an ordinary part of everyday life, and necessary changes became easier.

VI

In contrast to its surprising development of strength in other areas, the gay community has been less successful and shown more ambivalence in relation to medicine. Ronald Bayer perceptively suggests that gay men find themselves "between the specter and the promise of medicine."[24] Medicine offers both the promise of solutions to AIDS and the danger that physicians and medical researchers will reassume control over the gay community or work with the state to do so. After all, physicians and psychiatrists "medicalized" homosexuality in the first place, and only in 1973, after a long struggle, was the

American Psychiatric Association "persuaded"—some would say forced—to remove homosexuality from its list of mental disorders.[25]

The AIDS crisis has rekindled much of that long history of hostility between the homosexual community and the medical world. Physicians and other health workers have not been notably sympathetic to gay men or persons with AIDS. Homophobia has risen dramatically as AIDS threatens to link gay men permanently with a specific, deadly disease, thought to be of their own making.[26] The association of gay men with other "medicalized" risk groups, such as illicit drug users and prostitutes, has similar effects. In addition, gay male health—like that of other oppressed or poor groups in our society—has been neglected by the medical community. When gay men receive excellent medical treatment, they do so as middle-class men, rarely as gays. Not until the mid-to-late 1970s did anyone recognize that the gay male community might suffer from specific diseases that would best be treated in the context of the community. This recognition, which grew out of the liberation movement and the sense of community that accompanied it, was largely confined to gays.[27]

Before the 1970s most physicians and psychologists refused to recognize the existence of a gay community, insisting that gay people were merely individuals with certain medical or psychological characteristics. For the medical establishment gay men were mentally or physically sick and their other diseases were the expected byproduct of their homosexuality. With a heavy dose of moralism, the medical profession acted as if—and often stated that—it preferred to see people suffer from venereal disease than commit acts it did not approve.[28] Moreover, many gays feared that seeking treatment for particular sexually transmitted diseases would, in effect, be an admission of their homosexuality before a hostile world—precisely what happens when gay men are diagnosed as suffering from AIDS or AIDS-related complex (ARC). And, finally, the neglect of gay men's health derives also from gay men's willingness to accept the typical twentieth-century emphasis, especially in the age of antibiotics, on curative medicine rather than preventive measures, particularly condoms and other devices that might have prevented the spread of infectious disease through sexual contact. . . .

Gay persons with AIDS have undertaken to educate themselves in technical matters relating to their disease and its possible cures and treatments. They have actively demanded services from the medical and epidemiological establishments and have even occasionally provided them directly. Especially in San Francisco and New York, these services are, arguably, superior to those available for most other types of patients in the United States. As Richard Dunne, executive director of the New York-based Gay Men's Health Crisis, recently put it:

> What's happening today is something that has not happened before in modern medicine—maybe never before. The patient walks in to see the doctor and says, "I'm on AZT and I'm having some side effects. What do you think of my taking cyclovir?" Physicians are overwhelmed that the P.W.A. [person with AIDS] is the expert. A P.W.A. said to me recently that his physician is someone he consults but that he makes the decisions.[29]

In a closely related development, the gay community has been forced to pay particular attention to the private health insurance industry. In America, unlike almost every other industrial nation, private health insurance represents health care access to all but the elderly or very poor. Gay men, never in favor with the insurance industry,[30] are in danger of being denied adequate insurance against illness. The gay community, largely through its legal organizations, has actively combated discrimination that would deny them health insurance or employee benefits. While not uniformly successful, its efforts have helped ensure that gay men and others thought to be at risk for AIDS will not automatically be denied access to insurance and, therefore, to health care itself. . . .

The gay community, however, has tended to remain aloof from other groups (many of the poor, the aged, and a substantial number of health care workers) who are pressing for a national health care system. This is a surprising development in that it is precisely minority-group members who are, in addition to gay men, those most at risk for AIDS and AIDS-related conditions.

. . . Perhaps the reason the gay movement has been unable to work fully with minority and other protest groups for common aims is the continuing racism among white, middle-class gays and homophobia among other minorities as well as belief on each side that they compete for limited resources. . . .

In health-related activities the gay community has begun to move closer to the heterosexual world. . . . Many gay health organizations, caring agencies, and related institutions have become large and successful bureaucracies, dispensing hundreds of thousands of dollars and becoming attentive to public fund-raising and government grants. In addition, they are becoming integrated into the wider health-care world. Some radical gay political groups, however, see such mainstreaming as political timidity, an overemphasis on the purely clinical aspects of the epidemic, and too great a willingness to cooperate with governmental bodies, which are thereby allowed to escape justified charges of having done too little to meet the crisis.[31] Such tensions are probably inevitable as the gay community's position on the complicated problems of medicine and health care involved with AIDS continues to evolve.

VII

The detailed responses to AIDS coalesce into two complex and parallel sets of metaphors—two types of discourse, two varieties of practice—built on the relationship between AIDS and homosexuality. Both sets of metaphors accept the reality of that relationship, resting upon the complicated ideological and institutional history of American homosexuality. But they draw on different aspects of that heritage, reconfigurating them in response to the AIDS crisis and creating significantly different amalgams of past and present. In fact, they form the materials for two sides of a massive political and ideological struggle that extends far beyond the reality of AIDS as a disease.

The two discourses do not speak directly to one another, but they "echo" each other strongly. Both regard AIDS as a moral crisis, but where the first defines AIDS as the breakdown of the dominant "traditional" morality of sexual behavior and social organization, the second considers it the breakdown of a new "social contract" that includes all people as full members of the body politic. Where the first seeks to exclude homosexuality from public view and shroud it in silence, the second argues that "silence is death" with respect to both AIDS and homosexuality.[32] Where the first finds sexual irresponsibility and lack of self-control at the heart of homosexuality, the second demonstrates an unusual capacity on the part of gay people to alter their own sexual behavior. Where the first sees victims and villains, the second sees actors and heroes.

The gay community, in an ironic reversal of popular views of AIDS, constructed the AIDS crisis and its metaphors in the image of its own history. But the metaphors we find here, unlike those of homophobic approaches, are socially, institutionally, and ideologically rich, because they build on the history of a real community coming to terms with its past and its need to take control of its present and future. They are also useful metaphors, in that they allow for a rational intervention in the management of the disease while they let us remove at least some of the hysteria that has prevented us from treating AIDS like any other disease.

The gay community has largely succeeded in removing AIDS from the category of the new, the terrifying, and the special and made it more ordinary, normal, and therefore, manageable. In the same manner, people with AIDS have shown that it is possible—indeed necessary—to live with AIDS and not merely die from it.

The best summation of this story can perhaps be found in the moving remarks of Paul Monette in the preface to a cycle of poems on the death of his lover from AIDS:

> The story that endlessly eludes the decorum of the press is the death of a generation of gay men. What is written here is only one man's passing and one man's cry, a warrior burying a warrior. May it fuel the fire of those on the front lines who mean to prevail, and of their friends who stand in the fire with them. We will not be bowed down or erased by this. I learned too well what it means to be a people, learned in the joy of my best friend what all the meaningless pain and horror cannot take away—that all there is is love. Pity us not.[33]

NOTES

1. Institute of Medicine, National Academy of Sciences, *Mobilizing Against AIDS: The Unfinished Story of a Virus* (Cambridge, Mass.: 1986).
2. Karl Marx, *The 18th Brumaire of Louis Bonaparte* (New York: 1963), 15. I have altered the customary translation from "men" to "human beings," a change I feel is warranted by the German "menschen."
3. Susan Sontag, *Illness As Metaphor* (New York: 1978).
4. Cf. Ronald Bayer, "AIDS and the Gay Community: Between the Specter and the Promise of Medicine," *Social Research* 52 (Autumn 1985): 581–606 at 587ff.
5. Jeffrey Weeks, *Sexuality and its Discontents* (London: 1985), 45–46.
6. On the process in general, see Jeffrey Weeks, *Coming Out* (London: 1977); Jonathan Katz, *Gay American History* (New York: 1976) 129–207; and *Gay/Lesbian Almanac* (New York: 1983), 1–19. For further details: Vernon L. Bullough, "Homosexuality and the Medical Model," *Journal of Homosexuality* 1 (1974): 99ff; George Chauncey, Jr., "From Sexual Inversion to Homosexuality: The Changing Medical Conceptualization of Female 'Deviance,'" in this volume; Georges Lanteri-Laura, *Lecture des perversions: histoire de leur appropriation médicale* (Paris: 1979); Peter Conrad and Joseph W. Schneider, *Deviants and Medicalization: From Badness to Sickness* (St. Louis: 1980), ch. 7. Cf. Michel Foucault, *The History of Sexuality*, vol. I, *An Introduction* (New York: 1978), 44.
7. Cf. Philippe Ariès, "Thoughts on the History of Homosexuality," in *Western Sexuality: Practice and Precept in Past and Present Times*, ed. Philippe Ariès and André Béjin (Oxford: 1985), 62–75.
8. Cf. Patrick Carnes, *The Sexual Addiction* (Minneapolis: 1983); Craig Rowland "Reinventing the Sex Maniac," *The Advocate*, 21 Jan. 1986, 43–49; Daniel Goleman, "Some Sexual Behavior Viewed as an Addiction," *New York Times*, 16 October 1984, C1. Not surprisingly in light of the historical associations of this concept, it turns out that women and gay men are the groups most at risk for this supposed "addiction."
9. *New York Post,* 24 May 1983.
10. Cf. the material collected in Jonathan Katz, *Gay American History and Gay/Lesbian Almanac.*
11. Elizabeth Fee, commentary on the session "AIDS in Historical Perspective," annual meeting of the American Historical Association, Chicago, Ill., December 1986.
12. Nancy Krieger and Rose Appleman, *The Politics of AIDS* (Oakland, Cal.: 1986), 18. See most recently Randy Shilts, *And the Band Played On: Politics, People, and the AIDS Epidemic* (New York: 1987); and Daniel M. Fox, "AIDS and the American Health Polity: The History and Prospects of a Crisis of Authority," *Milbank Quarterly* 64 (1986), Supplement I, "AIDS: The Public Context of an Epidemic," ed. Ronald Bayer, Daniel M. Fox, and David P. Willis, 7–33.
13. See William H. McNeill, *Plagues and Peoples* (New York: 1976), for a good historical overview.
14. See the works cited in note 6 and Dennis Altman, *The Homosexualization of America and the Americanization of the Homosexual* (New York: 1982); John D' Emilio, *Sexual Politics, Sexual Communities: The Making of a Homosexual Minority in the United States, 1940–1970* (Chicago: 1983), and "Gay Politics, Gay Community: San Francisco's Experience," *Socialist Review* 55 (Jan.–Feb. 1981): 77–104; and Toby Marotta, *The Politics of Homosexuality: How Lesbians and Gay Men Have Made Themselves a Political and Social Force in Modern America* (New York: 1981).
15. On all these aspects and the reaction of the gay community in general to the AIDS epidemic, see Shilts, *And the Band Played On*; Dennis Altman, *AIDS in the Mind of America* (Garden City, N.Y.: 1986); "AIDS: The Politicization of an Epidemic,"

Socialist Review 78 (Nov./Dec. 1984): 93–109; "The Politics of *AIDS*," In *AIDS: Public Policy Dimensions,* ed. John Griggs (New York: 1987), 23–33; and, in some ways most perceptively, Cindy Patton, *Sex and Germs: The Politics of AIDS* (Boston: 1985). In addition, it is important to look at coverage of AIDS in the gay press from 1981 on in some detail, particularly the *New York Native*, the (national) *Advocate* and the San Francisco *Bay Area Reporter.*

16. On the gay ghettoes in general, see: Martin P. Levine, "Gay Ghetto," *Journal of Homosexuality* 4 (1979): 363–378; Martin P. Levine, ed., *Gay Men: The Sociology of Male Homosexuality* (New York: 1979); Manuel Castels, *The City and the Grassroots: a Cross-Cultural Theory of Urban Social Movements* (Berkeley: 1983).
17. An overview of the various gay liberation movements can be found in Barry Adam, *The Rise of a Gay and Lesbian Movement* (Boston: 1987).
18. Michael Bronski, "Death and the Erotic Imagination," *Gay Community News* (Boston), 7–13 Sept. 1986, 8–9 at 8.
19. Quoted by Mark Vandervelden, "Gay Health Conference," *The Advocate*, 28 April 1987, 12. The occasion for Rofes' statement was the eighth annual "National Lesbian and Gay Health Conference," held in Los Angeles on 26–29 March 1987, and attended by representatives of more than 250 organizations.
20. On the limits of voluntarism, cf. Peter S. Arno and Karyn Feiden, "Ignoring the Epidemic: How the Reagan Administration Failed on AIDS," *Health-PAC Bulletin* 17 (December 1986): 7–11, and Peter S. Arno, "The Contributions and Limitations of Voluntarism," in *AIDS: Public Policy Dimensions,* 188–192.
21. Arno and Feiden, "Ignoring the Epidemic," 11.
22. Cf. Dennis Altman, "Sex: The Frontline for Gay Politics," *Socialist Review* (Sept./Oct. 1982).
23. The increasingly large literature on changes in gay sexual behavior due to AIDS is summarized and discussed in Marshall H. Becker and Jill G. Joseph, "AIDS and Behavioral Change to Reduce Risk: A Review," *American Journal of Public Health* 78 (April 1988): 394–410.
24. Bayer, "AIDS and the Gay Community: Between the Specter and the Promise of Medicine."
25. See Ronald Bayer, *Homosexuality and American Psychiatry: The Politics of Diagnosis*, 2nd ed. (New York: 1988).
26. See, in general, Dan DeNoon, "AIDS Takes Its Psychological Toll on the Health-Care Community," *In These Times*, 14–20 October 1987, 6–7.
27. The first relatively thorough and scientific survey and analysis of diseases specific to the gay community that I am aware of appeared in 1981: William W. Darrow, Donald Barrett, Karla Jay, and Allen Young, "The Gay Report on Sexually Transmitted Diseases," *American Journal of Public Health* 71 (Sept. 1981): 1004–1011; cf. the accompanying editorial of H. Hunter Handsfield, 989–990, who cites other, less complete studies.
28. Cf. Allan M. Brandt, *No Magic Bullet*, 2nd ed. (Cambridge, Mass.: 1987).
29. Quoted in Anne-Christine D'Adesky, "Breaking the F.D.A Drugjam," *The Nation*, 17 October 1987, 405. See also the various PWA (persons with AIDS) newsletters, magazines such as "AIDS Treatment News," and the pages of the gay press, in particular the *New York Native,* all of which provide substantial coverage on AIDS treatments (in addition to AIDS politics and organizing). For work by PWAs themselves, see Michael Callen, ed., *Surviving and Thriving with AIDS* (Boston: 1978) and Peter Tatchell, *AIDSÚ A Guide to Survival* (Boston: 1987).
30. Cf. the 1906 remarks of Dr. William Lee Howard, "The Sexual Pervert in Life Insurance," quoted in Katz, *Gay/Lesbian Almanac*, 319f.
31. The most notable example of such an attack was by Larry Kramer, "An Open Letter to Richard Dunne and Gay Men's Health Crisis, Inc.," *New York Native*, 26 January 1987, 1ff., which in part led to a slightly more activist stand by the GMHC as well as to the formation of several direct action groups in New York City, most notably the civil disobedience-oriented group ACT-UP.
32. "Silence = Death" is the organizing slogan of the New York-based ACT-UP, an AIDS activist group that uses street demonstrations to demand more funding for AIDS-related treatment.
33. Paul Monette, *Love Alone: Eighteen Elegies for Rog* (New York: 1988), xii–xiii.

DISCUSSION QUESTIONS

1. Have media portrayals of gay men as villains and heroes of the AIDS epidemic changed in the years since this article was written (1989)? If so, how? If not, what themes are common to both periods?

2. Can you think of other historical examples where marginalized populations have born the brunt of blame and responsibility for epidemics or other social ills? Why is blame constructed in this way?
3. Why are lesbians who, epidemiologically speaking, are one of the lowest risk groups for AIDS transmission, at the forefront of the battle against AIDS?
4. How is the stigma of AIDS similar to and different from other forms of STDs?
5. What groups have interests in controlling or exerting influence over the definition of the AIDS crisis? What strategies have they used? Have they been successful? What is at stake?

1 5

Abortion, Conception, and Procreation

The Egg and the Sperm: How Science Has Constructed a Romance Based on Stereotypical Male-Female Roles

EMILY MARTIN

Anthropologist Emily Martin examines how cultural constructions of gender are naturalized and reified through scientific research. Using the case of the egg and the sperm Martin describes how scientists stubbornly cling to discourses privileging the contributions and activities of the sperm even when new data suggest that alternate interpretations are called for.

The theory of the human body is always a part of a world-picture. . . .
The theory of the human body is always a part of a fantasy.

—JAMES HILLMAN, *THE MYTH OF ANALYSIS*[1]

From "The Egg and the Sperm: How Science Has Constructed a Romance Based on Steriotypical Male-Female Roles" in *Signs* 16(3): 485–501, 1991. Reprinted by permission of University of Chicago Press.

As an anthropologist, I am intrigued by the possibility that culture shapes how biological scientists describe what they discover about the natural world. If this were so, we would be learning about more than the natural world in high school biology class; we would be learning about cultural beliefs and practices as if they were part of nature. In the course of my research I realized that the picture of egg and sperm drawn in popular as well as scientific accounts of reproductive biology relies on stereotypes central to our cultural definitions of male and female. The stereotypes imply not only that female biological processes are less worthy than their male counterparts but also that women are less worthy than men. Part of my goal in writing this article is to shine a bright light on the gender stereotypes hidden within the scientific language of biology. Exposed in such a light, I hope they will lose much of their power to harm us.

EGG AND SPERM: A SCIENTIFIC FAIRY TALE

At a fundamental level, all major scientific textbooks depict male and female reproductive organs as systems for the production of valuable substances, such as eggs and sperm.[2] In the case of women, the monthly cycle is described as being designed to produce eggs and prepare a suitable place for them to be fertilized and grown—all to the end of making babies. But the enthusiasm ends there. By extolling the female cycle as a productive enterprise, menstruation must necessarily be viewed as a failure. Medical texts describe menstruation as the "debris" of the uterine lining, the result of necrosis, or death of tissue. The descriptions imply that a system has gone awry, making products of no use, not to specification, unsalable, wasted, scrap. An illustration in a widely used medical text shows menstruation as a chaotic disintegration of form, complementing the many texts that describe it as "ceasing," "dying," "losing," "denuding," "expelling."[3]

Male reproductive physiology is evaluated quite differently. One of the texts that sees menstruation as failed production employs a sort of breathless prose when it describes the maturation of sperm: "The mechanisms which guide the remarkable cellular transformation from spermatid to mature sperm remain uncertain. . . . Perhaps the most amazing characteristic of spermatogenesis is its sheer magnitude: the normal human male may manufacture several hundred million sperm per day."[4] In the classic text *Medical Physiology,* edited by Vernon Mountcastle, the male/female, productive/destructive comparison is more explicit: "Whereas the female *sheds* only a single gamete each month, the seminiferous tubules *produce* hundreds of millions of sperm each day" (emphasis mine).[5] The female author of another text marvels at the length of the microscopic seminiferous tubules, which, if uncoiled and placed end to end, "would span almost one-third of a mile!" She writes, "In an adult male these structures produce millions of sperm cells each day." Later she asks, "How is this feat accomplished?"[6] None of these texts expresses such intense enthusiasm for any female processes. It is surely no accident that the "remarkable" process of making sperm involves precisely what, in the medical view, menstruation does not: production of something deemed valuable.[7]

One could argue that menstruation and spermatogenesis are not analogous processes and, therefore, should not be expected to elicit the same kind of response. The proper female analogy to spermatogenesis, biologically, is ovulation. Yet ovulation does not merit enthusiasm in these texts either. Textbook descriptions stress that all of the ovarian follicles containing ova are already present at birth. Far from being *produced,* as sperm are, they merely sit on the shelf, slowly degenerating and aging like overstocked inventory: "At birth, normal human ovaries contain an estimated one million follicles [each], and no new ones appear after birth. Thus, in marked contrast to the male, the newborn female already has all the germ cells she will ever have. Only a few, perhaps 400, are destined to reach full maturity during her active productive life. All the others degenerate at some point in their development so that few, if any, remain by the time she reaches menopause at approximately 50 years of age."[8] Note the "marked contrast" that this description sets up between male and female: the male, who continuously produces fresh germ cells, and the female, who has stockpiled germ cells by birth and is faced with their degeneration.

Nor are the female organs spared such vivid descriptions. One scientist writes in a newspaper article that a woman's ovaries become old and worn out from ripening eggs every month, even though the woman herself is still relatively young. "When you look through a laparoscope . . . at an ovary that has been through hundreds of cycles, even in a superbly healthy American female, you see a scarred, battered organ."[9]

To avoid the negative connotations that some people associate with the female reproductive system, scientists could begin to describe male and female processes as homologous. They might credit females with "producing" mature ova one at a time, as they're needed each month, and describe males as having to face problems of degenerating germ cells. This degeneration would occur throughout life among spermatogonia, the undifferentiated germ cells in the testes that are the long-lived, dormant precursors of sperm.

But the texts have an almost dogged insistence on casting female processes in a negative light. The texts celebrate sperm production because it is continuous from puberty to senescence, while they portray egg production as inferior because it is finished at birth. This makes the female seem unproductive, but some texts will also insist that it is she who is wasteful.[10] In a section heading for *Molecular Biology of the Cell,* a best-selling text, we are told that "Oogenesis is wasteful." The text goes on to emphasize that of the seven million oogonia, or egg germ cells, in the female embryo, most degenerate in the ovary. Of those that do go on to become oocytes, or eggs, many also degenerate, so that at birth only two million eggs remain in the ovaries. Degeneration continues throughout a woman's life: by puberty 300,000 eggs remain, and only a few are present by menopause. "During the 40 or so years of a woman's reproductive life, only 400 to 500 eggs will have been released," the authors write. "All the rest will have degenerated. It is still a mystery why so many eggs are formed only to die in the ovaries."[11]

The real mystery is why the male's vast production of sperm is not seen as wasteful.[12] Assuming that a man "produces" 100 million (10^8) sperm per day (a conservative estimate) during an average reproductive life of sixty years, he would produce well over two trillion sperm in his lifetime. Assuming that a woman "ripens" one egg per lunar month, or thirteen per year, over the course of her forty-year reproductive life, she would total five hundred eggs in her lifetime. But the word "waste" implies an excess, too much produced. Assuming two or three offspring, for every baby a woman produces, she wastes only around two hundred eggs. For every baby a man produces, he wastes more than one trillion (10^{12}) sperm.

How is it that positive images are denied to the bodies of women? A look at language—in this case, scientific language—provides the first clue. Take the egg and the sperm.[13] It is remarkable how "femininely" the egg behaves and how "masculinely" the sperm.[14] The egg is seen as large and passive.[15] It does not *move* or *journey,* but passively "is transported," "is swept,"[16] or even "drifts"[17] along the fallopian tube. In utter contrast, sperm are small, "streamlined,"[18] and invariably active. They "deliver" their genes to the egg, "activate the developmental program of the egg,"[19] and have a "velocity" that is often remarked upon.[20] Their tails are "strong" and efficiently powered.[21] Together with the forces of ejaculation, they can "propel the semen into the deepest recesses of the vagina."[22] For this they need "energy," "fuel,"[23] so that with a "whiplash-like motion and strong lurches"[24] they can "burrow through the egg coat"[25] and "penetrate" it.[26]

At its extreme, the age-old relationship of the egg and the sperm takes on a royal or religious patina. The egg coat, its protective barrier, is sometimes called its "vestments," a term usually reserved for sacred, religious dress. The egg is said to have a "corona,"[27] a crown, and to be accompanied by "attendant cells."[28] It is holy, set apart and above, the queen to the sperm's king. The egg is also passive, which means it must depend on sperm for rescue. Gerald Schatten and Helen Schatten liken the egg's role to that of Sleeping Beauty: "a dormant bride awaiting her mate's magic kiss, which instills the spirit that brings her to life."[29] Sperm, by contrast, have a "mission,"[30] which is to "move through the female genital tract in quest of the ovum."[31] One popular account has it that the sperm carry out a "perilous journey" into the "warm darkness," where some fall away "exhausted." "Survivors" "assault" the egg, the successful candidates "surrounding the prize."[32] Part of the urgency of this journey, in more scientific terms, is that "once released from the supportive environment of the ovary, an egg will die within hours unless rescued by a sperm."[33] The wording stresses the fragility and dependency of the egg, even though the same text acknowledges elsewhere that sperm also live for only a few hours.[34]

In 1948, in a book remarkable for its early insights into these matters, Ruth Herschberger argued that female reproductive organs are seen as biologically interdependent, while male organs are viewed as autonomous, operating independently and in isolation:

> At present the functional is stressed only in connection with women: it is in them that ovaries, tubes, uterus, and vagina have endless interdependence. In the male, reproduction would seem to involve "organs" only.
>
> Yet the sperm, just as much as the egg, is dependent on a great many related processes. There are secretions which mitigate the urine in the urethra before ejaculation, to protect the sperm. There is the reflex shutting off of the bladder connection, the provision of prostatic secretions, and various types of muscular propulsion. The sperm is no more independent of its milieu than the egg, and yet from a wish that it were, biologists have lent their support to the notion that the human female, beginning with the egg, is congenitally more dependent than the male.[35]

Bringing out another aspect of the sperm's autonomy, an article in the journal *Cell* has the sperm making an "existential decision" to penetrate the egg: "Sperm are cells with a limited behavioral repertoire, one that is directed toward fertilizing eggs. To execute the decision to abandon the haploid state, sperm swim to an egg and there acquire the ability to effect membrane fusion."[36] Is this a corporate manager's version of the sperm's activities—"executing decisions" while fraught with dismay over difficult options that bring with them very high risk?

There is another way that sperm, despite their small size, can be made to loom in importance over the egg. In a collection of scientific papers, an electron micrograph of an enormous egg and tiny sperm is titled "A Portrait of the Sperm."[37] This is a little like showing a photo of a dog and calling it a picture of the fleas. Granted, microscopic sperm are harder to photograph than eggs, which are just large enough to see with the naked eye. But surely the use of the term "portrait," a word associated with the powerful and wealthy, is significant. Eggs have only micrographs or pictures, not portraits.

One depiction of sperm as weak and timid, instead of strong and powerful—the only such representation in western civilization, so far as I know—occurs in Woody Allen's movie *Everything You Always Wanted to Know About Sex** But Were Afraid to Ask*. Allen, playing the part of an apprehensive sperm inside a man's testicles, is scared of the man's approaching orgasm. He is reluctant to launch himself into the darkness, afraid of contraceptive devices, afraid of winding up on the ceiling if the man masturbates.

The more common picture—egg as damsel in distress, shielded only by her sacred garments; sperm as heroic warrior to the rescue—cannot be proved to be dictated by the biology of these events. While the "facts" of biology may not *always* be constructed in cultural terms, I would argue that in this case they are. The degree of metaphorical content in these descriptions, the extent to which differences between egg and sperm are emphasized, and the parallels between cultural stereotypes of male and female behavior and the character of egg and sperm all point to this conclusion.

NEW RESEARCH, OLD IMAGERY

As new understandings of egg and sperm emerge, textbook gender imagery is being revised. But the new research, far from escaping the stereotypical representations of egg and sperm, simply replicates elements of textbook gender imagery in a different form. The persistence of this imagery calls to mind what Ludwik Fleck termed "the self-contained" nature of scientific thought. As he described it, "the interaction between what is already known, what remains to be learned, and those who are to apprehend it, go to ensure harmony within the system. But at the same time they also preserve the harmony of illusions, which is quite secure within the confines of a given thought style."[38] We need to understand the way in which the cultural content in scientific descriptions changes as biological discoveries unfold, and whether that cultural content is solidly entrenched or easily changed.

In all of the texts quoted above, sperm are described as penetrating the egg, and specific substances on a sperm's head are described as binding to the egg. Recently, this description of events was rewritten in a biophysics lab at Johns Hopkins University—transforming the egg from the passive to the active party.[39]

Prior to this research, it was thought that the zona, the inner vestments of the egg, formed an impenetrable barrier. Sperm overcame the barrier by mechanically burrowing through, thrashing their tails and slowly working their way along. Later research showed that the sperm released digestive enzymes that chemically broke down the zona; thus, scientists presumed that the sperm used mechanical *and* chemical means to get through to the egg.

In this recent investigation, the researchers began to ask questions about the mechanical force of the sperm's tail. (The lab's goal was to develop a contraceptive that worked topically on sperm.) They discovered, to their great surprise, that the forward thrust of sperm is extremely weak, which contradicts the assumption that sperm are forceful penetrators.[40] Rather than thrusting forward, the sperm's head was now seen to move mostly back and forth. The sideways motion of the sperm's tail makes the head move sideways with a force that is ten times stronger than its forward movement. So even if the overall force of the sperm were strong enough to mechanically break the zona, most of its force would be directed sideways rather than forward. In fact, its strongest tendency, by tenfold, is to escape by attempting to pry itself off the egg. Sperm, then, must be exceptionally efficient at *escaping* from any cell surface they contact. And the surface of the egg must be designed to trap the sperm and prevent their escape. Otherwise, few if any sperm would reach the egg.

The researchers at Johns Hopkins concluded that the sperm and egg stick together because of adhesive molecules on the surfaces of each. The egg traps the sperm and adheres to it so tightly that the sperm's head is forced to lie flat against the surface of the zona, a little bit, they told me, "like Br'er Rabbit getting more and more stuck to tar baby the more he wriggles." The trapped sperm continues to wiggle ineffectually side to side. The mechanical force of its tail is so weak that a sperm cannot break even one chemical bond. This is where the digestive enzymes released by the sperm come in. If they start to soften the zona just at the tip of the sperm and the sides remain stuck, then the weak, flailing sperm can get oriented in the right direction and make it through the zona—provided that its bonds to the zona dissolve as it moves in.

Although this new version of the saga of the egg and the sperm broke through cultural expectations, the researchers who made the discovery continued to write papers and abstracts as if the sperm were the active party who attacks, binds, penetrates, and enters the egg. The only difference was that sperm were now seen as performing these actions weakly.[41] Not until August 1987, more than three years after the findings described above, did these researchers reconceptualize the process to give the egg a more active role. They began to describe the zona as an aggressive sperm catcher, covered with adhesive molecules that can capture a sperm with a single bond and clasp it to the zona's surface.[42] In the words of their published account: "The innermost vestment, the *zona pellucida,* is a glycoprotein shell, which captures and tethers the sperm before they penetrate it. . . . The sperm is captured at the initial contact between the sperm tip and the *zona*. . . . Since the thrust [of the sperm] is much smaller than the force needed to break a single affinity bond, the first bond made upon the tip-first meeting of the sperm and *zona* can result in the capture of the sperm."[43]

Experiments in another lab reveal similar patterns of data interpretation. Gerald Schatten and Helen Schatten set out to show that, contrary to conventional wisdom, the "egg is not merely a large, yolk-filled sphere into which the sperm burrows to endow new life. Rather, recent research suggests the almost heretical view that sperm and egg are mutually active partners."[44] This sounds like a departure from the stereotypical textbook view, but further reading reveals Schatten and Schatten's conformity to the aggressive-sperm metaphor. They describe how "the sperm and egg first touch when, from the tip of the sperm's triangular head, a long, thin filament shoots out and harpoons the egg." Then we learn that "remarkably, the harpoon is not so much fired as assembled at great speed, molecule by molecule, from a pool of protein stored in a specialized region called the acrosome. The filament may grow as much as twenty times longer than the sperm head itself before its tip reaches the egg and sticks."[45] Why not call this "making a bridge" or "throwing out a line" rather than firing a harpoon? Harpoons pierce prey and injure or kill them, while this filament only sticks. And why not focus, as the Hopkins lab did, on the stickiness of the egg, rather than the stickiness of the sperm?[46] Later in the article, the Schattens replicate the common view of the sperm's perilous journey into the warm darkness of the vagina, this time for the purpose of explaining its journey into the egg itself: "[The sperm] still has an arduous journey ahead. It must penetrate farther into the egg's huge sphere of cytoplasm and somehow locate the nucleus, so that the two cells' chromosomes can fuse. The sperm dives down into the cytoplasm, its tail beating. But it is soon interrupted by the sudden and swift migration of the egg nucleus, which rushes toward the sperm with a velocity triple that of the movement of chromosomes during cell division, crossing the entire egg in about a minute."[47]

Like Schatten and Schatten and the biophysicists at Johns Hopkins, another researcher has recently made discoveries that seem to point to a more interactive view of the relationship of egg and sperm. This work, which Paul Wassarman conducted on the sperm and eggs of mice, focuses on identifying the specific molecules in the egg coat (the zona pellucida) that are involved in egg-sperm interaction. At first glance, his descriptions seem to fit the model of an egalitarian relationship. Male and female gametes "recognize one another," and "interactions . . . take place between sperm and egg."[48] But the article in *Scientific American* in which those descriptions appear begins with a vignette that presages the dominant motif of their presentation: "It has been more than a century since Hermann Fol, a Swiss zoologist, peered into his microscope and became the first person to see a sperm penetrate an egg, fertilize it and form the first cell of a new embryo."[49] This portrayal of the sperm as the active party—the one that *penetrates* and *fertilizes* the egg and *produces* the embryo—is not cited as an example of an earlier, now outmoded view. In fact, the author reiterates the point later in the article: "Many sperm can bind to and penetrate the zona pellucida, or outer coat, of an unfertilized mouse egg, but only one sperm will eventually fuse with the thin plasma membrane surrounding the egg proper (*inner sphere*), fertilizing the egg and giving rise to a new embryo."[50]

The imagery of sperm as aggressor is particularly startling in this case: the main discovery being reported is isolation of a particular molecule *on the egg coat* that plays an important role in fertilization. Wassarman's choice of language sustains the picture. He calls the molecule that has been isolated, ZP3, a "sperm receptor." By allocating the passive, waiting role to the egg, Wassarman can continue to describe the sperm as the actor, the one that makes it all happen: "The basic process begins when many sperm first attach loosely and then bind tenaciously to receptors on the surface of the egg's thick outer coat, the zona pellucida. Each sperm, which has a large number of egg-binding proteins on its surface, binds to many sperm receptors on the egg. More specifically, a site on each of the egg-binding proteins fits a complementary site on a sperm receptor, much as a key fits a lock."[51] With the sperm designated as the "key" and the egg the "lock," it is obvious which one acts and which one is acted upon. Could this imagery not be reversed, letting the sperm (the lock) wait until the egg produces the key? Or could we speak of two halves of a locket matching, and regard the matching itself as the action that initiates the fertilization?

It is as if Wassarman were determined to make the egg the receiving partner. Usually in biological research, the *protein* member of the pair of binding molecules is called the receptor, and physically it has a pocket in it rather like a lock. As the diagrams that illustrate Wassarman's article show, the molecules on the sperm are proteins and have "pockets." The small, mobile molecules that fit into these pockets are called ligands. As shown in the diagrams, ZP3 on the egg is a polymer of "keys"; many small knobs stick out. Typically, molecules on the sperm would be called receptors and molecules on the egg would be called ligands. But Wassarman chose to name ZP3 on the egg the receptor and to create a new term, "the egg-binding protein," for the molecule on the sperm that otherwise would have been called the receptor.[52]

Wassarman does credit the egg coat with having more functions than those of a sperm receptor. While he notes that "the zona pellucida has at times been viewed by investigators as a nuisance, a barrier to sperm and hence an impediment to fertilization," his new research reveals that the egg coat "serves as a sophisticated biological security system that screens incoming sperm, selects only those compatible with fertilization and development, prepares sperm for fusion with the egg and later protects the resulting embryo from polyspermy [a lethal condition caused by fusion of more than one sperm with a single egg]."[53] Although this description gives the egg an active role, that role is drawn in stereotypically feminine terms. The egg *selects* an appropriate mate, *prepares* him for fusion, and then *protects* the resulting offspring from harm. This is courtship and mating behavior as seen through the eyes of a sociobiologist: woman as the hard-to-get prize, who, following union with the chosen one, becomes woman as servant and mother.

And Wassarman does not quit there. In a review article for *Science,* he outlines the "chronology of fertilization."[54] Near the end of the article are two subject headings. One is "Sperm Penetration," in which Wassarman describes how the chemical dissolving of the zona pellucida combines with the "substantial propulsive force generated by sperm." The next heading is "Sperm-Egg Fusion." This section details what happens inside the zona after a sperm "penetrates" it. Sperm "can make contact with, adhere to, and fuse with (that is, fertilize) an egg."[55] Wassarman's word choice, again, is astonishingly skewed in favor of the sperm's activity, for in the next breath he says that sperm *lose* all motility upon fusion with the egg's surface. In mouse and sea urchin eggs, the sperm enters at the *egg's* volition, according to Wassarman's description: "Once fused with egg plasma membrane [the surface of the egg], how does a sperm enter the egg? The surface of both mouse and sea urchin eggs is covered with thousands of plasma membrane-bound projections, called microvilli [tiny "hairs"]. Evidence in sea urchins suggests that, after membrane fusion, a group of elongated microvilli cluster tightly around and interdigitate over the sperm head. As these microvilli are resorbed, the sperm is drawn into the egg. Therefore, sperm motility, which ceases at the time of fusion in both sea urchins and mice, is not required for sperm entry."[56] The section called "Sperm Penetration" more logically would be followed by a section called "The Egg Envelops," rather than "Sperm-Egg Fusion." This would give a parallel—and more accurate—sense that both the egg and the sperm initiate action.

Another way that Wassarman makes less of the egg's activity is by describing components of the egg but referring to the sperm as a whole entity. Deborah Gordon has described such an approach as "atomism" ("the part is independent of and primordial to the whole") and identified it as one of the "tenacious assumptions" of Western science and medicine.[57] Wassarman employs atomism to his advantage. When he refers to processes going on within sperm, he consistently returns to descriptions that remind us from whence these activities came: they are part of sperm that penetrate an egg or generate propulsive force. When he refers to processes going on within eggs, he stops there. As a result, any active role he grants them appears to be assigned to the parts of the egg, and not to the egg itself. In the quote above, it is the microvilli that actively cluster around the sperm. In another example, "the driving force for engulfment of a fused sperm comes from a region of cytoplasm just beneath an egg's plasma membrane."[58]

SOCIAL IMPLICATIONS: THINKING BEYOND

All three of these revisionist accounts of egg and sperm cannot seem to escape the hierarchical imagery of older accounts. Even though each new account gives the egg a larger and more active role, taken together they bring into play another cultural stereotype: woman as a dangerous and aggressive threat. In the Johns Hopkins lab's revised model, the egg ends up as the female aggressor who "captures and tethers" the sperm with her sticky zona, rather like a spider lying in wait in her web.[59] The Schatten lab has the egg's nucleus "interrupt" the sperm's dive with a "sudden and swift" rush by which she "clasps the sperm and guides its nucleus to the center."[60] Wassarman's description of the surface of the egg "covered with thousands of plasma membrane-bound projections, called microvilli" that reach out and clasp the sperm adds to the spiderlike imagery.[61]

These images grant the egg an active role but at the cost of appearing disturbingly aggressive. Images of woman as dangerous and aggressive, the femme fatale who victimizes men, are widespread in Western literature and culture.[62] More specific is the connection of spider imagery with the idea of an engulfing, devouring mother.[63] New data did not lead scientists to eliminate gender stereotypes in their descriptions of egg and sperm. Instead, scientists simply began to describe egg and sperm in different, but no less damaging, terms.

Can we envision a less stereotypical view? Biology itself provides another model that could be applied to the egg and the sperm. The cybernetic model—with its feedback loops, flexible adaptation to change, coordination of the parts within a whole, evolution over time, and changing response to the environment—is common in genetics, endocrinology, and ecology and has a growing influence in medicine in general.[64] This model has the potential to shift our imagery from the negative, in which the female reproductive system is castigated both for not producing eggs after birth and for producing (and thus wasting) too many eggs overall, to something more positive. The female reproductive system could be seen as responding to the environment (pregnancy or menopause), adjusting to monthly changes (menstruation), and flexibly changing from reproductivity after puberty to nonreproductivity later in life. The sperm and egg's interaction could also be described in cybernetic terms. J. F. Hartman's research in reproductive biology demonstrated fifteen years ago that if an egg is killed by being pricked with a needle, live sperm cannot get through the zona.[65] Clearly, this evidence shows that the egg and sperm *do* interact on more mutual terms, making biology's refusal to portray them that way all the more disturbing.

We would do well to be aware, however, that cybernetic imagery is hardly neutral. In the past, cybernetic models have played an important part in the imposition of social control. These models inherently provide a way of thinking about a "field" of interacting components. Once the field can be seen, it can become the object of new forms of knowledge, which in turn can allow new forms of social control to be exerted over the components of the field. During the 1950s, for example, medicine began to recognize the psychosocial *environment* of the patient: the patient's family and its psychodynamics. Professions such as social work began to focus on this new environment, and the resulting knowledge became one way to further control the patient. Patients began to be seen not as isolated, individual bodies, but as psychosocial entities located in an "ecological" system: management of "the patient's psychology was a new entrée to patient control."[66]

The models that biologists use to describe their data can have important social effects. During the nineteenth century, the social and natural sciences strongly influenced each other: the social ideas of Malthus about how to avoid the natural increase of the poor inspired Darwin's *Origin of Species*.[67] Once the *Origin* stood as a description of the natural world, complete with competition and market struggles, it could be reimported into social science as social Darwinism, in order to justify the social order of the time. What we are seeing now is similar: the importation of cultural ideas about passive females and heroic males into the "personalities" of gametes. This amounts to the "implanting of social imagery on representations of nature so as to lay a firm basis for reimporting exactly that same imagery as natural explanations of social phenomena."[68]

Further research would show us exactly what social effects are being wrought from the biological imagery of egg and sperm. At the very least, the imagery keeps alive some of the hoariest old stereotypes about weak damsels in distress and their strong male rescuers. That these stereotypes are now being written in at the level of the *cell* constitutes a powerful move to make them seem so natural as to be beyond alteration.

The stereotypical imagery might also encourage people to imagine that what results from the interaction of egg and sperm—a fertilized egg—is the result of deliberate "human" action at the cellular level. Whatever the intentions of the human couple, in this microscopic "culture" a cellular "bride" (or femme fatale) and a cellular "groom" (her victim) make a cellular baby. Rosalind Petchesky points out that through visual representations such as sonograms, we are given "*images* of younger and younger, and tinier and tinier, fetuses being 'saved.'" This leads to "the point of visibility being 'pushed back' *indefinitely*."[69] Endowing egg and sperm with intentional action, a key aspect of personhood in our culture, lays the foundation for the point of viability being pushed back to the moment of fertilization. This will likely lead to greater acceptance of technological developments and new forms of scrutiny and manipulation, for the benefit of these inner "persons": court-ordered restrictions on a pregnant woman's activities in order to protect her fetus, fetal surgery, amniocentesis, and rescinding of abortion rights, to name but a few examples.[70]

Even if we succeed in substituting more egalitarian, interactive metaphors to describe the activities of egg and sperm, and manage to avoid the pitfalls of cybernetic models, we would still be guilty of endowing cellular entities with personhood. More crucial, then, than what *kinds* of personalities we bestow on cells is the very fact that we are doing it at all. This process could ultimately have the most disturbing social consequences.

One clear feminist challenge is to wake up sleeping metaphors in science, particularly those involved in descriptions of the egg and the sperm. Although the literary convention is to call such metaphors "dead," they are not so much dead as sleeping, hidden within the scientific content of texts—and all the more powerful for it.[71] Waking up such metaphors, by becoming aware of when we are projecting cultural imagery onto what we study, will improve our ability to investigate and understand nature. Waking up such metaphors, by becoming aware of their implications, will rob them of their power to naturalize our social conventions about gender.

NOTES

Portions of this article were presented as the 1987 Becker Lecture, Cornell University. I am grateful for the many suggestions and ideas I received on this occasion. For especially pertinent help with my arguments and data I thank Richard Cone, Kevin Whaley, Sharon Stephens, Barbara Duden, Susanne Kuechler, Lorna Rhodes, and Scott Gilbert. The article was strengthened and clarified by the comments of the anonymous *Signs* reviewers as well as the superb editorial skills of Amy Gage.

1. James Hillman, *The Myth of Analysis* (Evanston, Ill.: Northwestern University Press, 1972), 220.
2. The textbooks I consulted are the main ones used in classes for undergraduate premedical students or medical students (or those held on reserve in the library for these classes) during the past few years at Johns Hopkins University. These texts are widely used at other universities in the country as well.
3. Arthur C. Guyton, *Physiology of the Human Body,* 6th ed. (Philadelphia: Saunders College Publishing, 1984), 624.
4. Arthur J. Vander, James H. Sherman, and Dorothy S. Luciano, *Human Physiology: The Mechanisms of Body Function,* 3d ed. (New York: McGraw Hill, 1980), 483–484.
5. Vernon B. Mountcastle, *Medical Physiology,* 14th ed. (London: Mosby, 1980), 2:1624.
6. Eldra Pearl Solomon, *Human Anatomy and Physiology* (New York: CBS College Publishing, 1983), 678.
7. For elaboration, see Emily Martin, *The Woman in the Body: A Cultural Analysis of Reproduction* (Boston: Beacon, 1987), 27–53.
8. Vander, Sherman, and Luciano, 568.
9. Melvin Konner, "Childbearing and Age," *New York Times Magazine* (December 27, 1987), 22–23; esp. 22.
10. I have found but one exception to the opinion that the female is wasteful: "Smallpox being the nasty disease it is, one might expect nature to have designed antibody molecules with combining sites that specifically recognize the epitopes on smallpox virus. Nature differs from technology, however: it thinks nothing of wastefulness. (For example, rather than improving the chance that a spermatozoon will meet an egg cell, nature finds it easier to produce millions of spermatozoa.)" (Niels Kaj Jerne, "The Immune System," *Scientific American* 229, no. 1 [July 1973]: 53). Thanks to a *Signs* reviewer for bringing this reference to my attention.
11. Bruce Alberts et al., *Molecular Biology of the Cell* (New York: Garland, 1983), 795.
12. In her essay "Have Only Men Evolved?" (in *Discovering Reality: Feminist Perspectives on Epistemology, Metaphysics, Methodology, and Philosophy of Science,* ed. Sandra Harding and Merrill B. Hintikka [Dordrecht: Reidel, 1983], 45–69, esp. 60–61), Ruth Hubbard points out that sociobiologists have said the female invests more energy than the male in the production of her large gametes, claiming that this explains why the female provides parental care. Hubbard questions whether it "really takes more 'energy' to generate the one or relatively few eggs than the large excess of sperms required to achieve fertilization." For further critique of how the greater size of eggs is interpreted in sociobiology, see Donna Haraway, "Investment Strategies for the Evolving Portfolio of Primate Females," in *Body/Politics,* ed. Mary Jacobus, Evelyn Fox Keller, and Sally Shuttleworth (New York: Routledge, 1990), 155–156.
13. The sources I used for this article provide compelling information on interactions among sperm. Lack of space prevents me from taking up

this theme here, but the elements include competition, hierarchy, and sacrifice. For a newspaper report, see Malcolm W. Browne, "Some Thoughts on Self Sacrifice," *New York Times* (July 5, 1988), C6. For a literary rendition, see John Barth, "Night-Sea Journey," in his *Lost in the Funhouse* (Garden City, N.Y.: Doubleday, 1968), 3–13.

14. See Carol Delaney, "The Meaning of Paternity and the Virgin Birth Debate," *Man* 21, no. 3 (September 1986): 494–513. She discusses the difference between this scientific view that women contribute genetic material to the fetus and the claim of long-standing Western folk theories that the origin and identity of the fetus comes from the male, as in the metaphor of planting a seed in soil.
15. For a suggested direct link between human behavior and purportedly passive eggs and active sperm, see Erik H. Erikson, "Inner and Outer Space: Reflections on Womanhood," *Daedalus* 93, no. 2 (Spring 1964): 582–606, esp. 591.
16. Guyton (n. 3 above), 619; and Mountcastle (n. 5 above), 1609.
17. Jonathan Miller and David Pelham, *The Facts of Life* (New York: Viking Penguin, 1984), 5.
18. Alberts et al., 796.
19. Ibid., 796.
20. See, e.g., William F. Ganong, *Review of Medical Physiology,* 7th ed. (Los Altos, Calif.: Lange Medical Publications, 1975), 322.
21. Alberts et al. (n. 11 above), 796.
22. Guyton, 615.
23. Solomon (n. 6 above), 683.
24. Vander, Sherman, and Luciano (n. 4 above), 4th ed. (1985), 580.
25. Alberts et al., 796.
26. All biology texts quoted above use the word "penetrate."
27. Solomon, 700.
28. A. Beldecos et al., "The Importance of Feminist Cirtique for Contemporary Cell Biology," *Hypatia* 3, no. 1 (Spring 1988): 61–76.
29. Gerald Schatten and Helen Schatten, "The Energetic Egg," *Medical World News* 23 (January 23, 1984): 51–53, esp. 51.
30. Alberts et al., 796.
31. Guyton (n. 3 above), 613.
32. Miller and Pelham (n. 17 above), 7.
33. Alberts et al. (n. 11 above), 804.
34. Ibid., 801.
35. Ruth Herschberger, *Adam's Rib* (New York: Pelligrini & Cudaby, 1948), esp. 84. I am indebted to Ruth Hubbard for telling me about Herschberger's work, although at a point when this paper was already in draft form.
36. Bennett M. Shapiro, "The Existential Decision of a Sperm," *Cell* 49, no. 3 (May 1987): 293–294, esp. 293.
37. Lennart Nilsson, "A Portrait of the Sperm," in *The Functional Anatomy of the Spermatozaon,* ed. Bjorn A. Afzelius (New York: Pergamon, 1975), 79–82.
38. Ludwik Fleck, *Genesis and Development of a Scientific Fact,* ed. Thaddeus J. Trenn and Robert K. Merton (Chicago: University of Chicago Press, 1979), 38.
39. Jay M. Baltz carried out the research I describe when he was a graduate student in the Thomas C. Jenkins Department of Biophysics at Johns Hopkins University.
40. Far less is known about the physiology of sperm than comparable female substances, which some feminists claim is no accident. Greater scientific scrutiny of female reproduction has long enabled the burden of birth control to be placed on women. In this case, the researchers' discovery did not depend on development of any new technology. The experiments made use of glass pipettes, a manometer, and a simple microscope, all of which have been available for more than one hundred years.
41. Jay Baltz and Richard A. Cone, "What Force Is Needed to Tether a Sperm?" (abstract for Society for the Study of Reproduction, 1985), and "Flagellar Torque on the Head Determines the Force Needed to Tether a Sperm" (abstract for Biophysical Society, 1986).
42. Jay M. Baltz, David F. Katz, and Richard A. Cone, "The Mechanics of the Sperm-Egg Interaction at the Zona Pellucida," *Biophysical Journal* 54, no. 4 (October 1988): 643–654. Lab members were somewhat familiar with work on metaphors in the biology of female reproduction. Richard Cone, who runs the lab, is my husband, and he talked with them about my earlier research on the subject from time to time. Even though my current research focuses on biological imagery and I heard about the lab's work from my husband every day, I myself did not recognize the role of imagery in the sperm research until many weeks after the period of research and writing I describe. Therefore, I assume that any awareness the lab members may have had about how underlying metaphor might be guiding this particular research was fairly inchoate.
43. Ibid., 643, 650.

44. Schatten and Schatten (n. 29 above), 51.
45. Ibid., 52.
46. Surprisingly, in an article intended for a general audience, the authors do not point out that these are sea urchin sperm and note that human sperm do not shoot out filaments at all.
47. Schatten and Schatten, 53.
48. Paul M. Wassarman, "Fertilization in Mammals," *Scientific American* 259, no. 6 (December 1988): 78–84, esp. 78, 84.
49. Ibid., 78.
50. Ibid., 79.
51. Ibid., 78.
52. Since receptor molecules are relatively *immotile* and the ligands that bind to them relatively *motile,* one might imagine the egg being called the receptor and the sperm the ligand. But the molecules in question on egg and sperm are immotile molecules. It is the sperm as a *cell* that has motility, and the egg as a cell that has relative immotility.
53. Wassarman, 78–79.
54. Paul M. Wassarman, "The Biology and Chemistry of Fertilization," *Science* 235, no. 4788 (January 30, 1987): 553–560, esp. 554.
55. Ibid., 557.
56. Ibid., 557–558. This finding throws into question Schatten and Schatten's description (n. 29 above) of the sperm, its tail beating, diving down into the egg.
57. Deborah R. Gordon, "Tenacious Assumptions in Western Medicine," in *Biomedicine Examined,* ed. Margaret Lock and Deborah Gordon (Dordrecht: Kluwer, 1988), 19–56, esp. 26.
58. Wassarman, "The Biology and Chemistry of Fertilization," 558.
59. Baltz, Katz, and Cone (n. 42 above), 643, 650.
60. Schatten and Schatten, 53.
61. Wassarman, "The Biology and Chemistry of Fertilization," 557.
62. Mary Ellman, *Thinking about Women* (New York: Harcourt Brace Jovanovich, 1968), 140; Nina Averbach, *Woman and the Demon* (Cambridge, Mass.: Harvard University Press, 1982), esp. 186.
63. Kenneth Alan Adams, "Arachnophobia: Love American Style," *Journal of Psychoanalytic Anthropology* 4, no. 2 (1981): 157–197.
64. William Ray Arney and Bernard Bergen, *Medicine and the Management of Living* (Chicago: University of Chicago Press, 1984).
65. J. F. Hartman, R. B. Gwatkin, and C. F. Hutchison, "Early Contact Interactions between Mammalian Gametes *In Vitro,*" *Proceedings of the National Academy of Sciences* (U.S.) 69, no. 10 (1972): 2767–2769.
66. Arney and Bergen, 68.
67. Ruth Hubbard, "Have Only Men Evolved?" (n. 12 above), 51–52.
68. David Harvey, personal communication, November 1989.
69. Rosalind Petchesky, "Fetal Images: The Power of Visual Culture in the Politics of Reproduction," *Feminist Studies* 13, no. 2 (Summer 1987): 263–292, esp. 272.
70. Rita Arditti, Renate Klein, and Shelley Minden, *Test-Tube Women* (London: Pandora, 1984); Ellen Goodman, "Whose Right to Life?" *Baltimore Sun* (November 17, 1987); Tamar Lewin, "Courts Acting to Force Care of the Unborn," *New York Times* (November 23, 1987), A1 and B10; Susan Irwin and Brigitte Jordan, "Knowledge, Practice, and Power: Court Ordered Cesarean Sections," *Medical Anthropology Quarterly* 1, no. 3 (September 1987): 319–334.
71. Thanks to Elizabeth Fee and David Spain, who in February 1989 and April 1989, respectively, made points related to this.

DISCUSSION QUESTIONS

1. Create your own account of the interactions of the egg and the sperm using a model that privileges the role of the female and marginalizes the role of the male.
   OR Create an account that might speak to a world in which sex and gender were not dichotomized.
2. Is it in any way problematic to give sperm and eggs active intentionality? Why or why not?
3. To what extent do cultural biases color scientific discoveries? What is the best way to avoid the kind of cultural biases exposed in this article?
4. What, if any, effects do you think scientific accounts such as those described in this article have on the gender and sexual socialization of boys and girls?

Motherhood and Morality in America

KRISTIN LUKER

Kristin Luker's classic study of female abortion activists in California reveals striking differences in the social characteristics of the activists on each of the two sides of the abortion debate. Her analysis suggests that their contrasting profiles represent fundamental differences in the activists' views of the role of women in society. For pro-life women, motherhood is given primacy while for pro-choice women, careers, or some combination of career and family are central. Luker concludes that these women's views on abortion constitute a referendum on the cultural meaning of motherhood and, more immediately, represent efforts to validate their life choices and the value of their own existence.

According to interested observers at the time, abortion in America was as frequent in the last century as it is in our own. And the last century, as we have seen, had its own "right-to-life" movement, composed primarily of physicians who pursued the issue in the service of their own professional goals. When abortion reemerged as an issue in the late 1950s, it still remained in large part a restricted debate among interested professionals. But abortion as we now know it has little in common with these earlier rounds of the debate. Instead of the civility and colleagueship that characterized the earlier phases of the debate, the present round of the abortion debate is marked by rancor and intransigence. Instead of the elite male professionals who commanded the issue until recently, ordinary people—and more to the point, ordinary women—have come to predominate in the ranks of those concerned. From a quiet, restricted technical debate among concerned professionals, abortion has become a debate that seems at times capable of tearing the fabric of American life apart. How did this happen? What accounts for the remarkable transformation of the abortion debate?

. . . This article will argue that all the previous rounds of the abortion debate in America were merely echoes of the issue as the nineteenth century defined it: a debate about the medical profession's right to make life-and-death decisions. In contrast, the most recent round of the debate is about something new. By bringing the issue of the moral status of the embryo to the fore, the new round focuses on the relative rights of women and embryos. Consequently, the abortion debate has become a debate about women's contrasting obligations to themselves and others. New technologies and the changing nature of work have opened up possibilities for women outside of the home undreamed of in the nineteenth century; together, these changes give women—for the first time in history—the option of deciding exactly how and when their family roles will fit into the larger context of their lives. In essence, therefore, this round of the abortion debate is so passionate and hard-fought *because it is a referendum on the place and meaning of motherhood.*

Motherhood is at issue because two opposing visions of motherhood are at war. Championed by "feminists" and "housewives," these two different views of motherhood represent in turn two very different kinds of social worlds. The abortion debate has become a debate among women, women with different values in the social world, different experiences of it, and different resources with which to cope with it. How the issue is framed, how people think about it, and, most importantly, where the passions come from are all related to the fact that the battlelines are increasingly drawn (and defended) by women. While on the surface it is the embryo's fate that seems to be at stake, the abortion debate is actually about the meanings of *women's* lives.

To be sure, both the pro-life and the pro-choice movements had earlier phases in which they were dominated by male professionals. Some of these men are still active in the debate, and it is certainly the case that some men continue to join the debate on both sides of the issue. But the data in this study suggest that by 1974 over 80 percent of the activists in both the pro-choice and the pro-life movements in California were women, and a national survey of abortion activists found similar results.

Moreover, in our interviews we routinely asked both male and female activists on both sides of the issue to supply information on several "social background variables," such as where they were born, the extent of their education, their income level, the number of children they had, and their occupations. When male activists on the two sides are compared on these variables, they are virtually indistinguishable from one another. But when female activists are compared, it is dramatically clear that for the women who have come to dominate the ranks of the movement, the abortion debate is a conflict between two different social worlds and the hopes and beliefs those worlds support.

WHO ARE THE ACTIVISTS?

On almost every social background variable we examined, pro-life and pro-choice women differed dramatically. For example, in terms of income, almost half of all pro-life women (44 percent) in this study reported an income of less than $20,000 a year, but only one-fourth of the pro-choice women reported an income that low, and a considerable portion of those were young women just starting their careers. On the upper end of the income scale, one-third of the pro-choice women reported an income of $50,000 a year or more compared with only one pro-life woman in every seven.

These simple figures in income, however, conceal a very complex social reality, and that social reality is in turn tied to feelings about abortion. The higher incomes of pro-choice women, for example, result from a number of interesting factors. Almost without exception pro-choice women work in the paid labor force, they earn good salaries when they work, and if they are married, they are likely to be married to men who also have good incomes. An astounding 94 percent of all pro-choice women work, and over half of them have incomes in the top 10 percent of all working women in this country. Moreover, one pro-choice woman in ten has an annual *personal* income (as opposed to a family income) of $30,000 or more, this putting her in the rarified ranks of the top 2 percent of all employed women in America. Pro-life women, by contrast, are far less likely to work: 63 percent of them do not work in the paid labor force, and almost all of those who do are unmarried. Among pro-life married women, for example, only 14 percent report any personal income at all, and for most of them, this is earned not in a formal job but through activities such as selling cosmetics to groups of friends. Not surprisingly, the personal income of pro-life women who work outside the home, whether in a formal job or in one of these less-structured activities, is low. Half of all pro-life women who do work earn less than $5,000 a year, and half earn between $5,000 and $10,000. Only two pro-life women we contacted reported a personal income of more than $20,000. Thus pro-life women are less likely to work in the first place, they earn less money when they do work, and they are more likely to be married to a skilled worker or small businessman who earns only a moderate income.

These differences in income are in turn related to the different educational and occupational choices these women have made along the way. Among pro-choice women, almost four out of ten (37 percent) had undertaken some graduate work beyond the B.A. degree, and 18 percent had an M.D., a law degree, a Ph.D., or a similar postgraduate degree. Pro-life women, by comparison, had far less education: 10 percent of them had only a high school education or less; and another 30 percent never finished college (in contrast with only 8 percent of the pro-choice women). Only 6 percent of all pro-life women had a law degree, a Ph.D., or a medical degree.

These educational differences were in turn related to occupational differences among the women in this study. Because of their higher levels of education, pro-choice women tended to be employed in the major professions, as administrators, owners of small businesses, or executives in large businesses. The pro-life women tended to be housewives or, of the few who worked, to be in the traditional female jobs of teaching, social work, and nursing. (The choice of home life over public life held true for even the 6 percent of pro-life women with an advanced degree; of the married women who had such degrees, at the time of our interviews only one of them had not retired from her profession after marriage.)

These economic and social differences were also tied to choices that women on each side had made about marriage and family life. For example, 23 percent of pro-choice women had never married, compared with only 16 percent of pro-life women; 14 percent of pro-choice women had been divorced, compared with 5 percent of pro-life women. The size of the families these women had was also different. The average pro-choice family had between one and two children and was more likely to have one; pro-life families averaged between two and three children and were more likely to have three. (Among the pro-life women, 23 percent had five or more children; 16 percent had seven or more children.) Pro-life women also tended to marry at a slightly younger age and to have had their first child earlier.

Finally, the women on each side differed dramatically in their religious affiliation and in the role that religion played in their lives. Almost 80 percent of the women active in the pro-life movement at the present time are Catholics. The remainder are Protestants (9 percent), persons who claim no religion (5 percent), and Jews (1 percent). In sharp contrast, 63 percent of pro-choice women say that they have no religion, 22 percent think of themselves as vaguely Protestant, 3 percent are Jewish, and 9 percent have what they call a "personal" religion.

. . . [A]lmost 20 percent of the pro-life activists in this study are converts to Catholicism, people who have actively chosen to follow a given religious faith, in striking contrast to pro-choice people, who have actively chosen not to follow any.

Perhaps the single most dramatic difference between the two groups, however, is in the role that religion plays in their lives. Almost three-quarters of the pro-choice people interviewed said that formal religion was either unimportant or completely irrelevant to them, and their attitudes are correlated with behavior: only 25 percent of the pro-choice women said they *ever* attend church, and most of these said they do so only occasionally. Among pro-life people, by contrast, 69 percent said religion was important in their lives, and an additional 22 percent said that it was very important. . . . [H]alf of those pro-life women interviewed said they attend church regularly once a week, and another 13 percent said they do so even more often. Whereas 80 percent of pro-choice people never attend church, only 2 percent of pro-life advocates never do so.

Keeping in mind that the statistical use of average has inherent difficulties, we ask, who are the "average" pro-choice and pro-life advocates? When the social background data are looked at carefully, two profiles emerge. The average pro-choice activist is a forty-four-year-old married woman who grew up in a large metropolitan area and whose father was a college graduate. She was married at age twenty-two, has one or two children, and has had some graduate or professional training beyond the B.A. degree. She is married to a professional man, is herself employed in a regular job, and her family income is more than $50,000 a year. She is not religiously active, feels that religion is not important to her, and attends church very rarely if at all.

The average pro-life woman is also a forty-four-year-old married woman who grew up in a large metropolitan area. She married at age seventeen and has three children or more. Her father was a high school graduate, and she has some college education or may have a B.A. degree. She is not employed in the paid labor force and is married to a small businessman or a lower-level white-collar worker; her family income is $30,000 a year. She is Catholic (and may have converted), and her religion is one of the most important aspects of her life; she attends church at least once a week and occasionally more often.

INTERESTS AND PASSIONS

To the social scientist (and perhaps to most of us) these social background characteristics connote lifestyles as well. We intuitively clothe these bare statistics with assumptions about beliefs and values. When we do so, the pro-choice women emerge as educated, affluent, liberal professionals, whose lack of religious affiliation suggests a secular, "modern," or (as pro-life people would have it) "utilitarian" outlook on life. Similarly, the income, education, marital patterns, and religious devotion of pro-life women suggest that they are traditional, hard-working people ("polyester types" to their opponents), who hold conservative views of life. We may be entitled to assume that individuals' social backgrounds act to shape and mold their social attitudes, but it is important to realize that the relationship between social worlds and social values is a very complex one.

Perhaps one example will serve to illustrate the point. A number of pro-life women in this study emphatically rejected an expression that pro-choice women tend to use almost unthinkingly—the expression *unwanted pregnancy.* Pro-life women argued forcefully that a better term would be a *surprise* pregnancy, asserting that although a pregnancy may be momentarily unwanted, the child that results from the pregnancy almost never is. Even such a simple thing—what to call an unanticipated pregnancy—calls into play an individual's values and resources. Keeping in mind our profile of the average pro-life person, it is obvious that a woman who does not work in the paid labor force, who does not have a college degree, whose religion is important to her, and who has already committed herself wholeheartedly to marriage and a large family is well equipped to believe that an unanticipated pregnancy usually becomes a beloved child. Her life is arranged so that for her, this belief is true. This view is consistent not only with her values, which she has held from earliest childhood, but with her social resources as well. It should not be surprising, therefore, that her world view leads her to believe that everyone else can "make room for one more" as easily as she can and that therefore it supports her in her conviction that abortion is cruel, wicked, and self-indulgent.

It is almost certainly the case that an unplanned pregnancy is never an easy thing for anyone. Keeping in mind the profile of the average pro-choice woman, however, it is evident that a woman who is employed full time, who has an affluent lifestyle that depends in part of her contribution to the family income, and who expects to give a child as good a life as she herself has had with respect to educational, social, and economic advantages will draw on a different reality when she finds herself being skeptical about the ability of the average person to transform unwanted pregnancies into well-loved (and well-cared-for) children.

The relationship between passions and interests is thus more dynamic than it might appear at first. It is true that at one level, pro-choice and pro-life attitudes on abortion are self-serving: activists on each side have different views of the morality of abortion because their chosen lifestyles leave them with different needs for abortion; and both sides have values that provide a moral basis for their abortion needs in particular and their lifestyles in general. But this is only half the story. The values that lead pro-life and pro-choice women into different attitudes toward abortion are the same values that led them at an earlier time to adopt different lifestyles that supported a given view of abortion.

For example, pro-life women have *always* valued family roles very highly and have arranged their lives accordingly. They did not acquire high-level educational and occupational skills, for example, because they married, and they married because their values suggested that this would be the most satisfying life open to them. Similarly, pro-choice women postponed (or avoided) marriage and family roles because they chose to acquire the skills they needed to be successful in the larger world, having concluded that the role of wife and mother was too limited for them. Thus, activists on both sides of the issue are women who have a given set of values about what are the most satisfying and appropriate roles for women, and they have made *life commitments that now limit their ability to change their minds.* Women who have many children and little education, for example, are seriously handicapped in attempting to become doctors or lawyers; women who have reached their late forties with few children or none are limited in their ability to build (or rebuild) a family. For most of these activists, therefore, their position on abortion is the "tip of the iceberg," a shorthand way of supporting and proclaiming not only a complex set of values but a given set of social resources as well.

To put the matter differently, we might say that for pro-life women the traditional division of life into separate male roles and female roles still works, but for pro-choice women it does not. Having made a commitment to the traditional female roles of wife, mother, and homemaker, pro-life women are limited in those kinds of resources—education, class status, recent occupational experiences—they would need to compete in what has traditionally been the male sphere, namely, the paid labor force. The average pro-choice woman, in contrast, is comparatively well endowed with exactly these resources; she is highly educated, she already has a job, and she has recent (and continuous) experience in the job market.

In consequence, anything that supports a traditional division of labor into male and female worlds is, broadly speaking, in the interests of pro-life women because that is where their resources lie. Conversely, such a traditional division of labor, when strictly enforced, is against the interests of pro-choice women because it limits their abilities to use the valuable "male" resources that they have in relative abundance. It is therefore apparent that attitudes toward abortion, even though rooted in childhood experiences, are also intimately related to present-day interests. Women who oppose abortion and seek to make it officially unavailable are declaring, both practically and symbolically, that women's reproductive roles should be given social primacy. Once an embryo is defined as a child and an abortion as the death of a person, almost everything else in a woman's life must "go on hold" during the course of her pregnancy: any attempt to gain "male" resources such as a job, an education, or other skills must be subordinated to her uniquely female responsibility of serving the needs of this newly conceived person. Thus, when personhood is bestowed on the embryo, women's nonreproductive roles are made secondary to their reproductive roles. The act of conception therefore creates a pregnant woman rather than a woman who is pregnant; it creates a woman whose life, in cases where roles or values clash, is defined by the fact that she is—or may become—pregnant.

It is obvious that this view is supportive of women who have already decided that their familial and reproductive roles are the major ones in their lives. By the same token, the costs of defining women's reproductive roles as primary do not seem high to them because they have already chosen to make those roles primary anyway. For example, employers might choose to discriminate against women because they might require maternity leave and thus be unavailable at critical times, but women who have chosen not to work in the paid labor force in the first place can see such discrimination as irrelevant to them.

It is equally obvious that supporting abortion (and believing that the embryo is not a person) is in the vested interests of pro-choice women. Being so well equipped to compete in the male sphere, they perceive any situation that both practically and symbolically affirms the primacy of women's reproductive roles as a real loss to them. Practically, it devalues their social resources. If women are only secondarily in the labor market and must subordinate working to pregnancy, should it occur, then their education, occupation, income and work become potentially temporary and hence discounted. Working becomes, as it traditionally was perceived to be, a pastime or hobby pursued for "pin money" rather than a central part of their lives. Similarly, if the embryo is defined as a person and the ability to become pregnant is the central one for women, a woman must be prepared to sacrifice some of her own interests to the interests of this newly conceived person.

In short, in a world where men and women have traditionally had different roles to play and where male roles have traditionally been the more socially prestigious and financially rewarded, abortion has become a symbolic market between those who wish to maintain this division of labor and those who wish to challenge it. . . .

Thus, the sides are fundamentally opposed to each other not only on the issue of abortion but also on what abortion *means.* Women who have many "human capital" resources of the traditionally male variety want to see motherhood recognized as a private, discretionary choice. Women who have few of these resources and limited opportunities in the job market want to see motherhood recognized as the most important thing a woman can do. . . .

To the extent that women who have chosen the larger public world of work have been successful, both legally and in terms of public opinion and, furthermore, are rapidly becoming the numerical majority, pro-life women are put on the defensive. Several pro-life women offered poignant examples of how the world deals with housewives who do not have an official payroll title. Here is what one of them said:

> I was at a party, two years ago—it still sticks in my mind, you see, because I'm a housewife and I don't work—and I met this girl from England and we got involved in a deep discussion about the English and the Americans and their philosophies and how one has influenced the other, and at the end of the conversation—she was a working gal herself. I forget what she did—and she says, "Where do you work?" and I said, "I don't." And she looked at me and said, "You don't work?" I said "No." She said, "You're just a housewife . . . and you can still think like that?" She couldn't believe it, and she sort of gave me a funny look and that was the end of the conversation for the evening. And I've met other people who've had similar experiences. [People seem to think that if] you're at home and you're involved with children all day, your intelligence quotient must be down with them on the floor someplace, and [that] you really don't do much thinking or get yourself involved.

Moreover, there are subtle indications that even the pro-life activists we interviewed had internalized their loss of status as housewives. Only a handful of married pro-life activists also worked at regular jobs outside the home; but fully half of those who were now fulltime homemakers, some for as long as thirty years, referred to themselves in terms of the work they had given up when they married or had their first child: "I'm a political scientist," "I'm a social worker," "I'm an accountant." It is noteworthy that no one used the past tense as in "I used to be a social worker"; every nonemployed married women who used her former professional identification used it in the present tense. . . . Ironically, by calling on earlier identifications these women may have been expressing a pervasive cultural value that they oppose as a matter of ideology. . . .

Because of their commitment to their own view of motherhood as a primary social role, pro-life women believe that other women are "casual" about abortions and have them "for convenience." There are no reliable data to confirm whether or not women are "casual" about abortions, but many pro-life people believe this to be the case and relate their activism to their perception of other people's casualness. For example:

> Every time I saw some article [on abortion] I read about it, and I had another friend who had her second abortion in 1977 . . . and both of her abortions were a matter of convenience, it was inconvenient for her to be pregnant at the time. When I talked to her I said, "O.K., you're married now, your husband has a good job, you want to have children eventually, but if you became pregnant now, you'd have an abortion. Why?" "Because it's inconvenient, this is not when I want to have my child." And that bothered me a lot because she is also very intelligent, graduated magna cum laude, and knew nothing about fetal development.

The assertion that women are "casual" about abortion, one could argue, expresses in a shorthand way a set of beliefs about women and their roles. First, the more people value the personhood of the embryo, the more important must be the reasons for taking its life. Some pro-life people, for example, would accept an abortion when continuation of the pregnancy would cause the death of the mother; they believe that when two lives are in direct conflict, the embryo's life can be considered the more expendable. But not all pro-life people agree, and many say they would not accept abortion even to save the mother's life. (Still others say they accept the idea in principle but would not make that choice in their own lives if faced with it.) For people who accept the personhood of the embryo, any reason besides trading a "life for a life" (and sometimes even that) seems trivial, merely a matter of "convenience."

Second, people who accept the personhood of the embryo see the reasons that pro-abortion people give for ending pregnancy as simultaneously downgrading the value of the embryo and upgrading everything else but pregnancy. The argument that women need abortion to "control" their fertility means that they intend to subordinate pregnancy, with its inherent unpredictability, to something else. As the pro-choice activists . . . have told us, that something else is participation in the paid labor force. Abortion permits women to engage in paid work on an equal basis with men. With abortion, they may schedule pregnancy in order to take advantage of the kinds of benefits that come with a paid position in the labor force: a pay-check, a title, a social identity. The pro-life women in this study were often careful to point out that they did not object to "career women." But what they meant by "career women" were women whose *only* responsibilities were in the labor force. Once a woman became a wife and a mother, in their view her primary responsibility was to her home and family.

Third, the pro-life activists we interviewed, the overwhelming majority of whom are full-time homemakers, also felt that women who worked *and* had families could often do so only because women like themselves picked up the slack. Given their place in the social structure, it is not surprising that many of the pro-life women thought that married women who worked outside the home were "selfish"—that they got all the benefits while the homemakers carried the load for them in Boy and Girl Scouts, PTA, and after school, for which their reward was to be treated by the workers as less competent and less interesting persons.*

*In fact, pro-life women, especially those recruited after 1972, were *less* likely to be engaged in formal activities such as Scouts, church activities, and PTA than their pro-choice peers. Quite possibly they have in mind more informal kinds of activities, premised on the fact that since they do not work, they are home most of the time.

Abortion therefore strips the veil of sanctity from motherhood. When pregnancy is discretionary—when people are allowed to put anything else they value in front of it—then motherhood has been demoted from a sacred calling to a job. In effect, the legalization of abortion serves to make men and women more "unisex" by deemphasizing what makes them different—the ability of women to visibly and directly carry the next generation. Thus, pro-choice women are emphatic about their right to compete equally with men without the burden of an unplanned pregnancy, and pro-life women are equally emphatic about their belief that men and women have different roles in life and that pregnancy is a gift instead of a burden.

The pro-life activists we interviewed do not want equality with men in the sense of having exactly the same rights and responsibilities as men do, although they do want equality of status. In fact, to the extent that *all* women have been touched by the women's movement and have become aware of the fact that society often treats women as a class as less capable than men, quite a few said they appreciated the Equal Rights Amendment (ERA), except for its implied stand on abortion. The ERA, in their view, reminded them that women are as valuable *in their own sphere* as men are in theirs. However, to the extent that the ERA was seen as downplaying the differences between men and women, to devalue the female sphere of the home in the face of the male sphere of paid work, others saw it as both demeaning and oppressive to women like themselves. . . .

It is stating the obvious to point out that the more limited the educational credentials a woman has, the more limited the job opportunities are for her, and the more limited the job opportunities, the more attractive motherhood is as full-time occupation. In motherhood, one can control the content and pace of one's own work, and the job is *intrinsically meaningful.* Compared with a job clerking in a supermarket (a realistic alternative for women with limited educational credentials) where the work is poorly compensated and often demeaning, motherhood can have compensations that far transcend the monetary ones. As one woman described mothering: "You have this little, rough uncut diamond, and you're the artist shaping and cutting that diamond, and bringing out the lights . . . that's a great challenge."

All the circumstances of her existence will therefore encourage a pro-life woman to highlight the kinds of values and experiences that support childbearing and childrearing and to discount the attraction (such as it is) of paid employment. Her circumstances encourage her to resent the pro-choice view that women's most meaningful and prestigious activities are in the "man's world."

Abortion also has a symbolic dimension that separates the needs and interests of homemakers and workers in the paid labor force. Insofar as abortion allows a woman to get a job, to get training for a job, or to advance in a job, it does more than provide social support for working women over homemakers; it also seems to support the value of economic considerations over moral ones. Many pro-life people interviewed said that although their commitment to traditional family roles meant very real material deprivations to themselves and their families, the more benefits of such a choice more than made up for it.

> My girls babysit and the boys garden and have paper routes and things like that. I say that if we had a lot of money that would still be my philosophy, though I don't know because we haven't been in that position. But it's a sacrifice to have a larger family. So when I hear these figures that it takes $65,000 from birth to [raise a child], I think that's ridiculous. That's a new bike every year. That's private colleges. That's a complete new outfit when school opens. Well, we've got seven daughters who wear hand-me-downs, and we hope that sometime in their eighteen years at home each one has a new bike somewhere along the line, but otherwise it's hand-me-downs. Those figures are inflated to

give those children everything, and I think that's not good for them.

For pro-life people, a world view that puts the economic before the noneconomic hopelessly confuses two different kinds of worlds. For them, the private world of family as traditionally experienced is the one place in human society where none of us has a price tag. Home, as Robert Frost pointed out, is where they have to take you in, whatever your social worth. Whether one is a surgeon or a rag picker, the family is, at least ideally, the place where love is unconditional.

Pro-life people and pro-life women in particular have very real reasons to fear such a state of affairs. Not only do they see an achievement-based world as harsh, superficial, and ultimately ruthless; they are relatively less well-equipped to operate in that world. . . .

It is relevant in this context to recall the grounds on which pro-life people argue that the embryo is a baby: that it is genetically human. To insist that the embryo is a baby because it is genetically human is to make a claim that it is both wrong and impossible to make distinctions between humans at all. Protecting the life of the embryo, which is by definition an entity whose social worth is all yet to come, means protecting others who feel that they may be defined as having low social worth; more broadly, it means protecting a legal view of personhood that emphatically rejects social worth criteria.

For the majority of pro-life people we interviewed, the abortions they found most offensive were those of "damaged" embryos. This is because this category so clearly highlights the aforementioned concerns about social worth. To defend a genetically or congenitally damaged embryo from abortion is, in their minds, defending the weakest of the weak, and most pro-life people we interviewed were least prepared to compromise on this category of abortion.

The genetic basis for the embryo's claim to personhood has another, more subtle implication for those on the pro-life side. If genetic humanness equals personhood, then biological facts of life must take precedence over social facts of life. One's destiny is therefore inborn and hence immutable. To give any ground on the embryo's biologically determined babyness, therefore, would by extension call into question the "innate," "natural," and biological basis of women's traditional roles as well.

Pro-choice people, of course, hold a very different view of the matter. For them, social considerations outweigh biological ones: the embryo becomes a baby when it is "viable," that is, capable of achieving a certain degree of social integration with others. This is a world view premised on achievement, but not in the way pro-life people experience the world. Pro-choice people, believing as they do in choice, planning, and human efficacy, believe that biology is simply a minor given to be transcended by human experience. Sex, like race and age, is not an appropriate criterion for sorting people into different rights and responsibilities. Pro-choice people downplay these "natural" ascriptive characteristics, believing that true equality means achievement based on talent, not being restricted to a "women's world," a "black world," or an "old people's world." Such a view, as the profile of pro-choice people has made clear, is entirely consistent with their own lives and achievements.

These differences in social circumstances that separate pro-life from pro-choice women on the core issue of abortion also lead them to have different values on topics that surround abortion, such as sexuality and the use of contraception. . . . [P]ro-life women believe that the purpose of sex is reproduction whereas pro-choice women believe that its purpose is to promote intimacy and mutual pleasure.

These two views about sex express the same value differences that lead the two sides to have such different views on abortion. If women plan to find their primary role in marriage and the family, then they face a need to create a "moral cartel" when it comes to sex. If sex is freely available outside of marriage, then why should men, as the old saw puts it, buy the cow when the milk is free? If many women are willing to sleep with men outside of marriage, then the regular sexual activity that comes with marriage is much less valuable an incentive to marry. And because pro-life women are traditional women, their primary resource for marriage is the promise of a stable home, with everything it implies: children, regular sex, a "haven in a heartless world."

But pro-life women, like all women, are facing a devaluation of these resources. As American society increasingly becomes a service economy, men can buy the services that a wife traditionally offers. Cooking, cleaning, decorating, and the like can easily be purchased on the open market in a cash transaction. And as sex becomes more open, more casual, and more "amative," it removes one more resource that could previously be obtained only through marriage.

Pro-life women, as we have seen, have both value orientations and social characteristics that make marriage very important. Their alternatives in the public world of work are, on the whole, less attractive. Furthermore, women who stay home full-time and keep house are becoming a financial luxury. Only very wealthy families *or families whose values allow them to place the nontangible benefits of a full-time wife over the tangible benefits of a working wife* can afford to keep one of its earners off the labor market. To pro-life people, the nontangible benefit of having children—and therefore the value of procreative sex—is very important. Thus, a social ethic that promotes more freely available sex undercuts pro-life women two ways: it limits their abilities to get into a marriage in the first place, and it undermines the social value placed on their presence once within a marriage.

For pro-choice women, the situation is reversed. Because they have access to "male" resources such as education and income, they have far less reason to believe that the basic reason for sexuality is to produce children. They plan to have small families anyway, and they and their husbands come from and have married into a social class in which small families are the norm. For a number of overlapping reasons, therefore, pro-choice women value the ability of sex to promote human intimacy more (or at least more frequently) than they value the ability of sex to produce babies. But they hold this view because they can afford to. When they bargain for marriage, they use the same resources that they use in the labor market: upper-class status, an education very similar to a man's, side-by-side participation in the man's world, and, not least, a salary that substantially increases a family's standard of living.

. . . These differences in social background also explain why the majority of pro-life people we interviewed were opposed to "artificial" contraception, and had chosen to use natural family planning (NFP), the modern-day version of the "rhythm method." To be sure, since NFP is a "morally licit" form of fertility control for Catholics, and many pro-life activists are very orthodox Catholics, NFP is attractive on those grounds alone. But as a group, Catholics are increasingly using contraception in patterns very similar to those of their non-Catholic peers. Furthermore, many non-Catholic pro-life activists told us they used NFP. Opposition to contraception, therefore, and its corollary, the use of NFP, needs to be explained as something other than simple obedience to church dogma.

Given their status as traditional women who do not work outside of the home, the choice of NFP as the preferred method of fertility control is a rational one because NFP enhances their power and status as women. The NFP users we talked with almost uniformly stated that men respect women more when they are using NFP and that the marriage relationship becomes more like a honeymoon. Certain social factors in the lives of pro-life women suggest why this may be so. Because NFP requires abstinence during the fertile period, one effect of using it is that *sex becomes a relatively scarce resource.* Rather than something that is simply there—and taken for granted—sex becomes something that disappears from the relationship for regular periods of time. Therefore, NFP creates incentives for husbands to be close and intimate with their wives. The more insecure a woman and the less support she feels from her husband, the more reasonable it is for her to want to lengthen the period of abstinence to be on the safe side. The increase in power and status that NFP affords a woman in a traditional marriage was clearly recognized by the activists who use NFP. . . .

> . . . You know, if you have filet mignon every day, it becomes kind of uninteresting. But if you have to plan around this, you do some things. You study, and you do other things during the fertile part of the cycle. And the husband and wife find out how much they can do in the line of expressing love for one another in other ways, other than genital. And some people can really express a lot of love and do a lot of touching and be very relaxed. Maybe others would find that they can only do a very little touching because they might be stimulated. And so they would have to find out where their level was. But they can have a beautiful relationship.

NFP also creates an opportunity for both husbands and wives to talk about the wife's fertility so that once again, something that is normally taken for granted can be focused on and valued. Folk wisdom has it that men and women use sexuality in different ways to express their feelings of caring and intimacy: men give love in order to get sex and women give sex in order to get love. If there is some truth to this stereotype (and both popular magazines and that rich source of sociological data, the Dear Abby column, suggest that there is), then it means that men and women often face confusion in their intimate dialogues with one another. Men wonder if their wives really want to have sex with them or are only giving it begrudgingly, out of a sense of "duty." Wives wonder if husbands really love them or merely want them for sexual relief. Natural Family Planning, by making sex periodically unavailable, puts some of these fears to rest. . . .

Furthermore, a few mutually discreet conversations during our interviews suggest that during abstinence at least some couples find ways of giving each other sexual pleasure that do not involve actual intercourse and hence the risk of pregnancy. Given traditional patterns of female socialization into sexuality and the fact that pro-life women are both traditional and devout women, these periods of mutual caressing may be as satisfying as intercourse for some women and even more satisfying than intercourse for others.

The different life circumstances and experiences of pro-life and pro-choice people therefore intimately affect the ways they look at the moral and social dilemmas of contraception. The settings of their lives, for example, suggest that the psychological side benefits of NFP, which do so much to support pro-life values during the practice of contraception, are sought in other ways by pro-choice people. Pro-choice people are slightly older when they marry, and the interviews strongly suggest that they have a considerably more varied sexual experience than pro-life people on average; the use of NFP to discover other facets of sexual expression is therefore largely unnecessary for them. Moreover, what little we know about sexual practices in the United States (from the Kinsey Report) suggests that given the different average levels of education and religious devoutness in the two groups, such sexual activities as "petting" and oral-genital stimulation may be more frequently encountered among pro-choice people to begin with.*

*Kinsey's data suggest that for males the willingness to engage in oral-genital or manual-genital forms of sexual expression is related to education: the more educated an individual, the more likely he is to have "petted" or engaged in oral sex (Alfred Kinsey, *Sexual Behavior in the Human Male,* pp. 337–81, 535–37). For females, the patterns are more complicated. Educational differences among women disappear when age at marriage is taken into account. But as Kinsey notes: "Among the females in the sample, the chief restraint on petting . . . seems to have been the religious tradition against it." The more devout a woman, the less likely she is to have ever petted (Kinsey, *Sexual Behavior in the Human Female,* pp. 247–48).

The life circumstances of the two sides suggest another reason why NFP is popular among pro-life people but not seriously considered by pro-choice people. Pro-choice men and women act on their belief that men and women are equal not only because they have (or should have) equal rights but also because they have substantially similar life experiences. The pro-choice women we met have approximately the same kinds of education as their husbands do, and many of them have the same kinds of jobs—they are lawyers, physicians, college professors, and the like. Even those who do not work in traditionally male occupations have jobs in the paid labor market and thus share common experiences. They and their husbands share many social resources in common: they both have some status outside the home, they both have a paycheck, and they both have a set of peers and friends located in the work world rather than in the family world. In terms of the traditional studies of family power, pro-choice husbands and wives use the same bargaining chips and have roughly equal amounts of them.

Pro-choice women, therefore, value (and can afford) an approach to sexuality that, by sidelining reproduction, diminishes the differences between men and women; they can do this *because they have other resources on which to build a marriage.* Since their value is intimacy and since the daily lives of men and women on the pro-choice side are substantially similar, intimacy in the bedroom is merely an extension of the intimacy of their larger world.

Pro-life women and men, by contrast, tend to live in "separate spheres." Because their lives are based on a social and emotional division of labor where each sex has its appropriate work, to accept contraception or abortion would devalue the one secure resource left to these women; the private world of home and hearth. This would be disastrous not only in terms of status but also in terms of meaning; if values about fertility and family are not essential to a marriage, what support does a traditional marriage have in times of stress? To accept highly effective contraception, which actually and symbolically subordinates the role of children in the family to other needs and goals, would be to cut the ground of meaning out from under at least one (and perhaps both) partners' lives. Therefore, contraception, which sidelines the reproductive capacities of men and women, is both useless and threatening to pro-life people.

THE CORE OF THE DEBATE

In summary, women come to be pro-life and pro-choice activists as the end result of lives that center around different definitions of motherhood. They grow up with a belief about the nature of the embryo, so events in their lives lead them to believe that the embryo is a unique person, or a fetus; that people are intimately tied to their biological roles, or that these roles are but a minor part of life; that motherhood is the most important and satisfying role open to a woman, or that motherhood is only one of several roles, a burden when defined as the only role. These beliefs and values are rooted in the concrete circumstances of women's lives—their educations, incomes, occupations, and the different marital and family choices they have made along the way—and they work simultaneously to shape those circumstances in turn. Values about the relative place of reason and faith, about the role of actively planning for life versus learning to accept gracefully life's unknowns, of the relative satisfactions inherent in work and family—all of these factors place activists in a specific relationship to the larger world and give them a specific set of resources with which to confront that world.

The simultaneous and ongoing modification of both their lives and their values by each other finds these activists located in a specific place in the social world. They are financially successful, or they are not. They become highly educated, or they do not. They become married and have a large family, or they have a small one. And at each step of the way, both their values and their lives have undergone either ratification or revision.

Pro-choice and pro-life activists live in different worlds, and the scope of their lives, as both adults and children, fortifies them in their belief that their own views on abortion are the more correct, more moral, and more reasonable. When added to this is the fact that should "the other side" win, one group of women will see the very real devaluation of their lives and life resources, it is not surprising that the abortion debate has generated so much heat and so little light.

DISCUSSION QUESTIONS

1. Discuss how cultural constructions of gender influence the abortion debate for women. Why does gender appear to be less salient for men regarding this issue?
2. What effects would you expect women's increasing participation in the workplace to have on the abortion debate? What effects would you expect if women's workplace participation decreased?
3. Beside motherhood, how do women establish their gender identity? How important to one's sense of self is the accomplishment of gender identity for women? How important to one's sense of self is parenthood? Does this differ for men and women? Why? How do institutional structures help or hinder men and women in pursuing these aspects of identity?
4. Given the identity issues outlined by Luker, generally speaking, where would you expect gay, lesbian, bisexual, and transgendered individuals to fall in terms of the abortion debate? Address each separately. Why? What factors other than sexual identity might be more important in predicting their stance (e.g., religiosity, class, education)?
5. Do you expect these same trends to hold today? If not, what social changes during the past 15 to 20 years may have affected the demographic composition of abortion activists?

Now You Can Choose! Issues in Parenting and Procreation

BARBARA KATZ ROTHMAN

Technological advances have made a wide array of reproductive choices available to prospective parents, and many more advances may be just on the horizon. In this cogent sociological analysis Barbara Katz Rothman reveals the constraints and limitations of the concept of choice *given existing structures of social inequality. Rothman also points to the potential dangers of choice that arise from reproductive technologies and the implications of these new choices for social justice.*

My oldest child is 23. I can date, with great precision, my concern with issues of procreation.

In 1973, when I was early-on pregnant, teaching the still-new "Sex Roles" course at Brooklyn College (a two-credit filler renamed from "Courtship and Marriage"), my friend and colleague Roslyn Weinman was teaching medical sociology. Roz's child was a toddler. We alone seemed to have missed the delayed childbearing message. Actually, at 26, I was oldest pregnant woman I knew, lots older than Roz, and I thought I had delayed childbearing.

Roz and I were graduate students, lecturers in those glory days when full-time lecturer jobs were still available. I lost my job that year, bumped to part-time adjunct: Visibly pregnant, I was told I was obviously not serious about my career. It was a very particular moment in feminist history. Women, WOMEN were all anyone could think about those days. Women, not as mothers, with a very small *m* indeed, but as SISTERS, fully equal to our brothers. We didn't really have a vocabulary of "gender studies" worked out yet: I remember long discussions in which the uses of *sex* and *gender* were negotiated. Feminism was everywhere, reborn radiant, if maybe a bit limited.

Roz and I talked endlessly. The boundaries, if there ever were any, between the personal—the physically intimate, bodily bound personal—and the political and between the intellectual and the political all vanished into the thin air from which they probably sprang in the first place. What was happening in my pregnancy and birth, what was happening in the politics of women's health and "reproductive rights," and what was happening in the scholarly explosion of women's studies—all that happened at once, again and again, as Roz and I talked and talked and talked.

And I decided I wanted a home birth. It was not particularly about feminism, that decision. But feminism gave me a vocabulary to talk about it, a set of motives to legitimate it, the power to accomplish it, and an audience with which to share it. . . .

I had not yet discovered midwives. I chose a feminist obstetrician—a phrase that now sounds suspiciously like "military intelligence" in its inherent contradiction. Not that the doctor was not both a feminist and an obstetrician, but I am not sure that the one identity really informed the other, that the two could ever be fully integrated. But at the time, both of us thought that they could. It was just a matter of being sensible. The simplest, most straightforward way to ground the home birth decision in a sensible, yet feminist way was to latch on to the concept of "choice." Choice covered a lot of territory. It had not yet been entirely folded into a code word for abortion, though one ought to have seen that coming.

So the issue of home birth distilled into choice—informed, sensible choice. Given the facts—the abysmal record of obstetrics, the untested and unproven nature of much of what they were doing in hospitals, the far more pleasant and emotionally supportive environment that home provides—home birth came to be a sensible choice. A home birth was a choice I made as an informed consumer, someone who weighed all the options, the safety records, the concerns of one sort or another that mattered to me. Having chosen, I then needed support for my choice, and where better to find it than with a feminist gynecologist-obstetrician, someone who supported "choice."

Choice radiated everywhere in those days, a one-size-fits-all philosophy of feminism. Home birth, abortion, Little League teams, the two-step biopsy procedure for breast cancer, shaving armpits and legs, engineering degrees, bras—you name it, we wanted *choice.* It took me years to realize that while that made for excellent politics at the time, it was poor sociology.

"Choice" is an excellent, expandable concept, offering an opportunity to challenge the status quo (someone's lack of choice) while entirely accepting the dominant way of thinking in Western culture. It took a quarter of a century for the profound limits of choice to become clear. As Roberts (1997) spells it out in *Killing the Black Body:*

> The dominant view of liberty reserves most of its protection only for the most privileged members of society. This approach superimposes liberty on an already unjust social structure, which it seeks to reserve against unwarranted government interference. Liberty protects all citizens' choices from the most direct and egregious abuses of government power, but it does nothing to dismantle social arrangements that make it impossible for some people to make a choice in the first place. Liberty guards against government intrusion; it does not guarantee social justice. (p. 294)

In 1973 I was a feminist with a job—for a bit longer anyway—and a checkbook. If I wanted to hire an obstetrician to come to my home, that was my choice and my right. For women without jobs and checkbooks, let alone women without comfortable homes with heat, running water, husband, and family who could take time off from work and lend a hand, for women who were not, as I was, among the most privileged, the right might have been there, but the choice most assuredly was not.

The democratic ideal with which we were working is that each person is free to make her own choices—the capitalist system puts the limiting clause on that sentence: given what she can afford. Anybody with money can buy anything that is for sale, and given the capitalist ideal it is extremely difficult to argue why any given thing should not be for sale. If there is a market for it, so be it. And there was—and even more so today is—a market for home birth.

This is a very American understanding of choice, an extremely useful starting place for political action, but ultimately limited. To take a completely different arena of life, think about the civil rights movement's lunch counter sit-ins. The right to be served, the right to buy, was instantly understandable to White America outside of the South. The larger economic picture was lost in the close-ups of Black would-be customers denied service. The fact is that many of the people of African descent living in the South at that time could no more afford a Woolworth's lunch than they could afford a trip up to New York for an integrated lunch. Americans tend to see social justice as utopian, unattainable, wishful thinking, while viewing liberty as our birthright, guaranteed to us as citizens. And so anything that interferes with one's ability to purchase an available good or service, other than a lack of money, is seen as a terrific affront.

America, I have heard it said, does not have a culture; it has an economy. In such a context, choice is a useful, even essential, political tool and concept. But it has a way of flattening everything out, reducing everything to the same level, all individual choice. . . .

[But, choices can get] complicated pretty quickly. I would, for example, like to choose the safest possible car. But I cannot afford a new car, and among used cars, I am limited to what is available, and what is in my price range. Am I sacrificing some safety quality of the brake system to get the lower-mileage car? I choose, and given the limited number of used cars available to me within my price range, I balance the factors that seem to matter most to me. Safety, mileage, size, cost—all get thrown into one pile, and somehow a decision has to emerge, and that decision will be counted as "my choice."

In procreation, the choices became *if, when* and *how.* Women had choices about whether or not to become mothers; to conceive at all or, having conceived accidentally, to continue the pregnancy or not. The decision about entering motherhood was quickly subsumed into the language of choice, and "choice" ultimately has come to represent primarily that one choice, the one we are "pro" as good feminists. Decisions about the timing of motherhood were similarly framed as choices, but mostly in one direction, namely, to postpone motherhood, to wait until we had finished school, established ourselves in careers, until we were ready. The choices about how to enter motherhood start with the relatively simple issue of where and with whom in attendance we would give birth, but over time have burgeoned into choices about a vast array of technological assists, substitutes, and impediments. The (liberal) feminist response to everything, from home birth to breast-feeding in public to egg harvesting to frozen embryos to prenatal diagnosis to surrogacy contracts to donor insemination for lesbian couples to IVF and the whole alphabet soup of infertility treatments, to *everything,* has been "choice."

My first indications of the ways the concept of choice was not working came when I was interviewing women about their decisions to use, or not to use, prenatal diagnosis (Rothman [1986] 1993). Testing was available to them, testing that would tell them if their fetuses had Down's syndrome, neural tube defects, or any of an increasingly large number of identifiable conditions. The women, deeply agonizing over this decision, used a phrase that continues to haunt me: "My only choice." Whatever they chose, to use the testing or not, to continue the pregnancy or to abort, the decision was often framed as an "only choice." What could that mean, that deep contradiction? It was a no-choice choice, a forced choice, a choice a woman makes when she is told she has a choice but sees only one way out.

A woman who lives in a fourth-floor walk-up in a city without curb cuts or services for people with disabilities and who terminates a pregnancy for neural tube defects is not exactly engaging in an exercise in free will, making a "choice." A woman who knows what state services will be like for her Down's syndrome child in the years after she has died, and who aborts rather than subject anyone to that treatment, is not experiencing a "choice."

Even access to the testing itself is not a matter of "choice." I did research in the Netherlands about how the midwives there were using, and not using, prenatal diagnosis in their practices. When I spoke to them about which women chose to use the testing, the differences between our two understandings of "choice" were all but laughable. In the United States, any women who can afford the testing and who wants it feels she has a right to it. In the Netherlands, the midwives earnestly informed me that they always offer women over the age of 36 the choice of amniocentesis. It is absolutely the woman's choice, and her right to have that choice. No woman who is over 36 who wants the test is denied it; it is available for every eligible woman. But women under 36, I asked, what of their choice? "Oh, they're not eligible." End of discussion. The Netherlands had made a decision, as a society, as a community, and as a state, about what is reasonable in light of risks, of costs, of potential benefits all around. The point of demarcation was established as 36; the test was freely available to every woman over 36 who wanted it, and unavailable to every woman under 36.

The American in me bristled: You mean a 35½-year-old woman who wants the test can be refused it and then give birth to a baby with Down's syndrome, and that's the way it goes? Yup. Or more accurately, "Ja." That's the way it goes.

Yet as an American, I recognize the sad inevitability that within this country a 38-year-old woman who wants the test may not be able to afford it, even though she just as surely cannot afford a disabled child, and that's the way it goes. Just as a young woman who wants an abortion, who wants very much not to have a baby now but to finish school, who wants to grow up first, may not be able to afford the abortion she has a legal right to purchase. And that's the way it goes. Life, even our more morally responsive presidents have informed us, is not fair.

"Choice" was, all things considered, a good place to *start.* Choice offered feminists concerned with procreation a point of entry, a wedge into the discussion. But in this, as in the civil rights movement, it was only a start: Seating in the front of a bus matters only if you have the bus fare. Questions of social justice provide the context in which issues of choice can unfold.

In the area of procreation this is most clear—and most deeply ironic—as choice expands from decisions about the ways we want to enter motherhood to those affecting the kinds of babies we want to mother, including choices about the sex of our children. It started with Down's syndrome and neural tube defects, but it is most assuredly not stopping there. Any study of the introduction of new technology has to confront the problem of the technological imperative. Once people know how to do something, it is very hard for them not to do it. And once people know how to do something they do it more and more, often with wider and wider applications. If here, why not there? If then, why not now?

In philosophical terms, this problem is often discussed as "the slippery slope," a long slide down from the acceptable to the (morally) unacceptable uses of technology. In the case of prenatal diagnosis, the slope is usually graphed as moving from diagnosis and abortion for conditions incompatible with life, passing through the firm but contested territory of Down's syndrome and neural tube defects, floundering on the rocky terrain of socially undesirable conditions like deafness on down to obesity, bouncing along the questionable areas of "gay genes" and "alcoholism genes," and finally crashing into the great moral abyss of sex selection.

Now *that,* the bioethicists generally agree, is wrong. Using prenatal diagnosis and selective abortion "just because" you want one sex and not the other is generally considered wrong—not only by bioethicists but by lots of ordinary folk, too. It has become, I think, a fairly standard place to draw the line. For people who are doing something that risks making them uncomfortable, morally edgy, having a line somewhere is reassuring: This here may be a bit tricky, but that there is *wrong.*

I have heard countless physicians, geneticists, and genetic counselors use sex selection as the line, the unacceptable choice, which by its very existence makes what they *are* doing more acceptable. I am very uncomfortable with that line. It bothers me because it seems to make two demarcations, neither of which feels right. One is the line between "medical" and "nonmedical" conditions. The argument is that prenatal diagnosis and selective abortion for a "medical condition" is morally acceptable. The original language, which one still hears, is that those are "therapeutic" abortions. The implication is that a medical decision is a scientific, rational decision, and one that is morally sound because it is in the interests of health.

But what exactly makes a given disability a "medical condition" and sex *not?* Sex is a diagnosable genetic condition, associated with variations in phenotype, health, longevity, life chances. Is it that sex, while it does make a difference physically, should not matter that much socially? Is it that both sexes should have what they need to have a good life? Is it that sex ought not to be a basis for valuing people?

Then what is the difference between sex and any other "genetic condition"? This is precisely what disability activists are saying: Deaf people, people in wheelchairs, people who are blind, people with retarded mental development, all ought to be given what they need to have a good life, and all ought to be valued for what and for who they are. The problem, the disability activists are saying, lies with the society. And so it should be, one can easily argue, with sex. We ought to value our girls and our boys, welcome both equally. There is something wrong with a society in which people feel the need to do sex selection.

And that leads to the second demarcation that makes me so uncomfortable: "us" and "them." Over and over again, I have heard American and European physicians and geneticists point out that sex selection is a "Third World" problem, something done "over there" and requested by "immigrants." *We,* the doctors tell conference audiences, use prenatal diagnosis and selective abortion for sound medical reasons. *They* misuse the technology.

Each and every woman who uses this extraordinarily difficult technology of selective abortion is making a decision based on what she knows about the baby-to-be and what she knows about the world into which she might bring that baby. An Indian woman who knows what faces her third daughter is not making a morally different decision, it seems to me, than an American woman who knows what faces her child with Down's syndrome. I have spoken with women in both of these situations, who spoke with great love and longing for the baby-that-might-have-been, and much regret about the world that is. Each has said to me, "It wouldn't be fair to the baby, or to my other children."

I am not trying to show you how selective abortion for sex and selective abortion for disability are similar in order to show you that either is right or is wrong. Rather, I remain the sociologist: We need to look not at the individual decision, but at the social context, the world in which that decision is made. We must move beyond liberty and choice to the question of social justice. I can understand and respect a woman who chooses to terminate a pregnancy based on what she knows about the fetus and the world. I cannot understand or respect a society that places her in that position when it is not inevitable.

Let us move beyond sex to consider race. Race, too, is a "genetic" condition expressing itself in phenotype, health, longevity, and life chances. Whatever characteristics of skin, hair, and bone that a particular society defines as "racial," those characteristics will write themselves upon the body of the child. Several years ago, as a guest lecturer feeling somewhat defensive on the issues of disability, I tried to explain that it was not that the women who terminated pregnancies for Down's syndrome or other diagnoses were themselves "antidisability," or "ablist," not that they necessarily felt any repugnance toward people with disabilities. But they knew all too well what happens to these children in America when they grow up, when their parents die. Groping for a comparison, I said, let us consider a South African White woman still living under apartheid, pregnant by her Black lover, who might choose an abortion rather than bring a mixed-race child into that situation. Does that make her a racist? I thought not; I thought she was doing what she felt she had to do as an individual living under impossible circumstances. If we were going to have discussions about regulating morality, it was not *her* morality with which we should start.

A young Black man came up to me afterward. But what about *Black* South African women? he asked. What were they supposed to do? They too are bearing their babies into that world. And indeed, what of those women? Haven't they sometimes made that same decision? It is not always about the otherness of the baby. . . . When a woman kills her baby rather than have it sold downriver to speculators, or starved, tortured, or experimented upon, isn't that a mercy killing?

And what of women with disabilities? They have not been spared these decisions. Some years back, in one of the excesses of "talk radio," a show focused on the situation of a television news anchor's pregnancy. The woman had a genetic condition that resulted in missing fingers and toes, and knew the possibility of passing this condition on to her child. Some callers felt she had no right to continue the pregnancy without testing. The woman herself was obviously not all that severely disabled—she was, after all, a television news anchor, and so, by whatever American standards one brings to bear, doing just fine. But other women have been yet more disabled, have found the world too hostile a place for people with their condition, or their condition too difficult in itself, and have indeed chosen not to bring children like themselves into the world.

When the genetic technology for making these decisions was the technology of prenatal diagnosis and selective abortion, there was inevitably a strong leaning toward continuing the pregnancy. You have to have a good reason to abort a wanted pregnancy, and to some extent, the later the testing, the better the reason had to be. Thus the moral questions that arose focused on what might warrant an abortion: which conditions were serious enough, which too frivolous. Thus it turns out not to be so straightforward after all to draw a line that neatly divides "medical" and "other" grounds, that divides us with our good decisions from them with their bad ones. All decisions are made in a context, and there is no objective place to stand and judge.

But the technology is not stable. It shifts, and with the shifts come different questions. When you are not selecting *against,* but selecting *for,* the issue changes. Selective abortion has been a "slippery slope" problem. But when we think about selective implantation, selective *creation,* then I prefer a different image—that of the "camel's nose." That argument goes that once you let the camel's nose into the tent, it is very hard to keep the rest of the camel out. I prefer that image because I do not see a terrain along which we societal explorers move at our own risk, but rather a very aggressive camel: the biotech industries, highly motivated to get the nose and the rest of that profitable camel entrenched in our tents.

Sex selection is a perfect example of this process. Sex selection clinics have opened up around the world, offering methods of selection, of choosing the sex of a baby, that do not involve abortion. One is sperm sorting, using the technology of "artificial insemination" after separating out X- and Y-bearing sperm. This is of limited value, only increasing odds from roughly 50/50 to roughly 70/30, but thousands of people have paid for it, finding it worth trying. There are other, more elaborate procedures, with higher success rates combining variations on in vitro fertilization. Embryos outside of the body can be sexed and only those of the "right sex" implanted. . . .

Sometimes people do sex selection to avoid the consequences of deep and profound sexism, and then it is, I think, comparable in every way to decisions women make about terminating pregnancies for disabilities. But, especially when abortion is not involved, sex selection is also marketed as a "consumer choice." Depending on how invasive a procedure she is willing to endure, and how much money she wants to spend, a woman can choose a method that at minimum increases her odds and at maximum virtually guarantees her having a child of chosen sex.

We are introducing choice into yet another arena of our lives. Choice always seems like a good thing to have, and from the point of view of the consumer, the purchaser, it probably is. But what about choice from the point of view of the—what? the consumed? purchased? Or let us be kind: the *chosen* child?

People in the adoption world have struggled with the idea of the chosen child. At first, it seemed such a satisfying bedtime tale: We chose you. It seemed, in adoption, a nice counterbalance to the implicit, understood but unstated fact of having been placed, made available in the first place. But it wasn't such a sweet tale after all. If a child is chosen, it is chosen *for* something. Why me? the child asks—What about me made you choose me? And suddenly parenthood becomes contingent. Chosen for being pretty, sweet, cute, for any given characteristic, implies that should you lose that characteristic, your chosen status is at risk. The child has to wonder, What if I get ugly, surly, stop being cute?

But if you choose a boy and get a boy, or choose a girl and get a girl, then what is the problem? The technology can be improved so that failure, the wrong sex, is not going to happen. So then, won't the child have been chosen for what it truly, indelibly is?

Who does such a thing? Who uses sex selection? A woman who has two sons and has always wanted a daughter? A family that has all girls and wants to "pass on the family name"? People do sex selection because they want a particular kind of child—or maybe, more accurately, because they want a particular kind of parenting experience, and they think sex selection will buy them that.

Sex is a diagnosable chromosomal condition. Choose for Y and you get a male; choose for X and you get a female. What exactly is it that you are choosing and getting?

Sometimes people call it "gender selection." I used to correct that, but I've stopped. *Sex* is the word social scientists use for the biological phenomenon of male/female. *Gender* we save for the social role, for being a boy, a girl, a man, a woman. A sperm cell or a zygote cannot possibly have "gender," but gender is what people are choosing when they select by sex. People who say they want a girl have something in mind—girlness, femininity, some set of characteristics that they expect will come in that girl-package that they think would not come with a boy. I have heard women say that they want the kind of relationship they had with their mothers; they think they cannot have that kind of relationship with a son. I have heard women talking about wanting to have the frills, the clothes, the manicures together, the pretty mother-daughter outfits, the fun of a prom gown and a wedding gown, that come with girls.

Can't you just see the disaster looming? That woman is not ready for a six-foot-tall, 300-pound daughter who wears nothing but denim and boots. People who want a son are probably none too pleased when he announces he wants ballet lessons. When people want a son or want a daughter, they want a host of characteristics that they believe are, and often believe *should be,* sex linked. Someone who wants a cuddly, warm, loving child who will remain close in adulthood chooses a girl. Someone who wants a child who will go out and accomplish great things in the world, bring glory to the family name, goes for a boy. The person is choosing sex, the chromosomes and the genitals, but he or she is also making a statement about personality, lifestyle, what he or she wants the child to be and to become. The person is opting for gender.

Sex is a very crude determinant of these personality and lifestyle traits. How much of gender is biological, "genetic," "nature," is a long-standing debate. But wherever you stand on that question, it is apparent that not every child is a living and breathing gender stereotype. Getting a girl or a boy does not guarantee a parent the characteristics he or she is seeking.

Well, lots of us are not what our parents had in mind, and so be it. What has changed with this technology is the implicit guarantee, offering the idea that parents can now choose, can hope to control the kind of child they will be parenting.

What are the characteristics that we think we can control when we plan our children? When we move past the list of diseases to be avoided, where are we? Sex is a crude genetic characteristic: It is writ large as a chromosome, but is only a loose indicator of what we might expect in a child. As the map of the human genome is unfolded and read, we expect finer and finer resolution. But we are looking at a map, not a crystal ball, and we are not dealing with three wishes from the blessing fairy. Our planning is limited to selecting among embryos, or selecting specific stretches of DNA to include in an embryo, and so our choices are limited to those things we believe are genetically determined.

But what exactly is that supposed to mean? The logic of genetic thinking, as Fox Keller (1995:3) sums it up, is that genes are the primary agents of life: They are the fundamental units of biological traits. In that logic, to read genes is to predict traits; and it follows then that to choose traits you have to choose the genes. Want a blue-eyed child? Select for the genes that cause blue eyes.

But are genes causes? Hubbard would never allow us to use the word *cause* for the action of a gene (see Hubbard and Wald 1993). Genes do not "do" anything; they are certainly not "for" anything. Genes are associated with, involved in, active in the production of proteins. And although I know Hubbard is correct, somehow I think that the language of a gene that causes something, like blue eyes or sickle-cell anemia, is a reasonable way to speak. What, after all, ever causes anything in this world?

I fell down the stairs and broke my ankle. But it is perhaps a bit more complicated. How did I come to be on the stairs? My Aunt Joan lent us the down payment for this house with its big staircase. I was carrying the Hanukkah presents at the time and couldn't see where I was going. Judah Maccabee fought some battle that Hanukkah commemorates. American Jews only make such a fuss over Hanukkah to compete with Christmas—actually, to compete with the commercialization of Christmas. And besides, some little child who shall remain nameless, lest godforbid she get a complex, left a plastic bag on the third-from-bottom step. And how can you break both bones in you ankle by falling three steps? Look at a skeleton sometime—the whole weight of the body tapers down to this absurdly thin point right above where the foot twists.

So what was the cause of my broken ankle? Aunt Joan, Judah Maccabee, Jesus Christ, American capitalism, a nameless child, or an orthopedic design flaw?

I do not know what ever causes anything. "Causality" in science is basically only a hypothesis you cannot disprove (yet). But with all of the hedging of my bets, I am still ready to sometimes use the word *cause* in connection with a gene. I know that genes only code for the production of proteins. They do not *do* anything, not even produce the protein, but still, in the more or less approximate way I am used to talking about causation, I feel comfortable saying that a gene causes, say, blue eyes. Sickle-cell anemia. What else?

We make this question hard to answer because we have painted ourselves into a corner with this nature/nurture thing; we have set up a dichotomy that exists nowhere but in our own heads, and then keep confronting it as if it were a fundamental truth of the universe. We make a list of characteristics, qualities, traits, states of being, and then see if we can assign them to the "nature" side by finding a "gene for" the characteristic. Is intelligence, sexual orientation, schizophrenia, the tendency to divorce, depression, or inability to spell genetic or not? The discussion too often seems to degenerate into "Is too!" "Is not!"

Two of the most public, vociferous, and politically important of these discussions have been the long-standing one focusing on the genetic component in intelligence and the more recent one focused on the "gay gene." Because the intelligence discussion has been hopelessly mired in the racism that surrounds it, let us take a look at the gay gene discussion: Is there a "gene for" being gay? For gay men, there does seem to be a genetic component, one of the pieces. Like every other gene, it speaks in probabilities, in odds. If an identical twin is gay, the chances of his twin being gay are 50 percent. That is considerably higher than chance, given that gay men make up less than 10 percent of all men, but very much lower than the odds on eye color matching in identical twins.

Burr (1996) has helpfully compared male sexual orientation with handedness. For both, we have a dominant and a minority orientation. About 92 percent of the population is right-handed. Left-handedness has at various times in history been treated as evil, sick, sinful, or an ordinary variation. Handedness is experienced as a very powerful given: It is not changeable by an act of will, though one can hide or pass if necessary. And so it seems to be with male homosexuality.

But is handedness "genetic"? If an identical twin is left-handed, the chances of his twin being left-handed are 12 percent. That is, the identical twin of a left-handed person is only one and a half times more likely to be left-handed than is the person sitting next to him on the bus. It is not a powerful argument for genetic causality. But neither does it make handedness a "lifestyle choice."

It seems as if what we are really talking about when we invoke genes is predestination versus free will. In our conversation, we talk as if the opposite of "genetic" were "a choice." Genes seem to function in our language and our thinking as inevitability, determination, predestination, *fate*. What the geneticists tell us is that genes work as probability factors that play a part in a causal equation. How can we think about human will, choice, agency, intentionality, if we are thinking about genes as causes? Go back to the stairs with me.

If you keep leaving things on the steps (and if I've said this once I've said it a thousand times), someone is going to get hurt. And if people march up and down the stairs day after day, year after year, it should be no surprise that eventually someone falls. And if you carry packages and cannot see where you are going, well, what do you expect? To each of these things, and probably a dozen more, maybe even one or two that are "genetic," having to do with bone structure, clumsiness, and distractibility, we can assign a probability rating. What are the odds of falling under each set of circumstances?

Now we are approaching some basic philosophical questions about determinism and inevitability. Was it inevitable that I break my leg? Given everything that happened in the world to that exact second—including the history of architecture, my relationship with Aunt Joan, the set of circumstances that brought that nameless child who shouldn't have a complex about this into my life, the invention of plastic, the evolution of the ankle—given all of it—was it inevitable? Did I have to put my foot there? Was it fate?

That is a fascinating philosophical question, but it is not terribly useful practically. For practical purposes, we focus on one or two of the factors that we think we can control. Do not leave things on the stairs. Watch where you are going. We act as if, and we have to act as if, we have control.

When genes become more and more important in our thinking, we start assigning them greater and greater causal power, moving them to more central positions. Sometimes that has meant giving up, that metaphorical throwing your hands up in the air and saying, "It's genetic," meaning "And that's that." Fate. Which is acceptable if the situation is one that we might want people to take their hands off of and leave be. So the "gay gene" might be useful as a political tool if invoking that gene becomes another way of saying, Give it up, you have to accept that some people are inevitably, determinedly, gay. But if the question we are looking at is not "Why are some men gay?" but "Why are more Black men in prisons than in colleges?" then saying "It's genetic" is quite dangerous. That of course is the underlying premise of racism: that there are genetic differences between categories of people. In the United States, such racist arguments resurface periodically, with books like *The Bell Curve* (Herrnstein and Murray 1994) making the best-seller lists as they invoke genetic determinist explanations for social problems.

In the context of procreation, from the perspective of would-be parents, what the developing technology of genetics might mean is that "It's genetic" is less a throwing-up-your-hands situation than a rollingup-your-sleeves kind of problem. "It's genetic" might come to mean "Let's fix it," engineer it, construct it to order. Let *us* make the determination, let us predetermine: Let us *choose*.

Take that highly publicized "gay gene," XQ, now officially recorded as GAY1. Individual prospective parents of privilege should soon be able to include that—or any other given gene—in their list of things to select for or select against. Many people would be considerably more distressed to learn that their child is gay than to learn that he has some disability. The idea that there is a genetic component to being gay leads quickly to either selecting against that gene or engineering to change it. I picture that aggressive camel at the side of our tent, wearing its sign advertising sex selection; won't sexual orientation, for a slight additional fee, be available in the next package?

Does that kind of selection assure parents that their child will not be gay? Certainly not: A person without the "gay gene" can grow up to be gay just as a person with it can grow up to be straight. But with the selection, one shifts the odds, the probabilities. People have demonstrated their willingness to pay, in time, money, and physical risks and pain, to shift those odds even for sex selection. Knowing that the technology cannot guarantee a child of chosen sex, or if it is a child of chosen sex, cannot guarantee the "kind of girl" or "kind of boy" the parents seek, still they enter the sex selection clinics. People using prenatal diagnostic technologies know that whatever diseases are screened for and against, there are still no guarantees of a healthy, bright, happy child, but they find it worthwhile to do what they can to shift the odds in their favor. They rule out what they can, exercise control where they can. Offer more technology promising yet more control, and people will use it.

Gay is a highly politicized trait. But every day seems to bring some other "gene for" some other quality, characteristic, trait. Can we control all of it? Can we test and select and read and decode and splice our way to what we really want in our children, for our children?

And have we any right to do that? I am not talking about our legal rights, our rights as citizens. American liberal legal scholars can show that whatever our discomforts with treating children like consumer products, it is not in our civil libertarian tradition, and probably not in our interests, to try to stop each other from doing so. So I will not argue against "choice," not even the very limited kind of choices that are available to some people and not to others in so profoundly unjust a society. Rather, I am thinking about our moral rights as parents in our relationship with our children. Do we even want to have them custommade? Would we have wanted our parents to have chosen our traits, predetermined whatever they could and wanted to about us? Whether a trait is what you like best or least about yourself, you probably would not like thinking about it as something your parents put on an order form.

As if they could. As if the traits and characteristics and parts of our being that we cared about were all separately and distinctively coded in "genes for," which in our determination we could choose for our children.

Parenthood does not come with guarantees. Motherhood, I have often said, is one more chance for a speeding truck to ruin your life. The world has plans for our children, and our children have plans for themselves; we will not be able to control this.

The demands of the information age drive us toward getting all the information, toward taking all the control we can. Perhaps wisdom lies in not always doing so, in making wise judgments about what information we want and what information we do not want; which choices we want to make and which choices are not ours to make.

Choices about if, when, and how to mother are ours to make. In a just and decent world, those choices would be available to all women and to all men who want to actively nurture in the way we call mothering. A good and just world, I believe, would provide people with opportunities to mother if that is what they want to do. A just world would provide women with genuine choices about pregnancy, and would support them in their choices about who will attend them and where as they labor to bring their children forth into the world. A good and just world would provide us with choices about *our* mothering. But it may go beyond what a good world, beyond what a just world should offer to give us choices about whom we will mother.

Our children are no more our property, subject to our consumerist choices, than we are the property of our parents. The geneticists like to talk about having unlocked the secrets of life, found the bible, the blueprint. The human genome, they tell us, is our book of life. We must not permit it to become a catalog.

REFERENCES

Burr, Chandler. 1996. *A Separate Creation: The Search for the Biological Origins of Sexual Orientation.* New York: Hyperion.

Herrnstein, Richard J. and Charles Murray. 1994. *The Bell Curve: Intelligence and Class Structure in American Life.* New York: Free Press.

Hubbard, Ruth and Elijah Wald. 1993. *Exploding the Gene Myth.* Boston: Beacon.

Keller, Elizabeth Fox. 1995. *Refiguring Life: Metaphors of Twentieth-Century Biology.* New York: Columbia University Press.

Roberts, Dorothy. 1997. *Killing the Black Body: Race, Reproduction and the Meaning of Liberty.* New York: Pantheon.

Rothman, Barbara Katz. 1976. "In Which a Sensible Woman Persuades Her Doctor, Her Family and Her Friends to Help Her Give Birth at Home." *Ms.,* December.

———. [1986] 1993. *The Tentative Pregnancy: Prenatal Diagnosis and the Future of Motherhood.* New York: W. W. Norton.

DISCUSSION QUESTIONS

1. In what ways do categories of privilege stratify choice?
2. Is selecting *for* ethically different than selecting *against?* Why or why not? Where would you draw the line—what kind of information would you want regarding the health and other characteristics of your own child?
3. What defines a "responsible" choice? *Who* defines "responsible"?
4. How much choice should individuals have in reproductive decisions? Which individuals (e.g., mothers, fathers, surrogates, etc.) should be involved in these choices? What limits, if any, should the state set?
5. How do social institutions shape, enable, and constrain choice in reproductive matters?